# Advances in Intelligent Systems and Computing

## Volume 1086

The series "Advances in Intelligent Systems and Computing" contains publications on theory, applications, and design methods of Intelligent Systems and Intelligent Computing. Virtually all disciplines such as engineering, natural sciences, computer and information science, ICT, economics, business, e-commerce, environment, healthcare, life science are covered. The list of topics spans all the areas of modern intelligent systems and computing such as: computational intelligence, soft computing including neural networks, fuzzy systems, evolutionary computing and the fusion of these paradigms, social intelligence, ambient intelligence, computational neuroscience, artificial life, virtual worlds and society, cognitive science and systems, Perception and Vision, DNA and immune based systems, self-organizing and adaptive systems, e-Learning and teaching, human-centered and human-centric computing, recommender systems, intelligent control, robotics and mechatronics including human-machine teaming, knowledge-based paradigms, learning paradigms, machine ethics, intelligent data analysis, knowledge management, intelligent agents, intelligent decision making and support, intelligent network security, trust management, interactive entertainment, Web intelligence and multimedia.

The publications within "Advances in Intelligent Systems and Computing" are primarily proceedings of important conferences, symposia and congresses. They cover significant recent developments in the field, both of a foundational and applicable character. An important characteristic feature of the series is the short publication time and world-wide distribution. This permits a rapid and broad dissemination of research results.

** Indexing: The books of this series are submitted to ISI Proceedings, EI-Compendex, DBLP, SCOPUS, Google Scholar and Springerlink **

More information about this series at http://www.springer.com/series/11156

Xiao-Zhi Gao · Shailesh Tiwari ·
Munesh C. Trivedi · Krishn K. Mishra
Editors

# Advances in Computational Intelligence and Communication Technology

Proceedings of CICT 2019

 Springer

*Editors*
Xiao-Zhi Gao
School of Computing
University of Eastern Finland
Kuopio, Finland

Shailesh Tiwari
Computer Science Engineering Department
ABES Engineering College
Delhi, India

Munesh C. Trivedi
Department of Computer Science
and Engineering
National Institute of Technology Agartala
Agartala, Tripura, India

Krishn K. Mishra
Motilal Nehru National Institute
of Technology
Allahabad, Uttar Pradesh, India

ISSN 2194-5357          ISSN 2194-5365   (electronic)
Advances in Intelligent Systems and Computing
ISBN 978-981-15-1274-2          ISBN 978-981-15-1275-9   (eBook)
https://doi.org/10.1007/978-981-15-1275-9

This Springer imprint is published by the registered company Springer Nature Singapore Pte Ltd.
The registered company address is: 152 Beach Road, #21-01/04 Gateway East, Singapore 189721, Singapore

# Preface

The International Conference CICT-2019 is an international forum for researchers, developers, and end-users to explore cutting-edge ideas and results for the problems involved in the general areas of Communication, Computational Sciences and Technology to disseminate and share novel research solutions to real life problems that fulfill the needs of heterogeneous applications and environments, as well to identify new issues and directions for future research and development. CICT also provides an international communication platform for educational technology and scientific research for the universities and engineering field experts, and professionals.

Nowadays, globalization of academic and applied research is growing with speedy pace. Computer, communication, and computational sciences are the heating areas with a lot of thrust. Keeping this ideology in preference, ABES Engineering College, Ghaziabad, India, has come up with an event—International Conference on Computational intelligence and Communication Technology (CICT 2019) during February 22–23, 2019.

ABES Engineering College, Ghaziabad, was established in the year 2000. The institute is now all set to take higher education to the masses that are still not getting benefits of various researches done for the socioeconomic and cultural values. This is the fifth time the institute is organizing the International Conference on Computational intelligence and Communication Technology (CICT 2019), with a foreseen objective of enhancing the research activities at a large scale. The Technical Program Committee and Advisory Board of CICT 2019 include eminent academicians, researchers, and practitioners from abroad as well as from all over the nation.

In this book, selected manuscripts have been subdivided into various tracks namely intelligent computational science and engineering, advanced security techniques, Web and informatics, intelligent hardware and software design, and computer graphics and vision. A sincere effort has been made to make it an immense source of knowledge for all and includes 46 manuscripts. The selected manuscripts have gone through a rigorous review process and are revised by authors after incorporating the suggestions of the reviewers.

CICT 2019 received around 200 submissions from around 235 authors of 6 different countries such as China, India, Egypt, Bangladesh, Taiwan, and Malaysia. Each submission has been gone through the plagiarism check. On the basis of plagiarism report, each submission was rigorously reviewed by atleast two reviewers with an average of 2.7 papers per reviewer. Even some submissions have more than two reviews. On the basis of these reviews, 48 high-quality papers were selected for publication in this proceedings volume, with an acceptance rate of 22.8%.

We are thankful to the keynote speakers—Prof. Andrew K. Ng, SIT, Singapore, Prof. T. V. Vijaykumar, JNU, Delhi, India, Mr. Aninda Bose, Senior Editor, Springer Nature, for enlightening the participants with their knowledge and insights. We are also thankful to the delegates and the authors for their participation and their interest in CICT 2019 as a platform to share their ideas and innovation. We are also thankful to the Prof. Dr. Janusz Kacprzyk, Series Editor, AISC, Springer, for providing guidance and support. Also, we extend our heartfelt gratitude to the reviewers and Technical Program Committee Members for showing their concern and efforts in the review process. We are indeed thankful to everyone directly or indirectly associated with the conference organizing team for leading it toward the success.

Although utmost care has been taken in compilation and editing, a few errors may still occur. We request the participants to bear with such errors and lapses (if any). We wish you all the best.

Ghaziabad, Uttar Pradesh, India

Xiao-Zhi Gao
Shailesh Tiwari
Munesh C. Trivedi
Krishn K. Mishra

# Contents

# Web and Informatics

# About the Editors

**Prof. Xiao-Zhi Gao** received his B.Sc. and M.Sc. degrees from the Harbin Institute of Technology, China, in 1993 and 1996, respectively. He obtained his D.Sc. (Tech.) degree from the Helsinki University of Technology (now Aalto University), Finland, in 1999. He has been working as a Professor at the University of Eastern Finland, Finland, since 2018. He is also a Guest Professor at the Harbin Institute of Technology, Beijing Normal University, and Shanghai Maritime University, China. Prof. Gao has published more than 400 technical papers on refereed journals and international conferences, and his current Google Scholar H-index is 30. His research interests are nature-inspired computing methods with their applications in optimization, data mining, machine learning, control, signal processing, and industrial electronics.

**Prof. Shailesh Tiwari** currently working as a Professor in Computer Science and Engineering Department, ABES Engineering College, Ghaziabad, India. He is an alumnus of Motilal Nehru National Institute of Technology Allahabad, India. His primary areas of research are software testing, implementation of optimization algorithms, and machine learning techniques in various problems. He has published more than 50 publications in international journals and in proceedings of international conferences of repute. He is editing Scopus, SCI, and E-SCI-indexed journals. He has organized several international conferences under the banner of IEEE and Springer. He is a senior member of IEEE, member of IEEE Computer Society, and Fellow of Institution of Engineers (FIE).

**Dr. Munesh C. Trivedi** currently working as an Associate Professor in the Department of Computer Science and Engineering, National Institute of Technology Agartala, Agartala, Tripura, India. He has published 20 textbooks and 80 research publications in different international journals and proceedings of international conferences of repute. He has received Young Scientist and numerous awards from different national as well as international forums. He has organized several international conferences technically sponsored by IEEE, ACM, and Springer. He is on the review panel of IEEE Computer Society, International Journal of Network Security, Pattern Recognition Letter, and Computer & Education (Elsevier's Journal). He is an executive committee member of IEEE India Council and IEEE Asia Pacific Region 10.

**Dr. Krishn K. Mishra** is currently working as Assistant Professor, Department of Computer Science and Engineering, Motilal Nehru National Institute of Technology Allahabad, India. He has also been worked as Visiting Faculty in the Department of Mathematics and Computer Science, University of Missouri, St. Louis, USA. His primary areas of research include evolutionary algorithms, optimization techniques and design, and analysis of algorithms. He has also published more than 50 publications in international journals and in proceedings of internal conferences of repute. He is serving as a program committee member of several conferences and also edited Scopus and SCI-indexed journals. He is also a member of reviewer board of Applied Intelligence Journal, Springer.

# Intelligent Computational Science

# Influence of Wind Speed on Solar PV Plant Power Production—Prediction Model Using Decision-Based Artificial Neural Network

Roshan Mohanty and Paresh G. Kale

**Abstract** An effort has been made to study the different factors influencing the output of a solar photovoltaic (PV) plant. Environmental factors play a significant role in planning the placement of PV modules. Various factors like irradiation, wind speed, atmospheric temperature and cloud coverage affect the power obtained from the plant. Establishing a relationship between the multiple factors and the output of the plant makes it easy to decide and plan the site of installation of the module and for further research purposes. The effect of wind speed on the power output of S. N. Mohanty solar plant located in Patapur, Cuttack (Odisha, IN) is studied. The research work has been carried out with data acquired every 15 min from January 2015 to December 2015 and keeping the irradiation constant. The monthly plots help establish a relationship between the wind speed and the power generated from the plant. The data is then used to create and train a neural network model which is used to predict the power output of the plant given the appropriate input data. This paper demonstrates the training of artificial neural network using environmental and PV module data (module temperature, wind speed and month).

**Keywords** Renewable · Photovoltaic · Wind speed · Gaussian · Artificial neural network

## Abbreviations

| | |
|---|---|
| PV | Photovoltaic |
| ANN | Artificial neural network |
| NOCT | Nominal operating cell temperature |
| PCC | Pearson's correlation coefficient |

R. Mohanty (✉) · P. G. Kale
Department of Electrical Engineering, National Institute of Technology,
Sundargarh, Rourkela, Odisha 769008, India
e-mail: mroshan54@gmail.com

P. G. Kale
e-mail: pareshkale@nitrkl.ac.in

© Springer Nature Singapore Pte Ltd. 2021
X.-Z. Gao et al. (eds.), *Advances in Computational Intelligence and Communication Technology*, Advances in Intelligent Systems and Computing 1086,
https://doi.org/10.1007/978-981-15-1275-9_1

3

MFFNNBP    Multilayer feed-forward with backpropagation neural networks
GRNN       General regression neural network
FFBP       Feed-forward backpropagation
RNN        Recursive neural network
MLP        Multilayer perceptron
GM         Gamma memory
RBFEF      Radial basis function exact fit
FFNN       Feed-forward neural network
DBNN       Decision-based neural network
RMSE       Root mean square error

# 1 Introduction

The increasing energy demand of the world is putting much pressure on the conventional energy source available in limited amount. The need for coal has increased in India with import dependency rising from negligible in 1990 to 23% in 2014. An alternative source of energy such as renewable energy source is essential which can provide power in a sustainable manner which would ease the pressure on these conventional energy sources. Renewable energy sources such as sunlight, wind, rain, tides, waves and geothermal heat can be used again and again without exhausting it. The demand for the sun's energy, which is abundant, has increased since it is convenient to use and freely available. Solar PV system has attracted the attention of many in recent years due to its convenience and reliability. Environmental crisis coupled with depletion of conventional energy sources has accelerated the use of solar PV systems. The amount of solar energy reaching earth every year is around $1.5 \times 10^{18}$ kWh/year which is captured by PV devices and converted into a useful form, i.e., electricity. The PV system consists of the PV module, the conversion system usually comprising of power electronics converters and a control unit to regulate the power supply. The fundamental unit of a PV system is solar cells which are combined to give PV module which is further expanded to form PV array. There are three types of PV systems, namely monocrystalline PV system, polycrystalline PV system and thin film PV system.

The PV module characteristics are given by maximum power, maximum power voltage, maximum power current, open-circuit voltage and short-circuit current [1]. The study conducted by Dinçer et al. showed that the energy conversion efficiency increases with the reduction in reflection of incident light. Another survey by Chikate and Sadawarte [2] showed that the open-circuit voltage increases logarithmically by increasing the solar radiation, whereas the short-circuit current increases linearly. Apart from the factors mentioned above, the performance is also affected by environmental factors such as solar flux, temperature and relative humidity. Usually, the specifications given for each PV panel are for Standard Test Conditions (STC). However, the performance of PV modules does not remain constant when deployed in outdoor conditions for a significant duration due to several environmental factors [3, 4]. The effect of wind speed on the power output for a solar PV plant was studied using data which comprised of primary environmental parameters that might affect the power output of S. N. Mohanty solar plant. The parameters include module temperature, ambient temperature, irradiation and wind speed. The other parameters are kept constant to study the effect of only wind speed. An effort has been made to investigate the relation between wind speed and power output.

## 2   Background Details and Related Work

Different studies have been conducted to study the PV module under different environmental conditions. Research has been done to establish a relationship between the output power, efficiency, etc. of a solar module and various factors. Solar PV plant is called an uncontrollable power source due to the uncertain nature of the factors affecting the output. Hence, there is a need for predictive tools such as ANN to study and optimize the performance of PV plants.

### 2.1   Effects of Environmental Factors on Solar Panel

Lay-Ekuakille et al. [5] have attempted to explore the PV plant site in depth by using tools to simulate the behavior of the module by varying the different environmental factors like ambient temperature peaks, module temperature, irradiance and external humidity. The plots of output power versus the parameters mentioned above for CdTe and CIS PV module showed a similar trend with irradiation affecting output power the most. Similar methods have been adopted by Ali et al. [3] to study the performance of photovoltaic panel under different colors of light, dust, shading and irradiance. Dust and shading decrease the amount of radiation reaching the solar cell and hence the output power. In an attempt to the study the effects of environmental parameters on solar PV performance, Garg [4] concluded in her research that the maximum power increases with increase in radiation and decreases with increase in humidity, ambient temperature and wind speed. Touati et al. have mentioned in their research that dust accumulation, panel's temperature and relative humidity affect the efficiency of the panel with dust reducing the efficiency by 10% [6].

Irradiance is the factor that majorly and directly affects the performance of the PV module. The module works best when the ambient temperature is minimum, and the irradiation is maximum. Other factors like module temperature and wind speed play a vital role too. A study was conducted at Lucknow, India, by Siddiqui et al. in an attempt to establish a relation of the efficiency of a PV module with ambient temperature and wind velocity. Equation 1 represents the relationship established by them where $\eta$ is the efficiency of the PV module; the $T$ is the average ambient temperature in °C, and $V$ is wind speed in m/s [1].

$$\eta = 14.9852 - 0.0866 \times T + 0.017647 \times V \qquad (1)$$

A study conducted in Greece by Kaldellis et al. highlighted the effect of cell temperature on efficiency. Wind cools the solar cell which in turn produces a change in efficiency, $\delta_n$ as shown in Eq. 2 [7] where the value of $\beta_{ref}$ is usually between 0.3 and 0.5%/°C, $T_C$ and $T_{NOCT}$ represent the cell temperature and the nominal operating cell temperature (NOCT), respectively, and $\eta_o$ represents the efficiency at NOCT.

$$\delta_n = \beta_{ref} \times (T_C - T_{NOCT}) \times \eta_o \qquad (2)$$

Manokar et al. have demonstrated the effect on wind speed on the power output of the PV module under constant solar irradiation and ambient temperature in the form of a graph which is monotonically increasing till a certain point that corresponds to the wind speed that gives maximum power output [8]. Mekhilef et al. have concluded in their study that wind helps in removing the heat from the cell surface and in turn increases the efficiency of the module [9].

## 2.2 Correlation Between Environmental Factors and the Output Power

Pearson's correlation coefficient (PCC), given by 'r,' can be used to identify the relationship between any of the environmental parameter and output power. Zhu et al. [10] have used PCC to study the influence of environmental parameters in different weather conditions such as clear, cloudy and overcast. The environmental parameters considered are irradiance, temperature, humidity and wind speed. The PCC, $r$, has been calculated between any one environmental factor and the output power. Mandal et al. [11] have also given PCC values for different factors such as solar irradiance, ambient and module temperature, wind speed and wind direction.

## 2.3 Prediction of Output Power Using Artificial Neural Network (ANN)

McCulloch and Pitts initially introduced ANN which helps to establish a complicated relationship between the inputs and outputs after training. Many prediction models given by researchers using different input parameters and training algorithm were studied. Yang et al. devised a weather-based hybrid model for day ahead hourly forecasting of PV power output with an aim to ensure good real-time performance and to minimize negative impacts of the whole system [12]. Qasrawi and Awad used multilayer feed-forward with backpropagation neural networks (MFFNNBP) to predict the power output of the solar cells in different places in Palestine for one day, one month and finally for the whole year [13]. Saberian et al. used general regression neural network (GRNN) feed-forward backpropagation (FFBP) with four inputs (maximum temperature, minimum temperature, mean temperature and irradiance) and one output to model a photovoltaic panel output power and approximate the generated power. They concluded that FFBP is the best way for neural network training [14]. Simon et al. have used similar day ahead prediction techniques to predict the profile of produced power of a grid-connected 20 kW solar power plant in Tiruchirappalli, India [15]. Brano et al. has obtained the short-term power output forecast using three different types of ANN, namely a hidden layer multilayer perceptron (MLP), a recursive neural network (RNN) and a gamma memory (GM) trained with backpropagation [16]. Similar practices have been adopted to predict the power output profile for different requirements using neural network. Other methods such as least square optimization of numerical weather prediction and insolation prediction have also been used for the same purpose [17, 18]. This paper demonstrates the use of radial basis function (exact fit) (RBFEF) and feed-forward neural network (FFNN) models for prediction.

RBFEF consists of three layers, namely the input layer, a hidden layer and an output layer. The transfer function used in the hidden layer of RBFEF is called radial basis function. Equation 3 represents the output of RBFEF where $X$ is the input vector, $\sigma$ is the Gaussian spread coefficient and $w_k$ is the weight to be tuned [13].

$$h(\bar{X}) = \sum_{k=1}^{N} w_k \exp(-\sigma \left\|\bar{X}_n - \bar{X}_k\right\|^2) \tag{3}$$

An FFNN, trained by a Levenberg–Marquardt backpropagation algorithm, has three or more layers out of which two are input and output layers. There can be more than one hidden layers with a different number of neurons and activation functions. The network is trained by comparing the error values, i.e., the difference between predicted value and actual value and using the error to tune the weights [19].

## 3 Methodology

The irradiance, module temperature, wind speed and power output data were obtained from S. N. Mohanty solar plant for the period January 2015–April 2015 every 15 min. The SCADA system used in S. N. Mohanty solar plant measures and records these meteorological data. An effort has been made to study the influence of wind speed on the output power using plots between the same and PCC. Neural network models have been built to predict the output power by taking some particular parameters as input.

## 3.1 The Dependency of Output Power on the Environmental Factors

Ranges of module temperature, irradiation, ambient temperature and wind speed reading are listed in Table 1 to have an idea of the range of data obtained from the solar power plant. Further analysis requires normalized data and removal of erroneous data points. The value of irradiance consists of three spans, 4.5–5, 4–4.5 and 3.5–4 kWh/m$^2$ represented by *I*, *II* and *III*, respectively. The curve of power output versus wind speed was plotted and fitted using the *cftool* toolbox provided in MATLAB using Gaussian fitting. The curve fitting details are listed in for irradiance of ranges *I*, *II* and *III*, respectively. The categorization is done to study the relationship between power output and wind speed keeping the irradiation constant. PCC, calculated between wind speed and power output for two cases, is used to further study the influence of wind speed on the output power. PCC was calculated with the original data in the first case and was computed keeping parameters other than wind speed fixed in the second case.

## 3.2 Prediction of Output Power Using FFNN and RBFEF

Neural network models were built to predict the output power of the solar plant. Module temperature and month influence the output power significantly. However, the effort has also been made to include the impact of wind speed for prediction purpose. Figure 1 shows the structure of FFNN with one hidden layer with eight neurons, module temperature, wind speed and month as inputs and output power as the output of the network. Inputs comprise the first 300 values of module temperature, wind speed and normalized month. Sample input of 61 values is taken from the dataset of January 2014 to verify the models. FFNN with one and two hidden layers was used with eight neurons in the hidden layer and log sigmoid activation function. A validation check was fixed before training, and 'spread constant' is set at 0.6 for RBFEF. The RMSE of all the methods were compared to conclude which network was the best at prediction. Decision-based neural network (DBNN) approach was used to get a lesser RMSE and hence a more accurate output. Comparison of the three output values from the three networks corresponding to one data point was carried out to build DBNN. The output value closest to the actual is considered to get a new set of output values which is expected to give a lesser RMSE value. Figures 2 and 3 show the steps of prediction using FFNN, RBFEF and DBNN.

**Table 1** Range of environmental factors taken from plant data for a year

| Month | Minimum module temperature (°C) | Maximum module temperature (°C) | Maximum irradiation (kWh/m²) | Minimum ambient temperature (°C) | Maximum ambient temperature (°C) | Minimum wind speed (km/h) | Maximum wind speed (km/h) |
|---|---|---|---|---|---|---|---|
| January | 10.19 | 52.57 | 5.77 | 14.39 | 32.58 | 0.51 | 19.90 |
| February | 12.71 | 62.98 | 6.05 | 16.56 | 38.46 | 0.51 | 21.39 |
| March | 17.27 | 61.43 | 6.50 | 21.54 | 40.84 | 0.52 | 23.21 |
| April | 22.05 | 63.13 | 6.62 | 23.79 | 42.86 | 0.52 | 25.63 |
| May | 22.99 | 62.48 | 6.24 | 24.96 | 47.17 | 0.53 | 25.01 |
| June | 25.39 | 61.83 | 5.93 | 26.10 | 43.48 | 0.53 | 29.15 |
| July | 23.95 | 59.47 | 5.67 | 24.32 | 39.25 | 0.52 | 39.39 |
| August | 24.39 | 59.23 | 6.04 | 24.41 | 41.03 | 0.52 | 33.07 |
| September | 25.55 | 61.03 | 6.32 | 25.46 | 39.19 | 0.52 | 20.74 |
| October | 21.81 | 65.13 | 6.00 | 23.87 | 39.37 | 0.52 | 16.12 |
| November | 16.54 | 55.94 | 5.82 | 19.83 | 37.16 | 0.51 | 21.00 |
| December | 10.39 | 55.44 | 5.44 | 14.05 | 33.84 | 0.51 | 17.33 |

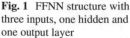
**Fig. 1** FFNN structure with three inputs, one hidden and one output layer

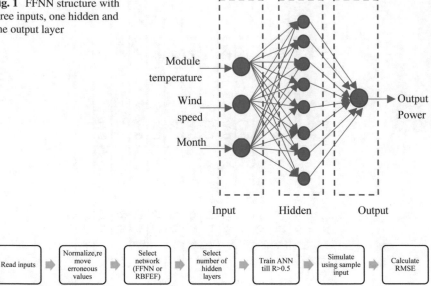

Module
temperature

Wind
speed

Month

Output
Power

Input          Hidden          Output

**Fig. 2** Process of prediction using ANN by selecting the type of network and validating it for a set of sample input

| Store FFNN and RBFEF outputs | Compare with actual output. | Find the network with minimum difference | Set this output as DBNN output | Repeat this process for every data point | Calculate RMSE |

**Fig. 3** Calculation of DBNN outputs from FFNN and RBFEF outputs and calculation of RMSE

## 4  Results and Discussion

The function and plots between wind speed and output power are studied to establish a relation between them. A model is built to calculate the output power when data of wind speed is available for a particular range of irradiation. PCC is calculated for different cases to study the influence of wind speed. Wind speed is expected to have a low impact on the output power. PCC is computed keeping other parameters like ambient temperature, module temperature and irradiation constant to study the change in the influence of wind speed. Prediction models (FFNN, RBFEF, DBNN) are developed to predict the output power.

## 4.1  Dependency of Output Power on Wind Speed

Gaussian fitting of output power results in Eq. 4 which represents the model to calculate the output power where $x$ is wind speed, $f(x)$ is the power output and $a$, $b$ and $c$ are constants.

$$f(x) = a \times e^{((-\frac{x-b}{c}))^2} \qquad (4)$$

The coefficients $a$, $b$, $c$, with their averages listed in Table 2, were observed to study its dependency on radiation for different months. The coefficient $a$ is found to decrease with an increase in the irradiation levels, but there is no particular pattern observed in the plots of coefficients $b$ and $c$. However, it is observed that the coefficient $c$ has a high value for lower ranges of radiation. Hence for smaller wind speeds and low value of irradiation, the output power approaches the constant $a$. Figure 4 shows the Gaussian fitted curve of normalized power output versus normalized wind speed for January where black dots represent the values of output power for particular wind speed. The model shown in Eq. 4 is obtained from the Gaussian fitting with the nonlinear least square method of fitting. Gaussian fitting is preferred over other methods as it gives a better fit with a higher value of $R$. PCC results are listed in Table 3 which shows low '$r$' value because the wind speed does not affect the output power significantly as expected. PCC-I is calculated taking the effect of other environmental factors, whereas PCC-II is computed keeping other environmental parameters constant. The PCC value increases considerably for certain months when other environmental factors are held constant. Hence, the wind speed shows a more significant impact on the output power when other parameters are fixed.

## 4.2  Comparison of RMSE of FFNN and RBFEF and Development of DBNN

The training is done using *nntool* with three models till '$R$' value is greater than 85% ensuring a good fit. The plot of predicted outputs of FFNN with one hidden layer, RBFEF, FFNN with two hidden layers as represented in Fig. 5 shows that the models can predict the output power with some minor error. The RMSE of each network, listed in Table 4, indicates that FFNN with two hidden layers has the least RMSE (0.0695) and gives the best prediction result among the three networks. Increasing hidden layers increases the accuracy of prediction as observed from RMSE values. The concept of DBNN is used to reduce the RMSE further to obtain a more accurate model. The output of the above-mentioned predictive are models used for DBNN, which is found to give even lesser RMSE, i.e., 0.0505 or a more accurate prediction as shown in Fig. 6.

**Table 2** Average values of coefficients $a$, $b$, $c$

| Coefficient | $a$ | | | $b$ | | | $c$ | | |
|---|---|---|---|---|---|---|---|---|---|
| Irradiation range | I | II | III | I | II | III | I | II | III |
| Average | 0.25 | 0.29 | 0.34 | 0.20 | 0.51 | 0.44 | 0.28 | 3.40 | 3.01 |

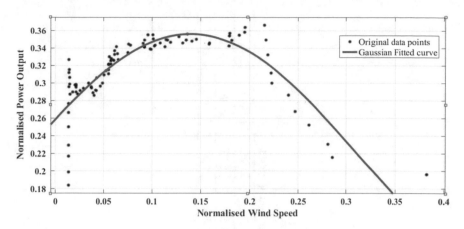

**Fig. 4** Smoothened curve of normalized power output plotted against normalized wind speed using *cftool*

## 5 Conclusion

Wind speed does not have a significant impact on the output power owing to the low PCC values. Hence, module temperature and month number are used along with wind speed for the day ahead prediction using neural network. The mathematical model established using a curve fitting tool in MATLAB provides a way to determine the output power from wind speed for fixed ranges of irradiation and for a particular month. Three ANN models: FFNN (one hidden layer), FFNN (two hidden layers) and RBFEF give the prediction of the output power of the plant under study. The comparison of the RMSE of the three networks shows that FFNN with two hidden layers has the least RMSE (0.0695). A more useful tool named DBNN compares the outputs of the three models with the original output to give modified outputs resulting in a lesser RMSE (0.0504) or a more accurate prediction. This model will be useful in the future part of this project which will involve a self-built data acquisition system followed by analysis and day ahead prediction of required parameter.

**Table 3** Pearson's correlation coefficients between output power and wind speed when other parameters and constant or variable

| Month | Jan | Feb | Mar | Apr | May | Jun | Jul | Aug | Sep | Oct | Nov | Dec |
|---|---|---|---|---|---|---|---|---|---|---|---|---|
| PCC-I | 0.04 | 0.39 | 0.32 | 0.28 | 0.29 | 0.24 | 0.28 | 0.28 | 0.32 | 0.27 | 0.29 | 0.32 |
| PCC-II | 0.19 | 0.43 | 0.14 | 0.19 | −0.23 | 0.06 | 0.85 | 0.08 | 0.68 | 0.24 | 0.49 | −0.08 |

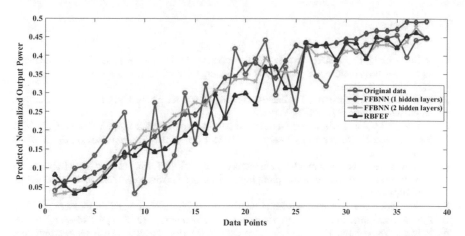

**Fig. 5** Plot of normalized output power predicted using FFNN and RBFEF network models

**Table 4** RMSE for training using different networks

| Neural network model | RMSE |
|---|---|
| FFNN (one hidden layer) | 0.0727 |
| FFNN (two hidden layers) | 0.0695 |
| RBFEF | 0.0738 |
| DBNN | 0.0504 |

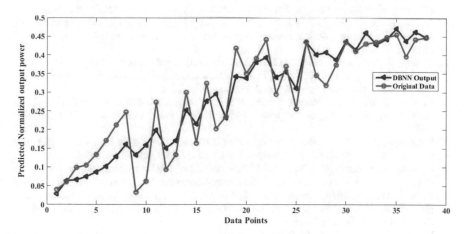

**Fig. 6** Plot of normalized output power predicted using DBNN model

# References

1. R. Siddiqui, U. Bajpai, Deviation in the performance of solar module under climatic parameter as ambient temperature and wind velocity in composite climate. Int. J. Renew. Energy Res. **2**(3), 486–490 (2012)
2. B.V. Chikate, Y. Sadawarte, The Factors affecting the performance of solar cell. Int. J. Comput. Appl. Sci. Technol. 975–8887 (2015)
3. R. Bhol, S.M. Ali, Environmental effect assessment on performance of solar PV panel, in *2015 International Conference on Circuits, Power and Computing Technologies* [ICCPCT-2015] (2015)
4. V. Garg, Effect of environmental parameters on solar PV performance with MPPT techniques on induction motor driven water pumping system. Int. J. Adv. Res. Ideas Innov. Technol. **3**, 996–1007 (2017)
5. P. Vergallo, Effects of environmental conditions on photovoltaic module measurements, in *2013 Seventh International Conference on Sensing Technology* (2013), pp. 944–947
6. F. Touati, M. Al-hitmi, H. Bouchech, Towards understanding the effects of climatic and environmental factors on solar PV performance in arid desert regions (Qatar) for various PV technologies, in *2012 First International Conference on Renewable Energies and Vehicular Technology* (2012), pp. 78–83
7. J.K. Kaldellis, M. Kapsali, K.A. Kavadias, Temperature and wind speed impact on the efficiency of PV installations. Experience obtained from outdoor measurements in Greece. Renew. Energy **66**, 612–624 (2014)
8. A.M. Manokar, D.P. Winston, A.E. Kabeel, S.A. El-Agouz, Integrated PV/T solar still—a mini-review integrated PV/T solar still—a mini-review. Desalination, no. January, pp. 0–1 (2018)
9. S. Mekhilef, R. Saidur, M. Kamalisarvestani, Effect of dust, humidity and air velocity on efficiency of photovoltaic cells. Renew. Sustain. Energy Rev. **16**(5), 2920–2925 (2012)
10. H. Zhu, X. Li, Q. Sun, L. Nie, J. Yao, G. Zhao, A power prediction method for photovoltaic power plant based on wavelet decomposition and artificial neural networks. Energies **9**(1), 1–15 (2016)
11. R.K. Mandal, P. Kale, Development of a decision-based neural network for a day-ahead prediction of solar PV plant power output, in *2018 4th International Conference on Computational Intelligence & Communication Technology,* September 2017
12. H. Yang, S. Member, C. Huang, Y. Huang, Y. Pai, A weather-based hybrid method for 1-day ahead hourly forecasting of PV power output. IEEE Trans. Sustain. Energy **5**(3), 917–926 (2014)
13. I. Qasrawi, Prediction of the power output of solar cells using neural networks: solar cells energy sector in palestine. Int. J. Comput. Sci. Secur IJCSS **9**, pp. 280–292 (2015)
14. A. Saberian, H. Hizam, M.A.M. Radzi, M.Z.A. Ab Kadir, M. Mirzaei, Modelling and prediction of photovoltaic power output using artificial neural networks. Int. J. Photoenergy **2014** (2014)
15. R.M. Ehsan, S.P. Simon, Day-Ahead prediction of solar power output for grid-connected solar photovoltaic installations using artificial neural networks, in *2014 IEEE 2nd International Conference on Emerging Electronics* (2014)
16. V. Lo Brano, G. Ciulla, M. Di Falco, Artificial neural networks to predict the power output of a PV panel. Int. J. Photoenergy **2014**, 12 (2014)
17. A. Yona et al., Application of neural network to 24-hour-ahead generating power forecasting for PV system (2008)
18. D.P. Larson, L. Nonnenmacher, C.F.M. Coimbra, Day-ahead forecasting of solar power output from photovoltaic plants in the American Southwest. Renew. Energy **91**, 11–20 (2016)
19. S.O. Haykin, *Neural Networks and Learning Machines,* 3rd edn. (Pearson, 2008)

# Prediction of Compressive Strength of Ultra-High-Performance Concrete Using Machine Learning Algorithms—SFS and ANN

Deepak Choudhary, Jaishanker Keshari and Imran Ahmed Khan

**Abstract** This paper presents machine learning algorithms based on back-propagation neural network (BPNN) that employs sequential feature selection (SFS) for predicting the compressive strength of ultra-high-performance concrete (UHPC). A database, containing 110 points and eight material constituents, was collected from the literature for the development of models using machine learning techniques. The BPNN and SFS were used interchangeably to identify the relevant features that contributed with the response variable. As a result, the BPNN with the selected features was able to interpret more accurate results ($r^2 = 0.991$) than the model with all the features ($r^2 = 0.816$). It is concluded that the usage of ANN with SFS provided an improvement to the prediction model's accuracy, making it a viable tool for machine learning approaches in civil engineering case studies.

**Keywords** ANN · SFS · UHPC · Compressive strength · Constituents

## 1 Introduction

Several types of machine learning algorithms such as artificial neural network (ANN) have been used in different fields for the development of models that predict response parameters (experimental dataset) using certain independent input parameters. However, an experiment could have a large number of independent parameters most of which are redundant and have negligible effects on the response parameters. Therefore, an artificially intelligent (AI) selection algorithm is required to overcome this shortcoming and identify the underlying parameters that improve the model's accuracy and simplify the computational complexity. The need for soft computing tools and models for the prediction of behavioral properties of engineering components, systems, and materials is continuously rising. ANN emerged as one of soft computing paradigms that have been successfully applied in several engineering fields [1].

D. Choudhary (✉) · J. Keshari
Electronics & Communication Department, ABES Engineering College, Ghaziabad, UP, India
e-mail: engg_deepak@yahoo.com

I. A. Khan
Instrumentation & Control Department, GCET, Gr. Noida, UP, India

© Springer Nature Singapore Pte Ltd. 2021
X.-Z. Gao et al. (eds.), *Advances in Computational Intelligence and Communication Technology*, Advances in Intelligent Systems and Computing 1086,
https://doi.org/10.1007/978-981-15-1275-9_2

Specifically, ANN has been used to solve a wide variety of civil engineering problems [2–4]. Mainly, ANN was utilized to model the nonlinear behavior of fatigue and creep of reinforced concrete (RC) members [5–8]. Recently, research interest has revolved around the development of ANN models to interpret the behavior of structural materials such as steel, concrete, and composites [9–14]. The utilization of ANN modeling made its way into the prediction of fresh and hardened properties of concrete based on given experimental input parameters, whereby several authors developed AI models to predict the compressive strength of normal weight, lightweight, and recycled concrete [14–17]. Afterward, several authors began developing ANN models for the prediction of compressive strength of high-performance concrete [18–21]. In this study ANN is employed with other machine learning techniques to identify the parameters that capture the compressive strength of UHPC using data collected from the literature.

## 2 UHPC and ANN Background

The evolution of UHPC has lead structural engineers to improve the compressive strength, ductility, and durability of heavy-loaded reinforced concrete (RC) structures. Several researchers have been investigating the mechanical behavior of UHPC and its applications over the last four decades, where it was founded that UHPC exhibits a compressive strength that would range from 150 to 810 MPa [22, 23]. The underlying material constituents that enable such a superior mix are cement (up to 800 kg/m$^3$), water/binder ratio that is lower than 0.20, high-range water-reducing (HRWR) admixture, very fine powders (crushed quartzite and silica fume), and steel fibers [24]. Other researchers proposed different mixtures by adding fly ash and sand to reduce the amount of cement and silica fume, and acquire an optimum mix, that is both economical and sustainable [25, 26]. However, most of the aforementioned mixtures result in exhausting a large amount of resources and performing tests on many batches, while barely predicting the strength of UHPC [19]. Therefore, researchers began conducting investigations on the utilization of machine learning techniques for the development of prediction models that could assist engineers and researchers to produce appropriate UHPC mixes.

Ghafari et al. [19] used the back-propagation neural network (BPNN) and statistical mixture design (SMD) in predicting the required performance of UHPC. The objective of the study was to develop an ANN and SMD model to predict both the compressive strength and the consistency of UHPC with two different types of curing systems (steam curing and wet curing). As a result, BPNN proved to be more accurate than SMD in the prediction of compressive strength and slump flow of UHPC. Despite the statistical advantages of ANN, it has been long regarded as a black box that evaluates functions using input covariates and yielding outputs. Meaning, the model does not produce any analytical model with a mathematical structure that can be studied. Therefore, ANN should be utilized in detecting the dominant input parameters that have direct association with the ANN model. This will reduce the amount of parameters in the model, which will improve the computation complexity of the ANN model and simplify the derivation strategies of a mathematical model used to predict the compressive strength of UHPC. In addition, prediction of compressive strength of high-strength and high-performance concrete was addressed by other researchers [20, 21].

There are several machine learning techniques, in the literature, that assist researchers in identifying the underlying covariates impacting the prediction model. Sequential feature selection (SFS) is a machine learning tool that sequentially selects features and inputs them into a fitting model (i.e., ANN) until the model's error function increases. This technique makes use of ANN's complex computation and allows the SFS tool to select and remove the influential and redundant parameters, respectively. The reduction in the covariate domain improves the accuracy of the fitting model, decreases its computation time, and facilitates a better understanding of the data processing [27]. There are two types of SFS classes—mainly filter method and wrapper method [28]—where Zhou et al. [29] used the Markov Blanket with a wrapper method to select the most relevant features of human motion recognition. Four sets of open human motion data and two types of machine learning algorithms were used. The total number of features was reduced rapidly, where this reduction helped the algorithm demonstrate better recognition accuracy than traditional methods. Moreover, Rodriguez-Galiano et al. [30] used SFS when tackling ground water quality problems, where 20 datasets of parameters were extracted from a GIS database. Four types of machine learning algorithms were used as wrappers for the SFS. As a result, the rain forest machine learning algorithm used with SFS showed promising results, where only three features were sufficient enough in predicting the most accurate results.

## 3 Methodology of Modeling

The steps that were are followed in developing a robust and accurate numerical model using SFS include (1) design and validation of ANN model by manipulating the number of neurons and hidden layers; (2) execution of SFS using ANN as a wrapper; and (3) analysis of selected features using both ANN and nonlinear regression. Table 1 presents the initial input variables together with their range (maximum and minimum values) and symbols for identifying them in this experimental program.

## 3.1 Artificial Neural Network

Artificial neural network (ANN) is a machine learning tool that imitates the learning functions of a human brain by providing a robust technique in classifying and predicting certain outcomes based on the model's objective. There are two types of ANN models: (1) feed forward; and (2) feed backward. In this study, the feed

**Table 1** Experimental program

| Variable | Symbols | Minimum | Maximum |
|---|---|---|---|
| Cement (kg/m$^3$) | C | 383 | 1600 |
| Silica fume (kg/m$^3$) | SI | 0 | 367.95 |
| Fly ash (kg/m$^3$) | FA | 0 | 448 |
| Sand (kg/m$^3$) | S | 0 | 1898 |
| Steel fiber (kg/m$^3$) | SF | 0 | 470 |
| Quartz powder (kg/m$^3$) | QP | 0 | 750 |
| Water (kg/m$^3$) | W | 109 | 334.5 |
| Admixture (kg/m$^3$) | A | 0 | 185 |
| fc (MPa) | fc | 95 | 240 |

backward ANN is used, where it is composed of input neurons, hidden neurons, bias units, wires containing randomly generated weights, and output neurons. The input neurons are responsible for containing the independent parameter presented by the user, the wires represent the randomly generated matrices called weights that manipulate the function's slope or steepness, the hidden neurons map the weights variables using an activation function, and the bias units control the output function's shift, either upward or downward. Equation (1) shows the linear combination of mapping weights from each input neuron, via wires, to the hidden neurons.

$$O_i = g\left( \sum_{i=0}^{n} \sum_{i=0}^{m} (X_i \times \theta_{ij}^T) \right) \tag{1}$$

where $X_i$ represents the first input parameter of size $R \times 1$ ($R$ is the number of data points), $\theta_{ij}$ is the weight of size $R \times (n + 1)$, $O_i$ is the value of the output neuron or prediction function $h_\theta(X)$, and $g(x)$ is the activation function. The bias unit is simulated by creating a column vector of size $R \times 1$ and assigning it with values of ones, where $X_0$ and $\theta_{i0}$ contain the bias values.

## 3.2 Sequential Feature Selection

SFS reduces the dimensionality of data by selecting only a subset of measured features to create a prediction model. SFS is composed of two components: the objective function, which is the criteria the algorithm follows when selecting the features (i.e., the NMSE), and the search algorithm, which are the methods of how the machine add/removes the features from the subset. There are two types of search algorithms: sequential forward selection and sequential backward selection. In this study, the previously verified ANN model was used as the objective function and the forward selection was used in selecting the relevant features. Figure 1 shows the algorithm SFS uses when performing forward selection.

**Fig. 1** Forward selection algorithm

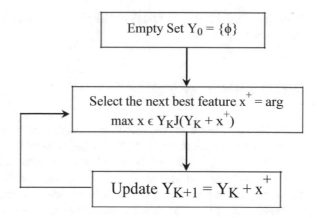

Empty Set $Y_0 = \{\phi\}$

Select the next best feature $x^+ = \arg$ $\max x \in Y_K J(Y_K + x^+)$

Update $Y_{K+1} = Y_K + x^+$

# 4 AI Modeling for Prediction of Compressive Strength of UHPC

## 4.1 Verification of ANN

The ANN numerical solver, Levenberg–Marquardt, was verified by testing different number of neurons using a basis like the normalized mean square error (NMSE) to measure the error. The increment started from one neuron and ended with 15 neurons, where the model was analyzed ten times, for each increment, because the Levenberg–Marquardt algorithm locates the local, and not the global, minimum of a function. Hence, for each neuron tested, ten NMSE values will be stored in a column vector, where each column vector will be averaged and plotted against its corresponding number of neuron(s). Figure 2 shows the plot of all the scenarios with the minimum point circled at 11 neurons. Therefore, 11 neurons is, approximately, the number of

**Fig. 2** Verification curve for ANN

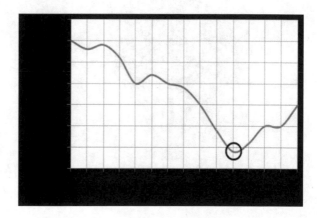

neurons that is sufficient enough for BPNN to facilitate an accurate ANN model for the collected dataset.

## 4.2 Execution of SFS

The SFS algorithm was run 200 times to capture all possible combinations of independent features when using ANN. Table 2 shows the percentage of features that were used during the 200 trials. Based on the results of these trials, the most abundant combination during the SFS analysis, within a 20% threshold, was selected as the important parameters that contribute mostly in the model. In this study, four variables (cement, silica fume, fly ash, and water) were selected as the most relevant features for the prediction model. Figure 3a, b presents the architecture of both ANN models, before and after selection.

**Table 2** SFS results

| Materials | Percentage used (%) |
| --- | --- |
| C | 24 |
| SI | 33 |
| FA | 24 |
| S | 14 |
| SF | 6 |
| QP | 5 |
| W | 31 |
| A | 13 |

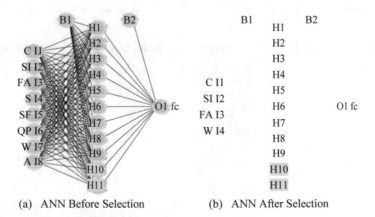

(a) ANN Before Selection          (b) ANN After Selection

**Fig. 3** ANN before and after selection

# 5 Results and Discussions

## 5.1 ANN Results

The selected features, using SFS, were analyzed by the previous BPNN model. As a result, the model that used the selected features showed stronger agreement with the experimental results in contrast with that prior to the selection. Table 3 shows the statistical measurements calculated for both cases. It was observed that the $r^2$ and NMSE before and after selection yielded 81.6% and 99.1%, respectively, and 0.0594 and 0.026, respectively.

The correlation plots between the predicted and experimental results for the ANN models, with and without selected features using SFS, summarized in Fig. 4a present the percent deviation, where an arbitrary percent deviation was plotted above and below the perfect fit line with a deviation value of $\pm20\%$. As a result, the ANN model with the relevant features was capable of predicting 89.6% of its values within the aforementioned boundaries, as opposed to the ANN model with all the features which predicted 58.7% of its values within the boundaries. Figure 4b summarizes the compressive strength ratio between both ANN models, where the model with the selected features demonstrated a higher percentage of values (90.6%) ranging between 0.8 and 1.2 than the other model (63.6%).

**Table 3** Statistical measurements before and after SFS

| Statistics measurements | Before selection | After selection |
|---|---|---|
| $r^2$ | 0.816 | 0.991 |
| NMSE | 0.0594 | 0.026 |

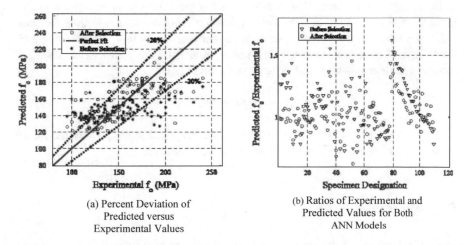

(a) Percent Deviation of Predicted versus Experimental Values

(b) Ratios of Experimental and Predicted Values for Both ANN Models

**Fig. 4** Comparison between experimental and predicted values

**Table 4** LSR coefficients

| UHPC constituents | Coefficient symbols | Coefficient values | $r^2$ | NMSE |
|---|---|---|---|---|
| C | $\theta_1$ | 115.56 | 0.938 | 0.0830 |
| SI | $\theta_2$ | 23.97 | | |
| FA | $\theta_3$ | 5.49 | | |
| W | $\theta_4$ | 27.85 | | |

**Fig. 5** Summary of the proposed model's performance

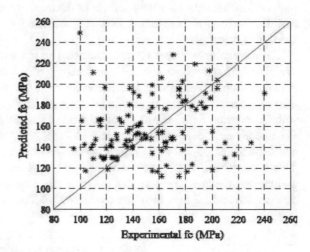

## 5.2 Linear Square Regression Analysis

A linear regression model was developed using the least square regression (LSR) method, where the analytical model consisted of the previously selected features. Table 4 shows the coefficient values, with their corresponding symbols, for each UHPC constituent with the statistical measurements of the LSR model. The LSR model is a linear function, and its form is shown in (2).

$$f_c = \theta_1 C + \theta_2 SI + \theta_3 FA + \theta_4 W \qquad (2)$$

Figure 5 summarizes the performance of the proposed model, where the model achieved an $r^2$ and NMSE of 0.938 and 0.0830, respectively.

## 5.3 Parametric Study

Since the developed LSR model is capable of accurately predicting the experimentally measured compressive strength, a parametric study was conducted, using this model, to study the effect of fly ash and silica fume on the compressive strength of

**Fig. 6** Variation of compressive strength as a function of fly ash and silica fume

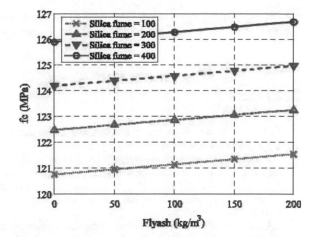

UHPC. Using fly ash quantities that range between 0 and 200 kg/m$^3$ and silica fume quantities that range between 40 and 160 kg/m$^3$ while fixing the quantity of cement at 1400 kg/m$^3$ and water at 175 kg/m$^3$, several plots showing the variation of strength of UHPC were generated as shown in Fig. 6. It is observed from Fig. 6 that there is noticeable increase in the compressive strength of UHPC with the increase in fly ash and more noticeable with the increase in silica fume.

# 6 Conclusion

This study was conducted to detect the correlation between the material constituents of UHPC and its compressive strength. BPNN was used, and three major steps were executed: (1) verification of ANN; (2) application of both SFS and NID, and (3) analysis of selected features using ANN and LSR. The SFS tool was used to select the relevant constituent that impacted have the most impact on the compressive strength of UHPC which are mainly cement, silica fume, fly ash, and water. It can be concluded from this study that:

- The use of ANN with SFS reduced the number of input parameters needed to accurately predict the compressive strength of UHPC. Mix for the prediction of compressive strength, making it less computationally expensive.
- The use of ANN with selected input parameters improved the accuracy of prediction of compressive strength of UHPC and reduced the computational effort. The correlation coefficient ($r^2$) before and after the use of SFS improved from 81.6 to 99.1%, while the NMSE improved from 0.0594 to 0.026, respectively.
- The ANN model with the selected relevant input parameters also showed a lower deviation (89.6%) than the ANN model with all the features (58.7%).

- LSR was implemented using the selected input parameters to develop an analytical model that can be used to accurately predict the compressive strength of UHPC.

# References

1. S. Haykin, *Neural Networks: A Comprehensive Foundation*, 2nd edn. (Prentice Hall PTR, Upper Saddle River, 1998)
2. I. Flood, N. Kartam, *Artificial Neural Networks for Civil Engineers: Advanced Features and Applications* (American Society of Civil Engineers, New York, 1998)
3. J.A. Abdalla, M.F. Attom, R. Hawileh, Prediction of minimum factor of safety against slope failure in clayey soils using artificial neural network. Environ. Earth Sci. **73**(9), 5463–5477 (2015)
4. J.A. Abdalla, M. Attom, R. Hawileh, Artificial neural network prediction of factor of safety of slope stability of soils, in *The 14th International Conference on Computing in Civil and Building Engineering* (Available at Moscow, Russia, 2012) pp. 27–29
5. L. Bal, F. Buyle-Bodin, Artificial neural network for predicting creep of concrete. Neural Comput. Appl. **25**(6), 1359–1367 (2014)
6. J.A. Abdalla, R. Hawileh, Modeling and simulation of low-cycle fatigue life of steel reinforcing bars using artificial neural network. J. Franklin Inst. **348**(7), 1393–1403 (2011)
7. J.A. Abdalla, R.A. Hawileh, Artificial neural network predictions of fatigue life of steel bars based on hysteretic energy. J. Comput. Civ. Eng. **27**(5), 489–496 (2013)
8. J. Abdalla, R. Hawileh, Energy-based predictions of number of reversals to fatigue failure of steel bars using artificial neural network, in *The 13th International Conference on Computing in Civil and Building Engineering* (2010)
9. J.A. Abdalla, A. Elsanosi, A. Abdelwahab, Modeling and simulation of shear resistance of R/C beams using artificial neural network. J. Franklin Inst. **344**(5), 741–756 (2007)
10. J.A. Abdalla, A. Abdelwahab, A. Sanousi, Prediction of shear strength of reinforced concrete beams using artificial neural network, in *The First International Conference on Modeling, Simulation and Applied Optimization* (Sharjah, U.A.E., 2005)
11. H.M. Tanarslan, M. Secer, A. Kumanlioglu, An approach for estimating the capacity of RC beams strengthened in shear with FRP reinforcements using artificial neural networks. Constr. Build. Mater. **30**, 556–568 (2012)
12. J.A. Abdalla, R. Hawileh, A. Al-Tamimi, Prediction of FRP-concrete ultimate bond strength using artificial neural network, in *International Conference on Modeling, Simulation and Applied Optimization (ICMSAO)* (Kuala Lumpur, Malaysia, 2011)
13. J.A. Abdalla, E.I. Saqan, R.A. Hawileh, Optimum seismic design of unbonded post-tensioned precast concrete walls using ANN. Comput. Concr. **13**(4), 547–567 (2014)
14. F. Deng, Y. He, S. Zhou, Y. Yu, H. Cheng, X. Wu, Compressive strength prediction of recycled concrete based on deep learning. Constr. Build. Mater. **175**, 562–569 (2018)
15. H. Naderpour, A.H. Rafiean, P. Fakharian, Compressive strength prediction of environmentally friendly concrete using artificial neural networks. J. Build. Eng. **16**(January), 213–219 (2018)
16. A. Heidari, M. Hashempour, D. Tavakoli, Using of backpropagation neural network in estimating of compressive strength of waste concrete. **1**, 54–64 (2017)
17. R. Gupta, M.A. Kewalramani, A. Goel, Prediction of concrete strength using neural-expert system. J. Mater. Civ. Eng. **18**(3), 462–466 (2006)
18. S.C. Lee, Prediction of concrete strength using artificial neural networks. Eng. Struct. **25**(7), 849–857 (2003)
19. E. Ghafari, M. Bandarabadi, H. Costa, E. Júlio, Prediction of fresh and hardened state properties of UHPC : comparative study of statistical mixture design and an artificial neural network model. **27**(11), 1–11 (2015)

20. G. Tayfur, T.K. Erdem, K. Önder, Strength prediction of high-strength concrete by fuzzy logic and artificial neural networks. J. Mater. Civ. Eng. **26**(11), 1–7 (2014)
21. A. Torre, F. Garcia, I. Moromi, P. Espinoza, L. Acuña, Prediction of compression strength of high performance concrete using artificial neural networks, in *VII International Congress of Engineering Physics, Journal of Physics: Conference Series,* p. 582 (2015)
22. A.E. Naaman, K. Wille, G.J. Parra-Montesinos, Ultra-high performance concrete with compressive strength exceeding 150 MPa (22 ksi): a simpler way. Mater. J. **108**(1)
23. Y.W.W. Zheng, B. Luo, Compressive and tensile properties of reactive powder concrete with steel fibres at elevated temperatures. Constr. Build. Mater. **41**, 844–851 (2013)
24. K.-Q. Yu, J.-T. Yu, J.-G. Dai, Z.-D. Lu, S.P. Shah, Development of ultra-high performance engineered cementitious composites using polyethylene (PE) fibers. Constr. Build. Mater. **158**, 217–227 (2018)
25. S. Abbas, M.L. Nehdi, M.A. Saleem, Ultra-high performance concrete: mechanical performance, durability, sustainability and implementation challenges. Int. J. Concr. Struct. Mater. **10**(3), 271–295 (2016)
26. N. Soliman, A. Tagnit-Hamou, Using glass sand as an alternative for quartz sand in UHPC. **145** (2017)
27. N. Gu, M. Fan, L. Du, D. Ren, Efficient sequential feature selection based on adaptive eigenspace model. Neurocomputing **161**, 199–209 (2015)
28. R. Kohavi, G.H. John, Wrappers for feature subset selection. Artif. Intell. **97**(1–2), 273–324 (1997)
29. H. Zhou, M. You, L. Liu, C. Zhuang, Sequential data feature selection for human motion recognition via Markov blanket. Pattern Recognit. Lett. **86**, 18–25 (2017)
30. V.F. Rodriguez-Galiano, J.A. Luque-Espinar, M. Chica-Olmo, M.P. Mendes, Feature selection approaches for predictive modelling of groundwater nitrate pollution: an evaluation of filters, embedded and wrapper methods. Sci. Total Environ. **624,** 661–672 (2018), in S. Haykin (ed.), *Neural Networks: A Comprehensive Foundation*, 2nd edn. (Prentice Hall PTR, Upper Saddle River, 1998)

# Classification of Plant Based on Leaf Images

Mayank Arya Chandra and S. S. Bedi

**Abstract** Plant identification has been important and complex task. Leaves are key part of the plant that distinguish, characterize, and classify. Every plant has unique leaves, and each leaf has a set of features that differentiate it from the other, inspire these features of plant leaves; here, this paper utilizes the idea of plant leaves for plant classification. Leaves are unique in relation to one another by attributes, for example, shape, shading, surface, and different other characteristics. This paper employs an algorithm for plant classification proposed based on plant leaves' image features' data through Linear Norms Decision Directed Acyclic Graph Least Square Twin Support Vector Machine (LN-DDAG-LSTSVM) classifiers. This proposed algorithm is demonstrated on leaf images from standard benchmarks database and compared with other methods where experimental results deliver higher accuracy.

**Keywords** Plant · Leaf · Classification · Linear norm · Recognition · Least square · SVM · Twin SVM · DDAG

## 1 Introduction

Plants are the backbone of life on earth and give oxygen, nourishment, medicine, fuel, and considerably more. Plants are the major source of oxygen; it is key factor of ecosystems and can decrease carbon dioxide by the process of photosynthesis [1]. Apart from this, plant plays vital role in production of food. Plant uses the energy of sunlight with water and carbon dioxide to make a food [2]. Today, the thousands of plant species are in danger of extinction or endangered caused by different anthropogenic activities. Thus, conserving of plant biodiversity, classification of plants is followed and the newly discovered species are distinguished and arranged. According to news report of nature journal, approximately 14% land species named and classified; the rest 86% remain undiscovered [3]. Scientist and botanist have named

M. A. Chandra (✉) · S. S. Bedi
MJP Rohilkhand University, Bareilly, UP, India
e-mail: machandra100@gmail.com

S. S. Bedi
e-mail: erbedi@yahoo.com

© Springer Nature Singapore Pte Ltd. 2021
X.-Z. Gao et al. (eds.), *Advances in Computational Intelligence and Communication Technology*, Advances in Intelligent Systems and Computing 1086,
https://doi.org/10.1007/978-981-15-1275-9_3

and classified millions of plants. It is not an easy task for any botanist to remember in excess of a minor division of the aggregate number of named plant species. The identification of the recently discovered species requires a specialist who has vast knowledge about it. The conventional techniques require significant amount of effort and time. So, it is demand of time to build up a system of automatic recognition of plant species.

Plant species identification and classification based on plant leaves are the fastest and facile approach to recognize a plant [4]. There are a few reasons why leaves are normally selected for classifying the plant species. Leaves are easily found and collected in any seasons. Leaf is 2-D shaped as compare to other parts of plant species (as blooms, seeds, and so on.) [5]. A lot of plants have remarkable leaves that are unique in relation to one another dependent on various attributes, e.g., shape, color, texture, and margin [6, 7]. Some researchers utilize more features for more appropriate strategies [8, 9]. Shape features deal with geometrical structure of leaf (aspect ratio, compactness, roundness, etc.) [10]. Color features are utilized by a few schemes [11] because most of the plant species have comparative leaf colors property; this highlights are not of key consequentiality. Textures features can be utilized to describe overall portray structure of the leaf or venation [12].

SVM is a successful mathematical supervised learning tool to resolve the classification and regression problems [1]. In the last decades, various modifications emerged in SVM like Lagrangian SVM [14], least square SVM [15, 16], proximal SVM [17]. Twin SVM (TWSVM) is proposed by Jayadeva [18]. Twin SVM is a mathematical modification of traditional SVM, and it solves pair of relative reduced QPP rather than large one [19]. TSVM computational cost is approximately four times faster than original SVM. A new version of twin SVM, known as least square Twin SVM, was proposed in which QPP problem is solved by equality constraints rather than inequality constraints of TSVM [20]. LS-TSVM has better performance than TSVM. Current scenario LS-TSVM was also modified by different researchers in order to achieve the better generalization capability and computational performance [21–24]. Motivation of improved classification accuracy of LSTSVM in classification domain, here, this paper utilizes the idea of least square twin support vector 56 machine for classifying the different plant species. Here we extend the LSTSVM by using decision directed acyclic graph with linear norm concept. It is analyzed that Linear Norm Decision Directed Graph Acyclic Graph Least Square Twin Support Vector Machine (LN-DDAG-LSTSVM) has present the better execution result with good prediction accuracy and running time complexity. This paper proposes a scheme that recognizes the different plants through their leaf images and their characteristics like aspect ratio, eccentricity, elongation maximal indentation depth, etc., and texture features like intensity, moment, contrast, entropy, etc. The scheme LN-DDAG-LSTSVM classifier for plant leaf recognition system is tested and compared to other machine learning approaches.

This paper is formatted as follows. The next section deals with review of previous works. Section 3 applied methodology and formulation of LN-DDAG-LSTSVM. Fourth section deals with experiment results and their comparison and finally concludes the paper at the end.

## 2 Previous Works

Plant recognition and classification are depending on images of leaf. Plant leaf images contain visual data, and their visual qualities like its shape, its color, its surface structure, its vein, etc., can be acclimated to describe plant species. Im et al. designed a system to identify the plant using leaf shape features [25]. Wang uses the concept of fuzzy on different leaf shape features [26]. The concept of hyper-sphere classifiers was introduced to recognition leaf image based on shape features [27]. A new method is proposed for leaf image that has complex background where a special feature Hu geometric and Zernike moments calculated from segmented binary images. Moving center hyper-sphere classifier is used to classification leaf images [28]. The concept of histogram of oriented gradients is used to identify the shape feature of leaf image, and a new dimensional reduction technique is utilized in this paper, known as Maximum Margin Criterion (MMC). HOG plays a crucial role with MMC in order to achieve a better performance on leaf classification [29]. Morphological features and Zernike moments are best choice for recognizing and categorizing the plant leaves. The process of feature extraction not depends upon the leaf growth as well geometric image transformation. These all terms build up an enhanced methodology that creates the best classification approach [30].

A new scheme has been proposed for identifying and characterizing plant leaves using texture and shape features. Texture descriptor of the leaf is calculated by Gabor filter and gray-level co-occurrence matrix (GLCM), while shape descriptor is determined by curvelet transform with invariant moments [10]. A system application known as Leaves is build up for plant classification dependent on the leaf's shape and venation. In this paper, the different types of machine learning procedures and different types of image filter are discussed like canny, moment invariance centroid, and ANN [31]. A new method which automatically identifies the different plant leaves has been proposed in which texture features are used for identifying and recognize the plant leaf [32]. An approach which is used the different color, shape, texture, and vein descriptor for the purpose of classification and recognizing plant leaf images [33, 34]. The concept of convolutional neural network is used to design a new algorithm known as AlexNet, in which different features were extracted by different CNN-based models [34].

However, plants are the major source of oxygen and its ability to absorb carbon forms around 66%. Without plants, it is difficult to think of existence of human life on the earth. So classifying plants helps at ascertaining the protection survival of all regular life. Plants can be classified on the basis of different parameters. This paper proposes a new scheme for plant classification based on leaf image.

# 3 Proposed Method

This proposed plant leaf classification scheme comprises four principles as appeared in Fig. 1. First are image acquisition, preprocessing, segmentation, feature extraction, and classification.

**Image acquisition**: In this paper, database of plant leaf image is used. A total of 52 leaf sample images from five different plant species are selected from open-source database [35]. Overall description of each plant with scientific name and total number of leaf samples are presented in Table 1 [35].

**Image preprocessing**: Image preprocessing is a preliminary step to reduce noise and upgrading the appearance of the leaf images which increase the overall performance of leaf identification. RGB images are converted into black and white (BW) images.

**Image segmentation**: The image segmentation is a method of partitioning a leaf images into region of interest (ROI). Here color slicing is used with Otsu's algorithm.

**Feature extraction**: In this paper, public image leaf database is used, and here we select only four plant species with all features.

**Classification**: We proposed a technique referred as Linear Norm Decision Directed Acyclic Graph Least Square Twin Support Vector Machine (LN-DDAG-LSTSVM) for classification and identification of plant leaf.

### Classification Algorithm

We proposed a technique referred as Linear Norm Decision Directed Acyclic Graph Least Square Twin Support Vector Machine (LN-DDAG-LSTSVM). In decision DAG, it has been observed that upper nodes of decision DAG have more effective over the performance of classifier. Therefore, building a predicted decision DAG in such a way that each node should be classified most separate group of classes from

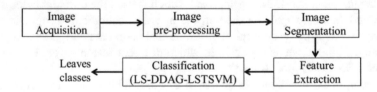

**Fig. 1** Proposed architecture of plant classification approach

**Table 1** Description selected plant leaf images

| Class | Scientific name | Instance |
|-------|-----------------|----------|
| 1 | Quercus suber | 12 |
| 2 | Salix atrocinerea | 10 |
| 3 | Populus nigra | 10 |
| 4 | Alnus sp. | 8 |
| 5 | Quercus robur | 12 |

the other one, i.e., split selection strategy at the root node is lowest error estimation between the two classes. DDAG always starts to compare between first and the last classes on the basis of decision function values. Repeat this process until remaining one class which has minimum criteria value. DDAG is generalizing the class of decision trees which allow repetition that occurs in different branch of tree.

These data values are normalized at the same time by using linear norm as given in formula Eq. (1).

$$x = (a - e * \min)/(e * (\max - \min)) \qquad (1)$$

where a data vector, min is data vector minimum values, max is data vector maximum values, and e is unit vector.

1. Let us input dataset $I = \{(x_1, y_1)(x_2, y_2) \ldots (x_n, y_n)\}$.
2. Compute the partition of the respective class versus rest at each node by using LS-TSVM as:

   (a) Let us two matrixes $A_i \& B_j$ and select penalty parameter $c_1$, $c_2$. Solve given Eqs. (2) and (3)

   $$\min_{w_1 b_1 \xi_1} \frac{1}{2}(Aw_1 + eb_1)^T (Aw_1 + eb_1) + \frac{c_1}{2}\xi_1^T \xi_1$$
   $$\text{Subject} \quad - (Bw_1 + eb_1) + \xi_1 = e, \quad \xi_1 \geq 0. \qquad (2)$$

   $$\min_{w_2 b_2 \xi_2} \frac{1}{2}(Aw_2 + eb_2)^T (Aw_2 + eb_2) + \frac{c_2}{2}\xi_2^T \xi_2$$
   $$\text{Subject} \quad - (Bw_2 + eb_2) + \xi_2 = e, \xi_2 \geq 0 \qquad (3)$$

   (b) Calculate hyper-plane parameters according to Eqs. (2) and (3)

   $$\begin{bmatrix} w_1 \\ b_1 \end{bmatrix} = -\left(B^T B + \frac{1}{c_1}A^T A\right)^{-1} B^T e \qquad (4)$$

   $$\begin{bmatrix} w_2 \\ b_2 \end{bmatrix} = -\left(A^T A + \frac{1}{c_2}B^T B\right)^{-1} A^T e \qquad (5)$$

3. Repeat Step 2, until all nodes have been trained by LS-TSVM.
4. Save the decision DAG and return.

Decision of the class 'k' for each new data sample decided by the given decision function based on respective perpendicular distance from the two planes

$$\text{Class } k = \min|x^T w_k + b_k| \quad \text{for } k = 1, 2 \qquad (6)$$

## 4    Experiment Result

For validating and examination of proposed method, we select the 40 leaf sample images from four different plant species, as Table 1, taken from open-source database provided leave recognition system [35]. The plant classification system implemented under the environment of MATLAB R2013a on Windows 10 with Intel Core i3 1.7 GHz, RAM 4 GB. The optimal values for penalty parameters are selected from the range $V_i \in \left(10^{-8}, 10^{-7}, \ldots, 10^{8}\right)$.

The proposed algorithm experimentally tested on binary class, 3-class, and 4-class problem. The performance in terms of accuracy has been compared with neural network (NN), SVM, and DAG-SVM. From Table 2 and Fig. 2, the fact can be extracted that proposed algorithm yields an average accuracy more than 90%. The quantitative comparison of the results shows that the proposed method has greater classification accuracy than existing systems.

Equation for calculating the accuracy is defined as

**Table 2** Quantitative measure for performance of different classifiers

| Class | NN Acc (%) | SVM Acc (%) | DAG-SVM Acc (%) | Proposed method acc (%) [running time (s)] |
|-------|-----------|-------------|-----------------|---------------------------------------------|
| 2 | 70 | 92.2 | 93.82 | 95 (0.000369) |
| 3 | 76.48 | 89.7 | 91.59 | 94.16 (0.000739) |
| 4 | 83.64 | 78.61 | 81.27 | 85.50 (0.001672) |
| 5 | 82.89 | 76.47 | 78.93 | 85.71 (0.002408) |
| Avg | 78.2525 | 84.245 | 86.4025 | 90.0925 (0.001297) |

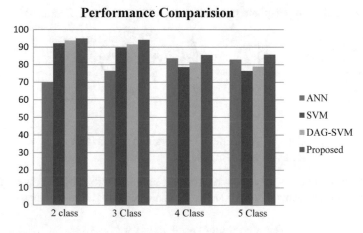

**Fig. 2** Accuracy comparison chart of proposed system and other classifiers

$$\text{Accuracy} = \frac{TP + TN}{FP + FN + TP + TN} \tag{7}$$

where $TP, TN, FP, FN$ are number of 'true positive,' 'true negative,' 'false positive,' and 'false negative,' respectively.

## 5 Conclusion

In this paper, we have proposed a LN-DDAG-LSTSVM method for classifying the plant leaf based on combining leaf shape and texture feature. The proposed plant leaf classification method is robust and computationally efficient, which takes into consideration 16 shapes and texture feature leaf. LN-DDAG-LSTSVM classifier is trained with these 16 leaf features. The proposed method was tested and evaluated on open-source leaf image database.

Performance comparison chart in Fig. 2 shows that LN-DDAG-LSTSVM is improved classification accuracy. Experimental results demonstrate that the proposed LN-DDAG-LSTSVM classifier-based plant recognition system achieves the best computation efficiency as compared to ANN, SVM, and DAG-SVM. As the number of leaves relating to each plant is small, the computational accuracy of the proposed method could be upgraded with large amount of leaves. This technique is very useful for tackling difficulties in traditional SVM, TSVM, and other classifiers. It is interesting to extract different features in future to improve the classification accuracy.

## References

1. Govindjee, A sixty-year tryst with photosynthesis and related processes: an informal personal perspective. Photosynth. Res. 1–29 (2018). https://doi.org/10.1007/s11120-018-0590-0
2. T. Beghin, J.S. Cope, P. Remagnino, S. Barman, Shape and texture based plant leaf classification, in *International Conference on Advanced Concepts for Intelligent Vision Systems (ACVIS)* (Springer, Sydney, Australia, 2010) Dec 13–16, pp. 345–353
3. L. Sweetlove, Number of species on Earth tagged at 8.7 million (Nature, Published online 23 Aug 2011). https://doi.org/10.1038/news.2011.498
4. Q.K. Man, C.H. Zheng, X.F. Wang, F.Y. Lin, Recognition of plant leaves using support vector machine, in *Advanced Intelligent Computing Theories and Applications with Aspects of Contemporary Intelligent Computing Techniques,* Springer (2008), pp. 192–199
5. A. Bhardwaj, M. Kaur, A. Kumar, Recognition of plants by leaf image using moment invariant and texture analysis. Int. J. Innov. Appl. Stud. 3(1), 237–248 (2013)
6. A. Kadir, L.E. Nugroho, A. Susanto, P.I. Santosa, Leaf classification using shape, color, and texture features. Int. J. Eng. Trends Technol. 2(1), 225–230 (2011)
7. J.X. Du, X.F. Wang, G.J. Zhang, Leaf shape based plant species recognition. Appl. Math. Comput. 185(2), 883–893 (2007)
8. P. Novotný, T. Suk, Leaf recognition of woody species in central Europe. Biosyst. Eng. 115(4), 444–452 (2013)

9. S.G. Wu, F.S. Bao, E.Y. Xu, Y.-X. Wang, Y.-F. Chang, Q.-L. Xiang, A leaf recognition algorithm for plant classification using probabilistic neural network, in *IEEE International Symposium on Signal Processing and Information Technology* (IEEE, Giza, 2007), pp. 11–16

10. J. Chaki, R. Parekh, S. Bhattacharya, Plant leaf recognition using texture and shape features with neural classifiers. Pattern Recogn. Lett. **58**(C), 61–68 (2015)

11. A. Caglayan, G. Oguzhan, A.B. Can, A plant recognition approach using shape and color features in leaf images, in *ICIAP 2013*. Lecture Notes in Computer Science, vol. 8157 (2013), pp. 161–170

12. A. Sabu, K. Sreekumar, Literature review of image features and classifiers used in leaf based plant recognition through image analysis approach, in *2017 International Conference on Inventive Communication and Computational Technologies (ICICCT)*, Coimbatore (2017) pp. 145–149

13. C. Cortes, V. Vapnik, Support vector network. Mach. Learn. **20**(3), 273–297 (1995)

14. O.L. Mangasarian, D.R. Musicant, Lagrangian support vector machines. J. Mach. Learn. Res. **1**, 161–177 (2001)

15. J.A.K. Suykens, J. Vandewalle, Least squares support vector machine classifiers. Neural Process. Lett. **9**, 293–300 (1999)

16. J.A. Suykens, T.V. Gestel, J.D. Brabanter, B.D. Moor, J. Vandewalle, *Least Squares Support Vector Machines* (World Scientific publishing Co., Singapore, 2002)

17. G. Fung, O.L. Mangasarian, Proximal support vector machine classifiers, in *KDD 01* (ACM, San Francisco CA USA, 2001)

18. Jayadeva, R. Khemchandani, S. Chandra, Twin support vector machines for pattern classification. IEEE Trans. Pattern Anal. Mach. Intell. **29**(5), 905–910 (2007)

19. M.A. Chandra, S.S. Bedi, Survey on SVM and their application in image classification. Int. J. Inf Technol. 1–11 (2018). https://doi.org/10.1007/s41870-017-0080-1

20. M.A. Kumar, M. Gopal, Least squares twin support vector machines for pattern classification. Expert Syst. Appl. **36**(4), 7535–7543 (2009)

21. Y. Xu, X. Pan, Z. Zhou et al., Structural least square twin support vector machine for classification. Appl. Intell. **42**(3), 527–536 (2015)

22. K. Mozafari, J.A. Nasiri, N.M. Charkari, S. Jalili, Hierarchical least square twin support vector machines based framework for human action recognition, in *2011 7th Iranian Conference on Machine Vision and Image Processing*, Tehran (2011), pp. 1–5

23. B. Richhariya, M. Tanveer, A robust fuzzy least squares twin support vector machine for class imbalance learning. Appl. Soft Comput. **71**, 418–432 (2018)

24. M.A. Chandra, S.S. Bedi, Benchmarking tree based least squares twin support vector machine classifiers. Int. J. Bus Intell. Data Min. https://doi.org/10.1504/ijbidm.2018.10009883

25. C. Im, H. Nishida, T.L. Kunii, Recognizing plant species by leaf shapes-a case study of the Acer family, in *Proceedings. Fourteenth International Conference on Pattern Recognition*, vol. 2, Brisbane, Queensland, Australia (1998), pp. 1171–1173

26. Z. Wang, Z. Chi, D. Feng, Fuzzy integral for leaf image retrieval, in *2002 IEEE World Congress on Computational Intelligence*, *2002 IEEE International Conference on Fuzzy Systems (FUZZ-IEEE'02)*, vol. 1, Honolulu, HI, USA (2002), pp. 372–377

27. X.F. Wang, J.X. Du, G.J. Zhang, Recognition of leaf images based on shape features using a hypersphere classifier, in *Advances in Intelligent Computing*, vol. 3644, Springer, ICIC (2005), pp 87–96

28. X.F. Wang, D.S. Huang, J.X. Dua, H. Xu, L. Heutte, Classification of plant leaf images with complicated background. Appl. Math. Comput. **205**, 916–926 (2008)

29. X.Y. Xiao, R. Hu, S.W. Zhang, X.F. Wang, HOG-based approach for leaf classification, in *Advanced Intelligent Computing Theories and Applications. With Aspects of Artificial Intelligence*, Springer, ICIC (2010), pp 149–155

30. B.S. Harish, A. Hedge, O. Venkatesh, D.G. Spoorthy, D. Sushma, Classification of plant leaves using morphological features and zernike moments, in *2013 International Conference on Advances in Computing, Communications and Informatics (ICACCI)*, Mysore (2013), pp. 1827–1831

31. A.L. Codizar, G. Solano, Plant leaf recognition by venation and shape using artificial neural networks, in *2016 7th International Conference on Information, Intelligence, Systems & Applications (IISA)*, Chalkidiki (2016), pp. 1–4
32. H.A. Elnemr, Feature selection for texture-based plant leaves classification, in *2017 Intl Conf on Advanced Control Circuits Systems (ACCS & 2017 Intl Conf on new Paradigms in Electronics & Information Technology (PEIT)*, Alexandria (2017), pp. 91–97
33. P. Mittal, M. Kansal, H.K. Jhajj, Combined classifier for plant classification and identification from leaf image based on visual attributes, in *2018 International Conference on Intelligent Circuits and Systems (ICICS)*, Phagwara (2018), pp. 184–187
34. D. Tomar, S. Agarwal, Leaf recognition for plant classification using direct acyclic graph based multi-class least squares twin support vector machine. Int. J. Image Graph. **16**(3), 1650012 (2016)
35. J.W. Tan, S. Chang, Binti, S. A. Kareem, H.J.Yap, K.Yong, Deep learning for plant species classification using leaf vein morphometric, in *IEEE/ACM Transactions on Computational Biology and Bioinformatics*
36. P.F.B. Silva, A.R.S. Marcal, R.M.A. de Silva, Evaluation of features for leaf discrimination. Springer Lect. Notes Comput. Sci. **7950**, 197–204 (2013)

# Controller Optimization for Boiler Turbine Using Simulated Annealing and Genetic Algorithm

Sandeep Kumar Sunori, Pradeep Kumar Juneja, Govind Singh Jethi, Abhijit Bhakuni and Mayank Chaturvedi

**Abstract** The present work takes up the boiler turbine process for control system design and its optimization. The control system is designed and optimized, to exhibit the best control performance, using two different optimization techniques, called simulated annealing (SA) and genetic algorithm (GA). Their control performance is also compared with that of controller designed, for the same process, using Ziegler–Nichol (ZN) technique. All the simulations have been done on MATLAB software.

**Keywords** Boiler turbine · Simulated annealing · Genetic algorithm · Optimization

## 1 Introduction

Simulated annealing is basically an imitation of a physical process, called annealing, in which first of all, a solid material is heated, and then, its temperature is gradually decreased, by which it starts melting, its defects are mitigated, and its internal energy is minimized. In the same manner, the SA algorithm decreases the temperature with every iteration until the minimum value is reached. The rate of lowering of temperature should be chosen to be small to increase the possibility of achieving the best solution.

At the start of algorithm, a test solution is generated randomly. The value of objective function is calculated at this current solution. Now, the current state is little perturbed, based on some probability distribution, to obtain a new state. If the value of objective function is better for this new state than the last current state, then it is considered as the current state, and next iteration of algorithm is run. This is done for preventing the algorithm from being stuck at a local minima point [1]. The worse point may also be accepted as the current state, after this comparison, with some

S. K. Sunori · G. S. Jethi · A. Bhakuni
Graphic Era Hill University, Bhimtal, Uttarakhand, India

P. K. Juneja · M. Chaturvedi (✉)
Graphic Era University, Dehradun, Uttarakhand, India

© Springer Nature Singapore Pte Ltd. 2021
X.-Z. Gao et al. (eds.), *Advances in Computational Intelligence and Communication Technology*, Advances in Intelligent Systems and Computing 1086,
https://doi.org/10.1007/978-981-15-1275-9_4

acceptance probability given by Eq. (1) [2].

$$P_A = e^{\frac{(U_1 - U_2)}{K\theta}} \tag{1}$$

where $U_1$ and $U_2$ represent objective function value in current and next state, respectively, such that $U_2 > U_1$, $K$ is Boltzmann constant, and $\theta$ represents temperature. Gelfand and Mitter proposed a modified annealing algorithm for a noise-corrupted inaccurate objective function [3].

Pawel Drag and Krystyn Styczen presented a solution to nonlinear optimization problem using simulated annealing for a two reactors system [4]. Young–Jae Jeon et al. minimized losses in electric power distribution system using SA [5]. Yogendra Kumar Soni and Rajesh Bhatt designed PID control system for a process with known transfer function model and optimized its performance using SA [6]. Alexander Hošovský optimized control system for temperature control of boiler water using SA [7]. J. S. Higginson et al. optimized performance of biomechanical system using SPAN [8]. Stanisław Mikulski performed optimization of parameters of PID controller using SA with integral squared error as the objective function and compared three different approaches of cooling [9].

Genetic algorithm (GA) is an imitation of how the progression of birth and reproduction takes place in human beings. This is very efficient and effective technique to solve an optimization problem which finds the most optimal solution in a very small amount of time. To initiate this algorithm, initial population of possible solutions, called chromosomes, is selected. Now, fitness of all the chromosomes is determined using a predefined fitness function. Finally, the two fittest chromosomes are selected, and crossover and mutation operations are performed on them to produce new pool of solutions (chromosomes). This sequence of steps represents one iteration, also called one generation of GA. This way a large number of generations are executed following the same sequence of steps on newly generated population until the global optimum solution is obtained [10].

## 2 Process Model and Controller Design

In the boiler turbine system, the high pressure steam, injected through a control valve, runs the turbine to produce electricity in large scale [11]. Here, a single-input single-output (SISO) model of boiler turbine process is considered. The corresponding state-space representation of this process is presented in Eqs. (2) and (3) with a time delay of 2 s. Here, the manipulated variable (mv) is control valve opening for steam flow. The control variable (cv) is electric power generated [12]. The state vector is represented by s.

$$\frac{ds}{dt} = As + B(mv) \tag{2}$$

$$(cv) = Cs + D(mv) \tag{3}$$

where the matrices $A$, $B$, $C$, $D$ are defined as under,

$$[A] = \begin{bmatrix} 0.4611 & 0.047 & 0.0015 & 0 & 0 & 0 \\ 1 & 0 & 0 & 0 & 0 & 0 \\ 0 & 1 & 0 & 0 & 0 & 0 \\ 0 & 0 & 1 & 0 & 0 & 0 \\ 0 & 0 & 0 & 1 & 0 & 0 \\ 0 & 0 & 0 & 0 & 1 & 0 \end{bmatrix} \tag{4}$$

$$[B] = \begin{bmatrix} 1 \\ 0 \\ 0 \\ 0 \\ 0 \\ 0 \end{bmatrix} \tag{5}$$

$$[C] = \begin{bmatrix} 0.0268 & 0.0757 & 0.0260 & 0.0008 & 0 & 0 \end{bmatrix} \tag{6}$$

$$[D] = 0 \tag{7}$$

Now, the formulated objective function, for the PI controller-based control system with the considered plant model, is the mean square error (MSE) as expressed in Eq. (4). Here, $K_P$ and $K_I$ are proportional constant and integral constant, respectively, of the PI controller. The corresponding objective function plot is displayed in Fig. 1.

$$\text{MSE} = 0.9959 - 1.063 K_P + 3.602 K_I + 0.8492 K_p^2 - 55.95 K_P K_I \tag{8}$$

The SA optimization algorithm is run to minimize this objective function with the parameters specified in Table 1. After running 100 iterations of SA algorithm, the objective function fitness (MSE) comes out to be 0.72 as depicted in Fig. 2. The optimized values are $K_P = 0.3$ and $K_I = 0.0018$.

Now, the GA optimization algorithm is executed to minimize this objective function with the parameters specified in Table 2. After running 100 iterations of GA algorithm, the objective function fitness (MSE) comes out to be 0.44 as depicted in Fig. 3. The optimized values are $K_P = 0.82$ and $K_I = 0.006$.

A clear performance comparison of ZN-, SA- and GA-based controllers is presented in Table 3, which is clearly showing that the SA and GA optimized controllers exhibit control performance much superior to that of ZN-based controller with smaller settling time and less peak overshoot. The corresponding set-point tracking responses are showcased in Figs. 4 and 5.

**Fig. 1** Objective function

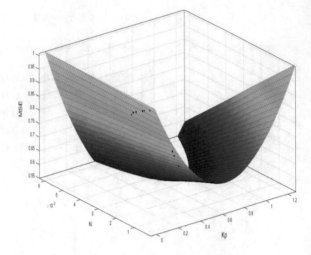

**Table 1** SA parameters

| Parameter | Value/type |
|---|---|
| Initial temperature | 50 |
| Re-annealing interval | 50 |
| Annealing function | Boltzmann |
| Temperature update function | Logarithmic |
| Iterations | 100 |

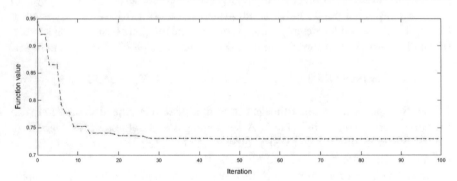

**Fig. 2** Fitness value plot in SA optimization

**Table 2** GA parameters

| Parameter | Value/type |
|---|---|
| Population size | 100 |
| Selection function | Roulette |
| Crossover probability | 0.6 |
| Mutation function | Gaussian |
| Crossover function | Single point |

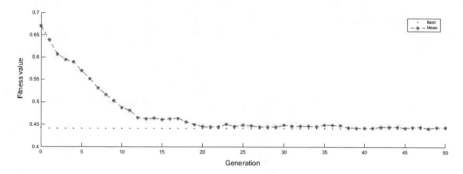

**Fig. 3** Fitness value plot in GA optimization

**Table 3** Comparison of performance

| Controller | $K_P$ | $K_I$ | Settling time (s) | Peak overshoot (%) |
|------------|-------|-------|-------------------|--------------------|
| ZN | 0.16 | 0.008 | 193 | 58.9 |
| SA | 0.3 | 0.0018 | 98.7 | 31.6 |
| GA | 0.82 | 0.006 | 81.3 | 48.8 |

**Fig. 4** Comparison of ZN- and SA-based set-point tracking responses

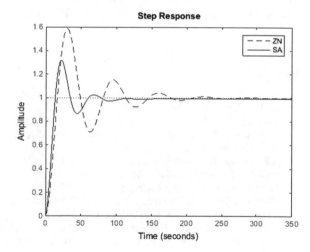

## 3 Conclusion

The present work highlights the optimization of PI controller parameters using simulated annealing and genetic algorithm optimization algorithm using MATLAB software. The controller is designed for the boiler turbine process with one input and one output variable. It has been observed that the SA- and GA-based controllers

**Fig. 5** Comparison of ZN-
and GA-based set-point
tracking responses

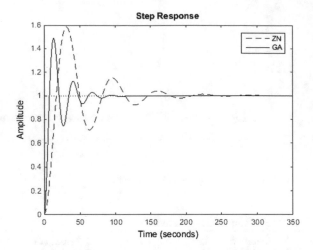

exhibit much better control performance than conventional ZN tuning-based controller. Mutual comparison of SA and GA reveals that the transient response of SA-based controller is better than that of GA-based controller with smaller peak overshoot. However, the steady-state response of GA-based controller is better than that of SA-based controller with smaller settling time.

# References

1. M. Najafi, Simulated annealing optimization method on decentralized fuzzy controller of large scale power systems. Int. J. Comput. Electr. Eng. **4**(4), 480–484 (2012)
2. Y. Nouraniy, B. Andresenz, A comparison of simulated annealing cooling strategies. J. Phys. A: Math. Gen. **31**, 8373–8385 (1998)
3. S.B. Gelfand, S.K. Mitter, Simulated annealing with noisy or imprecise energy measurements. J. Optim. Theory Appl. **62**(1), 49–62 (1989)
4. P. Drag, K. Styczen, Simulated annealing with constraints aggregation for control of the multistage processes, in *Proceedings of the Federated Conference on Computer Science and Information Systems,* vol. 5, ACSIS (2015), pp. 461–469
5. Y.J. Jeon, J.C. Kim, J.O. Kim, J.R. Shin, K.Y. Lee, An efficient simulated annealing algorithm for network reconfiguration in large-scale distribution systems. IEEE Trans. Power Deliv. **17**(4), 1070–1078 (2002)
6. Y.K. Soni, R. Bhatt, Simulated annealing optimized PID controller design using ISE, IAE, IATE and MSE error criteria. Int. J. Adv. Res. Comput. Eng. Technol. (IJARCET) **2**(7), 2337–2340 (2013)
7. A. Hošovský, Biomass-fired boiler control using simulated annealing optimized improved varela immune controller. Acta Polytech. Hung. **12**(1), 23–39 (2015)
8. J.S. Higginson, R.R. Neptune, F.C. Anderson, Simulated parallel annealing within a neighborhood for optimization of biomechanical systems. J. Biomech. **38**, 1938–1942 (2005)
9. M. Stanisław, The use of simulated annealing method for optimization of fractional order PID controller. Comput. Appl. Electr. Eng. **13**, 178–187 (2015)

10. S.K. Sunori, P.K. Juneja, M. Chaturvedi, P. Aswal, S.K. Singh, S. Shree, GA based optimization of quality of sugar in sugar industry. Ciencia e Tecnica Vitivinicola, ISSN: 0254-0223, **31**(4), 243–248 (2016)
11. Z.H. Guang, L. Cai, Multivariable fuzzy generalized predictive control. *Cybernetics and Systems: An International Journal* (Taylor & Francis, 2002), pp. 69–99
12. S.K. Sunori, P.K. Juneja, Controller design for MIMO boiler turbine process. Int. J. Control Theor. Appl. **2**, 477–486 (2015)

# Indian Language Identification for Short Text

Sreebha Bhaskaran, Geetika Paul, Deepa Gupta and J. Amudha

**Abstract** Language identification is used to categorize the language of a given document. Language identification categorizes the contents and can have a better search results for a multilingual document. In this work, we classify each line of text to a particular language and focused on short phrases of length 2–6 words for 15 Indian languages. It detects that a given document is in multilingual and identifies the appropriate Indian languages. The approach used is the combination of $n$-gram technique and a list of short distinctive words. The $n$-gram model applied is language independent whereas short word method uses less computation. The results show the effectiveness of our approach over the synthetic data.

**Keywords** $n$-gram · Language identification · Trigrams · Accuracy

## 1 Introduction

In an era of Internet, technologies and social media, there are many ways of communication. While communicating in these social medias through messaging or commenting or posting on Twitter [1], etc., people feel comfortable using their own language along with English. In a country like India which is rich in languages with 22 major languages, 13 different scripts and 720 dialects as per the survey reports [2]. Identification of language will have a major role to be played to identify the language used for communication in social medias.

S. Bhaskaran (✉) · G. Paul · D. Gupta · J. Amudha
Department of Computer Science & Engineering, Amrita School of Engineering, Amrita Vishwa Vidyapeetham, Bengaluru, India
e-mail: b_sreebha@blr.amrita.edu

G. Paul
e-mail: geetika2604.95@gmail.com

D. Gupta
e-mail: g_deepa@blr.amrita.edu

J. Amudha
e-mail: j_amudha@blr.amrita.edu

© Springer Nature Singapore Pte Ltd. 2021                                          47
X.-Z. Gao et al. (eds.), *Advances in Computational Intelligence and Communication Technology*, Advances in Intelligent Systems and Computing 1086,
https://doi.org/10.1007/978-981-15-1275-9_5

A common tendency for people is to mix up two or three languages while commenting or messaging for the posts. When we have to translate such multilingual texts, first we have to realize that the sentences are in more than one language, then correctly identify the language used and finally, translate it to a required language. This is another situation where the text identification comes in. Also, text is highly researched area in computer vision applications to model smart self-learning system, so identifying text from images [3, 4] also play a vital role. That is the optical character recognition (OCR) technology [5] uses language information and dictionaries during the process of optical character recognition. Other application of language identification, during translation [6, 7] of text from one language to another language, identifying the text of the source language is a crucial one.

One of the major bottlenecks of language identification system is discriminating between similar languages (DSL). The task of text categorization becomes more difficult as the number of languages increases and in Indian scenario, there are many languages with similar texture. The length of the sentence in a document also has a major role in text categorization. As number of words in a given sentence is more, it makes identification of that particular language easy. As the length of the text decreases, also similarity between languages increases and the size of training data is less, and then identification of language is a difficult task.

In this paper, we choose for 15 Indian languages and are grouped based on the similarity of text for each language. The groups and the language under it are mentioned as follows:

Group 1—{Gujarati (Guj), Kannada (Ka), Malayalam (Mal), Gurmukhi (Gur), Tamil (Ta), Telugu (Te), Urdu (Ur), English (En)}
Group 2—{Assamese (As), Bengali (Be), Oriya (Or)}
Group 3—{Hindi (Hi), Marathi (Mar), Nepali (Ne), Sanskrit (Sa)}

Group 1 set of languages is classified as unambiguous text or no similar text or character. In Group 2, Assamese and Bengali are considered as Indo-Aryan Language and also have similar affinity with Oriya. Group 3 set of languages are using Devanagari scripts. So, Group 2 and 3 have got languages which have similar scripts and few words having different meanings in each language. Identification for such set of languages is very critical. Also, the identification of languages has been carried out on phrase length of 2–6 words. A hybridized approach has been taken to classify the phrases from a document to any of the mentioned languages.

This paper has been organized as follows: Sect. 2 describes the detailed study of language identification with different Indian languages which helped to form this problem. Also the techniques used for language identification. In Sect. 3, it describes the proposed approach which is a hybridized technique with the combination of $n$-gram-based and short words. The dataset description and the evaluation measures are discussed in Sect. 4. The achieved results and the analysis made by it was described in Sect. 5. Conclusion of the present work and its future scope are justified in Sect. 6, and finally, it is followed by references.

## 2 Related Work

The language identification is the task of categorizing given text as a particular language based on certain features. These works can be broadly classified as linguistic models and statistical models. Linguistic model makes use of the linguistic rules. The developer of the system must be aware of all the rules associated with the languages under consideration. On the other hand, the statistical model is system dependent. The system is trained to automatically classify the languages based on features that are unique to a language like short word method, bag-of-words method, $n$-gram, support vector machine (SVM), etc.

One of the popular works in the text categorization has been done by Canvar and Trenkle [8] known as TextCat tool. The method uses $n$-gram profile to calculate a simple rank order statistical called as out-of place measure. An $n$-gram is a sequence of $n$ consecutive characters extracted from a given string. If we ignore the white spaces and all the special character then, the word "Hello World!" can be represented in terms of bi-gram and tri-gram as follows:

bi-gram: He, el, ll, lo, Wo, or, rl, ld
tri-gram: Hel, ell, llo, Wor, orl, rld

To calculate the out-of-place measure, they used finding the difference between the ranks of various matched $n$-grams and adding them up to find distance. The text belongs to the language with least distance. The method given by Carvar and Trenkle [8] has the advantages as: it is independent of the language and the alphabet, it does not require knowledge about the language and the effect of spelling or syntactical error in the document is limited.

The modification of the work [8] has been given by Keselj [9], in which they worked on word and character level $n$-grams. The algorithm used calculates the dissimilarity between two profiles and returns a positive value. This means $n$-gram which is present in less number of languages has a higher weight and could be decisive in classifying the text.

Another simple, effective and widely used method for text categorization is $k$-nearest neighbors (KNN) [10, 11]. The basic idea is to convert the test sample (text) into weighted feature vectors. It tries to find the KNN from the complete set of training samples using cosine similarity measure. The biggest drawback with this system is the large amount of computation that eventually hampers the speed and performance of categorization. There are multiple approaches to KNN method such as lazy learning and eager learning.

Certain linguistically motivated models have also been proposed. The work by Grefenstette [12] and Johnson [13] is based on generation of language model using the short words such as prepositions, pronouns, determiners and conjunctions to determine the language of the document. A list of unique words is generated by tokenizing the sample document, selecting the words with 5 or less characters and retaining the ones with frequency greater than 3. The probability of the test data belonging to a particular language is calculated as the product of the probabilities of occurrence of tokens in the list of that language. This is faster than the $n$-gram-based technique, since the training list is very small. But the quality of result is highly dependent on the occurrence of words from the list in the test data.

Lins and Goncalves [14] considered the syntactic structure of language (adverbs, articles, pronouns and so on) rather than the word structure and sequences. They have analyzed which set of words could be used to differentiate between the languages. Such a list of unique word model is highly dependent on the set of languages begin considered. The efficiency of this model may suit for one set of languages but may not work for similar set of languages.

Prager [15] proposed a hybridized approach for language identification by applying the vector space model based on the similarity computation with respect to the distance between the trained and test language model. The recent trend of language identification is with transliteration, where it replaces the words of the source language with an appropriate phonetic or spelling equivalent in target language. References [16–18] refers transliteration on very few languages and has been applied on an unambiguous text.

The $n$-gram model is slow but language independent, whereas short word method is less time consuming as it uses less complex computation but is language dependent. This leads to hybridized approach for Indian language identification. Most of the works done for Indian languages [19–21] have considered either few languages or languages with dissimilar text. The proposed work is based on combination of $n$-gram model and short word method applied on 15 different Indian languages having similar text patterns, and is classified as Group 2 and Group 3.

## 3 Proposed Approach

The work for the proposed model has started off by collecting the corpus for all these 15 languages from different sources such as newsgroups, Wikipedia, dictionaries, etc. The approach used is a hybridized language detection, which combines the advantages of $n$-gram method and short word method as discussed in the related work section.

The detailed description about the proposed work is shown in Fig. 1. This model has been classified into two phases, one is considered as training phase and other as testing phase. As the required data has been collected from different sources like newsgroups, Wikipedia's, etc., the data needs to be cleaned, as it consist of unwanted characters like digits, extra spaces, punctuation marks, words of some other language, etc., along with the source language text. So, the collected data is processed through the cleaning and filtering stage, where it removes all unwanted characters and gives us the required text in a document. This stage is also required to the data used for testing, as it may contain unwanted characters like digits, punctuation marks or extra space.

The data after cleaning and filtering stage, it then generates trigrams and its corresponding frequencies for both training and testing documents. In training phase,

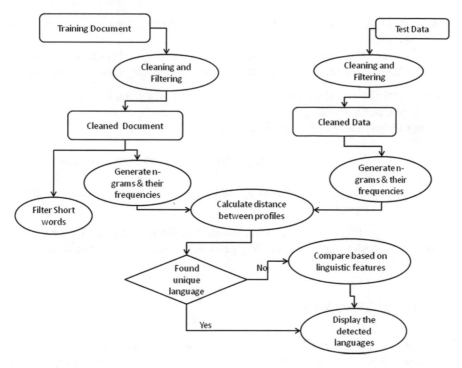

**Fig. 1** Flow of the proposed work

it also identifies unique short words having frequencies greater than the threshold frequency. Based on the distance calculated between the profiles, if the unique language could be identified then display that language. Otherwise, it compares with the short words collected for each language and displays the detected language.

## 3.1 N-gram Technique

This technique uses two phases called training phase and testing phase. Training phase consists of the following steps:

- Collect documents from different sources and concatenate them.
- The document is filtered to get rid of digits, special characters, extra white spaces and punctuation.
- Select the $n$ value. We have used $n = 3$, i.e. trigrams.
- Scan through the document to generate the $n$-grams.
- Store the $n$-grams in hash map. Hash maps are key-value pair, in this case keys are the trigrams and values are their frequencies. Keep incrementing the value with each occurrence of the particular $n$-gram.
- Sort them in decreasing order of their frequencies.

Repeat the above steps for all the languages and store the *n*-grams for all the languages in separate files.

The second phase, testing phase, uses the files produced during the training phase to determine the language of the given document. It consists of the following steps:

- Generate trigrams and compute their frequencies for the given input text following the steps described above.
- Compute the dissimilarity between the test document profile and the category's profile.
- Find the cumulative distance category wise adding the differences calculated in the previous step.
- The category with the smallest distance measure has highest probability of being the correct category.

If there are multiple languages with similar distance measure, then we apply the short words algorithm for the list of languages identified by the *n*-gram method.


## 3.2   List of Short Words

When dealing with set of similar languages such as Group 2 and Group 3, there are high possibilities to have trigrams that are common in two or more languages in this set. In such cases, there might be some ambiguity as to which language does the *n*-gram belong to. Even the cumulative distance might be very close. This could affect the accuracy of the result. To overcome this problem, we need language-specific features. Intuitively unique words such as determiners, conjunctions, and prepositions are good clues for guessing language.

Every Indian language follows a particular pattern which differentiates it from the other languages. A list of such differentiating words or features is collected and stored for each language in separate files. The test document is looked upon for pattern match in the set of similar languages identified.

We derive the attributes of language using the same corpus. This involves extracting the unique features for all the languages. The filtered corpus used for *n*-gram extraction is tokenized to get a list of words. The frequency of all the words is calculated. The words that have length less than or equal to four and frequency greater than threshold are retained.

Initially, the given document is tokenized and the tokens are searched in the list of the languages identified by the *n*-gram method. The sentence belongs to a particular category if the token in the test document is found in the corresponding list. In case of no match, the language-specific patterns are used to categorize the text. Our approach has been tested with 30–50 short phrases for Group 2 and Group 3 set of languages. Tables 1 and 2 show sample short words used to categorize text for Group 3 and Group 2 language sets.

Each language has its own pattern or style of writing, for example, most of the *Ma* phrases end with आहे. A list of such differentiating words or features are collected

**Table 1** Sample short words for Group 3 languages

| Tokens | Hi | Mar | Ne | Sa |
|---|---|---|---|---|
| है | ✔ | ✘ | ✘ | ✘ |
| और | ✔ | ✘ | ✘ | ✘ |
| छ | ✘ | ✘ | ✔ | ✘ |
| पनि | ✘ | ✘ | ✔ | ✘ |
| आहे | ✘ | ✔ | ✘ | ✘ |
| झाले | ✘ | ✔ | ✘ | ✘ |
| किम् | ✘ | ✘ | ✘ | ✔ |
| त्या | ✘ | ✔ | ✘ | ✘ |
| स | ✘ | ✘ | ✘ | ✔ |

**Table 2** Sample short words for Group 2 languages

| Tokens | As | Be | Or |
|---|---|---|---|
| ভাতো | ✘ | ✘ | ✔ |
| থেকো | ✘ | ✔ | ✘ |
| শুনুন | ✘ | ✔ | ✘ |
| কৰি | ✔ | ✘ | ✘ |

and stored for Group 2 and Group 3 set of languages. The test document is looked up for any such pattern match in the set of languages identified.

## 4 Data Description and Evaluation Measures

The proposed model was trained for 15 Indian languages classified as Group 1, Group 2 and Group 3, by collecting data randomly from different source of newsgroups, Wikipedia, dictionaries, books, etc. So, the size of the training dataset for each languages varied from 25 to 63,495 (kilobytes) based on the availability. The training dataset in terms of number of words, number of unique words and number of unique trigrams and testing dataset for each language is briefed in Table 3. The size of the training dataset used is not uniform for all languages because of the unavailability of data for those set of languages.

Testing has been done including and excluding similar languages. The evaluation has been calculated using accuracy measure (1) in percentage for analyzing the results.

$$\text{accuracy} = \frac{\text{total number of correctly identified phrases(words)}}{\text{total number of phrases(words)}} \quad (1)$$

For small phrase of 2–6 words, it is possible to have those words to be unique or not. In case of unique words, our tool can detect the language in which the phrase belongs too. But $n$-gram on the other hand is language independent. It is not restricted by a set of words. For example, consider the trigram "act" which could be present in "subtract," "enact," "acting," "actor," "fact" and so on. Although most of these words are not interrelated, but contain a common $n$-gram. All these words can be identified to belong to English language using just single trigram.

As we used hybrid model, it gave better results when executed for Group 2 and Group 3 Indian languages, where different languages have similar script used, but the meaning of the word in each language is different. For example, शिक्षा—Education in Hindi and Punishment in Marathi, अभ्यास—Practice in Hindi, study in Marathi, चेष्टा—attempt in Hindi and make fun of in Marathi, etc.

**Table 3** Training and testing dataset

| Language | Training dataset | | | Testing dataset |
|---|---|---|---|---|
| | # Words | # Unique words | #Unique trigrams | # Of words |
| As | 1,508 | 804 | 1465 | 127 |
| Be | 286,619 | 16,847 | 11,535 | 120 |
| En | 930,730 | 37,823 | 11,048 | 172 |
| Guj | 58,555 | 16,020 | 11,725 | 150 |
| Gur | 22,596 | 5585 | 6330 | 184 |
| Hi | 707,247 | 21,239 | 15,874 | 160 |
| Ka | 6685 | 3242 | 4651 | 94 |
| Mal | 298,267 | 32,976 | 16,299 | 100 |
| Mar | 11,971 | 4799 | 5967 | 184 |
| Ne | 2,860,354 | 220,734 | 43,171 | 164 |
| Or | 1439 | 814 | 1719 | 162 |
| Sa | 20,475 | 8918 | 10,788 | 176 |
| Ta | 339,463 | 31,527 | 8427 | 112 |
| Te | 416,306 | 35,202 | 15,695 | 132 |
| Ur | 583,476 | 14,689 | 10,150 | 186 |

# 5 Experimental Results and Analysis

Main objective of our proposed work is to identify the language for short text with the phrase length to be 2–6 words. Table 4, shows the results for Group 1 set of languages with phrase length of two words, where the accuracy is almost 100% but for Mal and Ur there was error because of unavailability of few words in training dataset.

In Table 5, the result for Group 2 and Group 3 set of languages with phrase length of two words each is shown. The error rate for Be in Group 2 is high, as the text pattern is similar to As. The results for Group 2 is better with our proposed approach, as same words with different meanings are less when compared to Group 3. Group 3 is a set of language with same text pattern and same words with different meaning for example, पैसा, खाना, अन्न, दण्ड, शिक्षा, चन्द्र, गीत, etc., so the error rate with two words phrase length is high as most of the words are getting mapped with other language in that set.

Table 6 shows the results of Group 3 set of languages with phrase length as 4 and 6 words, this shows a drastic improvement in accuracy as the word length increases. For Group 2 as the results were better with two word phrase length, it is prominent that it will show substantial improvement as the phrase length is increased. The accuracy can also be increased when all the unique words for each language are identified.

**Table 4** Result for Group 1 set of language

| Language | Accuracy (%) |
|----------|--------------|
| En | 100.00 |
| Guj | 100.00 |
| Gur | 100.00 |
| Ka | 100.00 |
| Mal | 98.00 |
| Ta | 100.00 |
| Te | 100.00 |
| Ur | 97.85 |

**Table 5** Result for Group 2 and Group 3 set of languages

| | Language | Accuracy (%) (Phrase length = 2 words) |
|---------|----------|------------------------------------------|
| Group 2 | As | 96.88 |
| | Be | 56.67 |
| | Or | 82.72 |
| Group 3 | Hi | 58.75 |
| | Mar | 66.67 |
| | Ne | 23.66 |
| | Sa | 28.18 |

Table 6 Result for Group 3
languages with 4 and 6 words

| Language | Accuracy (%) (phrase length = 4 words) | Accuracy (%) (phrase length = 6 words) |
|---|---|---|
| Hi | 67.50 | 81.37 |
| Mar | 69.57 | 83.87 |
| Ne | 42.71 | 54.71 |
| Sa | 52.27 | 63.33 |

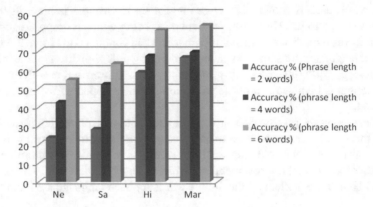

Fig. 2 Accuracy for Group 3 set of languages with different phrase length

While analyzing the results, Group 1 languages were giving 100% accuracy keeping two words as phrase length. Group 2 set of languages were also giving better accuracy keeping phrase length as two words. For Group 3 set of languages which has very similar text and words with different meanings as shown in Fig. 2, shows drastic improvement in accuracy for shorter phrase lengths. The model proposed gives better results with two words as phrase length. As the number of words increases in the phrase the accuracy is also increased.

## 6  Conclusion and Future Works

Though $n$-gram technique is the one which is most widely used to identify the languages, it has got certain drawback as the accuracy of the result depends on the quality of the training dataset and the execution time is higher when compared with other techniques like bag-of-words. So, in our model, it overcomes these drawbacks of $n$-gram by proper filtering of the collected data, before generating trigrams. As it is random collection of data from various different sources on many different topics, and by smartly selecting the threshold frequency to get rid of useless or erroneous trigrams. There is no way to reduce the execution time of $n$-gram technique. To overcome this issue, another famous technique list of short words is used in combination with the traditional $n$-gram technique.

Words are the privileged features in all the languages, it conveys more information than a sequence of characters ($n$-grams), and they should be treated as privileged units of the language detection tool. This is generally a small-sized file. Thus, matching through it is very fast as compared to scanning through $n$-grams.

In our work, we have developed a system to identify 15 Indian languages. The accuracy shows the system works fine with similar scripts when the minimum phrase length given is 2. This accuracy can be improved by increasing the number of words given for testing dataset but our goal is to identify the language for smaller phrase length and in the test results shows better accuracy. As a future work, the words which were unavailable in training dataset could be updated whenever a new word occurs.

# References

1. M. Venugopalan, D. Gupta, Exploring sentiment analysis on twitter data, in *2015 Eighth International Conference on Contemporary Computing (IC3)* (IEEE, 2015)
2. mhrd.gov.in/sites/upload_files/mhrd/files/upload_document/languagebr.pdf
3. P. Salunkhe, et al., Recognition of multilingual text from signage boards, in *International Conference on Advances in Computing, Communications and Informatics (ICACCI) (IEEE, 2017)*
4. J. Amudha, N. Kumar, Gradual transaction detection using visual attention system. Adv. Int. Inform. 111—122 (2014)
5. D. Gupta, M.L. Leema, Improving OCR by effective pre-processing and segmentation for devanagari script: a quantified study. J. Theor. Appl. Inf. Technol. (ARPN), **52**(2), 142—153 (2013)
6. K. Jaya, D. Gupta, Exploration of corpus augmentation approach for English-Hindi bidirectional statistical machine translation system. Int. J. Electr. Comput. Eng. (IJECE), **6**(3), 1059–1071 (2016)
7. D. Gupta, T. Aswathi, R.K. Yadav, Investigating bidirectional divergence in lexical-semantic class for English-Hindi-Dravidian translations. Int. J. Appl. Eng. Res. **10**(24), 8851–8884 (2015)
8. W.B. Cavnar, J.M. Trenkle, N-gram–based text categorization, in *Proceedings of the 3rd Annual Symposium on Document Analysis and Information Retrieval* (Las Vegas, Nevada, USA, 1994), pp. 161—175
9. V. Keselj, F. Peng, N. Cercone, C. Thomas, N-gram based author profiles for authorship attribution, in *Proceedings of the Pacific Association for Computational Linguistics* (2003), pp. 255–264
10. P. Soucy, G.W. Mineau, A simple KNN algorithm for text categorization, in *Proceedings 2001 IEEE International Conference on Data Mining* (San Jose, CA, 2001), pp. 647—648
11. W. Zheng, Y. Qian, H. Lu, Text categorization based on regularization extreme learning machine. Neural Comput. Appl. **22**(3–4), 447–456 (2013)
12. G. Grefenstette, Comparing two language identification schemes, in *3rd International Conference on Statistical Analysis of Textual Data* (1995)
13. N. Hwong, A. Caswell, D.W. Johnson, H. Johnson, Effects of cooperative and individualistic learning on prospective elementary teachers' music achievement and attitudes. J. Soc. Psychol. **133**(1), 58–64 (1993)

14. R.D. Lins, P. Goncalves, Automatic language identification of written texts, in *Proceedings of the 2004 ACM Symposium on Applied Computing, SAC '04* (ACM, New York, NY, USA, 2004), pp. 1128–1133

15. J.M. Prager, Linguini, language identification for multilingual documents, in *Proceedings of the 32nd Hawaii International Conference on System Sciences* (1999)

16. P.M. Dias Cardoso, A. Roy, Language identification for social media: short messages and transliteration, in *Proceedings of the 25th International Conference Companion on World Wide Web (International World Wide Web Conferences Steering Committee, 2016)*, April 11, pp. 611–614

17. S. Banerjee, A. Kuila, A. Roy, S.K. Naskar, P. Rosso, S. Bandyopadhyay, A hybrid approach for transliterated word-level language identification: CRF with post-processing heuristics, in *Proceedings of the Forum for Information Retrieval Evaluation* (ACM, 2014) Dec 5, pp. 54–59

18. D.K. Gupta, S. Kumar, A. Ekbal, Machine learning approach for language identification & transliteration, in *Proceedings of the Forum for Information Retrieval Evaluation*, 2014 Dec 5 (ACM), pp. 60–64

19. B. Sinha, M. Garg, S. Chandra, Identification and classification of relations for Indian languages using machine learning approaches for developing a domain specific ontology, in *International Conference on Computational Techniques in Information and Communication Technologies (ICCTICT)*, New Delhi, 2016, pp. 415–420

20. R. Bhargava, Y. Sharma, S. Sharma, Sentiment analysis for mixed script Indic sentences, in *2016 International Conference on Advances in Computing, Communications and Informatics (ICACCI)*, Jaipur, 2016, pp. 524–529

21. S.S. Prasad, J. Kumar, D.K. Prabhakar, S. Tripathi, Sentiment mining: an approach for Bengali and Tamil tweets, in *2016 Ninth International Conference on Contemporary Computing (IC3)*, Noida, 2016, pp. 1–4

# Improved Whale Optimization Algorithm for Numerical Optimization

**A. K. Vamsi Krishna and Tushar Tyagi**

**Abstract** In this paper, an Improved Whale Optimization Algorithm which is intended towards the better optimization of the solutions under the category of meta-heuristic algorithms is proposed. Falling under the genre of nature-inspired algorithms, the Improved Whale Optimization delivers better results with comparatively better convergence techniques used. A detailed study and comparative analysis have been made between the principal and the modified algorithms, and a variety of fitness functions has been used to confirm the efficiency of the improved algorithm over the older version. The merits with nature-inspired algorithms include distributed computing, reusable components, network processes, mutations and crossovers leading to better results, randomness and stochasticity.

**Keywords** Optimization algorithms · Nature-inspired algorithms · Whale Optimization Algorithm · Bubble-net forging · Meta-heuristic algorithm · Randomization · Fitness function · Spiral updating · Encircling mechanism

## 1  Introduction

The nature around us provides us with many efficient and optimal ways to solve problems. The nature-inspired algorithms are based on the optimization techniques offered by the nature, and in a sort of way, they imitate or mimic the processes available in nature where artificial intelligence and man's intellectual abilities fail. Nature-inspired algorithms along with fuzzy logic systems and neural networks fall under the category of computational intelligence. The use of nature in problem solving and optimizing real-world problems can be attributed to the design and construction of aeroplane wings based on the wings of eagle, turbine blade design based on the swimming of dolphins, etc. [1].

A. K. Vamsi Krishna (✉) · T. Tyagi
School of Electronics and Electrical Engineering, Lovely Professional University,
Jalandhar, India
e-mail: krish.0724@gmail.com

T. Tyagi
e-mail: tushar.20586@lpu.co.in

© Springer Nature Singapore Pte Ltd. 2021
X.-Z. Gao et al. (eds.), *Advances in Computational Intelligence and Communication Technology*, Advances in Intelligent Systems and Computing 1086,
https://doi.org/10.1007/978-981-15-1275-9_6

## *1.1 Nature-Inspired Algorithms*

The nature-inspired algorithms can be basically divided into two concepts, one being evolutionary algorithms and the other being swarm optimization. Evolutionary algorithms include genetic algorithm, differential evaluation algorithms and many others where the algorithm can be constantly improved with newer solutions available [1, 2].

**Examples**. A few of the nature-inspired algorithms are Firefly Algorithm, Migrating Birds Optimization, Ant Lion Optimization, etc.

## 2 Whale Optimization Algorithm

Whale Optimization Algorithm abbreviated as WOA is an optimization algorithm based on the hunting process or mechanism of humpback whales. This is a recently developed algorithm in the year 2016. It falls under the category of meta-heuristic algorithm and uses a compromise between the local search and randomization. This is a better approach to move from a localized scale of search to global-level search. WOA was proposed by Lewis and Mirjalili based on the simulation of bubble-net method of humpback whales during the hunt for their prey [3, 4].

## *2.1 About WOA*

Introduced by Mirjalili in the year 2016, WOA is simulation-based algorithm predominantly characterized depending on hunting technique or hunting style or humpback whales. The interesting aspect of their hunting technique is that the whales create distinctive bubbles along a spiral or circular path as the whales continue to encircle the krill or prey. This distinct technique is referred to as 'bubble-net foraging method'. The path traced by the bubbles can also be in the number nine-shaped path or a simple circular path [3–5].

## *2.2 Mathematical Modelling*

This technique is forged into a mathematical model which consists of three aspects of encircling the prey, bubble-net forging and search for the prey [3–5].

(1) Encircling the Prey:

Humpback whales first encircle the prey and then refresh and renew their position moving towards the best search from start to a maximum number of specified iterations given by Eqs. (1) and (2). The number of iterations is specified during the

algorithm, and the more number of iterations generally means much closer to the optimal solution [4].
Mathematically,

$$\vec{E} = \left| \vec{D} \cdot \vec{Y^*}(t) - \vec{Y}(t) \right| \tag{1}$$

$$\vec{Y}(t+1) = \vec{Y^*}(t) - \vec{B} \cdot \vec{E} \tag{2}$$

where

$\vec{B}$ and $\vec{D}$ are coefficient vectors,
$t$ indicates the present iteration,
$Y^*$ is the position vector of the best solution obtained so far,
$\vec{Y}$ is the position vector,
$\|$ is the absolute value, and
$\cdot$ is an element-by-element multiplication.

The vectors $\vec{B}$ and $\vec{D}$ are given by Eqs. (3) and (4) as follows:

$$\vec{B} = 2\vec{b} \cdot \vec{r} - \vec{b} \tag{3}$$

$$\vec{D} = 2 \cdot \vec{r} \tag{4}$$

where

$\vec{b}$ is linearly decreased from 2 to 0 as iterations progress,
$\vec{r}$ is a random vector in [0, 1].

(2) Bubble-Net Attacking Method:

For mathematical modelling of bubble-net attack method, there are two mechanisms to be approached [4]:

(a) Shrinking Encircling Mechanism:

This mechanism is achieved by reducing the value of $\vec{b}$ from 2 to 0 in Eq. (3) as the number of iterations progress. Now, the new position of a search agent can be set or defined anywhere in between or amongst the original position of the agent and the position of the current best agent by assigning arbitrary randomized values for $\vec{B}$ in [−1, 1].

(b) Spiral Updating Position:

The spiral equation between the position of the whale and the prey which mimics or imitates the helix-shaped movement of the whales is mathematically described as given:

$$\vec{Y}(t+1) = \vec{E'} \cdot e^{cl} \cdot \cos(2\pi l) + \vec{Y^*}(t) \tag{5}$$

From Eq. (5), the assumed probability is 50% to make a choice between either the shrinking encircling method or the spiral model so as to update the position of whales. The mathematical model is given by Eq. (6) as follows:

$$\vec{Y}(t+1) = \begin{cases} \vec{Y}(t+1) = \overrightarrow{Y^*(t)} - \vec{B}.\vec{E} & \text{if } p \leq 0.5 \\ \vec{E'} \cdot e^{cl} \cdot \cos(2\pi l) + \vec{Y^*}(t) & \text{if } p \geq 0.5 \end{cases} \tag{6}$$

where
$\vec{E} = \left| \vec{Y^*}(t) - \vec{Y}(t) \right|$ represents the distance between prey and whale,

$c$   is constant for spiral shape,
$l$   is a random number in $[-1, 1]$ and,
$p$   describes a randomized number in $[0, 1]$.

(3)   Search for Prey:

The variation of $\vec{B}$ vector can be used to search for prey, for during the exploration phase. Hence, $\vec{B}$ can be used with the random values greater than 1 or less than $-1$ to force the search agents to move far or away from a reference whale. This is represented mathematically as given by Eqs. (7) and (8) as follows [4]:

$$\vec{E} = \left| \vec{D} \cdot \overrightarrow{Y_{\text{rand}}} - \vec{Y} \right| \tag{7}$$

$$\vec{Y}(t+1) = \overrightarrow{Y_{\text{rand}}} - \vec{B} \cdot \vec{E} \tag{8}$$

where

$\overrightarrow{Y_{\text{rand}}}$   is defined as random position vector.

## 3   Improved Whale Optimization Algorithm

The improvement of the existing WOA and the development of new improved WOA known as IWOA is completely focused on the approach followed to generate new population and their updation as iterations progress. This is carried out in the bubble-net attacking method where shrinking encircling mechanism and spiral updating of position occurs. The idea here is to replace the existing method of spiral updation and search for prey. This is further carried out in three stages, namely diffusion, updation-I and updation-II. The new updation process adds the technique of ranking the points or positions such that the positions or the points with best rank make it out for the generation of new population of prey and the whales. The Gaussian Walk has an upper hand in generating random two-dimensional fractals compared to Levy flights, and hence, Gaussian Walk is preferred [4–7].

Gaussian Walk:

The Gaussian Walk is also referred to as random walk, is a stochastic process. This describes a path or positions in consecutive random sequence or steps based on simple or complex mathematical process [6].

## 3.1 Diffusion Process

The diffusion process used in IWOA implements the strategy of Gaussian Walk as it allows for exchanging information amongst all the involved points or positions. This means that the convergence to the minimum is accelerated alongside producing more favourable and better results. The Gaussian Walk is used for the generation of a random walk.

## 3.2 Updation-I

This is the first statistical procedure, and it ranks all the points individually obtained from the diffusion process. The ranking system is based on the value of the fitness functions. Here, the fitness functions are the 23 benchmark functions, and their description is tabulated in Table 1. The ranking method is based on uniform distribution and is applied to every point. Every position or point is first evaluated based on the fitness function, and a probability value is given or simply ranked based on the uniform distribution described by Eq. (9).

$$Ra_{01} = \frac{\text{rank}(D_i)}{U} \tag{9}$$

where

$\text{rank}(D_i)$    is the rank of the point $D_i$,
$U$           is the number of all the points in the group,
$Ra_{01}$      is the probability value.

## 3.3 Updation-II

The second updation process is aimed at modifying the position of point with respect to the change in the positions of other points. This updation process improves the

**Table 1** List of benchmark functions

| S. No. | Functions | Dimensions | Range |
|---|---|---|---|
| 1 | $F_1(y) = \sum_{i=1}^n y_i^2$ | 10 | $[-100, 100]$ |
| 2 | $F_2(y) = \sum_{i=1}^n |y_i| + \prod_{i-1}^n |y_i|$ | 10 | $[-10, 10]$ |
| 3 | $F_3(y) = \sum_{i=1}^n \left(\sum_{j-1}^i y_j\right)^2$ | 10 | $[-100, 100]$ |
| 4 | $F_4(y) = \max_i\{|y_i|, 1 \le i \le n\}$ | 10 | $[-100, 100]$ |
| 5 | $F_5(y) = \sum_{i=1}^{n-1}\left[100(y_{i+1} - y_i^2)^2 + (y_i - 1)^2\right]$ | 10 | $[-30, 30]$ |
| 6 | $F_6(y) = \sum_{i=1}^n ([y_i + 0.5])^2$ | 10 | $[-100, 100]$ |
| 7 | $F_7(y) = \sum_{i=1}^n (iy_i^2 + \text{random}[0, 1])$ | 10 | $[-1.28, 1.28]$ |
| 8 | $F_8(y) = \sum_{i=1}^D -y_i \sin(\sqrt{|y_i|})$ | 10 | $[-1.28, 1.28]$ |
| 9 | $F_9(y) = \sum_{i=1}^n [y_i^2 - 10\cos(2\pi y_i) + 10]$ | 10 | $[-5.12, 5.12]$ |
| 10 | $F_{10}(y) = -20\exp\left(-0.2\sqrt{\frac{1}{n}\sum_{i=1}^n y_i^2}\right) - \exp\left(\frac{1}{n}\sum_{i=1}^n \cos(2\pi y_i + 1) + 20 + e\right)$ | 10 | $[-32, 32]$ |
| 11 | $F_{11}(y) = \frac{1}{4000}\sum_{i=1}^n y_i^2 - \prod_{i=1}^{n-1}\cos\left(\frac{y_i}{\sqrt{i}}\right) + 1$ | 10 | $[-600, 600]$ |
| 12 | $F_{12}(y) = \frac{\pi}{n}\left\{10\sin(\pi z_1) + \sum_{i=1}^{n-1}(z_i - 1)^2[1 + 10\sin^2(\pi z_i + 1)] + (z_n - 1)^2\right\}$ | 10 | $[-50, 50]$ |

(continued)

**Table 1** (continued)

| S. No. | Functions | Dimensions | Range |
|---|---|---|---|
| 13 | $F_{13}(y) = 0.1\left\{\sin^2(3\pi y_1) + \sum_{i=1}^{n-1}(y_i - 1)^2[1 + \sin^2(3\pi y_i)]\right\}$ | 10 | $[-50, 50]$ |
| 14 | $F_{14}(y) = \left[\frac{1}{500} + \sum_{j=1}^{25}\frac{1}{j+\sum_{i=1}^{2}(y_i - a_{ij})^6}\right]^{-1}$ | 2 | $[-65.536, 65.536]$ |
| 15 | $F_{15}(y) = \sum_{i=1}^{11}\left[a_i - \frac{y_1(b_i^2+b_i y_i)}{b_i^2+b_i y_3+y_4}\right]^2$ | 4 | $[-5, 5]$ |
| 16 | $F_{16}(y) = 4y_1^2 - 2.1y_1^4 + \frac{1}{3}y_1^6 + y_1 y_2$ | 2 | $[-5, 5]$ |
| 17 | $F_{17}(y) = \left(y_2 - \frac{5.1}{4\pi^2}y_1^2 + \frac{5}{\pi}y_1 - 6\right)^2 + 10\left(1 - \frac{1}{8\pi}\right)\cos y_1 + 10$ | 2 | $[-5, 0], [10, 15]$ |
| 18 | $F_{18}(y) = [1 + (y_1 + y_2 + 1)^2(19 - 14y_1 + 3y_1^2 - 14y_2 + 6y_1 y_2 + 3y_2^2)]$ | 2 | $[-2, 2]$ |
| 19 | $F_{19}(y) = -\sum_{i=1}^{4}C_i \exp\left(-\sum_{j=1}^{3}a_{ij}(y_j - p_{ij})^2\right)$ | 3 | $[0, 1]$ |
| 20 | $F_{20}(y) = -\sum_{i=1}^{4}C_i \exp\left(-\sum_{j=1}^{6}a_{ij}(y_j - p_{ij})^2\right)$ | 6 | $[0, 1]$ |
| 21 | $F_{21}(y) = -\sum_{i=1}^{5}\left[(Y - a_i)(Y - a_i)^T + C_i\right]^{-1}$ | 4 | $[0, 10]$ |
| 22 | $F_{22}(y) = -\sum_{i=1}^{7}\left[(Y - a_i)(Y - a_i)^T + C_i\right]^{-1}$ | 4 | $[0, 10]$ |
| 23 | $F_{23}(y) = -\sum_{i=1}^{10}\left[(Y - a_i)(Y - a_i)^T + C_i\right]^{-1}$ | 4 | $[0, 10]$ |

probability of finding the solution, improves exploration quality and also satisfies the property of diversification. The points from the first updation process are considered and checked for the conditions and updated further. However, only the new ranks make it to the final process, and the latter are discarded [8, 9].

The procedure is two folded and proceeds as follows. Identical to the first updation process, if the condition $Ra_{01} < \varepsilon$ is satisfied for a new point say $D_i'$, the current position of $D_i'$ is modified according to the given equations to a new point $D_i''$. This is given by the two folded Eqs. (10) and (11) as follows:

$$D_i'' = D_i' - \hat{\varepsilon} \times \left(D_t' - BP\right) \quad \left|\varepsilon' \le 0.5\right. \tag{10}$$

and

$$D_i'' = D_i' - \hat{\varepsilon} \times \left(D_t' - D_r'\right) \quad \left|\varepsilon' > 0.5\right. \tag{11}$$

where

$D_r'$  and $D_t'$ are random points selected from the first updation process,
$\hat{\varepsilon}$   is random numbers by Gaussian distribution (Gaussian normal distribution).

The newly obtained point $D_i''$ from the second updation process will once again be replaced back to $D_i'$ if $D_i'$ function's fitness value is better than $D_i'$ (Fig. 1).


# 4   Results and Discussions

The assessment and evaluation of the obtained results have been based upon the standard benchmark functions [6–11].

The comparisons have been carried out on 23 classic benchmark functions. The functions from $F_1$ to $F_7$ are unimodal benchmark functions, the functions from $F_8$ to $F_{13}$ are multimodal benchmark functions, and the rest from $F_{14}$ to $F_{23}$ are composite benchmark functions. The composite benchmark functions are a combination of Ackley's function, Sphere function, Griewank's function, Rastrigin's function and Weierstrass function.

The evaluation is done in MATLAB, the number of iterations is 1000, and the number of runs is 20. From the obtained results from the minimum, maximum, mean and standard deviation, values are computed, and the values are tabulated. The respective graphs for every benchmark function are plotted. The results are compared to Chaotic Whale Optimization Algorithm (CWOA) [12], and only the best values from different chaotic maps that are included in both papers from the CWOA are considered here (Table 2).

From the results, it is obvious that the newly proposed IWOA surpasses WOA in optimizing the functions to their global minima'. Even compared to CWOA, IWOA outperforms it in most of the cases.

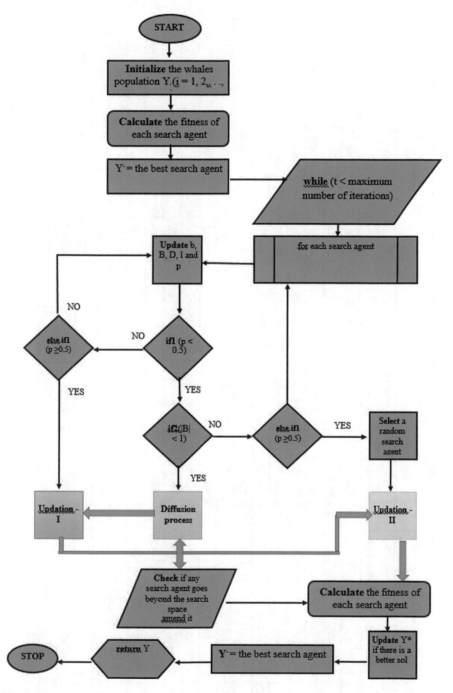

**Fig. 1** Flow chart of Improved Whale Optimization Algorithm

**Table 2** Results

| | WOA | | | IWOA | | | CWOA | |
|---|---|---|---|---|---|---|---|---|
| | Min | Avg | SD | Min | Avg | SD | Avg | SD |
| $F_1$ | 1.62e−200 | 1.20e−176 | 0 | 1.53e−217 | 6.57e−200 | 0 | 1.68e−70 | 2.58e−70 |
| $F_2$ | 5.47e−120 | 1.72e−113 | 6.64e−113 | 1.60e−128 | 2.66e−121 | 1.05e−120 | 3.31e−41 | 7.46e−41 |
| $F_3$ | 3.92e−6 | 1.5393 | 3.8817 | 1.74e−42 | 1.18e−17 | 5.28e−17 | 1.14e−292 | 0.00e+0 |
| $F_4$ | 1.86e−10 | 0.1007 | 0.3016 | 3.73e−53 | 1.58e−41 | 6.83e−41 | 1.07e−31 | 1.46e−31 |
| $F_5$ | 4.882 | 5.5273 | 0.335 | 4.5795 | 5.3413 | 0.7244 | 2.88e+01 | 4.59e−02 |
| $F_6$ | 1.23e−06 | 1.11e−05 | 9.87e−06 | 4.92e−07 | 3.05e−06 | 2.21e−06 | 0.00e+0 | 0.00e+0 |
| $F_7$ | 4.28e−05 | 7.77e−04 | 7.20e−04 | 5.84e−06 | 2.25e−04 | 2.29e−04 | n/a | n/a |
| $F_8$ | −4.19e+03 | −3.65e+03 | 638.0025 | −4.19e+03 | −3.87e+03 | 421.9218 | −1.14e+04 | 1.62e+03 |
| $F_9$ | 0 | 7.11e−16 | 3.18e−15 | 0 | 0 | 0 | 0.00e+0 | 0.00e+0 |
| $F_{10}$ | 8.88e−16 | 3.38e−15 | 2.03e−15 | 8.88e−16 | 2.49e−15 | 1.81e−15 | −1.44e−16 | 9.86e−21 |
| $F_{11}$ | 0 | 0.0349 | 0.085 | 0 | 0.0418 | 0.0995 | 0.00e+0 | 0.00e+0 |
| $F_{12}$ | 1.86e−06 | 2.48e−05 | 1.74e−05 | 1.21e−06 | 5.71e−06 | 3.83e−06 | n/a | n/a |
| $F_{13}$ | 1.54e−05 | 1.50e−04 | 1.83e−04 | 2.06e−06 | 5.81e−04 | 0.0025 | n/a | n/a |
| $F_{14}$ | 0.998 | 1.4445 | 0.8191 | 0.998 | 0.998 | 2.28e−16 | 1.91e−02 | 1.54e−02 |
| $F_{15}$ | 3.08e−04 | 6.71e−04 | 3.83e−04 | 3.08e−04 | 5.13e−04 | 3.20e−04 | n/a | n/a |
| $F_{16}$ | −1.0316 | −1.031 | 0 | −1.036 | −1.031 | | n/a | n/a |
| $F_{17}$ | 0.3979 | 0.3979 | 3.66e−07 | 0.3979 | 0.3979 | 3.66e−07 | n/a | n/a |
| $F_{18}$ | 3 | 3 | 2.24e−06 | 3 | 3 | 2.24e−06 | 0.00e+00 | 0.00e+00 |
| $F_{19}$ | −3.8628 | −3.861 | 0.0025 | −3.8628 | −3.8624 | 0.0015 | 8.94e−04 | 8.36e−04 |
| $F_{20}$ | −3.322 | −3.268 | 0.078 | −3.322 | −3.248 | 0.0868 | 0.00e+00 | 0.00e+00 |

(continued)

**Table 2** (continued)

| | WOA | | | IWOA | | | CWOA | |
|---|---|---|---|---|---|---|---|---|
| | Min | Avg | SD | Min | Avg | SD | Avg | SD |
| $F_{21}$ | −10.153 | −9.797 | 1.6001 | −10.1532 | −9.8982 | 1.1399 | n/a | n/a |
| $F_{22}$ | −10.402 | −9.236 | 2.1487 | −10.4029 | −9.8944 | 1.598 | n/a | n/a |
| $F_{23}$ | −10.536 | −8.086 | 3.3215 | −10.5364 | −9.7251 | 1.9811 | n/a | n/a |

Here, SD stands for Standard Deviation

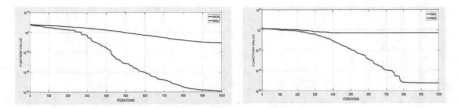

**Fig. 2** WOA versus IOA comparison graphs for function $F_3$ and $F_4$

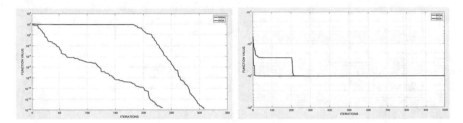

**Fig. 3** WOA versus IOA comparison graphs for function $F_9$ and $F_{22}$

## 4.1 Comparison Graphs

From the obtained graphs, it can be seen that for optimization of any solution, here the case being minimization, the Improved Whale Optimization Algorithm (IWOA) outperforms the Whale Optimization Algorithm (WOA). The noteworthy point is the fast rate of convergence of IWOA compared to WOA. The new ranking system allows for better results as it saves the best-obtained values and discards the least scoring values. This also allows for better exploration of the given search space and improves computational efficiency. From Fig. 2, it can be observed the computing efficiency of IWOA over WOA. The function here being to minimization yields the global minima quicker using IWOA. In Fig. 3, the speed of IWOA converging to the global minima is quite fast and requires further reduced iterations to achieve global minima compared to the WOA implementation.

## 5 Conclusion

In this paper, the performance of the existed algorithm that is WOA has been improved by including the updating process. Potential of IWOA is tested on the benchmark functions with varied complexity, and IWOA outperformed WOA in terms of convergence as well as in the function value. IWOA is performing better than WOA in the exploitation and explorations capabilities. Further, IWOA can be used in other optimization problems in different areas.

# References

1. X.S. Yang, *Nature-Inspired Metaheuristic Algorithms* (Luniver Press, 2010)
2. X.S. Yang, A new metaheuristic bat-inspired algorithm, in *Nature Inspired Cooperative Strategies for Optimization (NICSO 2010)* (Springer, Berlin, Heidelberg, 2010), pp. 65–74
3. S. Mirjalili, A. Lewis, The whale optimization algorithm. Adv. Eng. Softw. **95**, 51–67 (2016)
4. M.M. Mafarja, S. Mirjalili, Hybrid whale optimization algorithm with simulated annealing for feature selection. Neurocomputing (2017)
5. P.D.P. Reddy, V.V. Reddy, T.G. Manohar, Whale optimization algorithm for optimal sizing of renewable resources for loss reduction in distribution systems. Renew. Wind Water Sol. **4**(1), 3 (2017)
6. A.N. Jadhav, N. Gomathi, WGC: Hybridization of exponential grey wolf optimizer with whale optimization for data clustering. Alexandria Eng. J. (2017)
7. A. Kaveh, Sizing optimization of skeletal structures using the enhanced whale optimization algorithm. in Applications of metaheuristic optimization algorithms in civil engineering (Springer, 2017), pp. 47–69
8. T. Liao et al., Ant colony optimization for mixed-variable optimization problems. IEEE Trans. Evol. Comput. **18**(4), 503–518 (2014)
9. X.-S. Yang, S. Deb, Engineering optimisation by cuckoo search. Int. J. Math. Model. Numer. Optim. **1**(4), 330–343 (2010)
10. I. Aljarah, H. Faris, S. Mirjalili, Optimizing connection weights in neural networks using the whale optimization algorithm. Soft Comput. 1–15 (2016)
11. M.-Y. Cheng, D. Prayogo, Symbiotic organisms search: a new metaheuristic optimization algorithm. Comput. Struct. **139**, 98–112 (2014)
12. G. Kaur, S. Arora, Chaotic whale optimization algorithm. J. Comput. Des. Eng. **5**, 275–284 (2018)

# A Matrix-Based Approach for Evaluation of Vocal Renditions in Hindustani Classical Music

Kunjal Gajjar and Mukesh Patel

**Abstract** Indian or Hindustani Classical Music (ICM and HCM, respectively) is based on the Raga system of music. Unlike a Western Classical Music (WCM) composition (usually documented in a written score format), a Raga defines an overall melodic structure to which a vocal or instrumental rendition adheres to. HCM has around 200 Ragas, each of which is specified in terms of a subset of notes (that can be part of a composition in that particular Raga) as well as their sequencing (e.g. aroha–avaroha sequence), combinations and permutations. The rules and conventions (here under referred to syntax of Raga) together determine the general melodic structure (often referred to as a mood that the Raga aims to capture) of a rendition. Hence, a computational model for recognition or evaluation of HCM renditions would need to be able to handle a much higher degree of complexity than a similar system for WCM. Here we present a computational model that handles this complexity elegantly and efficiently for 8 Ragas (that are typically learned by novice singers). More specifically we show how a relatively simple note transition matrix-based approach incorporating key elements of a Raga's syntax results in highly reliable evaluation of songs and robust error identification (for feedback to novice learners as part of computer-based tutoring system).

**Keywords** Computational musicology · Hindustani classical music · Indian classical music · Raga evaluation · Transition matrix · Computational model · Note · Tutoring system · Novice · Early stage learners · Vocal Raga

K. Gajjar (✉)
Faculty, School of Computer Studies, Ahmedabad University, Ahmedabad 380009, India
e-mail: kunjal.gajjar@ahduni.edu.in

M. Patel
Charotar University of Science and Technology (CHARUSAT), Changa 388421, India
e-mail: mukeshpatel.mca@charusat.ac.in

© Springer Nature Singapore Pte Ltd. 2021
X.-Z. Gao et al. (eds.), *Advances in Computational Intelligence and Communication Technology*, Advances in Intelligent Systems and Computing 1086,
https://doi.org/10.1007/978-981-15-1275-9_7

# 1 Introduction

Computational musicology, an interdisciplinary research area is exciting and challenging field which motivates research for music information retrieval, music composition, recognition and analysis of extracted features [1]. Since 1960s, there has been reasonable amount of work done for Western Classical Music (WCM) in this field. However, Indian or Hindustani Classical Music (ICM and HCM) is still relatively unexplored [2, 3]. One main reason for this limited attention is the inherent complexity of HCM resulting from the fact that music is generated based on loosely defined Ragas.

Unlike Western Music, ICM songs or instrumental recitals conform to the structure of a Raga which is determined by rules and conventions about sequencing, combinations and permutations of notes which together determine the overall melodic structure by specifying general boundaries and constraints. It is important to note that these rules and conventions do not determine any specific rendition in a Raga but can be said to define its syntax. Hence, a singer/musician does not compulsorily recite a pre-composed piece of music but renders a version of it, typically by putting more emphasis on certain phrases or repeating some sequence of notes more often than others [4–6]. From the computational point of view, such latitude translates into a level of complexity that is relatively difficult to account for in computational models of HCM renditions.

Despite this complexity, attempt to develop computational models of HCM has been made by several researchers. Most such studies have focused on recognition of recitals in various Ragas and automated classification of the music for retrieval systems. The general approach has been to incorporate the rules and conventions as general heuristics. In the model presented here, we have extended this approach to a more systemic modelling of the syntax of eight selected Ragas such that it evaluates songs (according to the syntax of the Raga in which it is composed) and identifies any errors in the rendition.

The Raga syntax-based computer model of evaluation is part of a computer-based tutoring system designed for early stage learners (novices) of vocal HCM which also includes a note transcription module (described in Sect. 3.1). Typically during the early stage (defined as the first two years) of the learning process, a novice is confined to singing practice songs in the presence of tutor who engages in continuous evaluation of the novice's rendition and providing feedback on wrongly rendered notes/swars or violation of rule such as rendering the aroha sequence wrong. This is obviously a time-consuming process (particularly as rarely can it involve more than one learner at a time). Hence, the motivation for exploring the possibility of developing a computer-based tutoring system that automates as much of the initial stage of learning process is possible. In particular, it was felt that such a system needs to have robust and reliable process for evaluating whether a song has been rendered correctly and provide feedback on how to correct any errors with respect to the rules and conventions of the Raga.

In the next section, we describe the key elements of HMC that determine the melodic structure of Ragas, and how that translates into essentially non-deterministic and/or unconstrained nature of songs or musical renditions. In Sect. 2, we present a brief review of relevant previous work, and the scope of the tutoring system with particular emphasis on the evaluation model incorporated in the tutoring system.

## 1.1 Hindustani Classical Music and Raga Structure

Hindustani classical music (HCM) is one form of Indian Classical Music (ICM) that is primarily practised in northern and central parts of India. The uniqueness of HCM is in its melodic structure which is determined by a combination of a Raga—musical formalisms and melodic phrases—and beat that determines the tempo Rabunal [7]. Many of these aspects of Ragas have been codified in detail by Pandit Vishnu Narayan Bhatkhande [8, 9]. Here we describe the key musical elements, rules and conventions that collectively determine the structure of each Raga. Vocal and instrumental music (e.g. songs and solo recitals) are usually composed to conform to the structure or framework of specific Ragas.

The tonal material used in compositions known as swars (notes) of HCM scale are shadaj (Sa), rishabh (Re), gandhar (Ga), madhyam (Ma), pancham (Pa), dhaivat (Dha) and nishad (Ni). The scale contains these seven natural notes referred to as shuddha (literally pure) swars. However, each swar may also have two different forms which are komal (flat notes) and tivra (sharp notes) except 'Sa' (the tonic) and 'Pa' (the perfect fifth) which are rendered in only one form. Together, the following 12 notes make up the scale of HCM Ragas in which music and songs are composed or recited:

Shuddha swar—seven normal swars: Sa, Re, Ga, Ma, Pa, Dha and Ni.
Komal swar—four swars, Re, Ga, Dha and Ni can also be rendered in komal form, that is, their frequency is lower than as a shuddha swar.
Tivra swar—one swar, Ma' can also be rendered in tivra form, that is, its frequency is higher than shuddha Ma.

Collectively, these 12 swars are referred to as saptak. Unlike WCM, HCM uses a movable scale which means that there is no fixed pitch for a note. An octave can start from any pitch which is known as tonic frequency (frequency of 'Sa' of middle octave) [10, 11]. In a rendition, a singer/musician can shift across three saptak— lower (mandra), middle (madhyam) and upper (taar). In comparison to madhyam saptak, mandra saptak has lower pitch frequencies and taar saptak has higher pitch frequencies. All the notes in these three saptaks are defined in the relation to their tonic frequency. The frequency ratio with 'Sa' of middle octave (madhyam) is shown in Table 1. For more details, see Gajjar and Patel [12].

The basic structure of each Raga is determined by the selected swars from the saptak (listed in Table 1). Each Raga has minimum of five and maximum of seven swars. It is important to understand that a Raga is not just a sequence of various

**Table 1** WCM and HCM notes with frequency ratio [10]

| Western notation | Indian notation | Notation used in proposed model | Frequency ratio (natural) | Names |
|---|---|---|---|---|
| C | Sa | S | 1/1 | Tonic Sa |
| C# | Re | R | 16/15 | Komal Re |
| D | Re | R | 9/8 | Shuddha Re |
| Eb | Ga | G | 6/5 | Komal Ga |
| E | Ga | G | 5/4 | Shuddha Ga |
| F | Ma | M | 4/3 | Shuddha Ma |
| F# | Ma' | M | 7/5 | Tivra Ma |
| G | Pa | P | 3/2 | Perfect fifth Pa |
| Ab | Dh | D | 8/5 | Komal Dh |
| A | Dh | D | 5/3 | Shuddha Dh |
| Bb | Ni | N | 9/5 | Komal Ni |
| B | Ni | N | 15/8 | Shuddha Ni |
| C' | Sa' (next octave) | S_U | 2/1 | Sa of upper octave |

combinations of a set of swars: in addition, a Raga's structure is determined by rules and conventions that specify do's and don'ts regarding the sequencing and combination of swars as well as emphasis on certain subset (phrases) of notes.

The important conventions are summarized in the list below that collectively determines the melodic structure (or essence or mood) of a Raga [10, 13].

- Aroha–Avaroha—Aroha is the ascending order sequence of swars that describe how the Raga moves in its ascending order. Avaroha is the descending order sequence of swars that describe how the Raga moves in descending order. For example, in Raga Bhimpalasi:

    - Aroha sequence is: Sa, Ga, Ma, Pa, Ni, Sa'
    - Avaroha sequence is: Sa', Ni, *Dha*, Pa, Ma, Ga, *Re*, Sa
    - As can be seen swars, 'Re' and 'Dha', can be rendered in descending (avaroha) order; however, they cannot be used in ascending (aroha) order.

- Pakkad—phrase (or sequence) of notes that are specific to a Raga. Its usage or inclusion in a rendition is regarded as enhancing the quality of song. The pakkad is often used to determine the identity of a Raga, so can be regarded as a Raga's signature tune. For example, the pakkad "Pa, Ga, Dha, Pa, Ga, Re, Ga, Re, Sa" is specific to Raga Bhupali.
- Vadi swar—the highest frequency swar in a Raga. For example swar 'Ga' is Vadi swar of Bhupali. This is reflected in the higher frequency of 'Ga' of the Raga's pakkad, "Pa, Ga, Dha, Pa, Ga, Re, Ga, Re, Sa".
- Samvadi swar—the second highest frequency swar in Raga. For example 'Dha' is Samvadi swar of Raga Bhupali.

- Varjit swar—swars which are regarded as incompatible with the swars of a specified Raga, and therefore to be avoided in compositions of songs/recitals in that Raga. For example, swars 'Ma' and 'Ni' are varjit in Raga Bhupali, which is made up of swars, "Sa, Re, Ga, Pa, Dha".
- Vivadi swar—swars which are not part of aroha–avaroha sequence of the Raga but are also not varjit swar are known as vivadi swar. These swars should not be used in the Raga, hence also referred as "enemy" swar. However, experts might sometimes use them to enhance the beauty of Raga [14]. Usually, these are variations of swars in aroha–avaroha. For example, komal swars "Re, Ga, Dha" are vivadi swars for Raga Bhupali.

Hence, a set of swars combined with the above conventions (that determine emphasis on specific swars, sequencing of phrase, etc.) are the most significant determiners of the structure of a Raga which determines (literally) the mood of a Raga that the singer/musician is supposed to capture in a rendition of a song or instrumental piece in that Raga (and which in turn in evoked in the listener). It is the differences in the conventions applicable to each Raga that enables a listener (or learner) to distinguish them along melodic and/or aesthetic dimensions. So even when two Ragas have the same set of swars and aroha–avaroha sequences, the differences in features such as the vadi and samvadi swars and pakkad renditions in each will result in very different renditions in terms of the moods they invoke. For example, Raga Bhupali and Deshkar both use same set of swars—Sa, Re, Ga, Pa, Dha—yet they sound different because compositions of Bhupali have higher frequency for swars "Sa, Re, Ga", whereas in compositions of Deshkar, the swars "Pa, Dha, Sa" predominate [15].

There are more complicated and esoteric features of HCM, such as meend (sliding from 1 note to other) and which might even include a varjit (otherwise forbidden) swar for a very short duration; alap, the starting section of a song which is in slow rhythm compared to rest of the song, usually sung before the rendition of the Raga itself (in instrumental or vocal form) [16]; or, taan, compositions of Raga sung in very fast tempo [17]. These (and other such conventions), however, would typically be used by expert singers/musicians during renditions and as part of complicated improvisation.

Usually, early stage learners, apart from learning to render the correct notes of a specific Raga (which includes avoiding the varjit swar), focus on mastering aroha and avaroha sequences and learning when and how to launch into a pakkad. These aspects of a Raga have to be mastered before they can move on to a more advanced stage where the emphasis shifts to improvisation [6]. Hence, the tutoring system presented here has been designed to handle only these basic aspects of songs typically rendered by early stage learners. This approach hence reduces the complexity that needs to be handled by a computational model for automated recognition/evaluation of songs sung by novices.

In Sect. 2.1, we present a brief review of previous research focused on computational musicology models of ICM with an emphasis on findings that are specifically relevant to our research focus. It will be noted that much previous work has focused on exploring and/or modelling the nature of complexity of ICM (as part of a recognition or generation system) [12]. In Sect. 2.2, we provide a detailed description of the scope and objective of the proposed system for analysis and evaluation of songs (in one of the eight selected Ragas) typically sung by early stage learners of HCM.

# 2 Previous Findings and Scope and Objective of Proposed System

## 2.1 Previous Relevant Research

In this section, we review computational musicology models of aspect of ICM that are of relevance to the modelling approach and evaluation algorithms described here. A more comprehensive review and evaluation of computational modelling of ICM can be in found in Gajjar and Patel [12].

Bhattacharjee and Srinivasan [18] describe a transition probability model for Raga recognition. Their approach is based on the idea that in any Raga, the probability for transition between notes in ascending and descending order (i.e. aroha-avaroha) is a good indicator of a Raga's structure, and so can be reliably used for Raga identification. The authors carried out manual note transcription of renditions by two singers/musicians in ten Ragas. Based on this note transcription, they hand-crafted a Transition Probability Matrix (TPM) for each Raga. Each TM has 12 columns and 12 rows to represent 144 possible transitions between tuples of 12 swars. The TM has middle octave (madhyan) as the default, so swars rendered in the higher (taar) or lower (mandra) octave are shifted to the middle (madhyam) octave. The probability of transitions for each Raga was determined with a training set of ten clips of songs, each of 5 s duration.

While this approach provides insights into pattern of transition between notes in each Raga, it is highly constrained with respect to actual renditions. The primary focus on the middle octave (and shifting the notes from higher and lower octave to middle octave) means that critical information is lost as at times aroha–avaroha sequence can change with singer's shift between octaves. For instance, a sequence "Sa (madhyan octave), Ga (taar octave)" is considered aroha, whereas sequence "Sa (madhyam octave), Ga (mandra octave)" is considered to be avaroha. Hence, the proposed TM would not be able to handle this distinction which can prove to a significant shortcoming for an extended computer-based evaluation system for ICM renditions. However, as we will explain in Sect. 3, the TM approach to modelling note transition is a highly promising and efficient approach for an evaluation system for HCM renditions.

Das and Choudhury [19] focused on generation of HCM, for which they also considered aroha and avaroha features of Ragas. They have designed probabilistic Finite State Machine (FSM) for ascending and descending movements for a particular Raga. This FSM is used to generate a sequence of notes (i.e. a rendition of a Raga).

However, the authors conclude that the FSM generated renditions were not as good as those rendered by human musicians. The main reason for this being that while the FSM focused on modelling the aroha and avaroha features of a Raga, these two features were not sufficient for generating the complex melodic structure (i.e. richness) that is normally evident in renditions by human performers. This finding suggests that conventions such as pakkad, vadi and samvadi swars are critical in determining the essence of a Raga. Hence, an extended FSM that incorporates these additional features is likely to generate music of better quality; however, it would also result in the increased complexity.

Pandey, Mishra and Ipe [3] have proposed an approach for automatic Raga recognition. The note transcription is done using hill peak and note duration heuristics. For Raga identification, the notes identified from transcription module are used with hidden Markov model which is trained using Baum–Welch learning algorithm. It relies on the inherent flexibility of a Markov model, which unlike FMS's can handle probability based transition. This approach would be better at handling additional features such as pakkad, vadi and samvadi swars. However, the higher order of Markov model would result in an exponential increase in the number of combinations (of swars). Further, it is highly likely that a model using the same transition matrix will tend to generate music in similar style [20]. Overall, the conclusion is that this approach would still not be able to generate music that is aesthetically as creative and pleasing as would be by a trained human musician.

Overall computational models for music composition and recognition tend to rely on modelling the past sequence of note to predict next note or to identify patterns in the composition. This approach yields good results for recognition or composition where past sequence of notes can indeed be helpful in predicting the next note(s). However, for HCM at any point in a rendition, subsequent notes are as important as the preceding notes for deciding the correctness of a song. In order to model that requirement, it was decided to implement a transition matrix (using a Markov model approach). This matrix can be used to check the correctness of individual note by considering both, the preceding as well as the subsequent note.

Our approach of designing a tutoring system for novice learners hence is primarily focused on identifying the notes sung and assesses their appropriateness with respect to Raga's structure. In the next section, we describe scope and architecture of the tutoring system in more detail.


## 2.2   Scope of the HCM Tutoring System

The primary aim in developing this system was to enable a novice to do singing practice in the absence of a tutor (guru). For many students, the necessarily limited duration of one-to-one interaction with a tutor is the biggest reason for the relatively slow progress in mastering singing technique. The tutoring system was therefore designed to address this challenge; more specifically, it was focused on automated evaluation of rendered song by an early stage learner of HCM, and to provide feedback on how to correct any mistakes. Ideally, such a tutoring system should be designed to "listen" to a song being rendered in real time and provide feedback in real time as is often the case with a one-to-one learning session with a teacher. However, this would have raised the complexity of the system that would be beyond the scope of our research project; particularly so as the key challenge is to have a reliable evaluation system that can identify correctly sung notes (with respect to the Raga in which the song has been composed) and provide list of errors (if any). Hence, instead of designing a system robust enough to accept audio recordings of varying quality (depending on the quality of recording equipment), the system was designed to accept audio files in wav format.

Usually, performance by experts of HCM is accompanied by instruments like harmonium/sitar (to provide the tonic scale) and tabla (drums) which is used to set the tempo of the rendered song. However, an early stage learner usually practises in the absence of these instruments, thereby rendering song in a scale that he/she is comfortable with and keeping time with hand-claps. Tonic scale and tempo are used for identifying frequency of notes in octaves (lower, middle and upper) and segmentation of audio file, respectively, in note transcription module (described in detail in Sect. 3.2). Hence, the system requires the learner to select the tempo of song (in beats per minute) and tonic frequency (in hertz).

The tutoring system is designed to evaluate songs in eight Ragas—Bhupali, Durga, Sarang, Kafi, Bhimpalasi, Khamaj, Bhairavi and Desh—which are generally taught in the initial years of a singers learning process of HCM. Our selection of this subset was based on widely respected two leading proponents of syllabi for novice learners in HCM—first, Gandharv Mahavidyalaya founded by Pt. Vinayakbua Patwardhan in 1932 [21]; second, Akhil Bharatiya Gandharva Mahavidhyalaya Mandal [22] founded by Pt. Vishnu Digambar Paluskar in 1901. This selection is further supported by a survey of 50 singers of HCM based in Ahmedabad, Gujarat. Of these, 92% confirmed that these eight Ragas were the ones that they mastered in the initial phase of learning HCM. Even the remaining 8% agreed on at least six Ragas from this list (while suggesting alternative for the remaining one or two Ragas). In our system, the user (novice singer) is required to identify the Raga in which he/she has sung the (pre-recorded) song.

Evaluation of rendered song essentially means determining whether the transition between swars was permissible (with respect to the Raga's syntax). Hence, simple transition matrices were used to encode legitimate transition based on the aroha–avaroha sequence and varjit swar of each Raga. A matrix for the Raga in which the song has been rendered (by the novice) would be used to determine the correctness of a note with respect to both the previous note and next one. For instance, in Raga Bhimpalasi, swars 'Dha' and 'Re' are used only in avaroha (descending order). So, in the sequence "Sa, Dha, Pa, Ma, Re, Ga", the (initial) pair 'Sa, Dha' is moving in ascending order (aroha) which is not legitimate. But considering the subsequent (third) note 'Pa' which forms the pair 'Dha, Pa' which is avaroha (descending order) this transition would be considered legitimate. The generalized algorithm used for matrix generation is explained in Sect. 3.3.

Once the notes have been evaluated, users are presented with feedback that highlights the errors (if any) that they have made in the song. This includes indication of where in the song the error occurred with respect to the key aspects (described in Sect. 1.1) that determines a Raga's melodic structure.

However, as explained in Sect. 1.1, though aroha–avaroha and varjit swars are two most important conventions of Ragas, there might be scenarios where other Raga may follow the same convention. Hence, to differentiate between Ragas, vadi, samvadi and pakkad are also considered in second stage of evaluation process and which provides further feedback on whether the overall rendering of the song conform to the specified Raga by the novice (or is closer to another Raga's syntax).

In the next section, we present the overall architecture of the tutoring system with details of how the song evaluation process has been implemented. This is followed by the results of its performance when presented with 180 songs of eight Ragas with different types of errors listed in Sect. 4.

## 3 Architecture and Functionality of the HCM Tutoring System

The overall system architecture is illustrated in Fig. 1. The early stage learner would upload the audio file with relevant details of the Raga, tempo, gender and tonic frequency (octave). The input file is then processed for note transcription, the output of which is evaluated for errors. Each part of the system will be described in more detail in the rest of the sections.

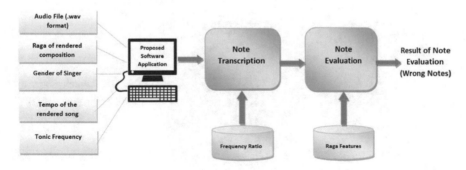

**Fig. 1** Architecture of proposed system

## 3.1  Note Transcription Process

Figure 2 shows the substeps for note transcription and inputs utilized in these steps. The note transcription module considers an audio file uploaded by user and generates list of notes. We have used a third-party wav file class by Andrew Greensted [23] for reading the sample values from uploaded wav file. Next, these samples are processed using algorithm from Praat to get the list of frequencies (pitch listing) [24] which are further segmented based on the tempo (as provided by early stage learner). The mean value of segmented frequency list is then compared with the note frequency table (generated by using tonic frequency listed in Table 1) to identify each note in the rendition. Figure 3 illustrated the format of information about notes displayed for the learner.

In general, musical note transcription from audio files is a complex process, particularly so for HCM renditions during which sliding between notes is considered a mark of excellence. However, this complexity can be tackled (to the extent required for our purposes) with the note duration heuristics approach described by Pandey, Mishra and Ipe [3]. It assumes that the minimum duration of each note is one-fourth of the duration of a beat. For example, if duration of a beat is 60 s, minimum duration of a note is then considered to be 15 s.

Figure 3 is an example of a typical output once a wav files has been segmented into individual notes. The figure displays identified notes in form of spectrogram and pitch contour. The identified list of notes along with their duration is stored in a separate text file. These files are then accessed by the evaluation module to check for any errors with respect to the Raga structure of the song.

The note transcription output is then displayed as shown in Fig. 3.

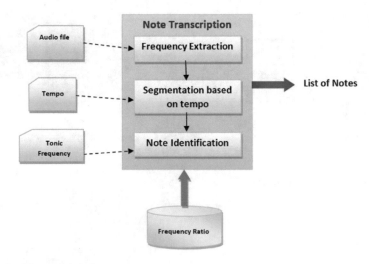

**Fig. 2** Note transcription module

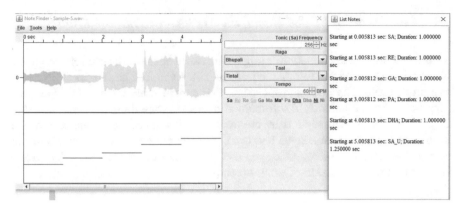

**Fig. 3** Notes listed through note identification module

## 3.2 Evaluation Process Overview

Figure 4 shows the substeps for note evaluation process. As explained in Sect. 2.2, the note evaluation module incorporates a two-stage evaluation process. In the first stage, the rendered notes are evaluated with respect to the aroha and avaroha sequences of the Raga and varjit swars: rendered notes are compared with a note TM (for the specific Raga of the rendition). The rendition is processed further (second stage) with respect to the vadi, samvadi and pakkad of the Raga. In the next Sect. 3.3, we present details of the algorithm used for generation of TM used in first stage. Sections 3.4 and 3.5 have detailed description of the evaluation process for the first and second stages, respectively.

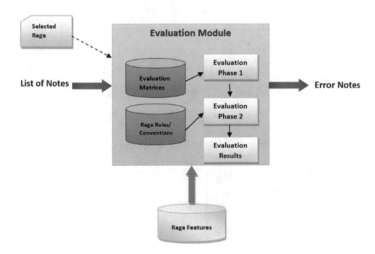

**Fig. 4** Note evaluation module

### 3.3 Computation of TM Based on Aroha–Avaroha of Raga

For the first step of the evaluation, an algorithm based on the aroha–avaroha sequence of each Raga is used for formulation of the TM. Each TM has an $36 \times 36$ array, containing 1,296 cells with values '1' or '0' as shown in Table 2, where '1' represents a legitimate transition and '0' a forbidden transition (based on) for that particular Raga. In our implementation, there are three sections in each TM; one each of the three octaves (mandra, madhyan and taar). The stepwise process for generation of each TM is described below. Once generated, the matrices are stored in the system for access during evaluation of songs uploaded by users (early stage learners).

- Step 1

  - Let the set of notes of three octaves (i.e. mandra, madhyan and taar: lower, middle and upper, respectively) be denoted as $S$.
  - $S = \{n_0, n_1, n_2 \ldots n_{35}\}$, where $n_0 = S =$ Sa of lower octave, $n_1 = r =$ Re of lower octave, $n_{12} = S =$ Sa of middle octave, $n_{35} = N =$ Ni of upper octave.

- Step 2

  - Let $S_A$ be set of notes in aroha (ascending order), where $S_A \subset S$.
  - And $S_D$ be set of notes in avaroha (descending order), where $S_D \subset S$.
  - Let $i$ and $j$ represent index of row and column of matrix, respectively, such that $i = 0 \ldots 35$ and $j = 0 \ldots 3$.

- Step 3

  - For $i \leq j$,

$$V_{ij} = \text{if} \quad n_i \in S_A \text{ and } n_j \in S_A$$
$$0, \quad \text{if } n_i \notin S_A \text{ or } n_j \notin S_A$$

  - For $i > j$,

$$V_{ij} = 1, \quad \text{if } n_i \in S_D \text{ and } n_j \in S_D$$
$$0, \quad \text{if } n_i \notin S_D \text{ or } n_j \notin S_D$$

Considering $S_A$ and $S_D$, matrix, denoted as $M$, is populated as

$$M = \left\{ \begin{array}{llll} V_{00} & V_{01} & \ldots & V_{0n} \\ V_{10} & V_{11} & \ldots & V_{1n} \\ \ldots & & & \\ V_{n0} & V_{n1} & \ldots & V_{nn} \end{array} \right\} \text{ where } n = 35 \text{ and } V_{ij} \in \{0, 1\}.$$

The example of TM for Raga Bhupali is shown in Table 2.

**Table 2** Example of transition matrix for Raga Bhupali

| Octave | | Lower | | | | Middle | | | | Upper | | | |
|---|---|---|---|---|---|---|---|---|---|---|---|---|---|
| | | S_L | r_L | M_L | N_L | S | R | M | N | S_U | r_U | M_U | N_U |
| Lower | S_L | 1 | 0 | 0 | 0 | 1 | 0 | 0 | 0 | 1 | 0 | 0 | 0 |
| | r_L | 0 | 0 | 0 | 0 | 0 | 0 | 0 | 0 | 0 | 0 | 0 | 0 |
| | M | 0 | 0 | 0 | 0 | 0 | 0 | 0 | 0 | 0 | 0 | 0 | 0 |
| | N_L | 0 | 0 | 0 | 0 | 0 | 0 | 0 | 0 | 0 | 1 | 0 | 0 |
| Middle | S | 1 | 0 | 0 | 0 | 1 | 0 | 0 | 1 | 0 | 0 | 0 | 0 |
| | r | 0 | 0 | 0 | 0 | 0 | 0 | 0 | 0 | 0 | 0 | 0 | 0 |
| | M | 0 | 0 | 0 | 0 | 0 | 0 | 0 | 0 | 0 | 0 | 0 | 0 |
| | N | 0 | 0 | 0 | 0 | 0 | 0 | 0 | 0 | 0 | 0 | 0 | 0 |
| Upper | S_U | 1 | 0 | 0 | 0 | 1 | 0 | 0 | 0 | 1 | 0 | 0 | 0 |
| | r_U | 0 | 0 | 0 | 0 | 0 | 0 | 0 | 0 | 0 | 0 | 0 | 0 |
| | M_U | 0 | 0 | 0 | 0 | 0 | 0 | 0 | 0 | 0 | 0 | 0 | 0 |
| | N_U | 0 | 0 | 0 | 0 | 0 | 0 | 0 | 0 | 0 | 0 | 0 | 0 |

## 3.4   Evaluation Process—Stage 1

First, the evaluation process compares sequences of notes (generated by the note identification module) with the TM to check for errors with respect to aroha–avaroha sequence of the Raga and varjit swars. Each bi-gram of note sequence is compared with TM in the following step-wise process:

Let the sequence of notes be denoted as:

$N = \{N_0, N_1 .... N_n\}$, where $N_0$ is starting note and $n$ represents total number of notes in the sequence.

Transition matrix ($M$) is used for evaluation (pre-generated) as shown in Sect. 3.3.

Apart from matrix, set of varjit swars of a Raga is also used for evaluation. Let it be denoted as $V$.

Evaluation algorithm is as follows:

1. For current note $N_i$ and next note $N_{i+1}$, where $0 \le i \le n$, find the value from $M[N_i][N_{i+1}]$.
2. If value of $M[N_i][N_{i+1}]$ is 1, increment $i$ by 1. If value of $M[N_i][N_{i+1}]$ is 0, check if $N_i$ or $N_{i+1}$ belongs to $V$.
3. If $N_i \in V$, mark $N_i$ as wrong note and increment value of $i$ by 1. Else if $N_{i+1} \in V$, mark $N_{i+1}$ as wrong note and increment value of $i$ by 2.
4. If $N_i \notin V$ and $N_{i+1} \notin V$ and $N_{i+1}$ is not last note in sequence, increment value of $i$ by 1. Check the value of $M[N_i][N_{i+1}]$. If this value is 1, increment $i$ by 2. Else mark $N_i$ as wrong note and increment the value of $i$ by 1.
5. If $N_{i+1}$ is last note and $M[N_i][N_{i+1}] = 0$, mark $N_{i+1}$ as wrong note.
6. Repeat from step 1 until $i$ reaches the last note of sequence $N$.

Figure 5 shows a section of the output of first stage of note evaluation of a rendition in Raga Bhupali. The notes in red colour are incorrect notes as per the TM based on the aroha–avaroha and varjit swar of that Raga.

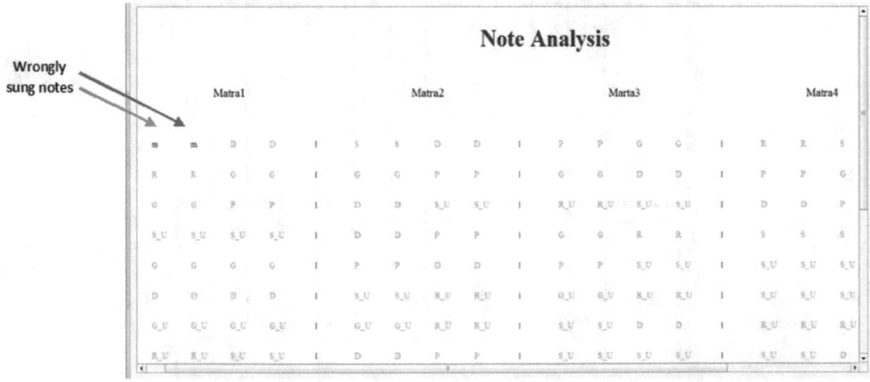

**Fig. 5**  Note evaluation result for Raga Bhupali

## 3.5   Evaluation Process—Stage 2

The second stage of evaluation considers the vadi, samvadi and pakkad features of the selected Raga to evaluate the overall extent to which the rendition confirms to the Raga's syntax. This stage is at a two-step process; first, a simple frequency count of different notes is calculated to check if the vadi and samvadi swars of the Raga are indeed to most and second most used swars in the rendition; and, second a string matching algorithm checks for the occurrences of any/all pakkads in the rendition. This process is implemented as follows:

Consider following sequence of notes which is evaluated for Raga Bhupali:

S_U S_U D D S_U S_U D D S_U S_U D D P P D D P P D D P P G R G P D P
G P D S_U D P G P D S_U D P G P D D D G P D S_U D P G G P D S_U D P G
D S D P G D S_U D P D S_U D P S_U S_U D P S_U S_U D P S_U D P S_U D P
S_U S_U D P G G P D S_U D P G G P D P G P D P G P G R S

The above sequence of notes conforms to the aroha–avaroha and varjit swars features of Raga Bhupali, so the first stage of evaluation has indicated not errors as such. However, the second-stage evaluation process has highlighted that the swar 'D' (Dha) occurs more frequently than the swar 'G' (Ga) which is the actual vadi swar of Raga Bhupali. In addition, the rendition has notably high frequency of the combination of "P(Pa), D(Dha), S_U(Sa_U)" swars which is a pakkad of Raga Deshkar and not Bhupali. Based on such observations, the tutoring system would provide the user qualitative feedback that some features of Raga Bhupali are not captured in the rendition.

## 4   Results

In this section, we present preliminary results of on the performance of the first stage of the evaluation process. Our primary aim is to validate the efficacy of the transition matrix approach for evaluation of HCM renditions by early stage learners. For this test, we compiled 180 renditions (with average 100 notes) of eight Ragas. The renditions collectively had 380 errors that comprehensively reflected all types of errors that can be typically made by early stage learners. These included:

– Varjit swar as first note, varjit swar as intermediate note in the sequence, three consecutive varjit swars in the sequence
– Vivadi swar as first note in sequence, vivadi swar as intermediate note in the sequence, three consecutive vivadi notes in the sequence
– Correct note between two varjit swars in the sequence, correct note between two vivadi swars in the sequence
– Five consecutive wrong notes in the sequence (combination of varjit and vivadi swars)
– Correct note between sequences of wrong notes
– Varjit swar as last note in sequence, varjit swar as last three notes in the sequence

**Table 3** Test results for TM-based evaluation (stage 1)

| Total number of sequences | Average number of notes in each sequence | Total number of notes in all the sequences | Total number of errors in the sequences | Number of errors identified through the implemented module | False positives |
|---|---|---|---|---|---|
| 180 | 100 | 18,000 | 380 | 380 | 0 |

- Vivadi swar as last note in the sequence, vivadi swar as last three notes in the sequence
- Swar allowed in aroha but not in avaroha used in avaroha sequence; swar allowed in avaroha but not in aroha used in aroha sequence.

Based on the types of errors, different testing sequences of notes were generated for eight Ragas. All the sequences had some combination of errors from the above list. Also, to check for false positives, there was at least one input sequence with zero errors.

As can be seen from the results presented in Table 3, the TM-based evaluation of HCM performs remarkable well in identifying errors (with respect to aroha–avaroha and/or varjit swar conventions of Ragas), as well as avoiding any false positives. These results confirm that algorithm to generate TM for each of the eight Ragas and their application as part of the evaluation process for HCM renditions is highly efficient and reliable in identifying typical errors while also avoiding false positives.

## 5  Discussion and Conclusion

Our primary objective was to develop a computational model that was capable of evaluating the correctness of vocal renderings (simple songs) in Ragas typically learnt by early stage learners of vocal HCM. The early stage of learning is highly interactive during which the main feedback is on getting the notes and prominent phrases correct. Our analysis and research in the early stage learning/tutoring process (i.e. the first two years) highlighted three focus areas of learning (apart from mastering the tonic scale). First, novices learnt how to sing by practising relatively simple songs (usually in 6–8 Ragas); second, the focus is primarily on rendering the correct notes with respect to aroha–avaroha, while avoiding the varjit swar; and third, to gain familiarity with the concept of pakkad. Hence, a computer–based tutoring system would need to be able to evaluate renditions with respect to these three aspects of the Ragas typically learnt by novices, and ideally also provide feedback on any errors.

The evaluation module of the tutoring system developed by us is constrained to consider only those elements of Raga's syntax that are expressed in renditions by novices. Given the relative importance of aroha–avaroha, the primary focus of the evaluation process was on assessing the appropriateness of each swar/note with respect to the preceding and subsequent swar/notes. For this assessment, transition matrices (for each of eight Ragas) based on aroha–avaroha and varjit swar features were generated. Test results show that this TM-based computational model is highly appropriate for assessing the correctness of songs sung by novices.

With respect to the note transcription part of the tutoring system, our implementation is a version that incorporates a note duration heuristics method that avoids the need for manual transcription of notes.

The computational model presented here has not considered other features of Raga that determines its overall melodic structure. Further research should focus on extending the evaluation and feedback process for more sophisticated renditions. To do so, the TM-based model presented here can be extended to consider features such as alap, taan and meend. This would necessarily increase the complexity of the transcription process which may affect performance time. Even further extensions of this approach can attempt top model improvisation which are very much a feature of expert singers/musician of HCM recitals. However, such models would probably need to include nonlinear approaches typically used in ML/AI modelling methods. Even so, we hope that the model and results presented here are indicative of the possibilities of developing computational musicology models that reflect and/or account for the richness, sophistication and complexity of HCM.

# References

1. A. Volk, F. Wiering, V.P. Kranenburg, Unfolding the potential of computational musicology, in Proceedings of the 13th International Conference on Informatics and Semiotics in Organisations (2011)
2. P. Agarwal, H. Karnick, B. Raj, A comparative study of Indian and Western music forms, in *ISMIR* (2013), pp. 29–34
3. G. Pandey, C. Mishra, P. Ipe, TANSEN: a system for automatic raga identification, in *IICAI* (2003), pp. 1350–1363
4. B. Mukerji, An analytical study of improvisation in Hindustani classical music (2014)
5. Improvisation in Indian Classical Music, Available at: https://auburnsangeet.wordpress.com/2012/08/10/improvisation-in-indian-classical-music/. Accessed 12 Dec 2018
6. Improvisation, Available at: https://www.aliakbarkhanlibrary.com/new_to_indian_music/tana.html. Accessed 12 Dec 2018
7. J.R. Rabuñal (ed.), *Artificial Neural Networks in Real-life Applications* (IGI Global, 2005)
8. P.V.N. Bhatkhande, A short historical survey of the music of upper India. J. Indian Musicolo. Soc. **1**, 2 (1970)
9. V.N. Bhatkhande. https://en.wikipedia.org/wiki/Vishnu_Narayan_Bhatkhande. Accessed 22 Mar 2018
10. S. Rao, P. Rao, An overview of Hindustani music in the context of computational musicology. J. new music Res. **43**(1), 24–33 (2014)
11. Indian Classical Music and Sikh Kirtan, http://fateh.sikhnet.com/sikhnet/gurbani.nsf/d9c75ce4db27be328725639a0063aecc/085885984cfaafcb872565bc004de79f!OpenDocument. Accessed 11 Feb 2018
12. K. Gajjar, M. Patel, Computational musicology for raga analysis in Indian classical music: a critical review. Int. J. Comput. Appl. **172**(9), 42–47
13. P. Dighe, H. Karnick, B. Raj, Swara histogram based structural analysis and identification of Indian classical ragas, in *ISMIR* (2013), pp. 35–40

14. Surgyan: Promoting Indian Classical Music, Available at: http://www.surgyan.com/vivadiswar. htm. Accessed 22 Mar 2018
15. ITC Sangeet Research Academy, http://www.itcsra.org/Bhupali-Deshkar-Shudh-Kalyan. Accessed 1 Mar 2018
16. Alap, Available at: https://en.wikipedia.org/wiki/Alap. Accessed 22 Mar 2018
17. Taan, Available at: https://en.wikipedia.org/wiki/Taan(music). Accessed 22 Mar 2018
18. A. Bhattacharjee, N. Srinivasan, Hindustani raga representation and identification: a transition probability based approach. IJMBC **2**(1–2), 66–91 (2011)
19. D. Das, M. Choudhury, Finite state models for generation of Hindustani classical music, in *Proceedings of International Symposium on Frontiers of Research in Speech and Music* (2005), pp. 59–64
20. J. McCormack, Grammar based music composition. Complex syst. **96**, 321–336 (1996)
21. Gandharva Mahavidyalaya, Pune. https://www.gandharvapune.org/cms/About-Bharatiya-Sangeet-Prasarak-Mandal,-Pune.aspx. Accessed 15 Mar 2018
22. Akhil Bhartiya Gandharva Mahavidyalaya, (2018) http://abgmvm.org/. Accessed 15 Mar 2018
23. A. Greensted, The lab book pages (2009), http://www.labbookpages.co.uk/audio/javaWavFiles. html. Accessed 22 Mar 2018
24. P. Boersma, D. Weenik, *Praat: doing phonetics by computer* (Institute of Phonetic Sciences of the University of Amsterdam, 2007)
25. Shruti, https://en.wikipedia.org/wiki/Shruti_(music). Accessed 11 Aug 2017

# Performance Analysis of ML Algorithms on Speech Emotion Recognition

**Pradeep Tiwari and Anand D. Darji**

**Abstract** Even though human–computer interface (HCI) applications such as computer-aided tutoring, learning and medical assistance have brought much changes in human lifestyle. This work has mainly focused on comparison of performance of five commonly used classifiers on emotion recognition. Since features are usually high-dimensional and structurally complex, the efficient classification has become more challenging particularly on low-cost processor and mobile (Android) environment. In this work, five machine-learning algorithms are implemented for speaker-independent emotion recognition and their performance is compared: (a) Logistic regression (LR), (b) K-nearest neighbour (KNN), (c) Naive Bayesian classifier (B), (d) Support vector machine (SVM), and (e) Multilayer perceptron (MLP) of neural network. The feature extraction techniques used to obtain features from speech are (a) Mel-scaled power spectrum; (b) Mel frequency cepstral coefficients. Naive Bayes classifier shows best results in speech emotion classification among other classifiers. Emotion data of happy and sad is taken from Surrey Audio-Visual Expressed Emotion (SAVEE) database.

**Keywords** Emotion recognition · MFCC · Logistic regression · K-nearest neighbour · Support vector machine

## 1 Introduction

Emotion recognition represents understanding and analysing mental state of a person from different bodily generated signal like facial expression, speech, etc. Emotion recognition is a complex process but it makes interaction easy as in [1]. In todays era,

P. Tiwari (✉)
NMIMS University, Mumbai, India
e-mail: pradeep.tiwari@nmims.edu

A. D. Darji
SVNIT, Surat, India
e-mail: addarji@gmail.com

© Springer Nature Singapore Pte Ltd. 2021        91
X.-Z. Gao et al. (eds.), *Advances in Computational Intelligence and Communication Technology*, Advances in Intelligent Systems and Computing 1086,
https://doi.org/10.1007/978-981-15-1275-9_8

when human-machine interface is solving many problems and improving human living standards, emotion recognition becomes important as in [2]. So, a novel prosody representation technique is required for higher accuracy of emotion recognition. Speech signal varies in accordance with different emotion [3]. Figures 1 and 2 consist of the waveforms and spectrograms of happy and sad emotions.

Ramamohan and Dandpat [4] have used sinusoidal features which can be characterized by its amplitude, frequency and phase as features. The emotion recognition performance is evaluated for four emotions neutral, anger, happiness and compassion with vector quantization (VQ) and Hidden Markov model (HMM) classifier with accuracy results as Amplitude: 76.4%, Frequency: 87.1% and Phase: 83.9%. A database consisting of five emotions, namely neutral, angry, happy, sad and Lombard using 33 words was considered by Shukla et al. [5]. 13 dimensional MFCC features were computed while VQ and Hidden Markov models (HMM) were used as classifiers. The performance was 54.65% for VQ and 56.02% for HMM while the human identification of stress was observed to give a performance of 59.44%. MFCC and Mel Energy Spectrum Dynamic Coefficients (MEDC) features with Lib-SVM

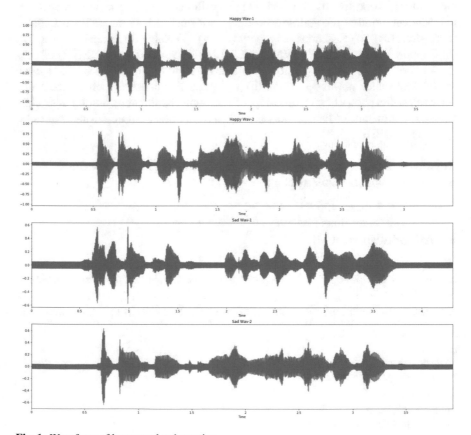

**Fig. 1** Waveform of happy and sad emotion

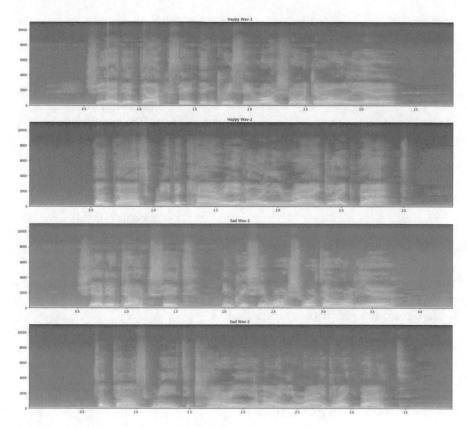

**Fig. 2** Spectrogram of happy and sad emotion

classifier was used by Chavhan et al. [6] for anger, happiness, sad, neutral and fear
from Berlin database with 93.75% accuracy. Sucharita et al. [7] have shown comparison of ML algorithms for classification of Penaeid prawn species. Berg et al. [8]
also employed different ML algorithms for automatic classification of sonar targets
and showed comparison. Fayek et al. [9] have used deep neural network (DNN)
classifier which is the latest trending classifier. System overview is shown in Sect. 2,
feature extraction algorithms are explained in Sect. 2.1 and classification algorithms
are explained in Sect. 2.2. Section 3 represents implementation and result discussion
followed by conclusion in Sect. 4.

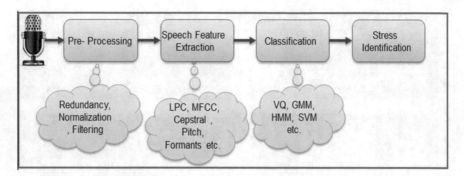

**Fig. 3** System set-up for Emotion Recognition

## 2 System Overview

The system set-up diagram shown in Fig. 3 provides the understanding of the steps involved in speech emotion recognition. It has three parts: (i) Preprocessing, (ii) Feature extraction and (iii) Classification.

The speech signal of happy and sad emotion is taken from standard SAVEE emotion database. The speech signal would contain silence, surrounding noise and dc offset values, so preprocessing is done to remove such redundant information.

*Preprocessing* of speech signal includes:

(a) Normalization: The formula shown in Eq. 1 represents normalization which removes DC offset for a speech signal $x$.

$$x = \frac{x - x_{\text{mean}}}{x_{\text{max}}} \tag{1}$$

(b) Pre-emphasis: A first-order finite impulse response (FIR) high-pass filter as shown in Eq. 2 is used to boost the high-frequency contents of the speech signal which gets suppressed during speech production. The value of $k$ is 0.96.

$$H(z) = 1 - kz^{-1}, \quad \text{where} \quad k \in [0.9, 1] \tag{2}$$

### 2.1 Feature Extraction

The feature extraction techniques used to obtain features from speech are (a) Mel-scaled power spectrum; (b) Mel frequency cepstral coefficients.

### 2.1.1  Mel-scaled Power Spectrum

This feature includes the multiplication of Mel filter bank [10] with power spectrum of speech signal. Mel-scale can be calculated for the given frequency $f$ in Hz, with Eq. 3.

$$S_k = \text{Mel}(f) = 2595 * \left( \log_{10} \left( 1 + \frac{f}{700} \right) \right) \tag{3}$$

The triangular Mel filter bank can be obtained using Eq. 3.

Power spectrum is the square of the absolute value of the discrete Fourier transform of the discrete time speech input signal $x[n]$. If $x[n]$ is the input signal, then the short-time Fourier transform for frame a is given in Eq. 4.

$$X_a[k] = \sum_{n=0}^{N-1} x[n] e^{\frac{-j2\pi nk}{N}}, \quad 0 \leq k \leq N \tag{4}$$

Now, $X_a[k]^2$ is called power spectrum. If it is passed through triangular filters of Mel frequency filter bank $H_m[k]$, the result is called Mel-scaled power spectrum and is given in Eq. 5.

$$S[n] = \sum_{n=0}^{N-1} X_a[k]^2 H_m[k], \quad 0 \leq m \leq M \tag{5}$$

### 2.1.2  Mel Frequency Cepstral Coefficients

This is widely used speech feature obtained with the multiplication of Mel filter bank with cepstrum of speech signal. Cepstrum [10] is the inverse discrete Fourier transform of the logarithm of the absolute value of the discrete Fourier transform of the discrete time input signal $x[n]$ given in Eq. 6.

$$\text{Cepstrum} = \text{IFT}[\text{abs}(\log(\text{FT}(x[n])))] \tag{6}$$

where, $\text{FT}(x[n])$ refers to the discrete Fourier transform of speech signal and IFT(signal) refers to the inverse Fourier transform of the speech signal. If $x[n]$ is the input signal, then the short-time Fourier transform for frame a is given in Eq. 4. As $X_a[k]^2$ is called power spectrum and if it is passed through triangular filters of Mel frequency filter bank $H_m[k]$, the result is called Mel frequency power spectrum and is given in Eq. 5. The filter bank which is triangular in shape is applied on power spectrum. Hence, the log-mel spectrum output transformed back to time domain by using a discrete cosine transform of the logarithm of $S[m]$. The MFCC calculated is given in Eq. 7.

$$\text{MFCC}[i] = \sum_{m=1}^{M} \log(S[m]) \cos\left[i\left(m - \frac{1}{2}\right)\frac{\pi}{M}\right] \quad i = 1, 2, \ldots, L \quad (7)$$

The value of $L$ represents MFCC coefficients for each frame whereas $M$ refers the length of the speech frames.

## 2.2 Classification

Machine-learning algorithms which are implemented for emotion recognition in this paper are: (a) Logistic regression (LR), (b) K-nearest neighbour (KNN), (c) Naive Bayesian classifier (B), (d) Support vector machine (SVM), and (e) Multilayer perceptron (MLP) of neural network.

### 2.2.1 Logistic Regression

Logistic regression is not a regression model, it is a classification model. Yet, it is closely related to linear regression. It is originally binary class classification.

LR model predicts the probability that a binary variable is 1, by applying a logistic function as shown in Eq. 8 to a linear combination of the variables where $k$ represents growth rate of function.

$$f(x) = \frac{1}{(1 + e^{-kx})} \quad (8)$$

### 2.2.2 K-Nearest Neighbour

It is one among widely used classification algorithm. The classification is done based on the distance between a test feature vector to the different classes feature points. Eucleadian distance is used to calculate the distance from the classifying feature to the other features.

$$D(x, y) = \sqrt{\sum_{i=1}^{N}(x_i - y_i)^2} \quad (9)$$

The number $K$ represents numbers of neighbours used for classification.

### 2.2.3 Naive Bayes Classifier

Naive Bayes classifier depends on the principle of famous Bayes theorem as shown in Eq. 10 which is used to find the probability of cause if the result is known.

$$P(c|X) = \frac{P(X|c)P(c)}{P(X)} \tag{10}$$

$P(c)$ = prior probability of test class $c$.
$P(X)$ = prior probability of training feature vector $X$.
$P(c|X)$ = Conditional probability of $c$ given $X$.
$P(X|c)$ = Conditional probability of $X$ given $c$.

The Naive Bayes classifier [11] technique is best suited when the dimensionality of the feature vectors is high. Even though it works very simple algorithm, still Naive Bayes mostly outperform in comparison to other sophisticated classification algorithms.

### 2.2.4 Support Vector Machine

SVM is binary classifier, which separates clustered datapoints into two classes. In SVM, supervised training is done by placing a line (for two-dimensional data) or a hyperplane between two different classes by maximizing the margin between all datapoints.

A hyper plane in an n-D feature space can be represented by Eq. 11.

$$f(\mathbf{x}) = \mathbf{x}^T \mathbf{w} + b = \sum_{i=1}^{n} x_i w_i + b = 0 \tag{11}$$

### 2.2.5 Multilayer Perceptron (MLP) of Neural Network

The multilayer perceptron describes the standard architecture of artificial neural networks and is based on the (single-layer) perceptron. A $(L+1)$ layer perceptron, depicted in Fig. 4, has $D$ input units, $C$ output units, and many hidden units. These units are set in layers, thus, named a multilayer perceptron. The $i$th unit within layer $l$ computes the output.

The training of the network is done by the highly popular algorithm know as error back propagation. This algorithm is based on the error-correcting Learning rule.

## 3 Implementation and Results

The speech samples used for emotion recognition in the experimentation are taken from standard database SAVEE [12] (English). The SAVEE database is multimodal (audiovisual) database with seven emotions: anger, disgust, fear, happy, neutral, surprise and sad. The sampling frequency of the database is 44,100 Hz (16 bit). The database consists of 480 audio–visual files recorded from four male speakers which are labelled categorically. The experimentation done here is part of the research project in which emotion is to be recognized particularly on low-cost processor and mobile (Android) environment where it is challenging to select less complex, more accurate and faster classifier. As there are many classifiers, we have selected five commonly used classifiers and compared the results. Also, considering the application like human health, depression, etc., where the happy and sad emotions can fulfil the desired criteria, the experimentation is carried on the two emotions (60 happy samples and 60 sad samples) out of the seven Emotions. The samples used for training (75%) and testing (25%) are totally different. Emotion recognition includes two steps: (a) Feature extraction; (b) Classification. The research has been carried either at feature extraction step or at classifier step. In feature extraction, 40 Mel-scaled power spectrum coefficients and 40 MFCC coefficient (whereas most of the papers take 13 MFCC coefficients). The work implemented here focuses mainly on performance comparison of the classifiers. The performance results of speech Emotion obtained from different machine-learning algorithms can be seen in Table 1.

The results for logistic regression and k-nearest neighbour is 90%. Naive Bayes classifier is giving accuracy of 100% since they use Bayes theorem, which finds the

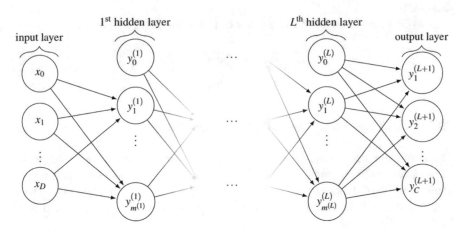

**Fig. 4** Multilayer perceptron

**Table 1** Performance comparison of emotion recognition for ML algorithms

| ML algo. | Acc. | Precision | | Recall | | No. of wav files | | F1-score | |
|---|---|---|---|---|---|---|---|---|---|
| | | Happy | Sad | Happy | Sad | Train | Test | Happy | Sad |
| LR | 0.9 | 1.0 | 0.8 | 0.83 | 1.0 | 90 | 30 | 0.91 | 0.89 |
| KNN | 0.9 | 0.94 | 0.85 | 0.89 | 0.92 | 90 | 30 | 0.91 | 0.88 |
| NB | 1.0 | 1.0 | 1.0 | 1.0 | 1.0 | 90 | 30 | 1.0 | 1.0 |
| SVM | 0.93 | 1.0 | 0.86 | 0.89 | 1.0 | 90 | 30 | 0.94 | 0.92 |
| MLP | 0.833 | 1.0 | 0.71 | 0.72 | 1.0 | 90 | 30 | 0.84 | 0.83 |

**Table 2** Performance comparison of emotion recognition between NB and DNN

| ML algo. | Acc. | Precision | | Recall | | No. of wav files | | F1-score | |
|---|---|---|---|---|---|---|---|---|---|
| | | Happy | Sad | Happy | Sad | Train | Test | Happy | Sad |
| NB | 1.0 | 1.0 | 1.0 | 1.0 | 1.0 | 90 | 30 | 1.0 | 1.0 |
| DNN [9] | 0.59 | 0.55 | 0.71 | 0.26 | 0.71 | 90 | 30 | 0.35 | 0.71 |

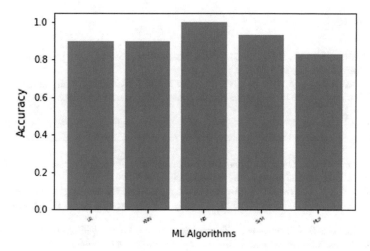

**Fig. 5** Comparison of accuracy of ML algorithms

probability of causes if results are known. NB also works better for high dimensionality features as it is true in this case. Support vector machine giving accuracy of 93%. Multilayer perceptron is neural network-based algorithm requires large amount of training data thus giving accuracy of 83.3% Thus, overall results show Naive Bayes classifier performs better than other classifiers. Now, the result obtained by Naive Bayes classifier is compared with the deep neural network (DNN) classifier [9]. The comparision result shown in Table 2 depicts that Naive Bayes classifier gives best performance among different ML algorithm discussed (Fig. 5).

## 4 Conclusion

This paper shows the performance evaluation of different machine-learning algorithms for speaker-independent emotion recognition. Five machine-learning algorithms are implemented for emotion recognition and their performance is compared, LR, KNN, NB, SVM and MLP gave accuracy of 90%, 90%, 100%, 93% and 83.3%, respectively. Speech features used are Mel-scaled power spectrum and

MFCC. Bayesian classifier shows best results of 100% in speech emotion classification among other classifiers. Emotion data of happy and sad for four speakers is taken from Surrey Audio-Visual Expressed Emotion (SAVEE) database.

Further, performance evaluation of these classifiers on all seven emotions of SAVEE database can be done. Other deep neural networks like CNN can also be used for recognition in future, thus their performances can also be compared.

# References

1. A. Chavhan, S. Dahe, S. Chibhade, A neural network approach for real time emotion recognition. IJARCCE **4**(3), 259–263 (2015)
2. C. Cameron, K. Lindquist, K. Gray, A constructionist review of morality and emotions: no evidence for specific links between moral content and discrete emotions. Pers. Soc. Psychol. Rev. **19**(4), 371–394 (2015)
3. J. Fredes, J. Novoa, S. King, R.M. Stern, N.B. Yoma, Locally normalized filter banks applied to deep neural-network-based robust speech recognition. IEEE Signal Process. Lett. **24**(4), 377–381 (2017)
4. S. Ramamohan, S. Dandpat, Sinusoidal model based analysis and classification of stressed speech. IEEE Trans. Speech Audio Process. **14**(3), 737–746 (2006)
5. S. Shukla, S.R.M. Prasanna, S. Dandapat, Stressed speech processing: human vs automatic in non-professional speaker scenario, in *National Conference on Communications* (2011), pp. 1–5
6. Y.D. Chavhan, B.S. Yelure, K.N. Tayade, Speech emotion recognition using RBF kernel of LIBSVM, in *2nd International Conference on Electronics and Communication Systems* (2015), pp. 1132–1135
7. V. Sucharita, S. Jyothi, V. Rao, Comparison of machine learning algorithms for classification of Penaeid prawn species, in *3rd International Conference on Computing for Sustainable Global Development (INDIACom)* (IEEE, 2016), pp. 1610–1613
8. H. Berg, K.T. Hjelmervik, D.H.S. Stender, T.S. Sastad, A comparison of different machine learning algorithms for automatic classification of sonar targets, in *OCEANS MTS* (IEEE Monterey, 2016), pp. 1–8
9. H. Fayek, M. Lech, L. Cavedon, Towards real-time Speech Emotion Recognition using deep neural networks, in *9th International Conference on Signal Processing and Communication Systems*, Cairns, QLD (2015), pp. 1–5
10. P. Tiwari, U. Rane, A.D. Darji, Measuring the effect of music therapy on voiced speech signal, in *Future Internet Technologies and Trends. ICFITT 2017. Lecture Notes of the Institute for Computer Sciences, Social Informatics and Telecommunications Engineering*, vol. 220. (Springer, 2017)
11. H. Atasoy, Emotion recognition from speech using Fisher's discriminant analysis and Bayesian classifier, in *Signal Processing and Communications Applications Conference (SIU)* (IEEE, 2015), pp. 2513–2516
12. P. Jackson, S. Haq, *Surrey Audio-visual Expressed Emotion (SAVEE) Database* (University of Surrey, Guildford, UK, 2014)

# EEPMS: Energy Efficient Path Planning for Mobile Sink in Wireless Sensor Networks: A Genetic Algorithm-Based Approach

Akhilesh Kumar Srivastava, Amit Sinha, Rahul Mishra
and Suneet Kumar Gupta

**Abstract** The area of wireless sensor networks (WSNs) has been highly explored due to its vast application in various domains. The main constraint of WSN is the energy of the sensor nodes. The use of mobile sink (MS) is one of the prominent methods to preserve the energy of sensor nodes. Moreover, use of mobile sink is also solving the hot-spot problem of wireless sensor network. In the paper, we propose a genetic algorithm-based approach to plan the path for mobile sink. All the basic intermediate operations of genetic algorithms, i.e., chromosome representation, crossover and mutation are well explained with suitable examples. The proposed algorithm is shown its efficacy over the randomly generated path.

**Keywords** Mobile sink · Energy efficacy · Wireless sensor network · Path planning · Data gathering

## 1 Introduction

Wireless sensor network (WSN) is the network of small tiny devices known as sensors [1]. Each sensor node has sensing capability, limited memory, and computational power [2]. The area of wireless sensor networks has been highly explored by the

A. K. Srivastava (✉) · A. Sinha
ABES Engineering College, Ghaziabad 201010, India
e-mail: akhilesh.srivastava@abes.ac.in; joinakhilesh@yahoo.com

A. Sinha
e-mail: sinha_mca@rediffmail.com

R. Mishra
CDAC, Noida 201309, India
e-mail: rahulmishra@cdac.in

S. K. Gupta
Bennett University, Gr. Noida 201310, India
e-mail: suneet.banda@gmail.com
URL: https://sites.google.com/view/ksuneet/

© Springer Nature Singapore Pte Ltd. 2021                                    101
X.-Z. Gao et al. (eds.), *Advances in Computational Intelligence and Communication Technology*, Advances in Intelligent Systems and Computing 1086,
https://doi.org/10.1007/978-981-15-1275-9_9

research community due to its vast applications, i.e., military surveillances, health care, security, and habitats monitoring [3, 4]. In WSNs, a sensor node senses the region for collecting the information and forwards the same information to Base Station directly or using multi-hop communication [5, 6]. As the sensor nodes have limited battery, energy consumption is one of the most important constraint in WSN [7]. Generally, the energy of sensor nodes consumed due to aggregation of the data, forwarding the data, and receiving the data. Moreover, in WSN the energy consumption is quadric proportional to distance between sender and receiver [8]. The lifetime of network is based on the energy consumption, so to improve the network lifetime, it is necessary that individual sensor node consumes minimum energy for forwarding the sensed data. To minimize the energy consumption, routing [9, 10] and clustering [11] are the prominent methods.

In WSNs, the sensor nodes close to the base station die quickly because these nodes forward the data of some other nodes which are so far from the BS. In this situation, there are many cases where nodes near to base station (BS) completely deplete the energy and partition the network. However, some of the node(s) have energy more than 90%. To resolve such kind of issue, many researchers have introduced the concept of mobile sink (MS) [12–14]. The pictorial representation of mobile sink is depicted in Fig. 1.

In Fig. 1, it is clearly seen that there is a sink which moves over a predefined trajectory and stay at some of the sensor nodes (SNs) for a predefined duration to collect the information. Generally, there are two methods to decide the path of mobile sink (1) random and (2) controlled. The random movement of mobile sink leads to uncontrolled behavior as well as buffer overflow. In controlled mobility, some researchers have proposed two methods for collecting the data (1) mobile sink (MS) visits all the sensor nodes and collect the data. However, in this method, the path of mobile sink is too long and it increases the data delivery latency. The other method is sink visits limited number of nodes known as rendezvous points. The sink

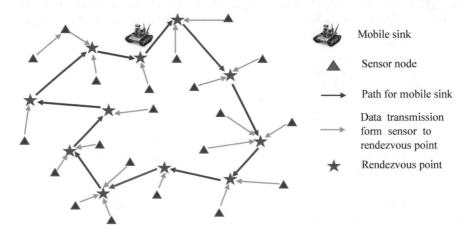

**Fig. 1** An example of wireless sensor networks with mobile sink

stops and stay at rendezvous points (RPs) [15] for limited duration. The sensor nodes either forward the data directly to rendezvous points or via multi-hop [16].

In the last decade, many algorithms have been proposed for the mobile sink, which may found in [17–19]. The mobility algorithms are categories in random and controlled mobility. The implementation of random mobility strategy is easy as only some of the RPs selected. However, this method increases the delay during the data gathering. In [15], authors presented a novel approach based on the controlled mobility. Moreover, the path is stationary for MS over RPs and SNs are also randomly organized on the basis of RPs. Ghafoor et al. [20] have applied one of the popular technique named as Hilbert curve to prepare the path of MS. In this technique, all the SNs forward the sensed data to RPs in one-hope communication. However, length of path is high and this method is not feasible for sensitive application such as military surveillances. A delay restricted method for mobile sink is proposed in [21]. However, the disadvantages of algorithm are that it has very high time complexity, i.e., $O(n^5)$.

In WSNs with mobile sink, there is a big challenge to take decision for selecting the number of RPs. As more number of RPs may alleviate the data delivery latency. However, the shorter number of RPs may increase the data delivery latency but there may be a possibility that sensor nodes forward the data to RP using multi-hop communication. Due to multi-hop communication energy of some of the sensor node(s) deplete quickly. Therefore, while designing the trajectory, it is desired that the length of the trajectory should be minimum and minimum number of sensor node(s) forward the data to RP by multi-hop transmission.

In this article, we address the same issue and propose the energy-efficient genetic algorithm (GA) [22]-based approach to find the trajectory of the MS. In GA-based approach, we optimize the number of RPs using some criteria function. The criteria function has three different parameters, i.e., the length of the MS tour, minimum number of multi-hop transmission, and load of RPs. In GA, there are some intermediate steps such as chromosome representation, selection, crossover, and mutation. All these steps are well explained with suitable examples.

## 2 System Model and Terminologies

In proposed work, we assume that network is formed using homogeneous sensor nodes and these nodes are randomly deployed in the predefined area. There is a sink, which moves in the area and stays at some predefined RPs for sufficient time to gather the data. It is also assumed that after the deployment, nodes are stationary and communication between two are in existence only when they are in the communication range of each other. Network is in functioning condition till some percentage of nodes alive.

- $S = \{s_1, s_2, s_3, \ldots, s_n\}$: represents the set of $n$ number of sensor nodes.
- $P = \{p_1, p_2, p_3, \ldots, p_m\}$: represents the set of $m$ number of position which may act as RPs and $m \leq n$.

- $DIST(s_i, s_j)$: represents the euclidean distance between node $s_i$ and $s_j$.
- $Tcost$: represents the cost of the tour.
- $NW1H$: It represents the numbers of nodes with 1-hop from RPs.
- $LoadRP(i)$: It stores the information that $i$th RP receives the information from how many sensor nodes.
- $Dif$: It represents the difference between maximum and minimum value $LoadRP(i)$.

In this research article, A GA-based algorithm for path planning of mobile sink is proposed. The detailed explanation of proposed algorithm is in following section.

## 3   Proposed Algorithm

All the intermediate steps, i.e., chromosome representation, crossover, and mutation are discussed in following subsection.

### 3.1   Chromosome Representation and Initial Population

In Fig. 2, a subgraph of WSN is depicted. There are 23 sensor nodes and out of 23, 11 act as RPs. The chromosome representation for this network is presented in Fig. 3. In proposed work, the chromosomes are randomly generated and the length of the chromosome is fixed which is equal to 10% of the deployed sensor nodes.

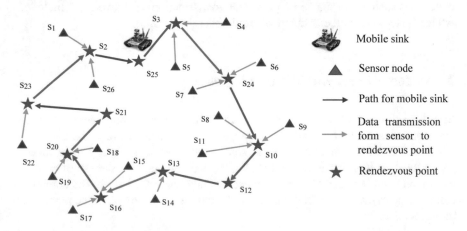

**Fig. 2** A subgraph of WSN having 23 number of sensor nodes. Out of 23, 11 sensors nodes act as RP

| Sensor node no. | 2 | 25 | 3 | 24 | 10 | 12 | 13 | 16 | 20 | 21 | 23 |
|---|---|---|---|---|---|---|---|---|---|---|---|

**Fig. 3** Chromosome representation for the graph given in Fig. 2

The collection of chromosomes is known as initial population. The chromosome representation of the subgraph (refer Fig. 2) is depicted in Fig. 3.

## 3.2 Fitness Function and Selection

The performance of chromosome is always judged by using the fitness function. Also fitness function decides that which chromosome is good for the next level operations, i.e., crossover, mutation, etc. In the proposed work, to design the fitness function three parameters has been used (1) length of the tour of the MS, (2) load of RP, i.e., it means that the RP receives the data from how many sensor nodes, and (3) number of nodes which forwards the data to RP using multi-hop. The proposed fitness function is represented as follows:

$$Minimize \ F = w_1 \times Tcost + w_2 \times \{n - NW1H\} + w_3 \times Dif$$

where $w_1$, $w_2$, $w_3$ represents the weights and $w_1 + w_2 + w_3 = 1$.

There are many methods for selection operation but in proposed algorithm roulette wheel method has been used.

## 3.3 Crossover and Mutation

During crossover operation, parent chromosomes interchange the information with each other and form two new chromosome named as child chromosomes. After the crossover operation, best two chromosomes replace the parent chromosomes. In proposed work, one-point crossover has been applied which is similar to crossover operation discussed in [23]. During mutation operation, we have tried to minimize the number of RP by flipping the gene value.

# 4 Experimental Result

A extensive simulation of proposed algorithm has been performed on an Intel CORE i5 processor with 2.53 GHz CPU and 8 GB RAM running on the platform Microsoft Windows 10. For the programming purpose, C and MATLAB (version 17.1) have

been used. For the experimental purpose, it is assumed that there is a scenario WSN#1
with randomly deployed 100 number of sensor nodes and the area of WSN#1 is
100 × 100. In this scenario, it is also assumed that only 10% of deployed sensor nodes
act as rendezvous points. Here, we first ran the algorithm to generate the trajectory
of mobile sink randomly. The pictorial representation of trajectory is represented in
Fig. 4.

From Fig. 4, it is clearly seen that we have randomly picked the ten nodes (10%
of total number of sensor nodes) as RP and mobile sink at coordinate (0, 0). The
mobile sink travels all the RPs one by one and stays for some time to collect the
information. Figure 5 represents the path of mobile sink using genetic algorithm.
From Figs. 4 and 5, it is clearly observed that GA-based approach is better than the
random approach. GA-based approach outperforms as its fitness function is based on
three parameters, i.e., distance covered in the tour, load of RPs and minimum number of nodes which transfer the data to RP via one-hop. However, other algorithm
randomly select some sensor nodes which act as RP. Moreover, random selection of
the nodes may develop a path which is in zigzag fashion and increase the overall
distance in the tour.

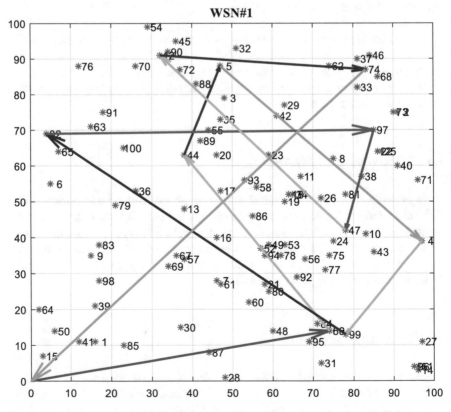

**Fig. 4** Path of the mobile sink using random algorithm with starting point from (0, 0)

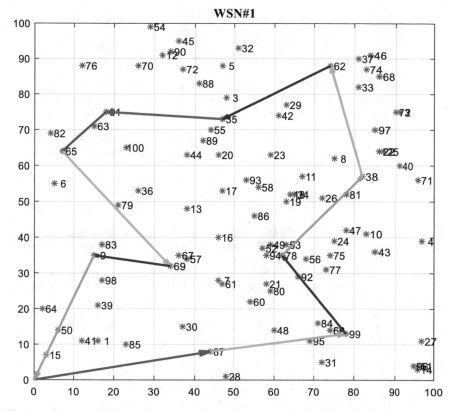

**Fig. 5** Path of the mobile sink using genetic algorithm with starting point from (0, 0)

## 5 Conclusion

In the research article, we have discussed GA-based algorithm for path planning of mobile sink. Moreover, the GA-based algorithm is well explained with suitable examples. The fitness function for GA-based approach is depended on tour length of MS, number of multi-hop nodes and load of RP. For the result comparison purpose, we have also executed the algorithm which randomly selects the 10% sensor node as RPs and it is observed that the GA-based approach outperforms with this approach. We are currently working on the node deployment and try to incorporate the issue of mobile sink during the deployment.

## References

1. I.F. Akyildiz, W. Su, Y. Sankarasubramaniam, E. Cayirci, Wireless sensor networks: a survey. Comput. Netw. **38**(4), 393–422 (2002)

2. G. Anastasi, M. Conti, M. Di Francesco, A. Passarella, Energy conservation in wireless sensor networks: a survey. Ad Hoc Netw. **7**(3), 537–568 (2009)
3. A. Gagarin, S. Hussain, L.T. Yang, Distributed hierarchical search for balanced energy consumption routing spanning trees in wireless sensor networks. J. Parallel Distrib. Comput. **70**(9), 975–982 (2010)
4. T. Arampatzis, J. Lygeros, S. Manesis, A survey of applications of wireless sensors and wireless sensor networks, in *Intelligent Control, 2005. Proceedings of the 2005 IEEE International Symposium on, Mediterranean Conference on Control and Automation* (IEEE, 2005), pp. 719–724
5. S.K. Gupta, P. Kuila, P.K. Jana, GAR: an energy efficient GA-based routing for wireless sensor networks, in *International Conference on Distributed Computing and Internet Technology* (Springer, 2013), pp. 267–277
6. S.K. Gupta, P.K. Jana, Energy efficient clustering and routing algorithms for wireless sensor networks: GA based approach. Wireless Pers. Commun. **83**(3), 2403–2423 (2015)
7. K. Akkaya, M. Younis, A survey on routing protocols for wireless sensor networks. Ad Hoc Netw. **3**(3), 325–349 (2005)
8. S.K. Gupta, P. Kuila, P.K. Jana, Delay constraint energy efficient routing using multi-objective genetic algorithm in wireless sensor networks, in *International Conference on Eco-friendly Computing and Communication Systems, TMH* (2013), pp. 50–59
9. S.K. Gupta, P. Kuila, P.K. Jana, GA based energy efficient and balanced routing in k-connected wireless sensor networks, in *Proceedings of the First International Conference on Intelligent Computing and Communication* (Springer, 2017), pp. 679–686
10. R.N. Shukla, A.S. Chandel, S.K. Gupta, J. Jain, A. Bhansali, GAE[3]BR: genetic algorithm based energy efficient and energy balanced routing algorithm for wireless sensor networks, in *2015 International Conference on Advances in Computing, Communications and Informatics (ICACCI)* (IEEE, 2015), pp. 942–947
11. S.K. Gupta, P. Kuila, P.K. Jana, E[3]BFT: energy efficient and energy balanced fault tolerance clustering in wireless sensor networks, in *2014 International Conference on Contemporary Computing and Informatics (IC3I)* (IEEE, 2014), pp. 714–719
12. Y.S. Yun, Ye Xia, Maximizing the lifetime of wireless sensor networks with mobile sink in delay-tolerant applications. IEEE Trans. Mob. Comput. **9**(9), 1308–1318 (2010)
13. M.I. Khan, W.N. Gansterer, G. Haring, Static vs. mobile sink: the influence of basic parameters on energy efficiency in wireless sensor networks. Comput. Commun. **36**(9), 965–978 (2013)
14. I. Papadimitriou, L. Georgiadis, Energy-aware routing to maximize lifetime in wireless sensor networks with mobile sink. J. Commun. Softw. Syst. (2006)
15. G. Xing, T. Wang, Z. Xie, W. Jia, Rendezvous planning in wireless sensor networks with mobile elements. IEEE Trans. Mob. Comput. **7**(12), 1430–1443 (2008)
16. D.J. Jezewski, J.P. Brazzel Jr., E.E. Prust, B.G. Brown, T.A. Mulder, D.B. Wissinger, A survey of rendezvous trajectory planning. Astrodynamics **1991**, 1373–1396 (1992)
17. T. Camp, J. Boleng, V. Davies, A survey of mobility models for ad hoc network research. Wireless Commun. Mob. Comput. **2**(5), 483–502 (2002)
18. G. Yu, Y. Ji, J. Li, F. Ren, Baohua Zhao, EMS: efficient mobile sink scheduling in wireless sensor networks. Ad Hoc Netw. **11**(5), 1556–1570 (2013)
19. T.-S. Chen, H.-W. Tsai, Y.-H. Chang, T.-C. Chen, Geographic convergecast using mobile sink in wireless sensor networks. Comput. Commun. **36**(4), 445–458 (2013)
20. S. Ghafoor, M.H. Rehmani, S. Cho, S.-H. Park, An efficient trajectory design for mobile sink in a wireless sensor network. Comput. Electr. Eng. **40**(7), 2089–2100 (2014)
21. H. Salarian, K.-W. Chin, F. Naghdy, An energy-efficient mobile-sink path selection strategy for wireless sensor networks. IEEE Trans. Veh. Technol. **63**(5), 2407–2419 (2014)
22. P. Kuila, S.K. Gupta, P.K. Jana, A novel evolutionary approach for load balanced clustering problem for wireless sensor networks. Swarm Evol. Comput. **12**, 48–56 (2013)
23. S.K. Gupta, P. Kuila, P.K. Jana, Genetic algorithm approach for k-coverage and m-connected node placement in target based wireless sensor networks. Comput. Electr. Eng. (2015)

# Mining Public Opinion on Plastic Ban in India

Nandini Tomar, Ritesh Srivastava and Veena Mittal

**Abstract** Every product available in our environment has a shelf life, but plastic is the only material that is non-degradable. The complex polymer present in the plastic makes it durable and non-degradable. As a result, it is found in different forms on the earth for a long time. People have become used to plastic-made product in day-to-day life like carrying bags, disposable cutlery, food packaging and many more. Extensive quantities of plastic waste have accumulated in the nature and landfills and have posed an alarming hazard to the environment, and now, it reached a crisis point. Currently, India is ranked as the top four producers of plastic waste in the world. Though there is a law against the use of plastic in India but the usage of plastic-made products is still high as the ban is not implemented completely and effectively. In this paper, we propose a framework for analyzing the opinion of Indian population on the plastic ban with the help of sentiment analysis technique on Twitter textual data. We train and test a machine learning classifier on different combination of datasets achieving 77.94% classification accuracy. The result obtained will help to understand how effective and successful polybags ban scheme will be when entirely implemented in India.

**Keywords** Sentiment analysis · Supervised machine learning · Natural language processing

N. Tomar (✉) · V. Mittal
Manav Rachna International Institute of Research and Studies (MRIIRS),
Faridabad, Haryana, India
e-mail: tomar.nandini@gmail.com

V. Mittal
e-mail: veena.mittal06@gmail.com

R. Srivastava
GCET, Greater Noida, India
e-mail: ritesh21july@gmail.com

© Springer Nature Singapore Pte Ltd. 2021                                         109
X.-Z. Gao et al. (eds.), *Advances in Computational Intelligence and Communication Technology*, Advances in Intelligent Systems and Computing 1086,
https://doi.org/10.1007/978-981-15-1275-9_10

# 1 Introduction

Twitter, one of the most popular micro-blogging websites among people all over the world, provides a podium to their registered users to express their feeling on the topic of interest effortlessly. People share their opinion on different social/political issues or product/services by writing a short message called as tweet having a maximum 140 characters. The informal structure and free format of tweets make it easy for a novice user to post a message with ease, and it results as an increasing number of users worldwide. All the tweets posted on Twitter wall carry user's emotions, feelings or feedback about the entity discussed which if can be mined appropriately can act as a crucial resource for different service industries and government organizations to track their status in the competitive market. Sentiment analysis process is a part of natural language processing which computationally identifies and categorizes the opinions expressed in a piece of text.

## 1.1 Plastic Pollution

Plastics also called as wonder material are cheap, lightweight, robust, durable, corrosion-resistant and long-lasting materials. They are now used in almost in every aspect of our day-to-day life including transport, telecommunications, clothing, footwear, medical equipment, food/product packaging, disposable cutlery, electric equipment, etc. As a result, the production and consumption of plastics have increased significantly over the last few years from around 0.8 million tons in 1955 to over 270 million tons today, and this rate of production is proliferating [1, 2].

Though plastics are convenient in our day-to-day use and help to gain societal benefits and offer future technological and medical advances; however, excessive consumption and careless disposal of plastic material result as the high accretion of plastic waste in the natural environment and in the landfills, which now has posed an alarming hazard to the environment.

Use of plastic and accumulation of plastic waste harm our environment in many ways:

- Mutilation Sanitary System: The residues of plastic like polybags, pet bottles, food wrappers, etc., block pipes causing the fear of many water contaminated diseases.
- Ocean Pollution: Throwing away the plastic wastes like plastic bags, bottles, etc., deposited in the ocean and pollute and disturb the natural ecosystem of the ocean.
- Environmental Imbalance: Excessive use of plastic-made goods and careless disposal of plastic waste trigger the water, soil and air pollutions disturbing various ecosystem of the earth.

Countries around the world are articulating ways to get rid of plastic waste disposal and plastic pollution. Some of the countries listed in Table 1, where the plastic use law is in place and strictly followed [2].

**Table 1** List of country and their law for the use of plastic

| Country | Law[a] |
|---------|--------|
| France | 'Plastic Ban' law is passed in 2016 which states all plastic plates, cups and utensils will be banned by 2020. France is the first country to ban all the plastic-made products used in daily life |
| Rwanda | Rwanda implemented a complete ban on plastic bags, and this country is plastic bag free since 2008 |
| Sweden | This country is known as one of the world's best recycling nations which follow the policy of 'No Plastic Ban, Instead More Plastic Recycling' |
| Ireland | In this country, a tax scheme is implemented in the use of plastic bags |
| China | In 2008, China passes a law to deal with plastic crisis. In China, it is illegal to give plastic bag to the customers for free |

## 1.2   The Ban on Plastic Bags in India

In August 2017 in India, the manufacturer, stock, sale and use of plastic bags have been banned in 17 states and Union Territories including Bengaluru, Maharashtra, Delhi, Punjab, Rajasthan, Goa, West Bengal, etc. However, use of plastic carry bags has been partially banned in 11 states which are pilgrimage centers, tourist and historical places. The name of states where the use of plastic carry bags is partially banned is Andhra Pradesh, Arunachal Pradesh, Assam, Goa, Gujarat, Karnataka, Odisha, Tamil Nadu, West Bengal, Uttar Pradesh and Uttarakhand [2].

National Green Tribunal in Delhi NCR announced a ban on disposable plastic items like food packaging, cutlery, polybags and other plastic items. The ban on plastic items came into effect on January 1, 2017. The ban on plastic affects the whole National Capital Territory (NCT) area of Delhi [3].

On March 23, 2018, in India's second-populous state Maharashtra, the government banned the manufacture, usage, sale, transport, distribution and storage of plastic bags with or without handle and disposable products made out of plastic and thermocol [4, 5].

Due to the massive dependence on plastic and lack of alternatives to the banned products, its effective enforcement is an issue. In this work, we try to find out people's reaction on the plastic ban by analyzing tweets posted by Twitter users so that government can get a clear picture of its effective implementation, and appropriate corrective actions can be taken on time.

## 1.3   Sentiment Analysis

Sentiment analysis is the area of natural language processing (NLP) and text mining which help to analyze the opinion of a person (called an opinion holder) toward an entity from given written text.

## *1.4   Techniques of SA*

Numerous methods have been devised so far to perform sentiment analysis process [6–12]. Each method has their applicability and strength, so the selection of suitable technique depends upon the type of textual data used in the analysis process and the area where analysis has to be performed. Techniques used to perform SA process are categorized into following two major categories [13]:

1.   Machine learning (ML) approach.
2.   Lexicon-based approach.

   Rest of this work is confined in the following sections; Sect. 2 gives the brief overview of the work done in the area of sentiment analysis by various researchers. Next, in Sect. 3, all the tools and techniques used to execute this work along with approach opted for data collection are specified. Section 4 graphically shows and defines the systematic flow and implementation of this work. Section 5 gives a detailed discussion of the results obtained by executing this work. Finally, in Sect. 6, we conclude this work along with future perspectives defining the scope of improvement.

## 2   Literature Review

Analyzing user's opinion from written text with improved accuracy is a quite challenging task. Researchers are unceasingly working on finding new methods and strategies to improve time, speed and efficiency of the sentiment analysis process. Some of the significant work done in this area by various researchers is given in subsequent part of this section.

   Tweets posted on Twitter are used as a corpus of positive and negative opinion words to perform sentiment analysis [14]. A method is proposed which collects an arbitrary large corpus of positive and negative opinion words automatically. It also collects a corpus of objective texts. With this collected corpora, sentiment analysis is performed. Domain-specific features extraction is one of the important tasks in sentiment analysis process. The work of [15] performed SA by using the strategy of selecting most frequent noun as a feature from collected blog or document with the help of association mining. They use nearby opinion words to find out infrequent features.

   A distance-based method [16] is suggested to identify compact features of a product, e.g., camera type, battery life, functions, etc., are the compact features of a mobile phone. In [17], the author defines the existing approaches for opinion-oriented information retrieval system including blogs and micro-blogging sites. A corpus of opinion words is created from web blogs [18]. Emoticons are also used to indicate the user's mood. In this work, SVM and CRF classifiers are trained using collected corpora of opinion words to perform sentiment analysis at the document level. A detailed discussion on sentiment analysis and its tools and techniques are described in [13].

To make analysis more precise and accurate, careful feature selection processes are also discussed. Use of emoticons smiley and sad symbolized as ':-)' and ': - (' is suggested to form a training dataset for training the classification model [19]. Emoticons are collected from Usenet newsgroups. Dataset to train the classifier contains two samples, one is text with positive opinion words and smiley emoticons, and another one is text with negative opinion words and sad emoticons.

OPINE [20] uses web-based search for calculating pointwise mutual information (PMI) score between the phrase and a part of differentiator related to product class using Eq. (1), where $f$ is fact and $d$ is a differentiator.

$$\text{PMI}(f, d) = \frac{(\text{hits}(f \wedge d))}{(\text{hits}(d)\text{hits}(f))} \tag{1}$$

The semantic orientation of adjectives is one of the important factors to consider carrying out sentiment analysis process. Remarkable work is done in [21], where they propose an algorithm to cluster all positive or negative objectives by their calculated semantic orientation. This algorithm is based on conjunction hypothesis and is designed for isolated adjectives.

# 3  Experimental Setup and Data Collection

## 3.1  Experimental Setup

Execution of this work is done on a machine with Intel Core i3, 2.4 GHz, 3 MB L3 cache processor with 4 GB of RAM. A script written in R language is executed to scrap out the tweets in which people express their feelings for a plastic ban on Twitter. We also utilize the online open resource SentiWordNet to perform subjectivity test on collected tweets. Finally, sentiment analysis on the collected dataset is performed using a powerful machine learning tool WEKA [24]. We can directly apply readily available machine learning algorithm on collected data or can use Java code to initiate the execution.

## 3.2  Tweets Extraction

A key search is performed by entering few plastic ban-related keywords mentioned in the harvest list given below. A script, written in 'R' language, is executed to scrap dawn the tweets from Twitter open API. Around 31,634 tweets posted on the Twitter platform from June 15, 2018 to June 22, 2018 (i.e., of one week) were extracted.

Harvest List = {Plastic Ban, Polythene Ban, Plastic Pollution}

### 3.3  Subjectivity Analysis

Tweets extracted from Twitter open API may contain some tweets with no sentiments or opinion within called as objective tweets. Subjectivity analysis filters out statements that include no opinion. To perform this analysis, an openly available lexicon resource, SentiWordNet [22] is used which contains positive/negative sentiment word. Each tweet is labeled manually according to the class they belong to, i.e., either positive or negative.

After subjectivity test and manual annotation on all the collected tweets (i.e., 31,634 tweets), we get 12,729 tweets with positive polarity and 11,943 tweets with negative polarity, and rest 6962 tweets were objective and so are discarded. To balance the collected, labeled dataset, we have selected 11,000 positive tweets and 11,000 negative tweets.

### 3.4  Training Data Preparation

The next step immediately after data extraction is training data preparation. In order to make our training data rich in opinion words and vast, here, we encapsulate in-domain dataset with out of domain dataset available online. The training dataset is constructed from the following two sources:

1. Manually Annotated Dataset.
2. International Movie Database (IMDb) [23].

**Manually Annotated Data**. The manually annotated dataset is the collection of labeled positive and negative tweets which are specific to the domain (in this work plastic ban). After applying subjectivity analysis and manual annotation on all collected tweets, we obtain 12,729 positive and 11,943 negative tweets out of which we have selected 11,000 positive polarity tweets and 11,000 negative polarity tweets to possess the uniform tweets selection throughout this work. The statistics of our dataset is given in Table 2.

**IMDb Reviews**. IMDb reviews are labeled collection of 1000 positive and 1000 negative movie reviews which are highly polar, that is all positive reviews comprise strictly positive opinion words, and in the same way negative reviews contain strictly negative opinion words. Encapsulating these reviews with our collected dataset will enrich the semantic knowledge base which will help in better polarity classification.

**Table 2** Training dataset statistics

| Sources | Positive instances | Negative instances |
| --- | --- | --- |
| IMDb reviews | 1000 | 1000 |
| Manually annotated dataset | 11,000 | 11,000 |
| Total instances | 12,000 | 12,000 |

# 4 Proposed Approach

The approach chosen to execute this work is given in Fig. 1. Following steps need to follow to perform sentiment analysis on collected data:

(a) Data Collection
(b) Preprocessing
(c) Training the Classifier
(d) Polarity Classification.

## 4.1 Data Collection

Tweets in which users have posted their views about plastic ban are collected from Twitter REST API from June 15, 2018 to June 22, 2018, i.e., of one week. A connection needs to be established between the user program and Twitter open API before the data extraction process. To build this connection first, we need to create a Twitter account and then get credentials (they are API Key, API Secret, Access Token and Access Secret) on the Twitter developer site.

A script written in R language is executed to get the authentication and related tweets from Twitter open API.

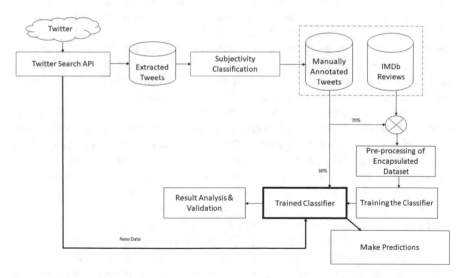

**Fig. 1** Proposed system architecture

## 4.2 Preprocessing

All the tweets collected from Twitter may contain some undesirable data including Twitter-specific terms like '@' tags, '#' tags, URLs, symbols, etc., and some other terms like punctuations, stop words, abbreviations, etc., called as noise which needs to be removed from the collected dataset before passing to the. Following preprocessing steps are followed to make the dataset clean and smooth:

– Tokenization
– Normalization
– Case Sensitivity
– Stemming.

## 4.3 Training Machine Learning Classifier

In this work, we opted for supervised machine learning classifier, Support Vector Machine (SVM).

Based upon the combination of training and testing dataset used to train the SVM classifier, the following three models are introduced:

**Model-1**. Model-1 is trained and tested over highly polar IMDb reviews using $k$-fold cross-validation process for $k = 10$. Cross-validation is a resampling process used to evaluate machine learning models on a limited data sample. In $k$-fold cross-validation testing, process starts by partitioning the given dataset into $k$ number of equal size sub-datasets, after that, a single subset is selected as the validation dataset, and rest $(k − 1)$ sub-datasets are used for training the classifier. This process is repeated $k$ times, with each $k$ subsamples used exactly once as the validation dataset. Final result is produced by averaging the $k$ calculated results.

It is a standard approach to follow to verify the correctness of the out-of-domain dataset used (IMDb reviews in this work) in combination with the in-domain dataset.

**Model-2**. Model-2 is trained with IMDb reviews and tested over manually annotated tweets using batch setting. This model is proposed to demonstrate that if we use a classifier trained solely with different domain dataset may misclassify few statements due to the absence of some domain-specific opinion words in classifiers knowledge base. Results shown in the subsequent section will validate this statement.

**Model-3**. Model-3 is our core model which is trained and tested over summed up dataset (i.e., domain-specific tweets + IMDb reviews) using ten-fold cross-validation process. We believe that the accuracy of the classifier can be improved by adding more opinion words in its knowledge base, and here, we achieve it by encapsulating data from two sources (IMDb reviews and domain-specific Twitter data) to train the machine learning classifier.

# 5   Results Analysis and Discussions

This section comprises a detailed discussion on the performance obtained with each proposed model to make it easy to compare and visualize the difference between them. This work of finding public opinion on plastic ban is carried out by using our core model-3 which is trained and tested with the dataset build by encapsulating data from two different sources. Classifier's result is received in the form of a matrix called as a confusion matrix which determines the correctness and accuracy of the SVM classification model on the labeled validation dataset. It is used for classification problem where the output can have two or more classes. The result is analyzed regarding precision value, recall value, $F$-curve, and overall accuracy attained.

## 5.1   Results Obtained with Model-1

Model-1 is trained and tested over IMDb dataset which contains 2000 highly polar reviews out of which 1000 are positive, and rest 1000 are negative reviews. Complete execution is performed in WEKA using $k$-fold cross-validation process where $k = 10$. Confusion matrix obtained with this model is shown in above-mentioned Table 3. With model-1, 78.95% classification accuracy is attained.

We propose this model to verify the correctness of the out-of-domain dataset used (IMDb reviews in this work) in combination with the in-domain dataset (collected tweets related to plastic ban).

## 5.2   Results Obtained with Model-2

Model-2 is trained with IMDb and validated over annotated domain-specific tweets (containing 22,000 labeled tweets) using batch setting. With this model, we attain

**Table 3** Confusion matrix obtained for model-1

| Actual | Predicted | | |
|---|---|---|---|
| | Negative tweets | Positive tweets | Accuracy obtained (%) |
| Negative tweets | 782 (TN) | 218 (FP) | 78.2 |
| Positive tweets | 203 (FN) | 797 (TP) | 79.7 |
| Total tweets | 985 | 1015 | 78.95 |

*TP* true positive; *TN* true negative; *FP* false positive; *FN* false negative instances

**Table 4** Confusion matrix obtained for model-2

| Actual | Predicted | | |
|---|---|---|---|
| | Negative tweets | Positive tweets | Accuracy obtained (%) |
| Negative tweets | 6784 (TN) | 4216 (FP) | 61.67 |
| Positive tweets | 4205 (FN) | 6795 (TP) | 61.77 |
| Total twects | 11,989 | 12,011 | 61.72 |

**Table 5** Confusion matrix obtained for model-3

| Actual | Predicted | | |
|---|---|---|---|
| | Negative tweets | Positive tweets | Accuracy obtained (%) |
| Negative tweets | 18,689 (TN) | 5311 (FP) | 77.87 |
| Positive tweets | 5264 (FN) | 18,736 (TP) | 78.06 |
| Total tweets | 23,953 | 24,047 | 77.97 |

61.72% classification accuracy. Accuracy with this model dropped because classifier is trained and validated over two different domain datasets which result as misclassification of few instances. Table 4 shows the predicted instances with this model.

## 5.3 Results Obtained with Model-3

In model-3, we enrich our training dataset by encapsulating data from two different sourced, i.e., one is manually annotated tweets (22,000 tweets) scrapped from Twitter API, and another one is IMDb reviews comprising 2000 highly polar reviews. This change in strategy results in improved classification accuracy. Results obtained with this model show the significant improvement in the performance of the classifier. Table 5 shows all the predicted values and accuracy obtained corresponding to the given dataset.

## 5.4 Performance Obtained with Each Suggested Model

Performance of the classifier is analyzed regarding precision value, recall value, *F*-curve, and overall accuracy attained. From Table 6, we can observe the perfor-

**Table 6** Performance measures obtained with model-1, model-2 and model-3

| Model | Model-1 | Model-2 | Model-3 |
|---|---|---|---|
| Model description | Trained and validated over IMDb reviews | Trained with IMDb reviews and validated over the domain-specific dataset | Trained and tested over encapsulated dataset (IMDb reviews + domain-specific labeled tweets) |
| Precision (%) | 50.47 | 50.04 | 50.06 |
| Recall (%) | 79.7 | 61.77 | 78.07 |
| F-measure (%) | 61.80 | 55.29 | 61.00 |
| Accuracy (%) | 78.95 | 61.72 | 77.97 |

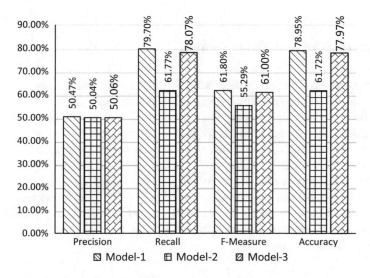

**Fig. 2** Graphical representation of performance measures obtained with each model

mance corresponding to each model, and it is evidently seen that our core model-3 outperforms when trained and tested with the encapsulated dataset (Fig. 2).

## 5.5 Performing SA on Domain-Specific Subjective Tweets

Finally, we use our core model-3 to perform classification on time, and domain-specific dataset (i.e., total 150,374 subjective tweets) in which issue related to plastic ban was discussed. The classifier classifies 94,735 tweets as positive polarity and remaining 55,639 tweets as the negative polarity. The pie chart given in Fig. 3 shows the percentage share of classified positive and negative tweets. Our observation on

**Fig. 3** Polarity classification
percentage

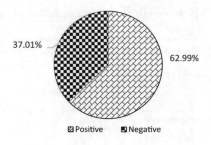

collected Twitter data shows that around 62.99% people carry positive opinion about plastic ban in India.

# 6 Conclusions and Future Work

We have successfully built a model with 77.94% classification accuracy and performed sentiment analysis on time-specific and domain-specific Twitter data posted on the Twitter wall. Our observation shows that 62.99% people are quite positive toward the law in polybags ban in India, and rest 37% people shows a negative opinion about it. Furthermore, we can explore other more accurate classification methods for getting more accurate results.

In this work, only two-class classification is performed, i.e., positive class and negative class. However, we can also extend this work by defining more classes according to the behavior people show about this law like neutral, anger, anticipation, etc. Also, we can improve the accuracy and efficiency of the classifier by applying more preprocessing steps.

# References

1. Plastic bag. Available at: https://en.wikipedia.org/wiki/Plastic_bag
2. Plastic ban: what india can learn from other countries. Available at: https://swachhindia.ndtv.com/plastic-ban-india-can-learn-countries-6161/
3. India just banned all forms of disposable plastic in its capital, independent. Available at: https://www.independent.co.uk/news/world/asia/india-delhi-bans-disposable-plastic-single-use-a7545541.html
4. Plastic ban in Maharashtra: what is allowed, what is banned. Available at: https://indianexpress.com/article/india/plastic-ban-in-maharashtra-mumbai-from-june-23-what-is-allowed-what-is-banned-all-you-need-to-know-5228307
5. Why have laws to completely ban plastic bags failed in India? Available at: https://scroll.in/article/872612/why-have-laws-to-completely-ban-plastic-bags-failed-in-india
6. R. Srivastava, M.P.S. Bhatia, Quantifying modified opinion strength: a fuzzy inference system for sentiment analysis, in *2013 International Conference on Advances in Computing, Communications and Informatics (ICACCI)* (IEEE, 2013), pp. 1512–1519

7. R. Srivastava, M. Bhatia, Ensemble methods for sentiment analysis of on-line micro-texts, in *International Conference on Recent Advances and Innovations in Engineering (ICRAIE)* (IEEE, 2016), pp. 1–6
8. R. Srivastava, M. Bhatia, Offline vs. online sentiment analysis: issues with sentiment analysis of online micro-texts. Int. J. Inf. Retr. Res. (IJIRR) **7**(4), 1–18 (2017)
9. R. Srivastava, M. Bhatia, Real-time unspecified major sub-events detection in the Twitter data stream that cause the change in the sentiment score of the targeted event. Int. J. Inf. Technol. Web Eng. (IJITWE) **12**(4), 1–21 (2017)
10. R. Srivastava, M. Bhatia, Challenges with sentiment analysis of on-line micro-texts. Int. J. Intell. Syst. Appl. **9**(7), 31 (2017)
11. R. Srivastava, M. Bhatia, H.K. Srivastava, C. Sahu, Exploiting grammatical dependencies for fine-grained opinion mining, in *International Conference on Computer and Communication Technology (ICCCT)* (IEEE, 2010), pp. 768–775
12. R. Srivastava, H. Kumar, M. Bhatia, S. Jain, Analyzing Delhi assembly election 2015 using textual content of social network, in *Proceedings of the Sixth International Conference on Computer and Communication Technology 2015* (ACM, 2015), pp. 78–85
13. V.B. Vaghela, B.M. Jadav, Analysis of various sentiment classification techniques. Analysis **140**(3), 22–27 (2016)
14. A. Pak, P. Paroubek, Twitter as a corpus for sentiment analysis and opinion mining. LREc **2010**(10), 1320–1326 (2010)
15. M. Hu, B. Liu, Mining and summarizing customer reviews, in *Proceedings of the Tenth ACM SIGKDD International Conference on Knowledge Discovery and Data Mining* (ACM, 2004), pp. 168–177
16. M. Hu, B. Liu, Mining opinion features in customer reviews. AAAI **4**(4), 755–760 (2004)
17. B. Pang, L. Lee, Opinion mining and sentiment analysis. Found. Trends Inf. Retr. **2**(1–2), 1–135 (2008)
18. C. Yang, K.H.-Y. Lin, H.-H. Chen, Emotion classification using web blog corpora, in *IEEE/WIC/ACM International Conference on Web Intelligence* (IEEE, 2007), pp. 275–278
19. J. Read, Using emoticons to reduce dependency in machine learning techniques for sentiment classification, in *Proceedings of the ACL Student Research Workshop* (Association for Computational Linguistics, 2005), pp. 43–48
20. A.-M. Popescu, O. Etzioni, Extracting product features and opinions from reviews, in *Natural Language Processing and Text Mining* (Springer, 2007), pp. 9–28
21. V. Hatzivassiloglou, K.R. McKeown, Predicting the semantic orientation of adjectives, in *Proceedings of the Eighth Conference on European Chapter of the Association for Computational Linguistics* (Association for Computational Linguistics, 1997), pp. 174–181
22. A. Esuli, F. Sebastiani, SENTIWORDNET: a publicly available lexical resource for opinion mining, in *Proceedings of LREC*, vol. 6 (Citeseer, 2006), pp. 417–422
23. B. Pang, L. Lee, A sentimental education: sentiment analysis using subjectivity summarization based on minimum cuts, in *Proceedings of the 42nd Annual Meeting on Association for Computational Linguistics* (Association for Computational Linguistics, 2004), p. 271
24. M. Hall, E. Frank, G. Holmes, B. Pfahringer, P. Reutemann, I.H. Witten, The WEKA data mining software: an update. ACM SIGKDD Explor Newsl **11**(1), 10–18 (2009)

# Whale Neuro-fuzzy System for Intrusion Detection in Wireless Sensor Network

**Rakesh Sharma, Vijay Anant Athavale and Sumit Mittal**

**Abstract**  A hybrid intrusion detection system (IDS) that depends on neuro-fuzzy system (NFS) strategy is proposed which identifies the WSN attacks. IDS which makes utilization of cluster-based engineering with upgraded the low-energy adaptive clustering hierarchy (LEACH) will be simulated for routing that expects to decrease energy utilization level by various sensor nodes. An ID utilizes anomaly detection and misuse detection dependent on NFS which will be changed by incorporating with meta-heuristic optimization strategies for ideally creating fuzzy structure. Fuzzy rule sets alongside the neural network are used to incorporate the location results and determine the attackers kinds of attacks, and the regular procedure of NFS is as per the following: Initially, fuzzy clustering strategy is used to produce distinctive training subsets; in light of unusual training subsets, divergent ANN models are prepared to devise unique base models and fuzzy aggregation module, which is being used to unite these result. The proposed WNFS is created by including the properties of the whale optimization algorithm (WOA) with the neuro-fuzzy architecture. The optimization algorithm selects the appropriate fuzzy rules for the detection.

**Keywords**  Intrusion detection · Wireless sensor network (WSN) · Routing · Neuro-fuzzy system · Whale optimization algorithm

R. Sharma
Department of Computer Science & Engineering, I. K. Gujral Punjab Technical University
Jalandhar, Kapurthala, Punjab, India
e-mail: rakeshsharma3112@gmail.com

V. A. Athavale (✉)
PIET, Panipat, Haryana, India
e-mail: vijay.athavale@gmail.com

S. Mittal
MMICT & BM, MMDU, Mullana, Haryana, India
e-mail: sumit.mittal@mmumullana.org

© Springer Nature Singapore Pte Ltd. 2021      123
X.-Z. Gao et al. (eds.), *Advances in Computational Intelligence and Communication Technology*, Advances in Intelligent Systems and Computing 1086,
https://doi.org/10.1007/978-981-15-1275-9_11

# 1 Introduction

Wireless sensor network (WSN) is a flow innovation and has customary massive thinking among researchers. WSNs have happened to a predominant and basic instrument because of their reasonable highlights and applications, for example, monitoring, domestic applications, surveillance systems, healthcare and disaster management [1]. Ordinarily, the WSN condition contains low power, ease, and a large amount of sensors that are dispersed haphazardly even the objective area or are redeployed physically. In WNSs, communicating message is a proficient and an acknowledged model that allows different clients to consolidate and disseminate message packets all through the system viably with the end goal to get information of their advantage (Fig. 1).

To build up productive system having full lifetime, proper sending of the energy in WSN nodes is immensely vital since this class of system has confined battery storage for the nodes, and along these lines, these sensor nodes are named lightweight and correlative equipment having the limits of communicating, detecting, and handling the information from one node to the goal node in a larger system. Information transmission to longer distance can be observed as a result middle person nodes, since WSNs are helpless against interior and outside commencements as they have a confined transmission extend and subsequently drive the information straightforwardly to the favored client with a transmission go restrict. In this way, they do not be able to deal with a sturdy assailant because of their resource-confined nature [2]. In this condition, a defense mechanism is required that is called intrusion detection system (IDS), to guard the framework from the attackers. An efficient IDS system ought to be created for legitimate recognizing different types of attacks [3].

As a result of WSN attributes, enemies can absolutely produce network traffic, which can likewise cause substantial packet drop amid broadcasting of the packets or change the unordinary substance of the message in the packets larger section of the sensor networks which are exceptionally susceptible toward attacks [4]. In WSNs, it is extremely fundamental to complete secure information transmission between the

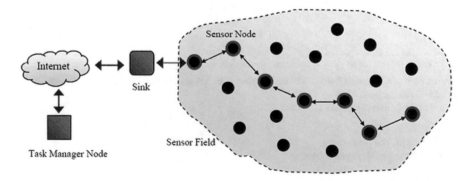

**Fig. 1** WSN model

nodes. For example, if WSNs are utilized in battlefield applications, sensor nodes are intruded with by the attackers and destroyed. Henceforth, security assumes a critical job for ensuring secure correspondence between the nodes; validation techniques are executed in the system. An evasion procedure is used to counter the all-around perceived assaults. Besides, a counteractive action technique cannot ensure close by every one of the attacks. The IDS is used generally to distinguish the packets in a system and determine which packet is harmed by the aggressors with the goal that these attacks ought to be recognized. What is more, IDS can help the anticipation framework through the created idea of attacks [5].

The misuse detection identifies the varied sorts of attackers by leveling or examination the current attack behavior and therefore the past attack behavior [6]; however, it is the very best accuracy but with low detection rate. It cannot sight unknown attackers that do not seem to be within the base of the model. The intrusion detection system (IDS) accomplishes the aim of getting the very best detection rate with low false positive rate [1]. Hybrid detection methodology will determine unknown attacks with the best accuracy of the misuse detection and therefore the greatest detection rate of anomaly detection.

Protocol dependent on the clustering algorithm for information accumulation in the WSN that works on the medium access control (MAC) is low-energy adaptive clustering hierarchy (LEACH) protocol [7]. Fundamental procedures of LEACH protocol incorporate algorithms for distributing cluster framing, adaptive cluster forming, and cluster header position evolving. The method of disseminating cluster shaping guarantees self-association of most target nodes. The adaptive cluster forming and cluster header position changing algorithms guarantee to share the energy scattering reasonably among all nodes and drag out the lifetime of the entire framework at last. In this manner, the energy stack associated with being a cluster head is uniformly appropriated with nodes for expanding the lifetime of the whole network. Misuse detection and anomaly identification are two imperative modules in the IDS [8]. Anomaly detection builds up a model to perceive the anomalous and typical conduct of the nodes, via completing the examination and correlation of the nodes conduct, and it has the most noteworthy location rate; in the meantime, it has the highest false positive rate. The misuse detection recognizes the different types of attackers by likening or looking at the present assault conduct and the past assault conduct [6]; however, it has the most elevated exactness yet with low recognition rate. It cannot recognize obscure aggressors, which are not in the base of the model. The intrusion detection system (IDS) achieves the point of getting the most astounding detection rate with low false positive rate [1]. Hybrid detection methodology can recognize obscure attacks with the best precision of the misuse detection and the best recognition rate of anomaly detection (Fig. 2).

Because of the resource-constrained condition of the WSN, customary security techniques had not been in employment, as they required energy and in recent years, the examination about the WSNs had thorough on the security of the sensor systems.

**Fig. 2** Hybrid intrusion
detection system

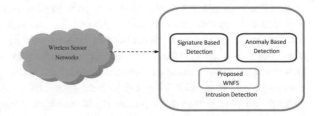

## 2 Background Details and Related Work

Selvakennedy et al. [9] present a peculiar purpose of aggregate social agents to
manage the development of clusters. Clustering is a methodology effectively sought
after by numerous gatherings in acknowledging more versatile data gathering and
routing. In any case, it is fairly testing to frame a proper number of clusters with all-
around adjusted participations. With the end goal to counter the standard issues of
such meta-heuristics, they propose a novel a run of the mill application that enables
their protocol to converge nimble with very defined overhead. An examination is
performed to decide the ideal number of clusters important to accomplish the most
astounding energy efficiency. With the end goal to consider a down to practical
assessment, a total test system connecting huge methods for the announcement stack
is utilized. Their protocol is set up to make beyond any doubt a fine division of cluster
heads all through a totally circulated methodology. To quantify persuaded clustering
properties, they too presented two fitness metrics that may be utilized to standard
different clustering algorithms.

Harold Robinson et al. [10] Energy Aware Clustering utilizing Neuro-fuzzy
methodology (EACNF) is anticipated by authors to shape optimum and energy aware
and the proposed scheme comprises of fuzzy subsystem and neural network. This is
primary concern in networks for enhancing network life span. Ordinary clustering
schemes are made with static cluster heads that pass on past than the typical nodes
that debase the network performance in routing. It is exceptionally imperative zone
to build up energy-aware clustering protocol in WSN to lessen energy utilization for
expanding network duration.

Magotra and Kumar [7] enhanced this mechanism and depended on the identification of the malicious node using the signal strength along with the distance between
the nodes. Nevertheless, when both of these parameters exhibit a certain threshold
value, then the test packet will increase the communication overhead, which affects
the transmission time.

Kapitanova et al. [11] in this paper demonstrate that by methods for fuzzy qualities
in its place of crisp ones extensively enhances the accuracy of occasion finding.
They likewise demonstrate that fuzzy logic approach gives event detection discovery
precision than two entrenched classification algorithms. Occasion location is a crucial
part of numerous wireless sensor network (WSN) applications. In any case, the area
of occasion account has not set up adequate mindfulness. The greater part of current
occasion clarification and location approaches depends on methods for exact qualities

to recognize occasion edges. Notwithstanding, they trust that fresh qualities cannot adequately deal with the regularly uncertain sensor readings. An impediment of utilizing fuzzy rationale is the exponentially developing size of the fuzzy logic rule base. As sensor nodes have constrained memory, putting away vast control bases may be a test. To address this issue, they have built up various strategies that assistance diminishes the span of the administer base by over 70%, while protecting the event detection accuracy.

Khan et al. [12] present a fault detection strategy for wireless sensor networks. The procedure depends on modeling a sensor node by Takagi–Sugeno–Kang (TSK) fuzzy inference system (FIS), where a sensor estimation of a node is approximated by a component of the sensor estimations of the neighboring nodes. They show a node by intermittent TSK-FIS (RFIS), where the sensor estimation of the node is exact as the capacity of genuine estimations of the neighboring nodes and the already approximated estimation of the node itself. Transitory mistakes in sensor estimations or potentially correspondences are overwhelmed by repetition of information gathering and node which has built up a defective sensor is not totally disposed of on the grounds that it is valuable for transferring the data among alternate nodes. Every node has its own fuzzy model that is prepared with contribution of neighboring sensors estimations and a yield of its real estimation. A sensor is proclaimed broken if the distinction between the result of the fuzzy model and the genuine sensor estimation is more noteworthy than the recommended sum contingent upon the physical amount being estimated. Reproductions are performed utilizing the fuzzy rationale tool stash of MATLAB and give an examination of acquired outcomes to those from a feedforward artificial neural network, recurrent neural network, and the median of measured values of the neighboring nodes.

Jabbar et al. [13] in their overview article display an entire contention on intelligent optimization of wireless sensor networks through bio-inspired computing. The brilliant flawlessness of biological frameworks and its distinctive viewpoints for upgraded answers for non-biological issues is available here in detail. In the current research inclination, employing biological solutions for explaining and upgrading diverse parts of counterfeit frameworks' issues has been formed into a vital field with the name of bio-motivated registering. They have arranged the abuse of key constituents of organic framework for creating bio-motivated frameworks to speak to its centrality and appearance in trouble settling patterns and introduced how the figurative relationship is produced between the two natural and non-natural frameworks by citing a case of connection between winning remote framework and the regular framework. Interdisciplinary research is playing an impressive commitment for different issues' tackling. The way toward consolidating the people's yield to frame a solitary critical thinking arrangement is delineated in three-organize troupe plan. Likewise, the hybrid solutions from computational insight-based streamlining are lengthened to exhibit the developing contribution of these propelled frameworks with rich references for the intrigued peruses. It is inferred that these flawless manifestations have solutions for the vast majority of the issues in non-biological framework.

Oliveira et al. [14] presented FLEACH, a protocol which is intended to give security to node-to-node correspondence in LEACH-based WSNs. It used arbitrary key pre-distribution system to improve the correspondence security in the LEACH protocol alongside symmetric-key cryptography in this protocol. FLEACH gives integrity, authenticity, confidentiality, and freshness in node-to-node transmission communication; however, it is dangerous to node distinguishing assault.

## 3 Intrusion Detection System

Intrusion detection system (IDS) employs a neural network (NN), which comprise feedforward neural network (FFNN) and backpropagation neural network (BPNN) of the supervised learning process that is based on the fuzzy logic method with anomaly and misuse detection technique to sense the various attacks with higher detection ratio and lower false alarm. Neural network (NN) is one of the extensively used techniques and has been successful in solving numerous composite sensible troubles and diverse meta-heuristic optimization algorithms for optimization because they (i) depend on to some degree direct origination and are anything but difficult to acknowledge; (ii) do not request angle data; (iii) have the capacity to sidestep local optima; (iv) can be used in a broad scope of issues covering differing disciplines. Nature-propelled meta-heuristic calculations settle enhancement issues by emulating organic or physical wonders. They can be gathered in three center classifications: advancement-based, material science-based, and swarm-based strategies. Advancement-based strategies are propelled by the laws of characteristic development. The pursuit procedure begins with a haphazardly created populace which is developed above ensuing ages. The quality purpose of these strategies is that the best people are forever joined altogether to frame the in this way age of people. This enables the populace to be upgraded in abundance of the method for ages. ANN-based IDS exists in two perspectives: (1) lower discovery accuracy, especially for low-visit attacks, e.g., remote to local, user to root, and (2) weaker detection stability [15]. For the more than two viewpoints, the significant reason is that the appropriation of different sorts of attacks is pointless. For low-visit attacks, the inclining test measure is excessively little contrasted with high-visit attacks. It makes ANN difficult to gain proficiency with the characters of these attacks, and in this manner, recognition accuracy is much lower. By and by, low-visit attacks do not infer; they are irrelevant. Rather, genuine outcome will be caused if these attacks succeeded. For instance, if the U2R attacks succeeded, the assailant can procure the specialist of root customer and do all that he jumps at the chance to the focused on PC frameworks or system gadget. Moreover in IDS, the low-visit attacks are regularly anomalies. We plan and created neuro-fuzzy-based model by incorporating a methodology dependent on whale enhancement which meta-heuristic improvement calculations (Fig. 3).

The proposed ID-WNFS system has two components, like sniffer and detector for effectively identifying the intruders in the WSN, and each component is explained below.

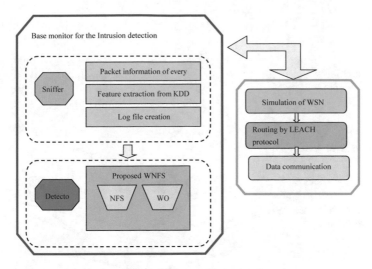

**Fig. 3** Architecture of the proposed ID-WNFS-based intrusion detection system

## 3.1 Sniffer: Creation of the Log File

The first component in the proposed ID-WNFS system is the sniffer, and as the name suggests, it carefully observes the packet information of the each node. The router nodes while interconnected with the other nodes for passing the information amidst the source and destination may undergo certain changes. Also, the nodes in the WSN are mobile and have limited energy resource. The information concerning the next node for which the packet needs to be send and the node from which the packet is received is there in each sensor node. The sniffer component in the proposed ID-WNFS observes the transmission pattern and stores all the packet information of the sensor node.

The packet information helps in finding a suitable log file for the intrusion detection. Each node in the WSN is registered within the server, and thus by observing the log information, the intruder node can be identified and eliminated from the communication.

## 3.2 Detector: Intrusion Detection Through Proposed WNFS

The detector component comprises the proposed WNFS network for identifying the intruder node. This work develops the WNFS network by including the optimization properties with the existing NFS. The existing NFS model developed in the literature combines both the neural network and fuzzy logic properties. The layers of the neural network are fed with the fuzzy rules for performing the classification as shown in Fig. 4.

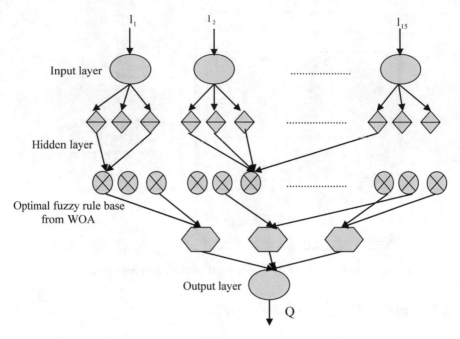

**Fig. 4** Diagrammatic representation of the WNFS

The fuzzy logic while implemented for any system provides large number of fuzzy rules in its rule base. It is practically impossible to identify the required rule base for the classification. For this purpose, this work has used the WOA algorithm for selecting the appropriate rules for the NFS system. The architecture of the deigned WNFS model is depicted in Fig. 3. The architecture of the deigned WNFS model is similar to the NFS, and the only change is the number of neurons provided for the training the log file created by the sniffer component.

As shown in the figure, the log files are provided as input to the input layer of the WNFS. The WNFS has different layers such as input, hidden, and output. As the aim is to find the node to be intruder or not, the WNFS has only one neuron at the output layer. The hidden layer provides optimal fuzzy rules for the modifying the input, and the output is observed from the output layer. The NFS model considered in this work is the Mamdani-type model, which uses the optimal fuzzy rule base. The input to the WNFS is the log file, and as there are 15 features in the log file, the total number of input neurons is considered as 15. The input layer of the WNFS is represented as follows:

$$L = \{l_1, l_2, \ldots, l_{15}\} \tag{1}$$

where $l_1$ refers to the first feature in the log file. For selecting the optimal rule base for the hidden layer, this work considers the WOA algorithm. Then, the hidden layer output is passed through the output layer and the final decision on the node is taken.

The output layer provides the value as 1 for the presence of intrusion and 0 for the normal node. The output layer of the WNFS is represented as follows:

$$Q = \begin{cases} 1; \text{ Intruder} \\ 0; \text{ Normal} \end{cases} \tag{2}$$

where $Q$ indicates the output of the WNFS model. The major task lies in identifying the optimal fuzzy rules. In the existing NFS model, the weights $X$ and $Y$ for fuzzy rules are tuned based on the backpropagation (BP) algorithm, and its tuning process is explained as follows.

In BP algorithm, the optimization procedure uses the minimal mean squared error (MSE) as the fitness parameter. The fitness criterion adopted by the BP is given as follows:

$$M = \frac{1}{2A} \sum_{a=1}^{A} (T_{\text{BP}} - O_{\text{BP}})^2 \tag{3}$$

where $A$ refers to the total number of iterations required for the BP algorithm, and the terms $T_{\text{BP}}$ signify the target output and the $O_{\text{BP}}$ actual output attained by the BP. Based on the minimal MSE error, the BP algorithm identifies the change in weights $X$ and $Y$ in the hidden layer, and it is expressed as,

$$\Delta X = -\kappa \frac{\partial M}{\partial X} \tag{4}$$

$$\Delta Y = -\kappa \frac{\partial M}{\partial Y} \tag{5}$$

where $\kappa$ refers to the learning rate used by the BP approach. Adopting basic fuzzy rule base complicates the learning task, as it is large in size. In this work, WOA selects the optimal rules for the proposed WNFS based on the prey-searching behavior of the whale. The steps involved in WOA [16] for identifying the optimal rules are explained as follows:

*Initialization*: The first step is the initialization of population, and thus, the solution space for the WOA is expressed as,

$$R = \{R_1, R_2, \ldots R_j, \ldots, R_F\} \tag{6}$$

where $R_j$ refers to the $j$th component in the rule base. WOA finds the best search agent $R*$ through the optimization.

Fitness: The next step identifies the best search agent $R*$ through the fitness criteria, and here, the fitness is considered to the detection accuracy.

Update phase: The WOA updates the position of the search agent in both the exploration and the exploitation phase. The update done based on the prey search behavior is expressed as follows:

$$\vec{Z} = |\vec{S}\,\vec{R}^*(t) - \vec{R}(t)| \tag{7}$$

$$\vec{R}(t+1) = \vec{R}^*(t) - \vec{U}\cdot\vec{Z} \tag{8}$$

where $\vec{Z}$ indicates the modified search agent. Now, the values of $\vec{U}$ and $\vec{S}$ are expressed as,

$$\vec{U} = \vec{a}\cdot\vec{b} - \vec{a} \tag{9}$$

$$\vec{S} = 2\cdot\vec{b} \tag{10}$$

where $a$ indicates the constant within the range of 2 to 0 and $b$ refers to the ranging vector [0, 1]. The update in the exploitation phase can be expressed as,

$$\vec{R}(t+1) = \vec{Z}'\cdot e^{st}\cdot\cos(2\pi x) + \vec{R}^*(t) \tag{11}$$

where

$$\vec{Z}' = |\vec{R}^*(t) - \vec{R}(t)| \tag{12}$$

Also, the exploration phase-based update depends on choosing random search agents, and it is expressed as,

$$\vec{R}(t+1) = R_{rand}^{\rightarrow} - \vec{U}\cdot\vec{Z} \tag{13}$$

$$\vec{Z} = \left|\vec{S}\,R_{rand} - \vec{R}\right| \tag{14}$$

Best search agent: Now, based on the fitness criteria, the best search $R^*$ is found, and it replaces the previous solution.

Termination: Up to T number of iterations, the WOA algorithm is carried out, and at the end of iteration, the optimal fuzzy rule base is selected and provided to the WNFS model.

### 3.2.1 Establishing the routing path based on LEACH protocol

After declaring the node as the 'intruder' or 'normal node,' the proposed ID-WNFS system passes the information to the LEACH protocol. The LEACH protocol [17] establishes the routing path by ignoring the intruder node. The LEACH protocol gets updated from time to time, and hence, the WSN is free from intrusion. This work chooses the LEACH protocol as the routing algorithm, since the LEACH establishes the routing path by constructing large-sized clusters. Also, the protocol updates the cluster head simultaneously, and hence, the chance of routing path failure is very low. The LEACH protocol can be explained in brief as follows:

LEACH protocol gets the information regarding the intruder node and carries out the simulation for creating the routing path amidst the source and destination. The LEACH protocol creates the routing path through the clustering, and it has two phases: (1) setup phase and (2) steady phase.

(1) Setup phase: The LEACH protocol creates the routing path by concentrating on the energy parameter and hence selects the node with the highest energy to the cluster head. The setup phase has three steps:

  (i)   Advertisement of the cluster head: After selecting the highest energy node as the cluster head, the protocol sends the information to the other nodes regarding the cluster head node. The selection of the cluster head is restricted for the single round, and the particular node can become the cluster head after each node in the WSN acts as the cluster head.
  (ii)  Cluster setup: The cluster is formed based on the nodes providing the response to the advertisement provided in the previous step. The non-cluster node only sends the information and hence switches off the transmission part.
  (iii) Creating a suitable transmission schedule: The cluster head creates the transmission path for the other nodes in the cluster.

(2) Steady phase: After establishing the connection, the head node collects the information from the node and sends to the base station and the transmission path to the destination is done through the various cluster heads in the WSN.

# 4   Experimental Setup and Results

Two attacks namely DoS and black hole are detected using the IDS ID-WNFS, and the WSN simulation is carried out in the MATLAB as the implementation tool. Here, the comparative analysis of the proposed ID-WNFS logic has been evaluated. Two kinds of attack such as DOS and the black hole attack have been introduced on the system. Total node count on the WSN is varied as 50 and 100. The metrics used to show the performance are network lifetime, energy, and detection accuracy. The results obtained are shown in Table 1.

From Table 1 and Fig. 5, it is indicated that ID-WNFS efficiently detects both DoS and black hole attacks. ID-WNFS achieves 99.9% detection accuracy in detecting DoS and 99.2% detection accuracy in detecting black hole attack.

**Table 1** Performance of ID-WNFS

| | DoS attack | | | | Black hole | | | |
|---|---|---|---|---|---|---|---|---|
| Number of nodes | 50 | | 100 | | 50 | | 100 | |
| Number of rounds | 500 | 1000 | 500 | 1000 | 500 | 1000 | 500 | 1000 |
| Network lifetime | 36 | 22 | 80.982 | 44 | 41 | 24 | 92.22 | 47.988 |
| Energy | 16.87 | 4.416 | 18.42517 | 5.515884 | 20.68005 | 6.007708 | 22.2291 | 7.106808 |
| Detection accuracy | 0.9699 | 0.50136 | 0.994435 | 0.787191 | 0.992307 | 0.50204 | 0.988165 | 0.778932 |

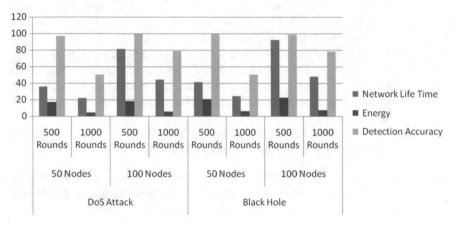

**Fig. 5** Performance of ID-WNFS

## 5 Conclusions

The IDS utilizes a fuzzy-neural network, which contains feedforward neural network and backpropagation neural network of the supervised learning process poisoned on the fuzzy logic operation with anomaly and misuse detection technique, and the combination of these two techniques is used to provide an intrusion detection system with a high detection rate and low false positive rate. IDS makes employ of cluster-based architecture through improved the low-energy adaptive clustering hierarchy (LEACH) which will be simulated for routing aim to lower energy consumption level of sensor nodes. An IDS will be modified by integrating with meta-heuristic optimization techniques for optimally generating fuzzy structure that uses anomaly detection and misuse detection based on NFS.

# References

1. Y. Maleh, A. Ezzatib, Y. Qasmaouic, M. Mbidac, A global hybrid intrusion detection system for wireless sensor networks. Procedia Comput. Sci. **52**, 1047–1052 (2015)
2. P. Sarigiannidis, E. Karapistoli, A.A. Economides, Detecting Sybil attacks in wireless sensor networks using UWB ranging-based information. Expert Syst. Appl. **42**(21), 7560–7572 (2015)
3. O. Depren, M. Topallar, E. Anarim, M.K. Ciliz, An intelligent intrusion detection system (IDS) for anomaly and misuse detection in computer networks. Expert Syst. Appl. **29**(4), 713–722 (2005)
4. Y. Shen, S. Liu, Z. Zhang, Detection of hello flood attack caused by malicious cluster heads on LEACH protocol. Int. J. Adv. Comput. Technol. **7**(2), 40–47 (2015)
5. S.K. Saini, M. Gupta, Detection of malicious cluster head causing hello flood attack in LEACH protocol in wireless sensor networks. Int. J. Appl. Innov. Eng. Manag. **3**(5), 384–391 (2014)
6. T.M. Rahayu, S.-G. Lee, H.-J. Lee, Security analysis of secure data aggregation protocols in wireless sensor networks, in *Proceedings of the 16th International Conference on Advanced Communication Technology*, 2014, pp. 471–474
7. S. Magotra, K. Kumar, Detection of HELLO flood attack on LEACH protocol, in *Proceedings of the IEEE International Advance Computing Conference (IACC '14)*, Gurgaon, India, 2014, pp. 193–198
8. S. Shamshirband, N.B. Anuar, M.L.M. Kiah et al., Co-FAIS: cooperative fuzzy artificial immune system for detecting intrusion in wireless sensor networks. J. Netw. Comput. Appl. **42**, 102–117 (2014)
9. S. Selvakennedy, S. Sinnappan, Y. Shang, A biologically-inspired clustering protocol for wireless sensor networks. Comput. Commun. **30**(14–15), 2786–2801 (2007)
10. Y. Harold Robinson, E. Golden Julie, S. Balaji, A. Ayyasamy, Energy aware clustering scheme in wireless sensor network using neuro-fuzzy approach, in *Wireless Personal Communication*, vol. 95 (Springer, 2017), pp. 1–19
11. K. Kapitanova, S.H. Son, K.-D. Kang, Using fuzzy logic for robust event detection in wireless sensor networks, in *Advances in Ad-hoc Network*, vol. 10 (2012), pp. 709–722
12. S.A. Khan, B. Daachi, K. Djouani, Application of Fuzzy Inference Systems to Detection of Faults in Wireless Sensor Networks, vol. 94 (Elsevier, 2012), pp. 111–120
13. S. Jabbar, R. Iram, A.A. Minhas, I. Shafi, S. Khalid, M. Ahmad, Intelligent optimization of wireless sensor networks through bio-inspired computing: survey and future directions. Int. J. Distrib. Sensor Netw. 1–13 (2013)
14. L.B. Oliveira, H.C. Wong, M. Bern, R. Dahab, A.A.F. Loureiro, SecLeach—a random key distribution solution for securing clustered sensor networks, in *Proceedings of the 5th IEEE International Symposium on Network Computing and Applications*, Washington, DC, USA, 2006, pp. 145–154
15. R. Beghdad, Critical study of neural networks in detecting intrusions. Comput. Secur. **27**, 168–175 (2008)
16. S. Mirjalili, A. Lewis, The whale optimization algorithm. Adv. Eng. Softw. **95**, 51–67 (2016)
17. W.R. Heinzelman, A. Chandrakasan, H. Balakrishnan, Energy-efficient communication protocol for wireless microsensor networks, in *Proceeding of the 33rd Hawaii International Conference on System Sciences* (IEEE, 2000), pp. 1–10

# Advanced Security Techniques

# The Threat Detection Framework for Securing Semantic Web Services Based on XACML

Nagendra Kumar Singh and Sandeep Kumar Nayak

**Abstract** In current scenario, the security issues are major concern for the proper functioning of semantic Web services. Whatever the facilities attained by the users or customers through Web services must have a concrete level of secure access, so that unauthorized access or usage may be prohibited to make their activities secure. The extant mechanism for access control restrains users from logging into data beyond their authorization, but they are incompetent to prohibit mishandling of the data made by them. This paper proposes a new framework for securing semantic Web services by applying anomaly detection mechanism with XACML. The main aim of the proposed framework is to identify the suspicious activities performed by a legitimate insider. In this research article, it is being tried to give the details of security framework on Web service which prevents information disclosure whether intentional or accidental.

**Keywords** Semantic Web · Access control · Threat detection · XACML · User profile

## 1 Introduction

In the present scenario, the semantic Web services are becoming the most important component for the Internet users and are responsible for the growth of information by enabling communication and management of individual's personal informative contents around the world. With the increasing usage of Internet, most of the services such as banking, shopping, bill payment, rail or air reservation, social networking and government services are accessible via the Web. Due to their vast applicability in many areas, semantic Web services are now emerging as a leading application domain in the open and complex Internet [1].

Semantic Web applications are progressively being transformed into a communication medium that is used to establish communication with the server that employs the database management system (DBMS) [2]. Such Web applications usually deploy the front-end Web server to facilitate user interface and a back-end server that stores

N. K. Singh · S. K. Nayak (✉)
Integral University, Dasauli, P.O. Bas-Ha Kursi Road, Lucknow 226026, India

© Springer Nature Singapore Pte Ltd. 2021      139
X.-Z. Gao et al. (eds.), *Advances in Computational Intelligence and Communication Technology*, Advances in Intelligent Systems and Computing 1086,
https://doi.org/10.1007/978-981-15-1275-9_12

files or data. The database server stores very crucial and sensitive information of an organization. These database servers are often susceptible to various types of outsider attacks as well as insider attacks such as guideline violation attack, elevation of privilege attack, SQL injection attack and privilege misuse attack [2–4]. Thus, the data stored in database must be protected from outsider attackers as well as from insider attackers.

Access control always plays a significant role in maintaining privacy, integrity and availability of SWS. Access control can be understood that the requesting user has to satisfy certain criterions to access the requested services. The main objective of access control is to secure the SWS from accidental or intentional attacks. The major goal of existing access control mechanism is to protect the organization from outsider attackers rather than from insider attackers. Many organizations including government agencies (e.g. defence, trade and judiciary), commercial enterprises, financial and banking institutions, etc. are now facing severe threats from malicious insiders. If a legitimate user has been authorized to access the database, then he can steal confidential information from the database and send this information to outside the organization.

Currently, most of the government and corporate organizations are storing their confidential data in relational databases. Therefore, the primary objective is to secure the data and files available in a database server. A threat detection mechanism enforced at the database server is an effective strategy for detecting malicious data extraction/behaviour by the trusted insider for the following reasons: First, the access in RDBMS is performed via standard query language (SQL) that has well-defined structure to operate the database. So, RDBMS is suitable for modelling user behaviour [5]. Secondly, it is more effective to monitor unauthorized access to confidential data stored on a database server because the RDBMS layer already has a standard system for implementing access control.

This paper proposes the threat detection framework for securing the semantic Web services from the anomalous database transactions done by the legitimate intruders. The remainder of the paper is organized as follows. Related research in semantic Web and access control domain is illustrated in Sect. 2. Section 3 briefly describes the threat detection framework for securing semantic Web services and core components of the profile manager module such as query signature format, signature creator (SC) and extract table row (ET). Section 4 explains the process for threat detection and significance of signature evaluator (SE) sub-module in detecting abnormal data extraction by a legitimate insider. Lastly, the conclusion and future work are given in Sect. 5.

# 2 Related Work

Lau et al. [6] use an oncology approach, an intelligent community transport service that has given semantic Web vision for the brokerage system. Jovanovic and Bagheri [7] have given a systematic analysis of semantic Web in the field of e-commerce.

They presented a technological stack for semantic Web-based e-commerce applications and briefly portray the challenges that are faced by the e-commerce applications. In [8], Lorenzo Bossi, Elisa Bertino and Syed Rafiul Hussain have proposed a formal model to detect malicious behaviour executed by the previously accredited application. The model stores the signature and constraints for each query submitted to the database but do not estimate the amount of data returned by the query. A model for controlling access in online social platforms has been developed by Cheng et al. [9] which uses regular expression notation for policy specification. The proposed algorithms are used to verify the necessary relationship path between users for a given access request. Workflow-based anomaly detection technique [2] identifies dissimilarity user behaviour in workflow-driven applications by converting Web sessions into a data-focused workflow.

## 3 The Threat Detection Framework for Securing Semantic Web Services

*eXtensible Access Control Markup Language* (XACML) is a platform-independent policy language that was developed by OASIS to control access to the Web environment. XACML permits the administrator to outline the policies for subjects in order to access the requested service and evaluates the incoming access requests from subjects. It is implemented in XML and follows the syntax and semantic of XML. XACML is an XML-based platform neutral security markup language for precisely defining the access control policy for Web services. XACML gives an architectonic model for implementing access control in Web services. XACML architecture has four major modules: Policy Enforcement Point (PEP), Policy Decision Point (PEP), Policy Administration Point and Policy Information Points (PIP) [10].

PEP is responsible for implementing the access control in the system. Each access request of subject is first transferred to PEP. First, PEP withdraws suitable traits from a request and regenerates the request in XACML format. In the end, it forwards the request to the PDP for evaluation. The PDP accepts an XACML request from PEP for evaluation. It extracts the applicable access policies from the PAP and retrieves the attribute values from XACML request and transfers its assessment to the PEP. The PAP creates access control policies for the resources. It also applies a set of permissions to block unauthorized access to these policies. PIP is considered as a midway storage that comprises various facets related to subjects, resources and environment.

Although it is believed that XACML has emerged as a major approach to address various requirements related to access control in a diversified environment such as semantic Web, there are some trust- and confidentiality-related problems in it. In the XACML-based system, PEP forwards an XACML request to the PDP. There is no mechanism available to ensure that the messages sent to the PDP are safe from attack. There is no mechanism to ensure the integrity of the message, which is transferred from one component to another. XACML allows the administrator to specify which data the user can access for which purpose. The main job of XACML is to protect data from external attacks but these mechanisms are unable to protect the data against legitimate intruders who have proper access to the database.

Since, the work of insiders is generated on a trusted domain within the organization, the insiders are not examined as the way external sources are examined.

Figure 1 shows the framework for detecting anomalous behaviour of a legitimate insider. XACML is specifically designed to address the requirement of access control in semantic Web environment. Since, the nature of semantic Web is changing continuously, it is required that the XACML must adapt to these changing environment. Generally, XACML secures semantic Web services from outsider attacks. As a number of attacks from legitimate insider are increasingly rapidly, it is required that the XACML should include provision for detecting anomalous behaviour of legitimate insider. The proposed framework integrates the threat detection framework with XACML. The proposed framework will alert the XACML in case of a possible data theft or anomalous behaviour of legitimate insider.

Legitimate intruders are well-skilled IT specialists with classified information regarding internal security and auditing control deployment. So, one possible solution to solving the above problem is to analyse the internal activities and data access patterns of these legitimate intruders to create a profile to expose the data extraction or anomalous activities performed by them.

The framework has two modules—Profile Manager and Threat Detector. The job of Profile Manager module is to create the normal profile of all legitimate insiders. The module creates the normal profile by including the access pattern of each legitimate insider in the profile when they are interacting with the database. The module will monitor each incoming query which is sent by the insider for execution. It creates

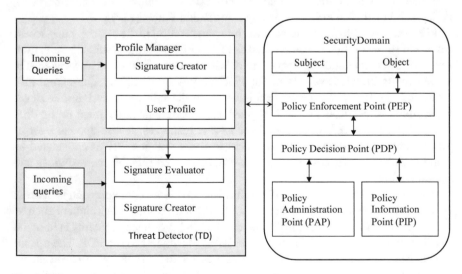

**Fig. 1** Threat detection framework

the normal profile by creating the query signature of each query submitted by each legitimate insider. As each organization stores their critical information in database, the objective is to secure the database from these legitimate insiders because these insiders hold valid credentials to access the database and may extract important data from the database.

The responsibility of Threat Detector module is to detect any anomalous behaviour of legitimate insider. The module monitors each incoming query that an insider is trying to execute. For this, it compares the current query signature with previous query signature that is created by the profile manager module. If both signatures are same, the query is found legitimate for execution, i.e. safe. Otherwise, not legitimate for execution, i.e. threat. The next section describes the above requirements.

## 3.1   Profile Manager

In the training phase, the user interacts with the database using SQL queries. The framework represents each input query in a special format known as query signature. The profile manager module builds normal profile for every individual user. The required training data has been collected when the user is performing regular function on the database, i.e. submitting queries for execution and waiting for result. The Profile Manager module transfers each query sent by the user to the database parser. The database parser generates possible parse tree of the query. The parsed tree of the query is sent to the signature creator sub-module of the Profile Manager to create the required signature of the input query. If a user sends a query, the query is first transferred to the Profile Manager module for processing and generating the signature of the query.

The major problem in designing the system is that which user should be supervised? One solution is to record the access patterns for each role available in the database. In some situations, roles are not used even if they are available. This situation may occur because currently available RDBMS allows users to login by authenticating himself and then the user can select a distinctive role. To solve this problem, the framework creates profile for every user of the database.

### 3.1.1   Signature Creator (SC)

This sub-module is responsible for creating the signature of each input query. It takes input from Oracle parse tree. The Oracle database transfers each input query to its parser to generate a query parse tree. SQL processing [11] is the parsing, optimization, row source generation and execution of a SQL statement. The problem is how the normal behaviour of a user is represented. The balance should be maintained when recording the access patterns of each user during profile creation stage. If profile includes very detailed access patterns, then it may improve the detection accuracy but may require more resources to execute. To resolve this problem, the SC sub-module represents the access pattern in the context of some query attributes for all queries submitted by the user.

*Queries* are the important element of *SQL* and are used to retrieve the data based on specific criteria. Each input query is represented in a special format knows as query signature. The query signature of each query includes: the command used in the query such as SELECT, INSERT INTO, DELETE and UPDATE the table IDs, and the IDs of attributes from each table which are referenced in the query. In some incidents, it is observed that the chunk of data retrieved through the query is a manifestation of unusual data elicitation from the database. Hence, the query signature includes the novel feature known as extract table row. This feature estimates the percentage of data extracted from each table referenced in the query. The estimated percentage of data extracted by the query can be obtained from the database's query optimizer. This feature helps in detecting abnormal data extraction before query is executed.

Whenever an SQL statement needs to be executed, the very first process that is initiated is known as parsing. In the parsing phase, SQL statements first separate into individual components and then formatted as data structures, so that other units can process those SQL statements. The first step of parsing is the Syntax check in which Oracle databases should check each SQL statement for syntax validity. If SQL statement is syntactically correct, the semantic check is performed on SQL statement. It determines that whether the table names and column names those are present in the statement actually exists in the database or not. The last step in the parsing process is known as the shared pool check which determines whether it can skip those steps of SQL statement that requires resource-intensive processing.

The signature creator collects required query data (command, table ID, attribute ID) from the query parse tree and generates the reciprocal signature of the submitted query. It also estimates the percentage of data extracted through the query. An execution plan of a query is an ordered set of steps used to access data in an Oracle RDBMS. The necessary steps of execution plans are articulated as a set of database operators that assemble rows from the database. The Oracle query optimizer decides what will be the order of operators and their implementation.

### 3.1.2 Extract Table row

The most challenging portion of the system is how to estimate the amount of data extracted by the query. This estimation should be done before sending a query to the targeted database for execution. The problem has been solved by using the *oracle optimizer* to estimate the percentage of data extracted from the database. The key function of Oracle optimizer is to determine an execution plan for input queries that can be executed efficiently. It builds optimal planning using statistical information about data and some database futures such as hash join, parallel query, partition, etc. [11]. The execution plan of a query describes brief details that are necessary to process SQL statements.

The optimizer creates a tree-structured plan that is used to execute a query. The each node in the query plan tree will represent the operators used in the query. When the optimizer reads a query plan tree, it starts from the bottom left, and after working at that level, it leads upwards and starts working at the next level. The optimizer then scans each table for determining the number of rows returned by the query and finally returns the output of the query to the user. The optimizer will create one or more query plan tree for each query. The optimizer first evaluates each query plan and then chooses the execution plan, which has the lowest cost of execution.

The estimated percentage of data returned by the query from the database can be obtained by processing the query plan tree of the query. The module then analyses the query plan tree in a schematic way to find out the expected number of rows returned by the query. After this, the cardinality of each table used in the query is calculated, which is the number of rows coming back from the table after executing the database operation on that table. For each table used in the query, the estimated number of rows coming back from the table has been computed as: dividing the minimum rows coming back after executing database operation by the total rows available in the table.

# 4 Threat Detection

The main idea of the threat detection module is to create the profile of normal behaviour of users when they interact with the database. These created profiles are then used to detect miscellaneous behaviour or abnormal data extraction by the user when they communicate with the database during the detection stage. If there are signs of threat during the detection phase, the module will take some necessary action against the user who communicates with the database.

The key modules of the threat detection process are the Threat Detector (TD), the signature creator and the signature evaluator. The data being secured is placed in the oracle database. The scrutinized user communicates with the Oracle database by using queries written in SQL. These queries are dispatched to the TD for potential threat detection. When Threat Detector module receives first query from the user, it loads all the query signatures of user profile, which are created by the profile manager. The TD then transfers the input query to signature creator to generate the corresponding query signature. Next, the signature creator generates the query signature which contains various information such as command used in the query, IDs of table used in the query, IDs of attributes from each table used in the query and extract table row. The signature creator forwards the signature of current query to the signature evaluator to compare the current query signature with previously generated query signatures of user profile.

## *4.1 Signature Evaluator (SE)*

The main function of the signature evaluator sub-module is to match the current query signature with the previously generated query signature. $Q_j$ stores the first query signature of user $U_i$ and $I_q$ stores the signature of current input query received from signature creator. The SE compares the current query signature with all previously generated query signatures of user $U_i$. If current signature matches with any previously generated signatures, the query is found legitimate and safe for execution. If current query signature is not matched with any query signature then query is considered as threat and will not be sent to database for execution.

## 5  Conclusion and Future Work

This paper proposed the threat detection framework for securing the semantic Web services based on XACML to identify anomalous database transactions by the insiders. The proposed framework effectively identifies the abnormal data extraction by the legitimate insider. This framework creates the profile of each authorized user and evaluates each incoming query against the previously created profiles. If there is a mismatch, the framework will prevent the query from execution. This framework, first, projects the number of rows to be extracted from each table and compares the result with previously generated query signatures. The main advantage of this framework is to evaluate the query before being sent to the database for execution. If there is any abnormality in query execution, the framework will revoke the query from execution.

In future work, focus will be on the development of an algorithm for the detection of anomalous behaviour performed by a legitimate insider. In future, the proper implementation of the various modules of the proposed framework will be addressed as well. The attention will also be given on defining some important parameters for assessing security and evaluating the effectiveness of proposed framework in semantic Web domain.

**Acknowledgements**  This work was supported by Integral University with Manuscript Communication Number (MCN)—IU/R&D/2019-MCN000540.

## References

1. P. Rodriguez-Mier, C. Pedrinaci, C. Lama, M. Mucientes, An integrated semantic web service discovery and composition framework. IEEE Trans. Serv. Comput. **9**, 537–550 (2016)
2. X. Li, Y. Xue, B. Malin, Detecting anomalous user behaviors in workflow-driven web applications. in *IEEE 31st Symposium on Reliable Distributed Systems* (IEEE, Irvine, CA, USA, 2012), pp. 1–10

3. Symantec internet security threat report, http://www.symantec.com/business/theme.jsp?themeid=threatreport
4. M. Le, A. Stavrou, B.B. Kang, Doubleguard: detecting intrusions in multitier web applications. IEEE Trans. Dependable Secure Comput. **9**, 412–425 (2012)
5. A. Sallam, E. Bertino, S.R. Hussain, D. Landers, R.M. Lefler, D. Steine, DBSAFE—an anomaly detection system to protect databases from exfiltration attempt. IEEE Syst. J. **11**, 483–493 (2017)
6. S. Lau, R. Zamani, W. Susilo, A semantic web vision for an intelligent community transport service brokering system. in *IEEE International Conference on Intelligent Transportation Engineering* (IEEE Press, Singapore, 2016), pp. 172–175
7. J. Jovanovic, E. Bagheri, Electronic commerce meets the semantic web. IEEE Comput. Society. **18**, 56–65 (2016)
8. L. Bossi, E. Bertino, S.R. Hussain, A system for profiling and monitoring database access patterns by application programs for anomaly detection. IEEE Trans. Software Eng. **43**, 415–431 (2017)
9. Y. Cheng, J. Park, R. Sandhu, An access control model for online social networks using user-to-user relationships. IEEE Trans. Dependable Secure Comput. **13**, 446–453 (2016)
10. OASIS eXtensible Access Control Markup Language. https://www.oasis-open.org/committees/tc_home.php?wg_abbrev=xacml
11. Oracle White Paper. https://www.oracle.com/technetwork/database/bi-datawarehousing/twp-explain-the-explain-plan-052011-393674.pdf

# Real-Time Detection of Fake Account in Twitter Using Machine-Learning Approach

Somya Ranjan Sahoo and B. B. Gupta

**Abstract** In day-to-day life for sharing of information and communication among others, online social networking has emerged as one of the eminent ways. Online social networking offers attractive means of social interaction and communication but it raises various concerns like privacy and security of user and its account. During the events, people share unreliable information and misleading content through social media platform such as Twitter and Facebook. Our objective is to detect the fake account by analyzing different characteristics that spread malicious contents in real-time environment. Fake profiles are the identity theft of genuine user's profile content and create a similar profile using users credential. At later stage, the profile is perverted for aspersing the legitimate profile owner as well as sending friend request to the user's friend. In this article, we discuss about our chrome extension-based framework that detects the fake accounts in Twitter environment by analyzing the different characteristics.

**Keywords** Fake profile · Identity theft · Online social networks · Chrome extension

## 1 Introduction

Online social networks (OSNs) such as Twitter [1], Facebook [2], LinkedIn [3], and Google+ [3] are becoming quite necessary platform for communication and information exchange among people, users of similar background, career interests, activities, or real-life connections in day-to-day life. Due to the popularity of OSN, it creates a new weapon and arena for cyber-crime. Fake accounts on OSN are labeled by its generality, interference, and trespassing with the WebPages key functionality. Fake accounts may cause of falling the credibility and reputation of these websites due to the militant's activity in social network, which may further harm their preface of a

S. R. Sahoo · B. B. Gupta (✉)
Department of Computer Engineering, National Institute of Technology, Kurukshetra, India
e-mail: gupta.brij@gmail.com

S. R. Sahoo
e-mail: somyaranjan.sahoo@gmail.com

© Springer Nature Singapore Pte Ltd. 2021     149
X.-Z. Gao et al. (eds.), *Advances in Computational Intelligence and Communication Technology*, Advances in Intelligent Systems and Computing 1086,
https://doi.org/10.1007/978-981-15-1275-9_13

pleasurable, secure, and alternative online existence. The existence of fake account can gradually decrease the experience of user's attraction on the OSN sites. The fake accounts can replace the content of genuine user account and communicate with friends of that account with harmful contents and convert the user accounts with their own experience to degrade the credibility of the user. Online social networking sites like Twitter is a microblogging site capable of spreading news and social networking services. For instance, Twitter owned 330 million active users as on January 2018 according to the report by Statista [4]. That report also suggests currently the user in India exceeds 30 million. According to the report by omnicore agencies, more than 500 million tweets can tweet by 100 million users per day. Due to the popularity, the business holder and kidders spread malicious contents in the network by using the services like tagging and short URLs. Modern statistics report indicated that Twitter is the desired target for fake account attacks, worm and spammer attacks due to the popularity. However, various defensive mechanisms have been proposed and investigated for relieving the impression of fake account attacks and spammer detection. For spreading malicious contents in the form of tweet, URLS, and other messages, Twitter platform spread spam messages in the form of direct message service or advertising media content. In addition to the above scenario, fake account follows the nameless Twitter users and sends fraudulent messages containing hash tag to obtain high exposure. Most of the time, fake profile creator creates the fake accounts by using the content of the genuine user that is available publicly on the social sites. Many times fake accounts may use embedded URLs to spread phishing attacks to still personal information of the user.

In this paper, we construct and evaluate an automated fake profile detector that detects the malicious content and account in chrome environment to protect the Twitter user account from attackers. We compare the different activities by using machine-learning approach to validate the user data gathered by using crawler and manual approach.

The remainder of our paper structured as follows. In Sect. 2, we describe the previous research solutions (related work) to detect fake account in Twitter. In Sect. 3, we describe the proposed framework-based data collection and analysis by our crawler and manual approach. We elaborate the different features and modeled the trained data in Sect. 4. In the next section, i.e., in Sect. 5, we describe the classification result with fake detection by using our chrome extension. In the final section, i.e., in Sect. 6, we conclude our work with feature directions.

## 2 Related Work

The demand of Twitter online social network has appealed the aid of security researchers. Since, these networks are trust based, people share their content with other users. Due to this, it raises significant issues in the network by the attackers in different scenarios. In the recent era, a lot of work was done to detect the fake activities by detecting fake accounts in Twitter environment. This portion of work furnishes an overall description of some related work with different approaches and methods that detect fake accounts. Amleshwaram et al. [5] used machine-learning approach to extract the features from different Twitter accounts based on the user behavior patterns like tweets and followers. The performance was analyzed based on the precession and recall of different features. But the generated accuracy was not so good due to the less number of features used. According to Dhingra et al. [6], fake profile creator spread malicious content in the form of tweets that contain the same malicious URLs in multiple tweets by the legitimate user. He also used some machine-learning classification techniques to classify the malicious URLs and genuine URLs. Other new features are added by Yang et al. [7], called bidirectional link ratio, to improve the reliability of the features like followers to following. This feature is very difficult to evade by the spammer. Yardi et al. [8] studied spam accounts features based on their behavior on Twitter and recognize that they behave differently in terms of number of followers, followings, rate of posting tweets, and so on from normal accounts. Wang et al. [9] proposed a spammer detection system for detecting fake user accounts which separate the legitimate behavior form fake accounts by implementing machine-learning classifier with the precision of 89% with Naïve Bayesian classifier.

Stringhini et al. [10] further examined spam user features by creating a dataset of more than 900 user profiles from highly prominent OSNs (MySpace, Twitter, and Facebook) and gather some feature/characteristics like number of friend, message sent in the form of tweet of direct message, similarity index of message, URL shared, and followee to follower ratio for identifying fake accounts on social networks. Lee et al. [11] proposed a social honey pots mechanism for identifying the spammer accounts on different social network platforms without human inspection for identification. The basic drawback of this mechanism is that for building a satisfactory dataset, it needed very long time. Venkatesh et al. [12] proposed a solution to detect the fake accounts by analyzing the short URLs in Twitter account. Based on the user activities, the profiles are decided to be genuine or fake. Aswani et al. [13] proposed a hybrid solution for identifying spammer accounts in Twitter by applying content spamming and the profile that enlists in spamming activity. Alowibdi et al. [14] proposed a framework-based solution for detecting deception in Twitter account by analyzing user data and combining multiple characteristics; he generated an accuracy measurement for detecting deception. In summary, the fundamental concept of detecting malicious account of the existing approaches using machine-learning approaches with public features does not provide good rate of detection. But our proposed framework is highly effective for detection of fake accounts in Twitter with public as well as private features of user accounts.

## 3  Proposed Framework for Fake Profile Identification

For our experiment, the basic steps for analysis are to collect and gather the Twitter dataset from different Twitter profiles by using our web crawler and Twitter API. The

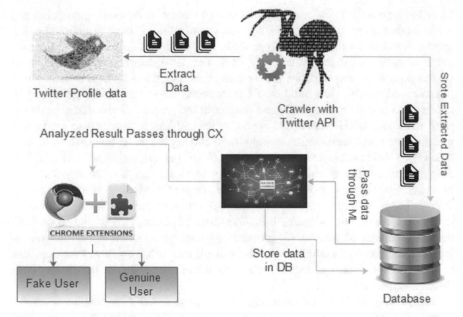

**Fig. 1** Framework for data extraction and process

gathering of data from different profiles is too critical in the current scenario due to the security principle used by different OSNs. We collected the public information that is shared by different users and also some personal information using manual approach by sending request. We collected more than 6500 Twitter user information with some malicious account that spreads malicious content in the form of tweet and URLs. The process of data collection and operation is shown in Fig. 1. On the basis of privacy and security purpose, we are not disclosing the details of user data. We collected profile information like profile id, profile name, status count, follower count, friends count, favorite count, listed count number of URLs in the tweet, and other profile contents. Most of the time the fake account holder spread malicious URLs in his/her tweet to gain the user personal information. The details of collected information about the users are described in the next section. Our goal is to extract the user information from various Twitter accounts and analyze the activity related to that account. For each user in the dataset, we collected followers, followings, friend request sent and accepted, number of tweets, and other spatial information.

## 4 Feature Selection and Analysis

The basic objective of creating fake profiles in Twitter is for financial benefit and degrades the performance of Twitter user, flow of information in the network. Since the characteristics and behavior of legitimate user and fake profile are different, we believe that to achieve their goal the behaviors of such account vary according to their uses like interaction rate with other users. In general, the interaction of the legitimate user with other users likes followers more active on Twitter. In order to identify the characteristics and behavior of different accounts in Twitter environment and classify them as fake or legitimate account, we extracted and examined tremendous set of characteristics and features from different Twitter profiles and analyzed the content to identify the actual motive of the user. On the basis of this analysis, in this section, we introduce a set of features from each Twitter user profile for the purpose of detecting fake accounts and separated from legitimate one.

## 4.1 Feature Analysis

In our proposed framework, we analyze different features and generated some trust score for detecting malicious account based on the performance of the account. The different features used for our analysis are shown in Table 1. After analyzing all these features, we generate a trust score that passes to our chrome extension to detect the fake accounts.

**Table 1** Different features used in our classifications

| Feature | | About |
|---|---|---|
| Profile ID ($P_{id}$) | | Twitter provides a unique profile ID to every user |
| Tweets ($U_{twt}$) | Total tweets $= (U_{twt}^{total})$ | Number of tweets posted by user from creation of account |
| | Average tweets $= (U_{twt}^{avg})$ | |
| Follower ($T_{follower}$) | Total follower $= (T_{follower})^{total}$ | Number of user following some specific ID |
| | Average follower $= (T_{follower})^{avg}$ | |
| Following ($T_{following}$) | Total following $= (T_{following})^{total}$ | Particular user following the number of people |
| | Average following $= (T_{following})^{avg}$ | |
| Likes ($T_{likes}$) | Total likes $= (T_{likes})^{total}$ | The number of tweets the user like |
| | Average likes $= (T_{likes})^{avg}$ | |
| Lists ($T_{lists}$) | Total lists $= (T_{lists})^{total}$ | List is the curated group of Twitter account We calculate the total subscribe lists of every user |
| | Average lists $= (T_{lists})^{avg}$ | |
| URL count ($T_{url}$) | Total URLs $= (T_{url})^{total}$ | We calculate the total number of URL tweet by every user |
| | Average URLs $= (T_{url})^{avg}$ | |
| User location ($T_{location}$) | | It shows the current geographic location of the user |

## *4.2   Modeling the Generated Dataset*

To model the generated dataset, we dissevered the data into two different groups called training data and testing data in the ratio of 80:20. All the trained data and testing data are passes through a machine-learning platform called WEKA. In that environment, we process the data by using different classification techniques and the classifier inspects the data for detecting malicious content.

## 5   Classification Results and Analysis

The accuracy and the different measures that are used to evaluate the performance of our chrome extension are shown in Table 2. The WEKA platform provides the result analysis in the form of confusion matrix. If the evaluated results give higher closer value near to 1, the observation satisfied with proper classification. In our observation, random forest and bagging approach gives satisfactory analysis with 99.4% as comparison to other classifiers in terms of receiver operating characteristics (ROCs) and correctly classified instance. The comparison-based analysis graph for our result is shown in Fig. 2.

## *5.1   Fake Detector Result Based on Chrome Extension*

The objective of our analysis is to detect the fake profile in Twitter by analyzing the characteristics and user behavior shown in Fig. 3. Our chrome extensions analyze the characteristics of different users and generate the result based on the suspiciousness by using the trust score. We calculated the trust score by using Beth et al. [15] formula which calculated the trust score of each node by analyzing the characteristics in online social network like Twitter by using Eq. (1).

$$\text{Transaction } T^n(\text{Node}^i, \text{Node}^j) = (1 - \alpha)^n \tag{1}$$

where $T^n$ ($\text{Node}^i$, $\text{Node}^j$) is the trust value of node $i$ for node $j$. Alpha ($\alpha$) is the communication and information sharing between different profiles in Twitter in a certain interval of time, and n is the number of successful communication of node $i$ with node $j$.

**Table 2** Performance analysis based on different classifier

| Different classifications | TP (true positive) rate | FP (false positive) rate | Precision | Recall | F-measure | MCC(Matthews correlation coefficient) | ROC area | PRC area | Correctly classified instance |
|---|---|---|---|---|---|---|---|---|---|
| Random Forest | 0.994 | 0.011 | 0.989 | 0.994 | 0.992 | 0.983 | 0.999 | 0.999 | 99.16 |
| Bagging | 0.994 | 0.012 | 0.989 | 0.994 | 0.991 | 0.982 | 0.998 | 0.998 | 99.12 |
| JRip | 0.992 | 0.014 | 0.986 | 0.992 | 0.989 | 0.978 | 0.991 | 0.986 | 98.88 |
| J48 | 0.991 | 0.013 | 0.988 | 0.991 | 0.990 | 0.979 | 0.991 | 0.987 | 98.93 |
| PART | 0.987 | 0.012 | 0.988 | 0.987 | 0.988 | 0.975 | 0.993 | 0.991 | 98.75 |

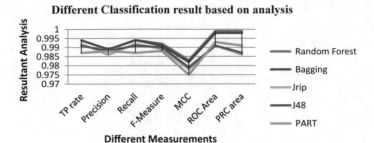

**Fig. 2** Comparison-based analysis graph

**Algorithm:** Trust score calculation
**Input**: Number of profile contents
**Output**: Trust score for each user profile
1.       Initial =0
2.          For(i=0; i<N;i++)
3.             {
4.                If(n>0)
5.                {
6.                   $T^n$ (Node$^i$, Node$^j$) = $(1-\alpha)^n$
7.                   goto step 12.
8.                } else
9.                {
10.                  $T^n$ (Node$^i$, Node$^j$) = 0
11.               }
12.            Initial = initial + ($T^n$(i , t)* Tn(t, j))/N
13            }
14       return initial

If the calculated trust score is higher than the threshold, then the profile activity is satisfactory; otherwise, the chrome extension displays a popup box with message 'Profile type—Fake' as shown in Figs. 4 and 5.

# 6 Conclusion and Future Work

In this paper, we have proposed a framework for detecting malicious profile and behavior of the Twitter account by using chrome extension-based machine-learning analyzer. The chrome extension evaluates the performance of any Twitter account and generates a trust score for detecting fake accounts with high accuracy. A major benefit is that it reduces the learning time and provides good scalable result by analyzing the features. To generalize the performance of our chrome extension, we use random

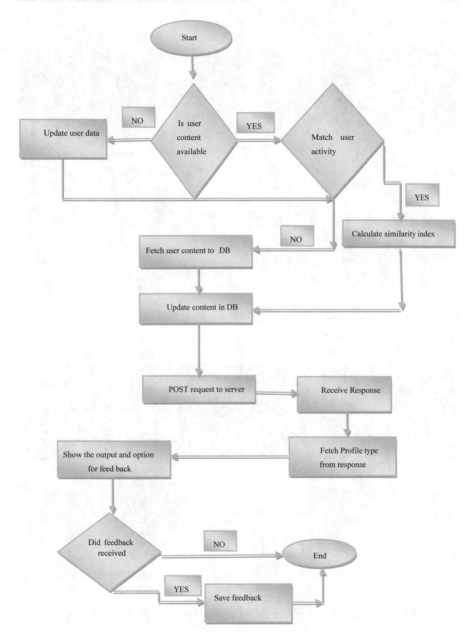

**Fig. 3** Flow control of detection on chrome extension

**Fig. 4** Fake account detected as per identity clone attack

**Fig. 5** Fake account detected as per its behavior

forest as well as bagging method for detection of fake and troll profiles. Some of the attributes may be pleonastic or pay slight to the identification task. Therefore, as a future work more features are required for better identification of Fake accounts by our chrome extension.

**Acknowledgements** This research work is funded by the Ministry of Electronics and Information Technology, Government of India, under YFRF scheme and under the project 'Visvesvaraya PhD Scheme' which is implemented by Digital India Corporation.

# References

1. Twitter. Available at: https://twitter.com/
2. Facebook. Available at: https://www.facebook.com/
3. LinkedIn. Available at: https://in.linkedin.com/
4. Statista report. Available at: https://www.statista.com/statistics/282087/number-of-monthly-active-twitter-users
5. A.A. Amleshwaram, N. Reddy, S. Yadav, G. Gu, C. Yang, Cats: characterizing automation of twitter spammers, in *Proceedings of the 5th International Conference on Communication Systems and Networks (2014)*, vol. 1, pp. 86–95 (Springer, Berlin, 2014)
6. A. Dhingra, S. Mittal, Content based spam classification in Twitter using multi-layer perceptron learning. Int. J. Latest Trends Eng. Technol. 5(4) (2015)
7. C. Yang, R.C. Harkreader, G. Gu, Die free or live hard? Empirical evaluation and new design for fighting evolving twitter spammers, in *RAID 2011*, ed. by R. Sommer, D. Balzarotti, G. Maier, LNCS, vol. 6961, pp. 318–337 (2011)
8. S. Yardi, D.M. Romero, G. Schoenebeck, D. Boyd, Detecting spam in a twitter network. First Monday **15** (2010)
9. A.H. Wang, Don't follow me: spam detection in twitter, in *Security and Cryptography (SECRYPT), Proceedings of the 2010, International Conference on Security and Cryptography, Athens, Greece*, July 26–28, 2010, pp. 142–151 (2010)
10. G. Stringhini, C. Kruegel, G. Vigna, Detecting spammers on social networks, in *Proceedings of the 26th Annual Computer Security Applications Conference (New York, NY, USA, 2010)*, ACSAC'10, pp. 1–9 (ACM, New York, 2010)
11. K. Lee, J. Caverlee, S. Webb, Uncovering social spammers: social honeypots + machine learning, in *Proceedings of the 33rd International ACM SIGIR Conference on Research and Development in Information Retrieval*, pp. 435–442 (ACM, New York, 2010)
12. R. Venkatesh, J.K. Rout, S.K. Jena, Malicious account detection based on short URLs in Twitter, in *Proceedings of the International Conference on Signal, Networks, Computing, and Systems. Lecture Notes in Electrical Engineering*, ed. by D. Lobiyal, D. Mohapatra, A. Nagar, M. Sahoo, vol 395 (Springer, New Delhi, 2017), pp. 34–41
13. R. Aswani, A.K. Kar, P. Vigneswara Ilavarasan, Detection of spammer in Twitter marketing: a hybrid approach using social media analytics and bio inspired computing. Inf. Syst. Front. **20**(issue 3), 515–530 (2017)
14. J.S. Alowibdi, U.A. Buy, S.Y. Philip, S. Ghani, M. Mokbel, Deception detector in Twitter. Soc. Netw. Anal. Min. **5**(1), 1–16 (2015)
15. T. Beth, M. Borcherding, B. Klein, Valuation of trust in open networks, in *ESORICS 94, Brighton, UK*, 2–9 Nov. 1994

# DPLOOP: Detection and Prevention of Loopholes in Web Application Security

Monika and Vijay Tiwari

**Abstract** Web application loopholes are related to different components. Defeat correctly sanitized users' given input is one of the prominent features that accompany to run illegal snippets in such type of programs. Due to the absence of proper input sanitization, common loopholes occur in web applications, such as SQL, Cross-site Scripting (XSS), XML, CSRF, and LDAP. Thus, research work presented in this paper deliberates possible methods to detect and mitigate vulnerabilities in order to prevent organizational websites against SQL and XSS loopholes. We have analyzed a dataset of URLs. SQL, XSS, and XML have the highest rate of detection and the least percentage of CSRF.

**Keywords** Web loopholes · SQLi · XSS · XML · Input validation

## 1 Introduction

The security of web application is always a significant and crucial task to protest different types of attacks like users' data stealing, password hacking, session hijacking, etc. The evolution of web attack is contingent upon continuous growth of web applications and web users. The popularity of applications attracts and invites hackers toward them. Nowadays, securing and maintaining the websites against attack are an arduous and challenging task—detection of loopholes in a web application, computer system, or network and exploiting them called hacking [1]. New approaches for web application attacks arise day by day, so the study of detection and prevention against web application attacks and the discovery of a solution is an integral part of the Internet world.

The ubiquity of the World Wide Web (WWW) makes websites and their visitor's attractive targets for various classification of cybercrime encompass data breaches, spear-phishing campaigns, ransomware, and fake technical support scams. As stated

Monika (✉) · V. Tiwari
CAS, Dr. APJ Abdul Kalam Technical University, Lucknow, India
e-mail: monikaahalawat07@gmail.com

V. Tiwari
e-mail: vktiwari@cas.res.in

© Springer Nature Singapore Pte Ltd. 2021                                                    161
X.-Z. Gao et al. (eds.), *Advances in Computational Intelligence and Communication Technology*, Advances in Intelligent Systems and Computing 1086,
https://doi.org/10.1007/978-981-15-1275-9_14

by Symantec's most begun Internet Security Threats are reported, more than 229,000 attacks against websites are increasing day by day, and more than 76% of sites contain unpatched loopholes.

Thus, research work presents some prominent and primary used web vulnerabilities, which may enlarge the security web exploits to increase the rate of loophole flow. These vulnerabilities are exploited by the attacker to steal the users' data. The OWASP 10 web application security threats were upgraded in 2017 to present instructions for the developers, and security experts for the guidance on the most significant loopholes in web applications are commonly found, easy to bypass them [2]. OWASP top 10 web loopholes are harmful because these loopholes are easily breached by attackers to implant malware, steal information of the users', or completely control your computers or web servers. To commit breaches, web application loopholes are currently the most recurrent pattern. These loopholes are extensively received as for being exploited, and rectifying them will significantly reduce your threat of infringement [3].

The essential contribution of this paper is to detect loopholes on a large dataset of organizational websites. Through the detection process, detect the presence of relative gaps.

Another significant contribution of the research work presented is the highest percentage of existing loopholes. The proposed model has been tested on web dataset, and we can detect 95.77% of SQL injection attacks, 89.02% of XML attack, and 83.54% of Cross-site Scripting attack.

Organization of this paper is as follows: The web loopholes is explained in Sect. 2. Methodology and data collection are described in Sect. 3. Results analysis presented in Sect. 4. Solutions for secure loopholes are defined in Sect. 5. Finally, conclusion of the work is defined in Sect. 6. And the future research directions are explained in Sect. 7.

## 2   Web Loopholes

### 2.1   OWASP Top 10 Web Application Security Threats

In this proposed paper, we will go through the top security loopholes according to the Open Web Application Security Project (OWASP) [4] such as SQL injection, XXS, XML attacks, and many more. Even though the programmer detects lots of software security threats, we propose a center of attraction on the detection of the issues that are "commonly known." The OWASP top 10 loopholes are a registry of shortcoming, so they can prevent an extreme argument that no web application can dispatch to users' without the proper sanitization, these types of faults are not carried out by any software. OWASP top 10 web application security threats are recognized, and it proposed a solution and the best implementation to mitigate them.

**Table 1** CIA Triad

| CIA Triad | |
|---|---|
| Confidentiality | Confidentiality is identical to the privacy, which usually prevents the sensitive data information from the attackers. Access of confidential data must be restricted for the attacker/unauthorized persons to maintain the security measures of the web. Network authentication, data encryption, and encapsulation are the primary example of confidentiality |
| Integrity | To maintain the integrity of the web, it contains the three parameters; accuracy, trustworthiness, and consistency of data. Integrity ensures that the data is same when transit, as per the sender and receiver's knowledge, and an unauthorized person cannot alter the data. To maintain integrity, some data may include cryptographic techniques like checksum |
| Availability | Availability is the third pillar of information security, it is not come up under the direct attention of web security, but still, it is essential to overcome the security flaws |

The top known server and application security risks regularly found online. OWASP 10 easy-to-handle security loopholes which are facing by site owners and web experts [5].

As per the security concern, Table 1 shows the three pillars—confidentiality, integrity, and availability that are also known as "CIA Triad."

Focus on the suggested approach, and question yourself that your site is frankly secure or not. (Suggestion: whenever you connect with the CMS "security" plugins that means you're not safe!)

### A1-Injection

Injection attacks, like SQL injection, Carriage Return and Line Feed (CRLF) injection, and Lightweight Directory Access Protocol (LDAP) injection, arise when an attacker sends some malicious snippet to an analyst that is executed as a command without any proper authentication. An injection of code happens when an attacker sends invalid data to the web application with the aim to make it do something unauthenticated from what the app was designed/programmed to do [6].

Maybe the most basic example of this security loophole is the SQL query takes untrusted data. You can see one of the SQL injection examples below:

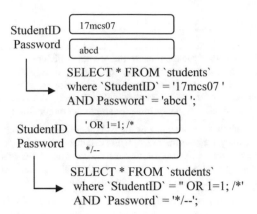

SQL injection example: To validate the username and password used "Basic SQL Injection" query "Request.Form" accepted by the user-supplied data as a query string.

**$query = "SELECT StudentID FROM `login` where `username` = '" + Request.Form("username") + "' AND `password` = '" + Request.Form("password") "'";**

SQL query construction, the attacker can inject malicious code as input and form an original SQL query statement executed by the attacker and some additional actions taken by the attacker [7]. In the SQL mentioned above statements, whenever the user enters the following

**`username`="admin' or 1=1--" AND `password`=" P@ssword1923"**

Let us see the original query

**SELECT ID FROM `Login` where `username`='admin' or 1=1 --' AND `password`='P@ssword1923'";**

The logic of the SQL query is changed because of malicious user-supplied data. 'OR 1=1 --', the 'OR 1=1' is returned as true and '--' as consider as a comment in the executed query and remains the analyst does not implement a portion of password='P@ssword1923'"of SQL query. In SQL query implementation, database will always return the first entry of the user login table in the stored database (in most of the examples, it will be an "admin" user), login credentials of an admin attacker will easily enter into the application. In primary SQL injection, the application retrieved the values as an input and given the responsibility as content. An attacker can identify its existence if comparing existing output for given standard input (malicious code as an input) [8].

SQL injection through SQL query:

```
$query = "SELECT * FROM `students` where `studentID` = '" +
request.getParameter("id") + "'";
```

This query can be exploited by calling up the web page executing it with the following URL: http://example.com/app/studentsView?id='or'1'='1, the return of all the rows stored on the database table.

The core of a code injection loopholes is the lack of validation and sanitization of the data consumed by the web application, which means that this loopholes can be present on almost any type of technology [9].

Anything that acquires specification as input can be vulnerable to a code injection attack. Injection loopholes can smartly be detected by web application security testing. To mitigate injection threats, experts should use parameterized query statements.

### A1.1 Lightweight Directory Access Protocol (LDAP) Injection

LDAP servers contain the vital information that is retrieved by the clients within the LDAP sessions. The most common operation that taken once in the whole session is launch for some actions like adding, deleting, and modifying entries. Other known activities that regularly perform, it includes:

1. Bind—To specify and validate LDAP and the LDAP protocol version.
2. Search—To retrieve and discover entries of LDAP directory.
3. Compare—To test that a names entry holds given attribute value.
4. Extended Operation—It contains an operation that is used to define a novel one.
5. Unbind—close the session.

LDAP injection examples:

In a user search field, the following code script is managed by the authorized user input value, and it produces an LDAP query statement that will be used in the LDAP database repository [10].

```
<input type="text" size=25 name="UserName">Insert the username</input>
```

The LDAP query statement is shrunk the execution time and the snippet for the following function:

```
String ldapSearchQuery = "(cn=" + $UserName + ")";
System.out.println(ldapSearchQuery);
```

If the mentioned variable $UserName is not properly authenticated, it could be done to execute an LDAP injection, as follows:

If an authenticated user puts "*" on the search field, the form may return all the user objects on the LDAP database. In a login form, for example, the following vulnerable code could be used:

If there is no input validation, the user can easily insert the following injection code in the "user" and "pass" fields:

```
user = *)(uid=*))(||(uid=*
pass = password
```

The search field would return the first user in the LDAP tree and allow the attacker to bypass or exploit authentication:

```
searchlogin="(&(uid=*)(uid=*))(||(uid=*)(userPassword={MD5}X03MO1qnZdYdgyfeuI
LPmQ==))";
```

## A2 Broken Authentication and Session Management

Amiss formation of the authenticated user and authenticated session could enable the attackers [10] to steal the passwords, authorized key, or session token authentication or grab charge of users' valid accounts to access their private information [11].

Multi-factor authentication reduced the threat of compromised accounts such as Fast ID Online (FIDO) or related apps.

## A3 Sensitive Data Exposure

Such type of applications and Application Program Interfaces (APIs) that do not appropriately cover confidential information like economic information, UserID, passwords or health-related data, could authorize attackers to access such type of information to execute unethical activities or steal valid id and personal information.

## A4 XML External Entity

Imperfectly formation of XML processors to analyze external entity recommends that lie within the XML documents. Attackers can access XML external entities for performing these types of attacks, and they include remote execution, and to reveal confidential files and shares Server Message Block (SMB) file [12].

Often, applications need to be obtained, current process and XML documents from users. Within XML documents, imperfectly configured XML documents can authorize as an XML attribute known as an external entity reference, and the attacker will insert the malicious content in the original file. An attacker can misuse these credentials to access internal system, read the confidential informations, and even shut down the application in a Denial-of-Service (DoS) attack [13].

XML document example is given below:

```
<!ENTITY xxe SYSTEM "file:///etc/passwd">]>&xxe;
```

Static application security testing (SAST), it can uncover these types of problem by inspecting format and its dependencies.

## A5 Broken Access Control

Amiss formation or miss validation allows an attacker to access the authorized users' data and approved accounts of them; they can obtain their rights to steal the confidential documents.

## A6 Security Misconfiguration

It introduces to the inappropriate execution of authority deliberates to maintain the security log of application, like misconfiguration of security headers, the error message that contains sensitive information (information breach), and not upgraded patches, structure, and elements [14].

Dynamic application security testing (DAST) can quickly recognize misconception, like cracked Application Program Interfaces (APIs).

## A7 Cross-Site Scripting

Cross-site Scripting (XSS) shortcoming gives attackers the capability to inject malicious script on client-side application, for example, to deviate user's page to malicious web applications. These types of attacks are classified into two categories: the one is persistent (or stored), and another is non-persistent (or reflected). Further, it is categorized into three types of XSS attacks; the third one is called as DOM-based XSS [15, 16]. The implementation process of XSS: the victim's browser receives the malicious link, and it automates the HTTP request to www.web.com and further receives mentioned HTML script. Then, the browser emerges translating the HTML script into Document Object Modeling (DOM) [17, 18]. To recognize an exploit, it is regularly studied in weeks or months. Unauthorized logging and inefficient combination with security [19] event systems response allow attackers to access to other systems response and keep undetermined threats.

## A8 CSRF (Cross-Site Request Forgery)

In one-click attack or CSRF, an attacker takes the victim's browsers credentials and forged HTTP request send by the attacker to an original website and start monitoring a state-changing activity at the web.

For example, whenever you enter into the login session on notable websites like timesofindia.indiatimes.com (an online news website), sbi.co.in (a famous banking website), etc., if a victim opens a malicious web page in browser which is controlled by the attacker, forged HTTP request is sent by the attacker to an original website and (1) victim's e-mail address hijacked by the attacker (2) attacker can easily transfer money from the victim's "(Some-bank)" bank account to its own account, etc. [20].

For example, $200 bank transfer through GET;

```
GET http://somebank.com/transfer.do?account=AnalystB&amount=$200 HTTP/1.1
```

Modified script by a hacker to transfer $200;

```
GET http://somebank.com/transfer.do?account=HackerA&amount=$200 HTTP/1.1
```

Further, without any known malicious activity behind it, a large number of bank customers circulate hyperlink through the e-mail, and they directly login into the bank account session and unintentionally originate the $200 transfer.

Note that, if an only POST request is using by the bank's website then using a <a href> tag it is impractical to frame malicious script. However, with the automatic execution of the embedded JavaScript, the attack can be delivered a <form> tag.

```
<body onload="documentation.forms.submit()">
    <form action="http://netbank.com/transfer.do" method="post">
        <input type="hidden" name="account" value="HackerA"/>
        <input type="hidden" name="amount" value="$200"/>
        <input type="submit" value="See my pictures!"/>
    </form>
</body>
```

## 3   Methodology and Data Collection

See Fig. 1.

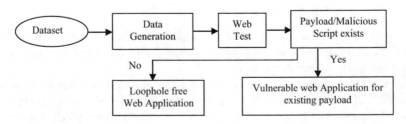

**Fig. 1**  Architecture of proposed model

A. *Data Generation*

The significance of the research, to analyze the organizational web applications in opposition to SQL injection, Cross-site Scripting, XXE, CSRF and LDAP vulnerabilities. Hence, the dataset [21] contains a large number of URLs. The case is relevant to the combinations of government organizations and individual organizations.

B. *Testing Methodology*

White-box and black-box testing are used to identify loopholes in web applications. In white-box testing, the formation of web application and framework is known by the analyst, but in the black box testing, the target web application is not identified by the analyst. To analyze the security measures and to contain relevant datasets of the web application using the black box testing. After generating the dataset, web testing is done by existing payloads/malicious script. If the payloads/malicious script are present in the web application, then the web application is vulnerable to the existing loopholes. Otherwise, web application is free from loopholes.

This work developed a loophole detection system by using python, shell, and PHP scripts and to identify SQL, XSS, XXE, and LDAP vulnerabilities in web applications. The proposed mechanism of a loophole in web application security and it supervise a test set on the relevant websites. All these have a high ratio of accuracy to detect SQL, XSS, XXE, CSRF, and LDAP loopholes.

# 4 Results Analysis

The web application security loopholes result in datasets of web applications show that the maximum rate of SQL injection loopholes and minimum rate of Cross-Site Request Forgery (CSRF) and rest vulnerabilities lie between the maximum and minimum range. The proposed model has been tested on web dataset, and we can also detect the data rate 95.77% of SQL injection attacks, 89.02% of XML attack, 83.54% of Cross-site Scripting attack, 77.44% of LDAP, and 68.6% of Cross-Site Request Forgery. Chart reflects our review on the dataset (Fig. 2).

The results signify that the web applications are superior to secure from XSS vulnerabilities than SQL injection loopholes. To verify that XSS loophole in web applications compromised users but, if an authorized user is trapped into the loopholes and an authorized user with a high user credential (i.e., administrator) entire web application is threatened. The validation of results obtained and compared with the existing OWASP 10 vulnerability.

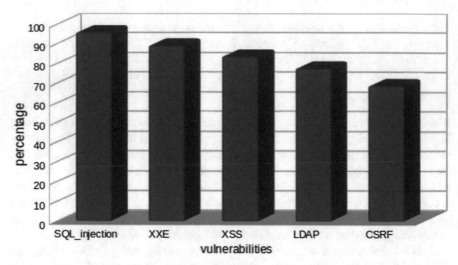

**Fig. 2** The percentage of most widespread loopholes

## 5 Solutions to Secure XSS Vulnerabilities

After identifying vulnerable web application, the initial step is to fix the detected loopholes. Therefore, we have proposed some techniques to remove the weaknesses which are recognized in the web:

(1) *Data Validation*

Web to running with validated data is known as data validation. After confirm receiving as valid data type every content of user's data should be organized robustly; it eliminates if it lies under the validation process.

*Example*, contact number validation, it must contain an only number (0–9).

(2) *Data Sanitization*

Modulate the data and remove additional tags which are present in it, is known as data sanitization process [22]. *Example*, form an everyday user's input text, and it must separate presented tags.

(3) *Output Encoding*

To convert all the information that has a particular definition to a client-side script translator into the description, quickly eliminate them known as output escaping/output encoding process. Before acceptance in web browser, skip all the raw data.

# 6 Conclusions

Research begins with by understanding the interconnection and complementary nature between privacy and security. We have a comprehensive structure of the web application loopholes or threats. OWASP 10 web application security threat gives a resource to identify and be attentive to the top ten web loopholes. Analyzed the advancement of existed web loopholes over the last decade, and also describe these loopholes where which construct widespread destruction for worldwide users. Collected five of the significant web loopholes indexed lie in OWASP 10, and smartly implant the technique to detect the loopholes and avoid them. This work developed a loophole detection system by using python and shell scripts. The proposed model has been tested web dataset, and we can identify 95.77% of SQL injection attacks, 89.02% of XML attack, and 83.54% of Cross-site Scripting attack. We later recommend the techniques to mitigate these loopholes.

# 7 Future Scope

As future work, we will try to demonstrate the rest loopholes and the mitigation techniques of all remains loopholes, and increase the size of the dataset.

# References

1. F.T. Pirvadlu, G. Sepidnam, Assessments Sqli and Xss vulnerability in several organizational websites of North Khorasan in Iran and offer solutions to fix these vulnerabilities, in *2017 3rd International Conference Web Research ICWR 2017* (2017), pp. 44–47
2. Y. Ruan, X. Yan, Research on key technology of web application security test platform, in *EMSS* (2018) pp. 218–223
3. A.M. Hasan, D.T. Meva, A.K. Roy, J. Doshi, Perusal of web application security approach, in *ICCT 2017—International Conference on Intelligent Communication and Computational Techniques*, vol. 2018 (2018), pp. 90–95
4. C. Borghello, Top 10-2017 Top 10. [Online]. Available: https://www.owasp.org/index.php/Top_10-2017_Top_10. Accessed: 25-Aug-2018
5. P.R. Kadam, *Stop Website Attack before It Attacks You—XSS And SQLi Detection*, vol. 2, no. 1 (2017), pp. 331–336
6. D.M. Varun, N.K. Mydhili, S. Dhrumil, *Major Web Application Threats for Data Privacy & Security–Detection, Analysis and Mitigation Strategies*, vol. 3, no. 7 (2017), pp. 182–198
7. M. Arianit, R. Ermir, J. Genc, Testing techniques and analysis of SQL injection attacks, in *2017 2nd International Conference on Knowledge Engineering and Applications,* vol. 91 (2017), pp. 399–404
8. A. Alzahrani, A. Alqazzaz, Y. Zhu, H. Fu, N. Almashfi, Web application security tools analysis, in *Proceedings 3rd IEEE International Conference on Big Data Security. Cloud, BigDataSecurity 2017, 3rd IEEE International Conference on High Performance Smart Computing. HPSC 2017, 2nd IEEE International Conference on Intelligence Data and Security* (2017), pp. 237–242

9. M.R.V. Bhor, H.K. Khanuja, Analysis of web application security mechanism and attack detection using vulnerability injection technique, in *Proceedings of 2nd International Conference on Computer and Communications, Control and Automation, ICCUBEA 2016* (2017)
10. J. Thome, L.K. Shar, D. Bianculli, L. Briand, An integrated approach for effective injection vulnerability analysis of web applications through security slicing and hybrid constraint solving. IEEE Trans. Softw. Eng. **5589**, 1–33 (2018)
11. M.M. Hassan et al., Broken authentication and session management vulnerability: a case study of web application. Int. J. Simul. Syst. Sci. Technol. **19**(2), 1–11 (2018)
12. S. Jan, C.D. Nguyen, A. Arcuri, L. Briand, A search-based testing approach for XML injection vulnerabilities in web applications, in *Proceedings of 10th IEEE International Conference on Software Testing, Verification and Validation, ICST 2017* (2017), pp. 356–366
13. I. Ilyas, M. Tayyab, A. Basharat, Solution to web services security and threats, in *2018 International Conference on Computing, Mathematics and Engineering Technologies Inven. Innov. Integr. Socioecon. Dev. iCoMET 2018*, vol. 2018 (2018), pp. 1–4
14. R.P. Adhyaru, Techniques for attacking web application security. Int. J. Inf. Sci. Tech. **6**(1/2), 45–52 (2016)
15. T.A. Taha, M. Karabatak, A proposed approach for preventing cross-site scripting, in *6th International Symposium on Digital Forensic and Security ISDFS 2018—Proceeding*, vol. 2018 (2018), pp. 1–4
16. M.R. Zalbina, T.W. Septian, D. Stiawan, M.Y. Idris, A. Heryanto, R. Budiarto, Payload recognition and detection of Cross Site Scripting attack, in *2017 2nd International Conference on Anti-Cyber Crimes, ICACC 2017* (2017), pp. 172–176
17. A.W. Marashdih, Z.F. Zaaba, Detection and removing cross site scripting vulnerability in PHP web application, in *2017 International Conference on Promising Electronic Technologies ICPET 2017* (2017), pp. 26–31
18. I. Dolnak, Content Security Policy (CSP) as countermeasure to Cross Site Scripting (XSS) attacks, in *ICETA 2017—15th International Conference on Emerging eLearning Technologies and Applications* (2017)
19. V. Dehalwar, A. Kalam, M.L. Kolhe, A. Zayegh, Review of web-based information security threats in smart grid, *2017 7th International Conference on Power Systems, ICPS 2017* (2018), pp. 849–853.
20. A. Sudhodanan, R. Carbone, L. Compagna, N. Dolgin, A. Armando, U. Morelli, Large-scale analysis & detection of authentication cross-site request forgeries, in *Proceedings—2nd IEEE European Symposium on Security and Privacy (EuroS&P) 2017* (2017), pp. 350–365
21. "OpenPhish." [Online]. Available: https://openphish.com/. Accessed: 07-Sept-2018
22. A. Nair, P. Chame, S. Gaikwad, S. Ethape, P.S. Agarwal, Prevention of Cross Site Scripting (XSS) and securing web application atclient side. Int. J. Emerg. Technol. Comput. Sci. **3**(2), 83–86 (2018)

# Security Enhancement Using Modified AES and Diffie–Hellman Key Exchange

Y. Bhavani and B. Jaya Krishna

**Abstract** In today's world, providing data security is a primary concern. For this purpose, many researchers have introduced asymmetric and symmetric algorithms to ensure security. But they are not resistant to many attacks. In this paper, we combine symmetric and asymmetric techniques to provide more security. Advanced Encryption Standard algorithm is modified by generating Dynamic S-Boxes (DS-Boxes) to provide a better attack-resistant algorithm. In our approach, Diffie–Hellman is used to generate and exchange both keys and random numbers. These random numbers create DS-Boxes used in Modified AES. The proposed algorithm is resistant to timing attacks, linear, and differential cryptanalysis attacks due to the usage of DS-Boxes.

**Keywords** AES · Diffie–Hellman key exchange · Cryptography · Dynamic S-Boxes · Attacks

## 1 Introduction

Cryptology is a vast subject where its primary objective is to provide security to users. It is a combination of cryptography and cryptanalysis. Cryptography concentrates on securing data through encryption and decryption techniques using a secured key. Cryptanalysis concentrates on performing decryption but without knowing the key. In cryptography, the main basis is on three terms encryption, decryption, and key exchange.

Encryption is a process of converting a known language text to unknown or unreadable text using an algorithm and a key. Decryption is a process of converting an unreadable or unknown text to a known text by using the same algorithm (reversible)

Y. Bhavani (✉) · B. Jaya Krishna
Department of Information Technology, Kakatiya Institute of Technology and Science, Warangal, India
e-mail: yerram.bh@gmail.com

B. Jaya Krishna
e-mail: jkbjk549@gmail.com

© Springer Nature Singapore Pte Ltd. 2021          173
X.-Z. Gao et al. (eds.), *Advances in Computational Intelligence and Communication Technology*, Advances in Intelligent Systems and Computing 1086,
https://doi.org/10.1007/978-981-15-1275-9_15

used for encrypting the text. Key exchange is a process of exchanging keys between sender and receiver which are used for encryption and decryption [1].

There are many types of cryptographic techniques, namely symmetric key cryptography, asymmetric key cryptography, and Hash function cryptography. Symmetric key cryptographic techniques are of two types, namely stream ciphers and block ciphers. Rivest Cipher 4 (RC4), A5/1, are some of the stream cipher examples. Data Encryption Standard (DES), Advanced Data Encryption Standard (AES) are block cipher examples. Similarly, there are different types in asymmetric key cryptography. They are RSA, Diffie–Hellman key exchange, Elliptic Curve Cryptography (ECC). Now to enhance the security system in cryptography, the symmetric and asymmetric cryptographic techniques are combined to create more secured algorithms. The combination of these techniques helps to achieve security services like confidentiality, data integrity, and key management [2, 3].

This paper is structured in the following way, Sect. 2 will depict about the related work and attacks to which algorithms are vulnerable, and Sect. 3 elucidates the proposed work. Section 4 describes experimentation and results, and Sect. 5 presents conclusion and future work.

## 2  Related Work

In this section, we discussed symmetric and asymmetric cryptographic techniques and also their combination. Combining asymmetric and symmetric cryptographic techniques more security could be provided than they used individually.

Diffie–Hellman [4] proposed a key exchange algorithm that allowed users to exchange keys securely. It is the first algorithm to introduce exponentiation for providing security and to overcome from attacks. This algorithm uses public keys of users, a prime number and its primitive root to exchange keys securely between users. But this algorithm is vulnerable to many attacks like known-plaintext attack and man in the middle attack.

Segal et al. [5] proposed the modification of Diffie–Hellman key exchange technique for a more secured key exchange. This algorithm introduced a random integer selection by users in key exchange process which made it resistant to known-plaintext attack and man in the middle attack. But the algorithm is only supportable to a key exchange and not for complete secured communication.

Daemen et al. [6] proposed a symmetric algorithm which can run on 128-, 192-, and 256-bit keys. This algorithm consists of sequence of steps like replacing of bits using S-Boxes called as S-Bytes, shifting of rows to change the sequence of bits called as Shift rows, multiplying of a constant using group multiplication to change the bits known as Mix columns. In the next step, the resultant state from Mix columns is X-ORed with a unique sub key. The unique subkeys are generated for each round through key expansion process for providing more security. These steps are repeated 10, 12, and 14 rounds based on the 128-, 192-, and 256-bit key length, respectively. The decryption process is the exact reverse process of the encryption process. Every step of encryption process is reversed with Add round key, Inverse Mix columns, Inverse Shift rows, and Inverse S-Bytes. For a particular round, key expansion process generates exact key used in encryption. In [7], the authors had mentioned some important features of AES and proved that AES is better than DES, 3DES, and Blowfish. The usage of static S-boxes makes the algorithm vulnerable to linear and differential cryptanalysis attacks.

Mathur et al. [8] proposed a text encryption algorithm using AES and elliptic curve cryptography. It is based on cache hits and misses during encryption. Due to the usage of ECC for key encryption, the proposed algorithm successfully resisted cache timing attack.

Alex et al. [9] proposed an algorithm to find the cost of cryptanalytic attacks. They are able to analyze different time cost trade-offs using special-purpose hardware in the multicore AES processors. This approach proves that AES may not provide long-term security.

Hadi et al. [10] proposed a differential attack on AES-128 with 7 rounds. This is the fastest attack from the point of time and computational complexity.

Kapoor et al. [11] proposed a hybrid algorithm which uses AES, MD-5, and RSA algorithms. The RSA and MD-5 are used for a double encryption of key, and AES is used for encryption of data. This algorithm is effective in providing services like confidentiality, integrity, authentication. The algorithm is restricted to 128-bit key only.

Alkady et al. [12] proposed a hybrid algorithm which is a combination of AES, Elliptic Curve Cryptography (ECC), and MD-5. ECC and MD-5 are utilized for double encryption of key, and AES performs encryption and decryption process. This algorithm is successful in overcoming the general problems of cryptosystems like short response time.

Henriques et al. [13] proposed a secured algorithm based on symmetric and asymmetric cryptographic techniques to provide security between IoT devices. They generated random keys using timestamps for symmetric encryption, and this reduced the encryption time and to overcome some threats.

The most common attacks in symmetric and asymmetric cryptographic algorithms are explained as follows [14]:

- **Brute Force attack**: In this attack, attacker tries to decrypt the ciphertext with every possible permutation of key and finally succeeds in identifying the key used for encryption. AES is immune to this attack because of its key length.
- **Timing attack**: It is a side-channel attack in which attacker can guess the key and plaintext based on time of encryption and decryption.
- **Differential cryptanalysis attack**: It is a chosen-plaintext attack which computes the differences between ciphertexts for a respective plaintext at S-Boxes, and these differences in turn help to find the main key by comparing the every possible key with a difference at every S-Box.
- **Linear cryptanalysis attack**: It is a known-plaintext attack which compares the specific bits of plaintext and ciphertext and generates a linear expression based on these comparisons. These expressions are generated from linear components

of S-Boxes. The generated expressions can be solved, and based on these results, key can be identified.

One of the reasons that differential and linear cryptanalysis attacks are successful is due to the usage of static S-Boxes. Our algorithm is resistant to all of the above attacks due to the usage of Dynamic S-Boxes (DS-boxes).

# 3  Proposed Work

This paper combines the asymmetric Diffie–Hellman key exchange and symmetric AES by modifying it. Firstly, we generate a shared secret key and a random number using Diffie–Hellman key exchange. Then using the random number generated in the above step, Dynamic S-Boxes (DS-Boxes) are generated. Finally, modified AES encryption and decryption processes are discussed.

## 3.1  Key and Random Number Generation Process

In Diffie–Hellman key exchange, User-X and User-Y agree upon two global elements $q$ and $\alpha$ where '$q$' is a prime number and '$\alpha$' is a primitive root of $q$. Each user selects one private key to generate a public key. This public key is shared with other user to generate a shared secret key K. Apart from this, in our algorithm user X selects a random number $R_X$ and user Y selects a random number $R_Y$. User X multiplies $R_X$ with shared secret key K and sends this value ($R_1$) to user Y. Now user Y can calculate $R_X$ using $R_1$ and K. Similarly, user X can generate $R_Y$ from $R_2$ and K. $R_X$ and $R_Y$ can be used in our modified AES to generate Dynamic S-Boxes.

**Key and random number generation algorithm**

Step 1: X selects a private key $KR_X = e, 0 < e < q$
Generates a public key $KU_X = \alpha^e \bmod q$
Step 2: Y selects a private key $KR_Y = f, 0 < f < q$
Generates a public key $KU_Y = \alpha^f \bmod q$
Step 3: X, Y exchanges public keys $KU_X$ and $KU_Y$, respectively.
Step 4: X calculates shared key $K = (KU_Y)^e \bmod q$
Selects a random number $Rx, 0 < Rx < q$
Calculate $R_1 = Rx * K$
Send $R_1$ to Y
Step 5: Y calculates shared key $K = (KUx)^f \bmod q$
Selects a random number $R_y, 0 < R_y < q$
Calculate $R_2 = R_y * K$
Send $R_2$ to X

Step 6: Y calculates $R_x$ as $R_x = R_1/K$
Step 7: X calculates $R_y$ as $R_y = R_2/K$

## 3.2 Dynamic S-Box Generation

The static S-Boxes of AES are converted into Dynamic S-Boxes using the algorithm given below, where even number S-Box uses $R_x$ and odd number S-Box uses $R_y$ as R in the below algorithm.

**Dynamic S-Box generation Algorithm**

Step 1: R value is to be converted into binary format.
Step 2: The first element of state is to be X-ORed with R, if the first bit of R is 1 else X-OR operation need not to be performed.
Step 3: Similarly, the next element of state is to be X-ORed with R if the respective bit of R is 1 else X-OR operation need not to be performed.
Step 4: Step 3 is repeated for all the elements of state. If number of elements in state exceeds the length of R, then R is to be circularly repeated until all elements of state undergo Step 3.

## 3.3 Encryption and Decryption Processes

**Encryption process**

The encryption process of modified AES generates a shared key K using key and random number generation algorithm explained in Sect. 3.1. This key K undergoes key expansion process and generates subkey for each round known as Round key.

Before the encryption process the entire data is to be divided into 128-bit blocks and if the last block is less than 128 bits, then it is padded with extra bits. Each block is converted into state and encrypted individually.

**Encryption Algorithm**
Step 1: The initial state is X-ORed with Round key.
Step 2: The output of Add round key undergoes S-Byte procedure where byte-by-byte substitution is performed using DS-Box.
Step 3: The output of S-Byte will undergo permutation in Shift rows.
Step 4: The output of Shift rows will be transformed into another state by using Mix columns to achieve a substitution by using mathematical transformations.
Step 5: The output of Mix columns is X-ORed with Add round key generated by key expansion process.

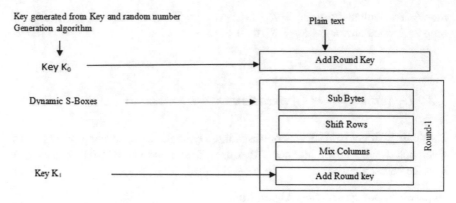

**Fig. 1** Modified AES single-round encryption process

The above step 2 to step 5 represents a single round of modified AES encryption process as shown in Fig. 1. The last round does not perform step 4 procedure, and the output from this round is considered as ciphertext.

**Decryption process**

Decryption is exactly the reverse process of encryption. The process is explained in the algorithm below.

**Decryption Algorithm**

Step 1: The cipher state is X-ORed with Round key.
Step 2: The output of Add Round key will undergo Inverse Shift rows first and gets de-permutated.
Step 3: The output of Inverse Shift rows will undergo Inverse S-Bytes using DS-Box.
Step 4: The output of Inverse S-Bytes is X-ORed with Round key generated for the specific round by key expansion process.
Step 5: The output of Add round key will be transformed by using Inverse Mix Columns.

The above step 2 to step 5 represents a single round of modified AES decryption process. The last round does not perform step 5 procedure, and the output from this round is plaintext.

## 4 Experimentation and Results

We implemented the Diffie–Hellman key exchange algorithm and modified AES using Java networking [15]. To consider the large size keys, plaintext and random number values BigInteger class of Java is used. Based on the results of our implementation, we can say that our proposed algorithm is resistant to timing attack, linear,

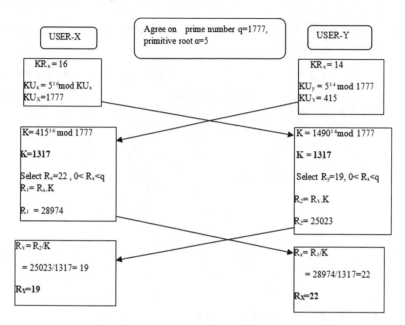

**Fig. 2** Key agreement and random numbers generation

and differential cryptanalysis attacks. To make it clear how our algorithm works, we explained an example below.

Consider that User-X and User-Y agree upon two global elements—a prime number $q = 1777$ and a primitive root $\alpha = 5$. Each user selects a private key to generate a public key. This public key is shared with other user to generate a shared secret key K. Apart from this, in our algorithm user X selects a random number $R_X = 22$ and calculates $R_1$ and user Y selects a random number $R_Y = 19$ and calculates $R_2$. These $R_1$ and $R_2$ are exchanged between X and Y to generate $R_Y$ and $R_X$, respectively, as shown in Fig. 2.

Now, by using these random numbers $R_X = 22$ and $R_Y = 19$ users generate Dynamic S-Boxes, by using Dynamic S-Box generation algorithm explained in Sect. 3.2.

Let $(R_X)_2 = 10110$, $(R_Y)_2 = 10011$

Let us consider the static S-Boxes:

$$S_1 = \begin{bmatrix} 0 & 1 & 2 & 2 \\ 2 & 0 & 1 & 3 \\ 3 & 2 & 1 & 0 \\ 1 & 0 & 3 & 2 \end{bmatrix}$$

Applying Dynamic S-Box generation algorithm on $S_1$ the following DS-Box$_1$ is generated as follows. The first element of $S_1$ box is X-ORed with $R_X$ as $R_X$ first bit is 1, and the second element is not X-ORed because $R_X$ second bit is 0, and similarly, all the other elements of DS-Box$_1$ are generated. The $R_X$ bits are considered circularly for performing XOR operation.

$$DS\text{-}Box_1 = \begin{bmatrix} 00000 \oplus 10110 & 1 & 00011 \oplus 10110 & 00010 \oplus 10110 \\ 2 & 00000 \oplus 10110 & 1 & 00011 \oplus 10110 \\ 00011 \oplus 10110 & 2 & 00001 \oplus 10110 & 0 \\ 00001 \oplus 10110 & 00000 \oplus 10110 & 3 & 00010 \oplus 10110 \end{bmatrix}$$

$$DS\text{-}Box_1 = \begin{bmatrix} 22 & 1 & 21 & 20 \\ 2 & 22 & 1 & 21 \\ 21 & 2 & 23 & 0 \\ 23 & 22 & 3 & 20 \end{bmatrix}$$

$$S_2 = \begin{bmatrix} 3 & 2 & 0 & 1 \\ 1 & 0 & 3 & 2 \\ 2 & 0 & 1 & 3 \\ 0 & 3 & 2 & 1 \end{bmatrix}$$

Applying Dynamic S-Box generation algorithm on $S_2$ and using $R_Y$ on $S_2$ the DS-Box$_2$ is as follows:

$$DS\text{-}Box_2 = \begin{bmatrix} 00011 \oplus 10011 & 2 & 0 & 00001 \oplus 10011 \\ 00001 \oplus 10011 & 00000 \oplus 10011 & 3 & 2 \\ 00010 \oplus 10011 & 00000 \oplus 10011 & 00001 \oplus 10011 & 3 \\ 0 & 00011 \oplus 10011 & 00010 \oplus 10011 & 00001 \oplus 10011 \end{bmatrix}$$

$$DS\text{-}Box_2 = \begin{bmatrix} 16 & 2 & 0 & 1 \\ 18 & 19 & 3 & 2 \\ 17 & 19 & 18 & 3 \\ 0 & 16 & 17 & 18 \end{bmatrix}$$

Similarly, we apply Dynamic S-Box generation algorithm on S-Box 3, 5, 7, 9, 11, 13 using $R_X$ and on S-Box 4, 6, 8, 10, 12, 14 using $R_Y$ and if the generated shared secret key length is less than required key length, then it is to be padded up to the required key length. Now these Dynamic S-Boxes and shared key are used by encryption algorithm in Sect. 3.3 to encrypt plaintext and by decryption algorithm in Sect. 3.4 to decrypt the ciphertext.

**DS-Box Analysis**

Ratio of independence plays a major role in S-box properties. In our proposed methodology, we analyzed generated DS-Boxes for ratio of independence. We used two random numbers for DS-Box generation; therefore, independence of DS-Boxes is better. By observing the results of DS-Box analysis, we can conclude that ratio of independence varies between 85%–100% which makes the proposed algorithm secure as shown in Fig. 3.

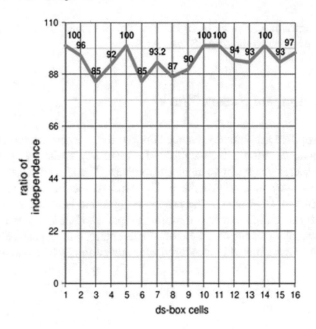

**Fig. 3** Analysis of DS-Boxes

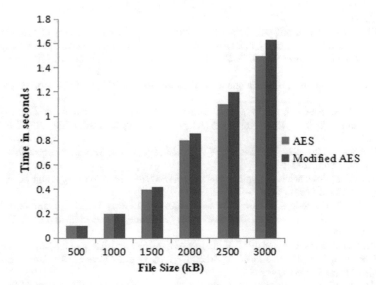

**Fig. 4** Comparison of AES and modified AES

When compared to AES our modified AES is a little bit slower as shown in Fig. 4, but it is more resistant to timing, linear, and differential cryptanalysis attacks. We observed that due to the usage of DS-Boxes, a small change in plaintext has created a significant amount of change in ciphertext. This indicates the avalanche effect is more effective in our modified AES than AES.

## 5 Conclusion and Future Work

This paper proposed a modified Advanced Encryption Standard which uses Dynamic S-Boxes instead of static S-boxes. Diffie–Hellman key exchange is used to generate shared secret key and two random numbers. The shared secret key is used as a symmetric key in modified AES. The random numbers are used to create Dynamic S-Boxes (DS-Boxes). The proposed algorithm is resistant to attacks like brute force attack, timing attack, differential cryptanalysis, and linear cryptanalysis attacks because of the usage of DS-Boxes. In the future, our proposed algorithm can be modified and implemented for secured communication in the cloud environment.

## References

1. A. Kahate, *Cryptography and Network Security* (Tata McGraw-Hill Companies, 2008)
2. A. Eskicioglu, L. Litwin, Cryptography. IEEE Potent. **20**(1) (2001). https://doi.org/10.1109/45.913211
3. S. Chandra, S. Paira, S.K. Alam, S. Bhattacharyya, A comparative survey of symmetric and asymmetric key cryptography, in International Conference on Electronics, Communication and Computational Engineering, ICECCE 2014, pp. 83–93. https://doi.org/10.1109/icecce.2014.7086640
4. W. Diffie, M. Hellman, New directions in cryptography. IEEE Trans. Inf. Theory **22**, 644–654 (1976)
5. P. Sehgal, N. Agarwal, S. Dutta, P.M. Durai Raj Vincent, Modification of Diffie-Hellman algorithm to provide more secure key exchange. Int. J. Eng. Technol. (IJET) **5**(3), 2498–2501 (2013)
6. J. Daemen, V. Rijmen, *The Design of Rijndael AES-The Advanced Encryption Standard* (Springer, Berlin, 2002)
7. A. Abdullah, Advanced Encryption Standard (AES) algorithm to encrypt and decrypt data. Cryptogr. Netw. Security, pp. 1–12 (2017)
8. N. Mathur, R. Bansode, AES based text encryption using 12 Rounds with dynamic key selection, in *International Conference on Communication Computing and Virtualization*, vol. 79 (2016), pp. 1036–1043
9. B. Alex, G. Johann, Cryptanalysis of the full AES using GPU-like special-purpose hardware. J. Fundam. Inf. **114**(3–4), 221–237 (2012). *Cryptology in Progress: 10th Central European Conference on Cryptology*
10. S. Hadi, S. Alireza, B. Behnam, A. Mohammadreza, Cryptanalysis of 7-Round AES-128. Int. J. Comput. Appl. **10**(23), 21–29 (2013)
11. V. Kapoor, A. Jain, Novel hybrid cryptography for confidentiality, integrity, authentication. Int. J. Comput. Appl. **171**(8), 35–40 (2017)
12. Y. Alkady, M.I. Habib, R.Y. Rizk, A new security protocol using hybrid cryptography algorithms, in *9th International Computer Engineering Conference (ICENCO)*, pp. 109–115 (2013). https://doi.org/10.1109/icenco.2013.6736485

13. M.S. Henriques, N.K. Vernekar, Using symmetric and asymmetric cryptography to secure communication between devices in IoT, in *International conference on IoT and Application(ICIOT)* (2017)

14. W. Stallings, *Cryptography and Network Security: Principles and Practice*, 5th edn. (Pearson Education, 2011). ISBN 10: 0-13-609704-9, ISBN 13: 978-0-13-609704-4

15. E.R. Harold, *Java Network Programming: Developing Networked Applications* (O'Reilly Media, Inc., 2013)

# Detection of Advanced Linux Malware Using Machine Learning

Nitesh Kumar and Anand Handa

**Abstract** The malware attacks targeting Linux are increasing recently, because the popularity of Linux has been growing by years, and many popular applications are also available for Linux. There are lots of research that has been done on detecting malicious programs for the Windows-based operating system. But identifying malicious programs for the Linux-based operating system are rarely present. Anand The methods that are present to detect malware are lacking to detect advanced malware effectively. This work shows a machine learning approach by extracting static as well as dynamic features to identify malicious Executable and Linkable Format (ELF) files that is a file format of the Linux operating system. This work uses the best features of benign executables and malware executables to build and train a classification model that can classify malicious and benign executable efficiently. And the classification results show 99.66% accuracy by using XGBoost classifier to distinguish between malicious and benign executable.

**Keywords** ELF · Feature extraction · Machine learning · Static analysis · Dynamic analysis

## 1 Introduction

Nowadays, security became a vital issue and securing data is essential to everyone. Now, malware is playing a significant role to compromise the security; Malware refers to a program or file which do some malicious activity on a target system like modify data, delete files, and download valuable information.

N. Kumar (✉)
CAS, Dr. APJ Abdul Kalam Technical University, Lucknow, India
e-mail: 17mcs08@cas.res.in

A. Handa
C3I Lab, IIT Kanpur, Kanpur, India
e-mail: ahanda@cse.iitk.ac.in

© Springer Nature Singapore Pte Ltd. 2021      185
X.-Z. Gao et al. (eds.), *Advances in Computational Intelligence and Communication Technology*, Advances in Intelligent Systems and Computing 1086,
https://doi.org/10.1007/978-981-15-1275-9_16

Malware became a severe threat to information security, and various reports [1] address these threats. In recent years, malware researchers have mentioned that a large number of new malware that is unknown malware is found. Linux is an open-source and free operating system. Because of the popularity of Linux, uses and development of Linux application become faster. There are different Linux releases present which can be installed on all kinds of computer hardware equipment such as routers and IoT devices. Due to an increase in the development and application of Linux, there is more disruptive malware that appeared for the Linux platform. The malware detection [2] analyses are seldom studied in Linux platforms when compared with Windows platforms, and also the current malware analysis and detection methods are not sufficient and also have some limitations.

There are some problems with the existing approaches first consider static analysis [3] approach this is faster because binary files do not need to run in a controlled environment like dynamic analysis. It has some limitations like polymorphic [4], and metamorphic [5] malware cannot be detected by using a static analysis approach. These type of methods are frequently used by malware authors to create malware, so when the hash of the malware is taken, each time it will return a different hash value which will help them to bypass signature-based detection system. Second is the dynamic analysis [6] approach. In this approach, an executable is executed in a controlled environment that is sandbox or VMs and then the behavior of the malware is observed to conclude whether it is a malware or not, and limitations of this approach are incomplete code coverage, VM detection.

So both static and dynamic analysis approaches have some restriction. Static analysis can be confused once some polymorphic or metamorphic malware is used while dynamic analysis is experiencing from low code coverage issue. So, in this paper, both static and dynamic features are used. Malware authors use the obfuscation techniques to confuse the signature-based static analysis features. They make the malware to do some random task like random access files, arbitrary system calls use, etc., to confuse the dynamic analysis. But to bypass both the analysis approach is so much challenging task. And by using both the features, this work detects the unknown malware (i.e., zero-day malware) and experimental results show the 99.66% accuracy of detecting the malware.

Organization of this paper is as follows: The related work part is explained in Sect. 2. Overview of the ELF file format described in Sect. 3. The experimental framework presented in Sect. 4. Experimental results and comparison with others results are described in Sect. 5. Finally, conclusion of the work is described in Sect. 6. And the future research directions are explained in Sect. 7.

## 2 Related Work

The authors in [7] have proposed a work in which they are using printable string information (PSI) that is extracted from the binary files. This information extracted is used as the static feature, and only API calls are using as a dynamic feature. Their dataset contains 997 malicious files and 490 clean files, that is decidedly less amount of dataset. They registered 98.7% detection accuracy. They also mentioned in their conclusion part that more features could be extracted in the future.

The authors in [8] have proposed a work in which they are using static analysis approach of Android malware detection. The dataset contains 2130 sample files, 1065 files used as clean file and 1065 files used as malicious file, and they are extracting features from the manifest.xml file and bytecode files (classes.dex) file. They are using features permissions, system event, permission rates, and some APIs. They registered 88.26% accuracy with 88.40% sensitivity.

The authors in [9] have proposed a work in which they were using Executable and Linkable Format for analysis and extracted 383 features from the ELF header. They have used information gain as a feature selection algorithm. They used four famous supervised machine learning algorithms C4.5 Rules, PART, RIPPER, and decision tree J48 for classification. Their dataset contains 709 benign executables scrapped from the Linux platform, and 709 malware executable downloaded from vxheaven and offensive computing. They registered nearly 99% detection accuracy.

The authors in [10] proposed a technique in which they were extracting system calls from the symbol table of Linux executable. Out of many system calls, they selected 100 of them as features. Their work proposed an accuracy of 98% in malware detection. Their dataset contains 756 benign executables scrapped from the Linux system and 763 malware executable from vx heavens.

The authors in [11] have proposed another approach based on system call features. They use "strace" to trace all the system calls of the executable running in a controlled environment. In this paper, authors have used two-step correlation-based feature reduction. They first calculated feature-class correlation by using information gain and entropy to rank the features than in the next step they removed redundant feature by calculating feature-feature correlation. They used three supervised machine algorithms J48, random forest AdaBoost for classification and have feature vector with 27 feature length. The authors of this paper used 668 files in the dataset, from the dataset 442 files are benign 226 files are malware . From this approach, they proposed the accuracy of 99.40%.

The authors in [12, 13] have proposed a concept of genetic footprints in which mined information of process control block (PCB) of the kernel is used to detect the runtime behavior of a process. In this approach, the authors have selected 16 out of available 118 parameters of the task _struct for each running process. To decide which parameters to select, authors have claimed to done forensics study on that. Authors believe that these parameters will define the semantic and the behavior of the executing process. These selected parameters are called genetics footprints of the process. Authors have then generated system call dump of all these parameters for 15 s with a resolution of 100 ms. All the instances of benign and malware process are classified using RBF network, SVM, J48 (decision tree), and J-Rip in WEKA environment. Authors have analyzed their result and shortlisted J-48 J-Rip classifiers having less class-noise as compared to others. In the end, authors have also listed the comparison with other system call based on existing solutions. Authors have also discussed evasion related to the robustness of their approach to evasion and access to task struct for modification. They used a dataset in which 105 files used as benign and 114 files used as malware process and proposed a result of 96% detection rate.

The authors in [14] have proposed a method for detecting unknown advanced malware using static analysis approach. The dataset contains 6010 malware and 4573 benign files. They used *objdump* utility installed in linux OS for disassembling the executables and also used opcode frequency as a feature. They have used various feature selection algorithms and select top 20 features. For classification, random forest, LMT, NBT, J48 Graft, and REPTree classifiers are used with the help of weka tool and also registered 100% accuracy.

## 3 Overview of ELF File Format

An Executable and Linkable Format (ELF) [15] file format is a standard binary file format for UNIX and UNIX-like system. A binary file contains various fields like headers, section, and segments. Headers include information like it describes the organization of file, read-only data, symbol table relocation table, etc. ELF object files participate in both program linking and program execution. Program linking concerns to building a program and program execution relate to running a program. The different views are shown in Fig. 1. An ELF header presents the starting and describes the file's organization. A program header table if present describes how to create a process image. A section header table holds the information representing the file's sections.

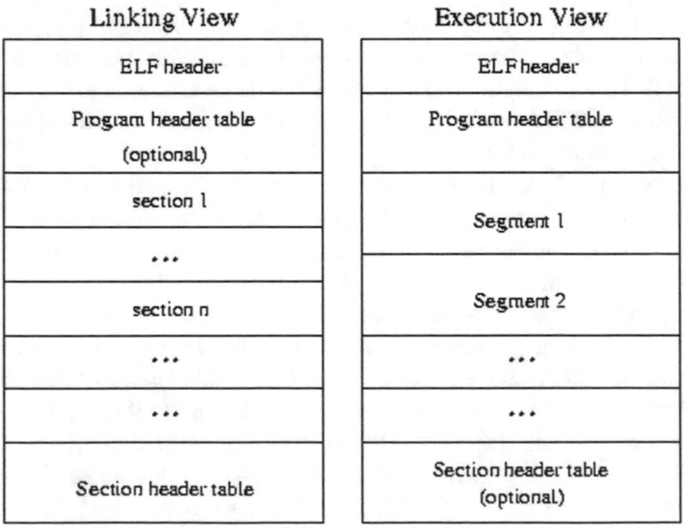

| Linking View | Execution View |
|---|---|
| ELF header | ELF header |
| Program header table (optional) | Program header table |
| section 1 | Segment 1 |
| ... | |
| section n | Segment 2 |
| ... | ... |
| ... | ... |
| Section header table | Section header table (optional) |

**Fig. 1** ELF file format [15]

# 4  Experimental Framework

As shown in Fig. 2 proposed malware detection model contains four steps:

1. First data is generated by using *readelf* for static analysis and by using Limon sandbox for dynamic analysis.
2. Perform feature extraction after data is generated from step 1. Feature selection will be done from the extracted features with the help of some feature selection techniques.
3. After feature selection, use classification algorithms to construct the classifier.
4. Finally, over constructed classifier classifying a file either malware or benign executable.

## 4.1  Dataset

In this section, dataset overview is described used in this work. The dataset contains malware and benign executable. This work collected 2462 benign executable 6456 malware executable. The malware ELF files are collected from VX heavens, Virustotal.com [16] and benign ELF files are collected from Linux operating system directories /bin, /sbin and /usr/bin and downloaded some of the open C C++ software and compiled them into our system to get the more benign executable.

**Fig. 2** Architecture of proposed model

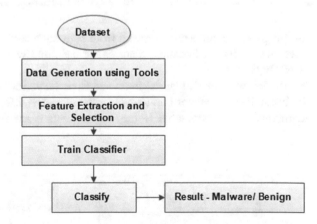

## 4.2 Data Generation Using Tools

This section shows how to generate static and dynamic analysis reports from the ELF executable for further analysis in the next step that is feature extraction. For static analysis report, readelf is used, and for dynamic analysis report, Limon sandbox in used.

**readelf** readelf [17] is a UNIX binary utility for displaying various information fields of one or more ELF files like program headers, sections, symbols, etc. In section II, ELF file format's fields are discussed and Fig. 1 shows the different views of ELF file.

**Limon Sandbox** Limon [18] is a tool which permits us to run an executable in a sand-boxed (controlled) environment and provide us the report about the runtime behavior of an executable. The configuration of limon sandbox involves a host machine that manages the guest machine. In this work, Ubuntu 18.04 used as the host machine, and Ubuntu 14.02 machine used as guest machine. For a better understanding of a le, the file executed in a full privileged mode in the guest machine. For running a le in limon sandbox, its path is given using a command line analysis of each file performed in a fresh virtual machine. While setting up a virtual machine, a current snapshot is taken so that after execution of the le, limon can revert. After execution completed, the limon sandbox saves a text le final report to the analysis report folder containing the full trace of the network, system calls, and functions. This work monitors the executable by default for 40 s. The architecture of limon sandbox has shown in Fig. 3.

## 4.3 Feature Extraction and Feature Selection

In this paper, features are extracted from both the analysis static and dynamic. This section described what static features are used and what are the dynamic features used for the model.

**Static Features** Static features extracted from the various field shown in Fig. 1, of an ELF file. Feature vector contains information derived from ELF headers, sections, segments, and symbol table. Some features' details are given below:

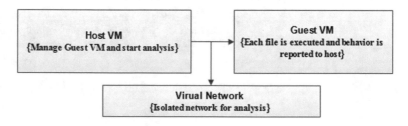

**Fig. 3** Architecture of Limon Sandbox [18]

1. **ELF headers**: Table 1 shows the ELF header information.
2. **Sections**: All the information required during the linking process to create a target object file a working executable resides in sections. Table 2 shows the information of the section in ELF file.
3. **Segments**: Segments are primarily known as program headers. Table 3 shows the ELF segments information.

**Dynamic Features** The dynamic features extracted from the reports are generated by the limon sandbox. Limon sandbox creates a full report in a text file which contains static analysis and dynamic analysis. In static analysis, Limon makes the structure information, and for dynamic analysis, system call traces and network analysis if a file is involved in some network activity. In this paper, dynamic features contains system calls and shell commands. System calls provide the information about what the process want to do or can be understood by it tells the behavior of the process. Shell commands used to interact with OS and using commands can use the service of OS. So, malware can use commands like reboot command or ufw command which is used to alter the firewall. In this work, some features optimization techniques [19] will be used during the features selection process, because during feature extraction process, tools extract a large number of features, but all the features extracted from the tools are not relevant for the work, so it is vital to select only those features

**Table 1** ELF headers information

| Header | Information |
| --- | --- |
| e _ident | This field contains various flags for identification of file. These flags helps in decoding and interpreting the file content |
| c _entry | Process entry point address |
| e _ehsize | ELF header's size |
| e _phentsize | Program header table entry's size |
| e_phnum | Sum of entries in the program header table |

**Table 2** ELF sections information

| Sections | Information |
| --- | --- |
| .text | Includes user executable codes |
| .data | Includes the uninitialized data |
| .dynamic | Includes information about dynamic linking |
| .dynsym | This is the runtime symbol table |

**Table 3** ELF segments information

| Sections | Information |
| --- | --- |
| NULL | Unassigned segment |
| LOAD | This segment gets a load into memory |
| DYNAMIC | .dynamic section in memory |
| INTERP | .interp section mapped to this segment |

that are discriminatory in nature. So, there is a need for some features optimization techniques for minimum features could get better accuracy.

## 4.4 Classification

Now, machine learning technology is widely used in the area of detecting malware. This paper uses decision tree [20], random forest [21], and XGBoost [22] classification learning methods to classify a file as malware or benign with the help of WEKA [23] tool. For classification, split the dataset into 70:30 ratio that means 70% dataset is used to train the classifier and rest 30% dataset is used to predict. This work achieved 99.66% accuracy using XGBoost classification method. The standard tenfold cross-validation processes are used, and this process helps in finding the effectiveness of this study to detect malware that are not known previously.

## 5  Experimental Results and Comparison with Other Work

This section shows the experimental results achieved by our approach and also shows the comparison with the other works results. This work uses both the analysis method static and dynamic. Figure 4a and Table 4 show the accuracy comparison between approaches this work proposed with the others. Figure 4b shows the different classifiers result comparison used in this work, and with our results, we achieved 99.66% accuracy score of detecting unknown malware.

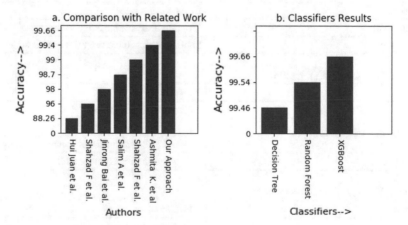

**Fig. 4**  Results comparison

**Table 4** Results comparison

| Authors | Method | Dataset | Accuracy% |
|---------|--------|---------|-----------|
| Salim et al. [7] | Static and dynamic | 490 benign and 997 malware | 98.7 |
| Shahzad et al. [9] | Static | 709 benign and 709 malware | 99 |
| Bai et al. [10] | Static | 756 benign and 763 malware | 98 |
| Ashmita et. al. [11] | Dynamic | 442 benign and 226 malware | 99.4 |
| Shahzad et. al. [12, 13] | Dyanmic | 105 benign and 114 malware | 96 |
| Zhu et al. [8] | Static | 1065 benign and 1065 malware | 88.26 |
| Our approach | Static and dynamic | 2462 benign and 6456 malware | 99.66 |

# 6 Conclusion

This work is using an approach in performing Linux malware analysis with the help of traditional static and dynamic analysis approaches. This work shows some great result 99.66 % of accuracy using XGBoost classification method, and this work uses a tenfold cross-validation to prove the better performance of the work. This work developed a malware detection system by using Python, shell, and PHP scripts. This detection system is running on the local server when any file submitted to the server it checks the file type only ELF files can be parsed, features are extracted, and at last, result shows after prediction using model developed whether it is a malware file or benign file.

# 7 Future Scope

This work focused on malware detection only. In the future, new modules will be added to perform classification of malware into their families.

# References

1. Internet security report. https://media.scmagazine.com/documents/306/wg-threat-reportq1-2017_76417.pdf (2017)
2. A. Azmoodeh, A. Dehghantanha, K.K.R. Choo, Robust malware detection for internet of (battlefield) things devices using deep eigenspace learning. IEEE Trans. Sustain. Comput. (2018)
3. J. Li, L. Sun, Q. Yan, Z. Li, W. Srisa-an, H. Ye, Significant permission identification for machine learning based android malware detection. IEEE Trans. Ind. Inf. (2018)
4. J. Drew, M. Hahsler, T. Moore, Polymorphic malware detection using sequence classification methods and ensembles. EURASIP J. Inf. Secur. **2017**(1), 2 (2017)

5. N. Runwal, R.M. Low, M. Stamp, Opcode graph similarity and metamorphic detection. J. Comput. Virol. **8**(1–2), 37–52 (2012)
6. A. Mohaisen, O. Alrawi, M. Mohaisen, Amal: high-fidelity, behavior-based automated malware analysis and classification. Comput. Secur. **52**, 251–266 (2015)
7. P. Shijo, A. Salim, Integrated static and dynamic analysis for malware detection. Procedia Comput. Sci. **46**, 804–811 (2015)
8. H.J. Zhu, Z.H. You, Z.X. Zhu, W.L. Shi, X. Chen, L. Cheng, Droiddet: effective and robust detection of android malware using static analysis along with rotation forest model. Neurocomputing **272**, 638–646 (2018)
9. F. Shahzad, M. Farooq: Elf-miner: using structural knowledge and data mining for detecting linux malicious executables
10. J. Bai, Y. Yang, S. Mu, Y. Ma, Malware detection through mining symbol table of linux executables. Inf. Technol. J. **12**(2), 380–383 (2013)
11. K. Asmitha, P. Vinod, A machine learning approach for linux malware detection, in *2014 International Conference on Issues and Challenges in Intelligent Computing Techniques (ICICT)* (IEEE, New Delhi, 2014), pp. 825–830
12. F. Shahzad, S. Bhatti, M. Shahzad, M. Farooq, In-execution malware detection using task structures of linux processes, in *2011 IEEE International Conference on Communications (ICC)* (IEEE, New York, 2011), pp. 1–6
13. F. Shahzad, M. Shahzad, M. Farooq, In-execution dynamic malware analysis and detection by mining information in process control blocks of Linux OS. Inf. Sci. **231**, 45–63 (2013)
14. S. Sharma, C.R. Krishna, S.K. Sahay, Detection of advanced malware by machine learning techniques, in *Soft Computing: Theories and Applications* (Springer, Berlin, 2019), pp. 333–342
15. T.I.S. Committee et al., Executable and linkable format (elf). Specification, Unix System Laboratories **1**(1), 1–20 (2001)
16. Virus total. https://www.virustotal.com/ (2018)
17. Readelf https://sourceware.org/binutils/docs/binutils/readelf.html
18. K. Monnappa, Automating linux malware analysis using limon sandbox. Black Hat Europe **2015** (2015)
19. M.F. Ab Razak, N.B. Anuar, F. Othman, A. Firdaus, F. Afifi, R. Salleh, Bio-inspired for features optimization and malware detection. Arab. J. Sci. Eng. **43**(12), 6963–6979 (2018)
20. K.M. Gunnarsdottir, C.E. Gamaldo, R.M. Salas, J.B. Ewen, R.P. Allen, S.V. Sarma, A novel sleep stage scoring system: combining expert-based rules with a decision tree classifier, in *2018 40th Annual International Conference of the IEEE Engineering in Medicine and Biology Society (EMBC)* (IEEE, New York, 2018), pp. 3240–3243
21. N. Dogru, A. Subasi, Traffic accident detection using random forest classifier, in *2018 15th Learning and Technology Conference (L&T)* (IEEE, New York, 2018), pp. 40–45
22. Y. Zhang, Q. Huang, X. Ma, Z. Yang, J. Jiang, Using multi-features and ensemble learning method for imbalanced malware classification, in *Trustcom/BigDataSE/I SPA, 2016 IEEE* (IEEE, New York, 2016), pp. 965–973
23. D. Jain, Improving software cost estimation process through classifcation data mining algorithms using weka tool. J. Commun. Eng. Syst. **8**(2), 24–33 (2018)

# Host-Server-Based Malware Detection System for Android Platforms Using Machine Learning

Anam Fatima, Saurabh Kumar and Malay Kishore Dutta

**Abstract** The popularity and openness of Android have made it the easy target of malware operators acting mainly through malware-spreading apps. This requires an efficient malware detection system which can be used in mass market and is capable of mitigating zero-day threats as opposed to signature-based approach which requires regular update of database. In this paper, an efficient host-server-based malicious app detection system is presented where on-device feature extraction is performed for the app to be analyzed and extracted features are sent over to remote server where machine learning is applied for malware analysis and detection. At server-end, static features such as permissions, app components, etc., have been used to train classifier using random forest algorithm resulting in detection accuracy of more than 97%.

**Keywords** Android malware · Machine learning · Decision tree · Random forest · Feature selection · Parameter tuning

## 1 Introduction

Android with its Google backing and open-source community has the world's highest global market share in the mobile computing industry with 85% in 2017-Q1 (Smartphone OS market share, 2017) due to its open-source distribution and sophistication [1]. Open-source software means that source code is freely available to users to use, understand, modify, and distribute. It gives huge scope for design perspective and support for a rich development community as it is freely available for everyone,

A. Fatima (✉) · M. K. Dutta
Centre for Advanced Studies, Dr. A.P.J. Abdul Kalam Technical University, Lucknow, India
e-mail: 17mcs01@gmail.com

M. K. Dutta
e-mail: malaykishoredutta@gmail.com

S. Kumar
Indian Institute of Technology, Kanpur, India
e-mail: skmtr@cse.iitk.ac.in

© Springer Nature Singapore Pte Ltd. 2021
X.-Z. Gao et al. (eds.), *Advances in Computational Intelligence and Communication Technology*, Advances in Intelligent Systems and Computing 1086,
https://doi.org/10.1007/978-981-15-1275-9_17

not necessarily professionals, to develop applications for Android phones as well as for Android users to download and install thousands of apps for free from various Android markets.

Its popularity and openness have made it a target of malware operators working majorly on malware-spreading apps. Compromised applications found majorly on third-party app stores pose a serious mobile security threat. These malicious apps can cause harm to users by obtaining their private data such as contact lists, email addresses, location information, and phone recordings sending it to third parties as well as taking control of the infected phones or subscribing it to premium SMS and other services.

Malware operators use various techniques such as code obfuscation to bypass conventional signature-based approaches requiring regular update of signature database. Malware analysis, also known as reverse engineering of malware, based on machine learning and data mining techniques have shown to be more effective to evade the ever-emerging malware varieties exploiting zero days. However, an important requirement for machine-learning classifier to detect malicious apps efficiently is feature selection during the data pre-processing while preserving the accuracy.

In this paper, the proposed framework will help users to identify such compromised applications performing malicious activities on his device using machine-learning techniques. Since mobile phones have limited resources performing end-to-end machine-learning detection/analysis on-device can be resource-draining. A trained classifier is deployed at server-end which uses machine-learning techniques to efficiently detect anomalous behavior of unknown apps to mitigate zero-day threats overcoming the short-comings of signature-based approaches. Additionally, depending on response from server, user is alerted and given option to act accordingly such as uninstalling app if detected as malicious.

The main contribution of the proposed system is to provide an efficient and lightweight framework where on-device feature extraction can be performed without the need for rooting device. Since classifier training and malware detection have been performed on server-side, it saves the host from hefty workload given in to the lack of resources on mobile device. Using random forest algorithm for training the classifier, a decent detection accuracy of more than 97% has been achieved. The remaining paper is structured in the following way: Section 2 give brief literature survey of related work. Section 3 discusses about the proposed methodology used for designing a host-server-based malware detection system. Section 4 presents results obtained by applying machine-learning algorithms at server-end on the extracted feature-set. Section 5 provides some conclusions which are derived from results obtained during the experiment.

## 2 Related Work

In MADAM [2], a complete framework for Android malware detection and prevention was proposed where in-depth analysis of Android Apps is performed at various levels such as kernel, application, user level, and effectively detecting more than 96% of apps; however, drawback for such in-depth analysis being that it required rooting of devices which means it cannot be used in mass market. Zhao et al. [3] proposed an approach for malware detection by extracting features such as permissions and APIs according to a set of predefined rules and authoring a feature selection algorithm based on difference in frequencies of features between malware and benign apps.

A faster and light-weight method has been proposed in Drebin [4] where as many as possible features are extracted statically achieving an accuracy of 94% using support vector machine-learning algorithm. Methodologies shown in [3, 4] perform malware analysis using static approach.

In [5], Xiaolei Wang, Yuexiang Yang, and Yingzhi Zeng proposed a hybrid approach for Android malware detection but requires to be deployed off-device on cloud to be provided as a service, considering resource-constrained mobile devices. Some approaches as in [6], make use of ensemble learning by running machine-learning classifiers in parallel to improve detection accuracy and harness their individual strengths. Recent trends in malware analysis are based on making use of deep learning frameworks for malware analysis such as those proposed in [7, 8].

Faruki et al. [9] have given survey of various Android security issues and defenses, a detail of Android security architecture, threats, analysis approaches, and various tools available for assessment and analysis of Android App. It gives brief about the Androguard tool used in the proposed architecture. SigPID [10] proposed a malware detection system based on permission usage by mining permissions data, pruning it at three levels for correctly identifying the significant permissions which will help in distinguishing between benign and malicious apps.

Zhu et al. in [11] use features extracted from APK file for malware detection using random forest classifier. In SAMADroid [12], a novel approach combining benefits of both static and dynamic analysis based on local host and remote server for efficient malware detection for resource-constrained mobile device has been devised. A comparative study of extraction of features and their selection methods for machine-learning-based techniques has been discussed in [13]. In [14], machine-learning-based approach, in combination with features dimension reduction methods, has been applied on permissions and their associated API calls feature-set to detect and classify malwares efficiently in terms of accuracy and performance.

## 3 Proposed Methodology

In this section, we first give a brief introduction of the overall architecture of proposed framework followed by in-detail working of each unit. The proposed framework consists of two sections: Server-end and mobile client-end. To save the resource constraint mobile devices from hefty computations, machine-learning-based classifier training and deployment for analysis and detection of malicious application are performed at server-end. At mobile client-end, user selects an app from the interface provided, features of the selected App are extracted on-device and sent over network

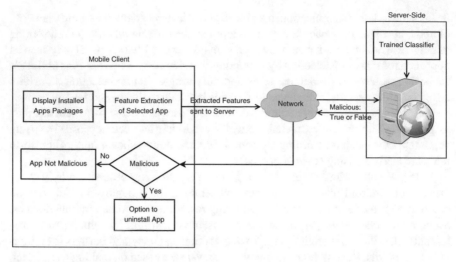

**Fig. 1** Proposed architecture

to server-side for detection. These features are used for detection by trained classifier as to whether the selected App is malicious or not and depending on response from server, the user is alerted if the selected app is detected as malicious.

As shown in Fig. 1, the proposed architecture provides a host-server-based malware analysis and detection approach where on-device feature extraction is performed at mobile client-end for the selected app while machine-learning-based malware analysis and detection is performed at server-end.

## 3.1 Server-End: Android Malware Analysis Using Machine Learning

Static analysis on large-set of APKs requires high computational power and hence, it has been performed on server-side. Unlike the conventional signature-based approach which requires regular update of signature database to detect unknown malware, machine-learning techniques provide an automated analysis process to detect these new variants of malware, also known as zero-day variants. The training dataset consists two set of APKs: one consisting of malicious apps and other consisting of benign apps, thereby labeled as malware and goodware. Since the training dataset is resource-rich, it has been kept in memory of remote server [12]. Figure 2 gives a brief of server-end work.

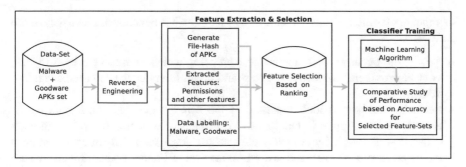

**Fig. 2** Flowchart for server-end work

### 3.1.1 Disassembling and Decompiling APKs

There are a number of tools available for reverse-engineering Android Apps. For the proposed methodology, Androguard, an open-source tool usable for static analysis and providing Python APIs to get access to disassembled resources and decompiled code, has been used [9]. With the help of Androguard, features have been extracted from the AndroidManifest.xml file [4] only. The extracted features are shown in Fig. 3.

### 3.1.2 Feature Extraction

The extracted features can be used to design a complete static analysis framework for detecting unknown malwares. For the proposed host-server-based architecture, application components and permissions are used for mapping to vector space, given

**Fig. 3** Information obtained from decompressed/decompiled APK

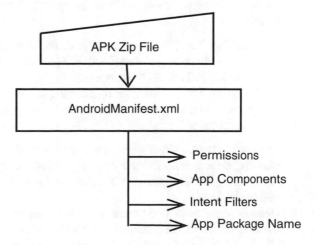

as input to classifier. Feature vector of 674 features has been used. It consists of the following:

- Firstly, the occurrence frequency of App components such as activity, services, content providers, broadcast receivers and intent filters have been used as feature vector.
- Secondly, the permissions requested by the App are mapped to a feature vector as follows: An App $x$ is mapped to $|S|$ dimensional vector space having each dimension as 0 or 1 by constructing a vector $\psi(x)$ such that for each feature $s$ extracted from $x$, the respective dimension is set to 1 and all other dimensions are 0 [12].

$$\psi : X \longrightarrow \{0, 1\}^{|S|} \tag{1}$$

In this way, a CSV file is generated with the above two feature vectors and class label as malware/goodware depending on whether the application is malicious or not, respectively. Hash of respective APKs has been used to uniquely identify them. This CSV is given as input to the machine-learning algorithm.

### 3.1.3   Learning-Based Classification

The generated CSV with labeled data is used to train classifier at remote host. The trained classifier once deployed at remote server can efficiently detect the unknown App whose features are extracted at client-end and sent over network from mobile-host to the server for detection. The output of classifier labeled as malware or goodware is used to classify the unknown App. Figure 4 shows the overall working of machine-learning unit.

The working process of the machine-learning unit is as follows:

1. **Machine-Learning Algorithms**: The feature vector generated is given as input to following two machine-learning algorithms to determine which performs better:

   a. **Decision Tree**: With large datasets, decision trees are proved to be very effective in malware classification. The tree-like structure makes it easy to visualize and simple to implement.
   b. **Random Forest**: Random forest mitigates the problem [15] of overfitting in decision tree by using an ensemble learning method [16] and aggregating the results.

   The models were trained with the above-mentioned algorithms and comparative study is performed for the above two algorithms and it is observed random forest performs better. The classifier is trained and kept at server-side for detection of malicious Apps.

2. **Detection**: Since the performance of random forest in terms of accuracy is better than decision tree, the classifier used for detection at server-side is random forest classifier. The features extracted at mobile client-end are converted into required

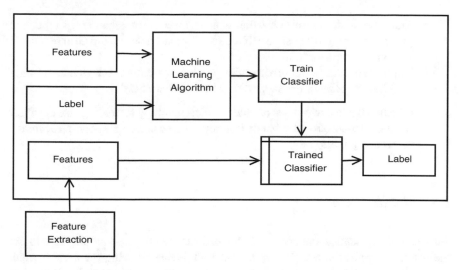

**Fig. 4** Machine-learning unit

feature vector format and fed to the classifier for detection. Depending on the class label (malware/goodware) predicted by the classifier, the response is sent back to mobile client.

3. **Feature Selection Method and Parameter Tuning**: Using a feature vector consisting of more than 600 features for training classifier, causes unnecessarily complicated computations and low efficiency. Further work is done to improve the overall accuracy while working on a low-dimensional feature vector.

   Random forest algorithm provides a way to identify the most important features by giving a relative score for each feature after training. Python's Scikit-Learn is being used for feature selection based on feature importance as it provides a good toolkit for this purpose.

   Hyperparameters of Scikit-Learn built-in random forest function can be used to increase the predictive power of the model. Mostly, there is a vague idea of best hyperparameters, hence to narrow the search of best parameters to a range of values, Scikit-Learn's GridSearchCV has been used [17].

**Feature Selection, Hyper-Parameter Tuning and Accuracy Improvement on Random Forest Algorithm**: Using Scikit-Learn Python library for feature selection and hyperparameter tuning, below steps are followed to make a comparative study of features selected and accuracy computed for selected features:

a. Feed the training dataset with all the features to random forest classifier with criteria set as "entropy" for computation of score of each feature using random forest's feature importance.

b. Sort the features in decreasing order of feature importance.

c. Train the random forest classifier with a subset of total features as input and providing a range of best possible values for parameters using GridSearchCV for hyperparameter tuning.
d. Compute the accuracy corresponding to subset of features selected for training with best hyperparameters predicted by GridSearchCV.

Repeat through last two steps: (c) and (d), incrementally increasing size of subset to maintain a record for number of features selected and corresponding accuracy obtained using them.

## 3.2  Mobile Client

An Android application interface is developed for the proposed architecture to get the list of all applications currently installed on device allowing the user to select any application. An efficient and fast way without the need to root the device has been used to extract the features of the selected App to be analyzed. Using Android PackageManager API, following features have been extracted and send to remote server for analysis:

- Requested Permissions List
- List of App Components as: Activities, Services, Content Providers, Broadcast Receivers
- Intent Filters.

The extracted features put in JSON format are sent to remote server for analysis. Depending on response from server whether app is malicious or not, the user is alerted to uninstall the App if detected as malware.

Figure. 5a shows a screenshot of the user interface designed to view list of Apps installed on device. Using the search option user can select the App to be analyzed, Fig. 5b shows the screenshot of second interface where an alert is generated if the App is detected as malware giving user option to uninstall it.

## 4  Experimental Results

For classification, the decision tree and random forest algorithms have been evaluated on dataset of 50,000 Android Apps consisting of 25,000 malicious Apps (Malware) and 25,000 benign Apps (Goodware). Evaluation criterion for the classifiers has been measured by the accuracy:

$$\text{Accuracy} = \frac{\text{Number of correct predictions}}{\text{Total number of predictions}} \tag{2}$$

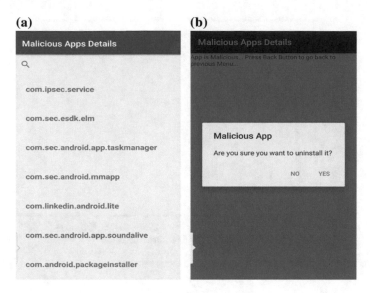

**Fig. 5** Screenshot of the client-end mobile interface: Left-side (**a**) to display list of installed Apps and Right-side (**b**) to give option to user to uninstall App

Table 1 shows results using 674 features for decision tree and random forest classifier. Further, work has been done to improve performance of the algorithm using best features while preserving accuracy.

It is observed that random forest performs better than decision tree in terms of accuracy. To improve performance, selecting good features is important and hence, random forest's feature importance criterion has been used for selecting best features while preserving accuracy. Further, hyper-parameter tuning of the random forest algorithm has been performed to improve accuracy. Following steps of feature selection, hyper-parameter tuning and accuracy improvement on random forest algorithm as given in proposed methodology, number of features selected, and respective accuracy obtained after hyper-parameter tuning have been noted.

Figure 6 shows the results of accuracy percent obtained corresponding to the number of features selected, post hyper-parameter tuning of random forest algorithm. As visible in the graph, feature selection helps in preserving accuracy with low-dimensional feature vector(with just 132 features accuracy of more than 97% is preserved).

**Table 1** Results of the algorithm on the dataset of 674 features

| Algorithm | Accuracy in percent |
| --- | --- |
| Decision Tree | 96.54 |
| Random Forest | 97.25 |

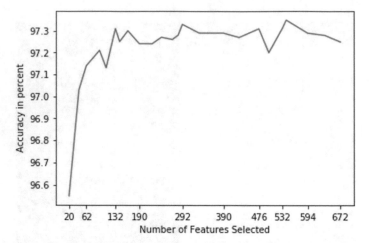

**Fig. 6** Percentage of accuracy with number of features selected

Using random forest feature importance and applying parameter tuning on them, it has been observed that using as less as 132 features, an improved accuracy of 97.31% has been achieved. Using less features while preserving accuracy will help in performance improvement of the overall framework.

## 5 Conclusion

Given in to the large number of malware-spreading apps available in the market, this paper aims to provide efficient and light-weight framework which will help users identify malicious apps installed on his device using machine-learning techniques. For resource-constrained mobile devices, it is important to provide an efficient way to perform on-device feature extraction without the need for rooting it while machine-learning-based malware analysis and detection at server-end are effective in mitigation of zero-day attacks as against signature-based approach. Further, to improve performance and accuracy of the classifier deployed at server-end, feature selection and parameter tuning have been done. The proposed framework has promising results in terms of accuracy of more than 97% on random forest classifier with much lesser features than in original dataset. In the future, work can be enhanced to design an efficient mechanism for capturing run-time behavior of apps as well and working toward improved performance of classifier in terms of accuracy.

**Acknowledgements** We would like to thank C3i Center (Interdisciplinary Center for Cyber Security and Cyber Defense of Critical Infrastructures), IIT Kanpur for sharing their Android Applications dataset.

# References

1. E.M.B. Karbab, M. Debbabi, A. Derhab, D. Mouheb, MalDozer: automatic framework for android malware detection using deep learning. Digit. Investig **24**, S48–S59 (2018 Mar)
2. A. Saracino, D. Sgandurra, G. Dini, F. Martinelli, MADAM: effective and efficient behavior-based android malware detection and prevention. IEEE Trans. Depend. Secure Comput. Digit. Invest. **15**(1), 83–97 (1 Jan-Feb 2018),
3. K. Zhao et al., Fest: a feature extraction and selection tool for android malware detection, in *2015 IEEE Symposium on Computers and Communication (ISCC'15)*, pp. 714–720
4. D. Arp, M. Spreitzenbarth, H. Gascon, K. Rieck, Drebin: effective and explainable detection of android malware in your pocket, in *Symposium on Network and Distributed System Security (NDSS'4)*
5. X. Wang, Y. Yang, Y. Zeng, Accurate mobile malware detection and classification in the cloud. Springer Plus **4**(1), 1–23 (2015)
6. S.Y. Yerima et al., Android malware detection using parallel machine learning classifiers, in *Eighth International Conference on Next Generation Mobile Apps Services and Technologies* (2014 Sept)
7. K. Xu et al., DeepRefiner: multi-layer android malware detection system applying deep neural networks, in *Conference Proceedings: 2018 IEEE European Symposium on Security and Privacy (EuroS&P '18)*
8. Z. Yuan, Y. Lu, Y. Xue, Droiddetector: android malware characterization and detection using deep learning. Tsinghua Sci. Technol. **21**(1), 114–123 (2016)
9. P. Faruki et al., Android security: a survey of issues, malware penetration, and defenses. IEEE Commun. Surv. Tutor. (2015)
10. L. Sun, Z. Li, Q. Yan, W. Srisa-an, Y. Pan, SigPID: significant permission identification for android malware detection, in *2016 11th International Conference on Malicious and Unwanted Software (MALWARE), Fajardo*, 1–8
11. H.J. Zhu, T.H. Jiang, B. Ma et al., HEMD: a highly efficient random forest-based malware detection framework for android. Neural Comput. Appl. **30**, 3353 (2018)
12. Arshad, S., Shah, M. A., Wahid, A., Mehmood, A., Song, H., Yu, H., SAMADroid: a novel 3-level hybrid malware detection model for android operating system. IEEE Access **6**, 4321–4339 (2018)
13. L.D. Coronado-De-Alba, A. Rodrguez-Mota, P.J.E. Ambrosio, Feature selection and ensemble of classifiers for android malware detection, in *2016 8th IEEE Latin-American Conference on Communications (LATINCOM), Medellin*, pp. 1–6 (2016)
14. M. Qiao, A.H. Sung, Q. Liu, Merging permission, A.P.I. features, for android malware detection, in *5th IIAI International Congress on Advanced Applied Informatics (IIAI-AAI), Kumamoto*, pp. 566–571 (2016)
15. Decision Trees and Random Forests. https://towardsdatascience.com/decision-trees-and-random-forests-df0c3123f991. Last accessed on 1st Nov 2018
16. Random forest, https://en.wikipedia.org/wiki/Random_forest. Last accessed on 2nd Nov 2018
17. Hyperparameter Tuning the Random Forest in Python, https://towardsdatascience.com/hyperparameter-tuning-the-random-forest-in-python-using-scikit-learn-28d2aa77dd74. Last accessed on 2nd Nov 2018

# Identification of Windows-Based Malware by Dynamic Analysis Using Machine Learning Algorithm

Areeba Irshad and Malay Kishore Dutta

**Abstract** One of the main concerns of the research community today is the continuously increasing new categories of malware which is a harmful threat to the Internet. Various techniques have been used but they are incapable of identifying unknown malware. To overcome them, the proposed work makes use of dynamic malware analysis techniques in conjunction with machine learning for windows-based malware identification and classification. It involves running the executables in cuckoo sandbox tool which provides a limited environment having an uncovered minimum of resources for execution and post-execution analyzing the behavior reports. The generated JSON report has been used to select the features and their count frequencies. Feature selection is done taking the most important features where the proposed framework's experimental result shows that five features were enough to distinguish malware from benign with the most effective accuracy. Further, in this paper, the top ten frequencies are considered for classification. Two classifiers have been used, the random forest classifier with an accuracy of 85% and decision tree classifier with an accuracy of 83%.

**Keywords** Malware identification · Dynamic malware analysis · Cuckoo sandbox · Feature selection · Machine learning

## 1 Introduction

Malware is a type of unknown activity. The purpose of the malware is to corrupt all files of the system and perform malicious activity [1]. There are different types of computer applications downloaded by users from the Internet at a large scale [2]. With the ever-increasing proliferation of this malicious activity, it is important to develop new techniques to detect and contain this malware [3, 4]. There are a number of anti-malware detection systems and anti-virus software available. However, they do not

A. Irshad (✉) · M. K. Dutta
Department of Computer Science and Engineering, Centre for Advanced Studies, Lucknow, India
e-mail: 17mcs04@gmail.com

M. K. Dutta
e-mail: malaykishoredutta@gmail.com

© Springer Nature Singapore Pte Ltd. 2021
X.-Z. Gao et al. (eds.), *Advances in Computational Intelligence and Communication Technology*, Advances in Intelligent Systems and Computing 1086,
https://doi.org/10.1007/978-981-15-1275-9_18

accurately identify malicious activity [5]. There are two methods of malware analysis: static and dynamic. In the proposed approach, dynamic malware identification using machine learning algorithm has been used as this captures the runtime behavior of the malware. Dynamic malware detection gives a better analysis of the malware and benign files [6]. Capturing the runtime behavior of the malware is one of the most complicated tasks which is achieved by dynamic analysis [7], and the other method of malware analysis is static analysis [8]. Static analysis provides only malicious information about its functionality [9] and does not give us the behavioral activity of malware [10]. Dynamic malware analysis records the behavioral activity of the malware providing more accurate detection results than static analysis [11]. There are various malwares which have affected data on the cloud, compromising its integrity and security by various techniques. By dynamic analysis, such attacks by malwares can be mitigated [12, 13].

In this approach, the cuckoo sandbox tool has been used to perform the dynamic malware analysis. Cuckoo sandbox generates the analysis report in JavaScript Object Notation (JSON) format [14–16] which contains behavior of the malware [17]. With the help of JSON report, the most significant features of malware are extracted and selected. In the proposed methodology, supervised learning approach has been used to detect the malware. Malware detection and feature selection are one of the most difficult operations in identifying malicious activity in the system using machine learning approach in supervised mode. Decision tree and random forest classifier have been used for detection of the malware. Machine learning plays a major role in computer science research, and such research has already had an impact on real-world applications [18].

The main contribution of the proposed work lies in the detection of the malware and benign files using random forest classifier and decision tree classifier which gives the accurate result and the further selection of the best features using the frequency count, and plotting the top ten frequency graph of malware and benign in the graph to pick up the top ten entries consisting of the executable files, dynamic link library, registry key of the windows system, application programming interface, and directories of the windows system.

The remaining paper consists of the following sections: Sect. 2 talks about the proposed methodology which is subcategorized as dataset collection, dataset generation, feature extraction, detection, and machine learning algorithm. Section 3 presents the experimental results and discussions including the confusion matrix and malware detection result. The conclusion derived during the experimentation process is presented in Sect. 4.

## 2 Proposed Methodology

Collecting dynamic data is more robust for obfuscated malware; however, it usually takes a long time. This section presents the architecture of the proposed methodology incorporating various feature engineering and machine learning techniques to obtain high classification accuracy as shown in Fig. 1. The basic outline is as follows:

- Dataset collection and generation from the cuckoo sandbox
- Feature extraction
- Detection and classification of windows-based malware.

In Fig. 1, the malware and benign data files are fed to cuckoo sandbox to examine runtime behavior of these applications. The JSON reports generated by cuckoo sandbox are used for feature extraction and selection. The selected features are fed as input to machine learning algorithms splitting it into training and testing set for effective malware/benign classification and detection.

A. Dataset collection

Windows OS-based malicious samples have been collected from VirusShare database, which is available in the public domain. For clean files, manual collection has been done from system directories of successive version of windows operating system.

B. Dataset generation

In this approach, a virtual and constrained environment provided by cuckoo sandbox is used to run malicious programs and analyze their runtime behavior. Cuckoo sandbox tool is an open-source automatic equipment malware analysis framework. The output created from the cuckoo sandbox is in JSON report file format.

Figure 2 describes the cuckoo sandbox architecture. Cuckoo sandbox, being the open-source software, has been used in automating the analysis of suspicious files. To achieve it, it uses custom components that monitor the behavior of malicious processes running them in an isolated/covered environment. A cuckoo sandbox makes use of minimum of two machines: One is the host machine and the other is the guest machine. All the behavior of the samples is recorded and stored in JSON format report on the host machine. In this approach, Ubuntu 16.04 on the host machine with 4 GB RAM and VirtualBox (latest version) have been used. For the guest machines, we have used 32-bit Windows 7. Installation of the guest operating systems is done on the VirtualBox while their network is configured as a host-only network interface separating the cuckoo host actual network from the virtual network. The cuckoo agent behaves as a communication medium between the cuckoo host and cuckoo guest. After all this, the results are recorded in a cuckoo internal database. Finally, it generates the report as shown in Fig. 3.

C. Feature extraction

Feature extraction is an important step for this proposed research work. One way is to select the smallest number of features that keep the detection and classification rate as the state of the art model in order to minimize the resources for the malware detection task. The JSON reports obtained from the cuckoo sandbox tool are further analyzed to get various malware and benign features. In this approach, five of the most significant features from JSON report have been selected. These features were

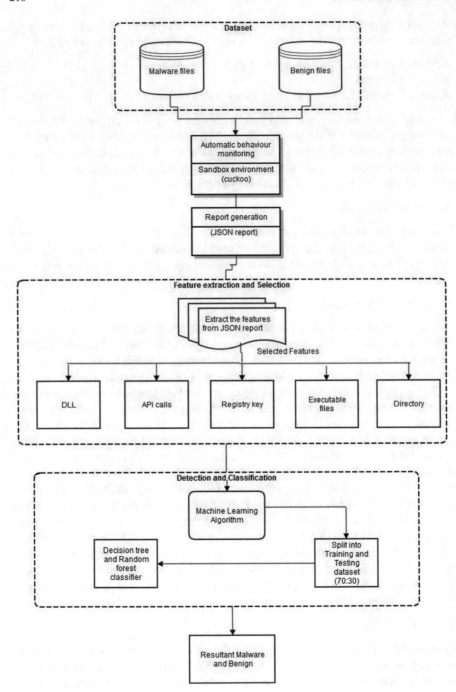

**Fig. 1** Flowchart of the proposed methodology

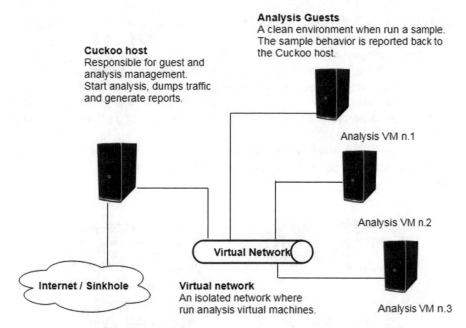

**Fig. 2** Cuckoo sandbox architecture [19]

**Fig. 3** JSON report

repeatedly called in the JSON report and are the ones that are most harmful to the system, and hence, they are used in this proposed work. They include .dll (dynamic link library), registry key, .exe (executable file), .api(), directories. They give better results than others. Then, the frequencies of these features are computed and sorted in descending order to pick top ten frequencies. Then, they are merged together removing duplicate entries to find resultant feature vector which is plotted in the graph for malware and benign. Figure 4 describes the process of the top ten frequency graph.

In Figs. 5, 6, 7, 8, and 9, the top ten frequency graph of five features that is an executable file, API, registry key,.dll, and directories of malware and benign file, are shown.

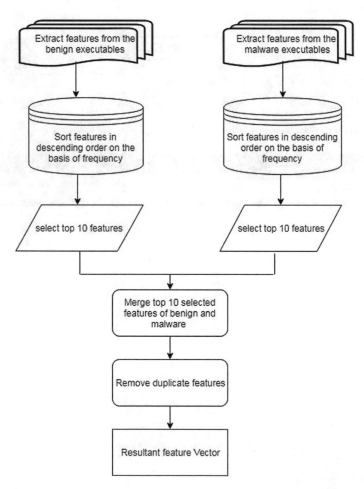

**Fig. 4** Process of top ten frequency graph

## D. Detection

In the proposed methodology, supervised learning approach has been used to detect the malware. The proposed system in Python extracts the raw features from JSON which is converted to a CSV file to perform analysis of malware and is the most important task in supervised machine learning-based malware analysis.

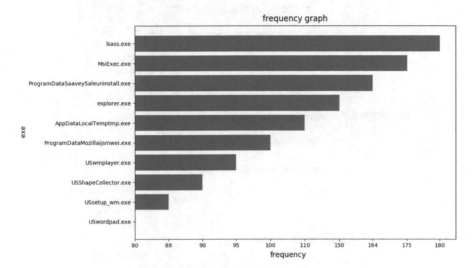

**Fig. 5** Frequent executable collected by malware

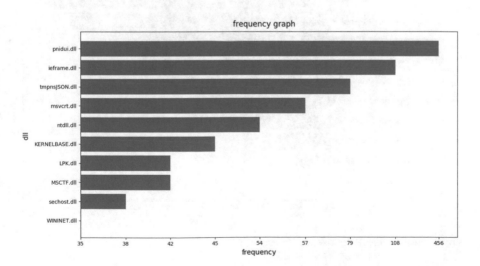

**Fig. 6** Frequent dll collected by malware

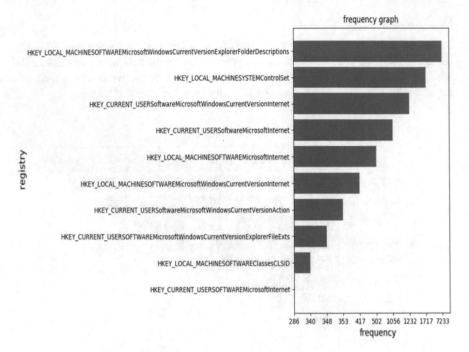

**Fig. 7** Frequent registry collected by malware

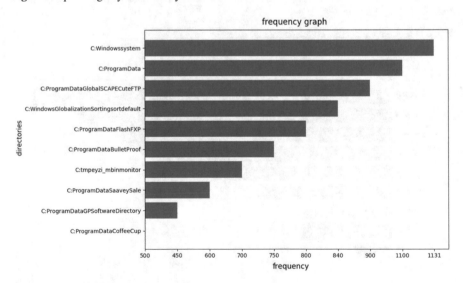

**Fig. 8** Frequent directories collected by malware

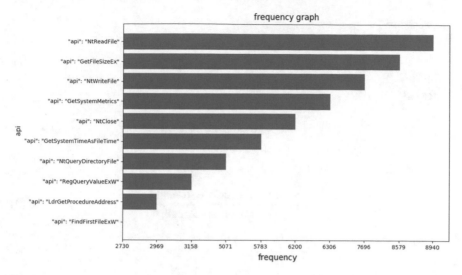

**Fig. 9** Frequent API collected by malware

E.   Machine learning algorithm

In this approach, random forest and decision tree classifier have been used. Both are supervised classification algorithms. Random forest algorithm gives the class of dependent variable based on the results obtained from collection of many decorrelated decision trees to get the best accuracy.

The other classifier algorithm is a decision tree which also comes in the supervised learning technique. Decision tree gives a simple way to understand, interpret, and visualize the trees.

## 3   Experimental Results and Discussions

In this section, the experimental results are presented where 236 samples of data containing malware as well as benign files have been used for this experiment. 115 samples of benign file and 121 samples of malware files have been used. The samples are divided into two sets: the training set consisting of 70% malware samples and the testing set for remaining 30% of malware samples. The results of the experimental evaluation for the following two supervised machine learning algorithms, decision tree and random forest classifiers, are described in the below subsection with the help of Tables 1, 2, and 3.

(a)   Confusion Matrix: The confusion matrix helps to assess models' accuracy. Table 1 shows the confusion matrix.

The following criteria have been used for comparative studies:

**Precision**

$$Precision = \frac{TP}{TP + FP} \tag{1}$$

**Recall**

$$Recall = \frac{TP}{TP + FN} \tag{2}$$

**F1-score**

$$f = \frac{2 * Precision * Recall}{Precision + Recall} \tag{3}$$

**Accuracy**

$$Accuracy = \frac{TP + TN}{TP + TN + FP + FN} \tag{4}$$

**Table 1** Confusion matrix

| Actual class | Predicted class | |
|---|---|---|
| | Negative | Positive |
| Negative | TN | FP |
| Positive | FN | TP |

TN or True Negative shows no. of benign files classified correctly as benign
FP or False Positive shows no. of benign files classified incorrectly as malware
FN or False Negative shows no. of malware files classified incorrectly as benign
TP or True Positive shows no. of malware files classified correctly as malware

**Table 2** Malware detection result using random forest classifier

| Random forest classifier | | | |
|---|---|---|---|
| Class | Precision | Recall | F1-score |
| Benign | 0.75 | 0.93 | 0.83 |
| Malware | 0.93 | 0.75 | 0.83 |
| Accuracy | 0.85 | 0.83 | 0.83 |

**Table 3** Malware detection result using decision tree classifier

| Decision tree classifier | | | |
|---|---|---|---|
| Class | Precision | Recall | F1-score |
| Benign | 0.84 | 0.84 | 0.84 |
| Malware | 0.82 | 0.82 | 0.82 |
| Accuracy | 0.83 | 0.83 | 0.83 |

(b)   Malware detection result

Details of the malware detection assessment obtained from the experimentations are described in Tables 2 and 3. Malware detection rate with an accuracy of 83% using decision tree classifier and 85% accuracy using random forest classifier has been obtained from the experiments.

# 4   Conclusion

In this paper, as against static analysis, in dynamic analysis the runtime behavior of the malware has been captured using cuckoo sandbox tool which generates a JSON report used to find the selected features of malware and benign files through frequency counts. Further, to determine the most significant features for improved malware analysis, the system of top ten frequency graph has been plotted giving better understanding results. Unlike signature-based approach, machine learning-based techniques have been used for malware analysis to detect new variants of malware. Using two classifiers for comparative studies, decent results have been obtained as follows: decision tree classifier with accuracy of 83% and random forest classifier with 85% accuracy. Further, work can be done with larger dataset and working toward accuracy improvement of the classifiers.

# References

1. A. Souri, R. Hosseini, A state-of-the-art survey of malware detection approaches using data mining techniques. Human-centric Comput. Inf. Sci. (2018)
2. T. Lee, B. Choi, Y. Shin, J. Kwak, Automatic malware mutant detection and group classification based on the n-gram and clustering coefficient. J. Supercomput. **74**(8), 3489–3503
3. M. Akiyama, T. Yagi, T. Yada, T. Mori, Y. Kadobayashi, Analyzing the ecosystem of malicious URL redirection through longitudinal observation from honeypots. Comput. Secur. **69**, 155–173 (2017)
4. A. Damodaran, F. Di Troia, C.A. Visaggio, T.H. Austin, M. Stamp, A comparison of static, dynamic, and hybrid analysis for malware detection. J. Comput. Virol. Hacking Tech. **13**(1), 1–12 (2017)
5. P. Burnap, R. French, F. Turner, K. Jones, Malware classification using self organising feature maps and machine activity data. Comput. Secur. **73**, 399–410 (2018)

6. S. Kilgallon, L. De La Rosa, J. Cavazos, L.D. La Rosa, J. Cavazos, Improving the effectiveness and efficiency of dynamic malware analysis with machine learning. 2017 Resil. Week, pp. 30–36 (2017)
7. C. Miller, D. Glendowne, H. Cook, D.M. Thomas, C. Lanclos, P. Pape, Insights gained from constructing a large scale dynamic analysis platform. Digit. Investig. **22**, S48–S56 (2017)
8. N. Udayakumar, S. Anandaselvi, T. Subbulakshmi, Dynamic malware analysis using machine learning algorithm. in *2017 International Conference on Intelligent Sustainable Systems*, ICISS, pp. 795–800 (2017)
9. P.V. Shijo, A. Salim, Integrated static and dynamic analysis for malware detection. Procedia Comput. Sci. **46**(ICICT 2014), 804–811 (2015)
10. K. Sethi, S.K. Chaudhary, B.K. Tripathy, P. Bera, A novel malware analysis framework for malware detection and classification using machine learning approach
11. S. Joshi, Machine learning approach for malware detection using random forest classifier on process list data structure, pp. 98–102
12. H. Dhote, M.D. Bhavsar, Practice on detecting malware in virtualized environment of cloud. SSRN **2014**, 603–608 (2018)
13. A. Sujyothi, S. Acharya, Dynamic malware analysis and detection in virtual environment. Int. J. Mod. Educ. Comput. Sci. **9**(3), 48–55 (2017)
14. S. Jamalpur, Y.S. Navya, P. Raja, G. Tagore, and G.R.K. Rao, Dynamic malware analysis using cuckoo sandbox, in *2018 Second International Conference on Inventive Communication and Computational Technologies (ICICCT)*, no. ICICCT 2018, pp. 1056–1060 (1974)
15. A. Pektas, T. Acarman, Classification of malware families based on runtime behaviors. J. Inf. Secur. Appl. **37**, 91–100 (2017)
16. N. Miramirkhani, M.P. Appini, N. Nikiforakis, M. Polychronakis, Spotless Sandboxes: Evading Malware Analysis Systems using Wear-and-Tear Artifacts, pp. 1009–1024
17. S. Huda, R. Islam, J. Abawajy, J. Yearwood, A hybrid-multi filter-wrapper framework to identify run-time behaviour for fast malware detection (2018)
18. H. Zhao, M. Li, T. Wu, F. Yang, Evaluation of supervised machine learning techniques for dynamic malware detection. Int. J. Comput. Intell. Syst. **11**, 1153–1169 (2018)
19. Cuckoo sandbox architecture, http://docs.cuckoosandbox.org/en/latest/introduction/what/. Last accessed on 3rd Oct 2018

# Secure E-mail Communications Through Cryptographic Techniques—A Study

Shafiya Afzal Sheikh and M. Tariq Banday

**Abstract** E-mail is one of the leading and most reliable modes of communications even after the emergence of many new methods of electronic communication systems. E-mail messages are transmitted from the senders' system to the recipients' system over the Internet with the help of some intermediary nodes including servers, switches, and gateways. E-mail communication relies on some of the oldest communication protocols which have not been modified ever enough since. Hence, e-mail communication is vulnerable to some security risks including eavesdropping, spoofing, tampering, and phishing. The e-mail communication system makes use of some security protocols, techniques, and data encryption methods in many ways to make this communication secure and reliable. This paper reviews the security of e-mail protocols and data encryption techniques that are in use in the e-mail system. It reports the results of a study of the roles and advantages of e-mail security techniques, protocols, and algorithms. This paper also highlights security vulnerabilities in an e-mail communication system and the possibility for improvement in the control measures.

**Keywords** Secure E-mail · Cryptography · Spam · Symmetric key · Asymmetric key · RSA

## 1 Introduction

E-mail communication starts with a sender composing a message on an e-mail client and attaching files to the e-mail message. The e-mail message is then submitted for transmission to a sending mail server which is a Mail Transfer Agent (MTA) or a Mail Submission Agent (MSA) which in turn hands it over to another MTA in the path. The MTA then sends it over to the recipients' mail server which is also an MTA. The e-mail message is sent from the sending MTA to the recipient MTA directly, or through one or more intermediate MTA nodes; thus, an e-mail message may take multiple

S. A. Sheikh · M. Tariq Banday (✉)
PG Department of Electronics and Instrumentation Technology, University of Kashmir, Srinagar, Jammu and Kashmir 190006, India
e-mail: sgrmtb@yahoo.com

© Springer Nature Singapore Pte Ltd. 2021
X.-Z. Gao et al. (eds.), *Advances in Computational Intelligence and Communication Technology*, Advances in Intelligent Systems and Computing 1086,
https://doi.org/10.1007/978-981-15-1275-9_19

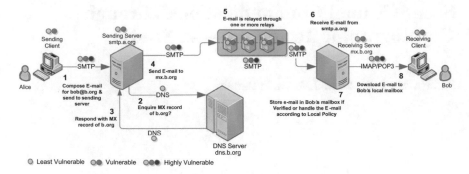

**Fig. 1** E-mail communication system highlighting message transmission along with its vulnerable spots

hops before arriving at the destination mail server. E-mail messages are submitted for sending and transferred between MTAs using Simple Mail Transfer Protocol (SMTP) or its extended version called ESMTP protocol. From the recipients' mail server, the e-mail message can be downloaded to the recipients' mail client (MUA) using IMAP or POP3 protocols. The protocols work with or without the use of any secure communication channels [1]. E-mail messages are vulnerable to security threats right from the spot where they are created, up to the recipients' system. Figure 1 depicts steps of s typical e-mail system and how a message travels from sending mail server to the recipient mail server and highlights the vulnerable spots in the path of communication.

## *1.1 E-mail Communication Protocols*

**Simple Mail Transfer Protocol (SMTP)**: SMTP is the standard communication protocol for transmission of e-mail messages from the sending client to the recipient server [2]. ESMTP or Extended SMTP [3], which works typically using ports 587 or 465, is a more widely used version of the SMTP protocol worldwide. The standard SMTP protocol is less secure than ESMTP because it does not support user authentication and hence can be misused to spread SPAM. ESMTP implements user authentication by username and password allowing only authorized users to send e-mail messages.

**Extended Simple Mail Transfer Protocol (ESMTP)**: ESMTP includes a setup extension to the classical SMTP protocol which enables features like SMTPUTF8 (UTF-8 characters in the headers and mailbox names), 8BITMIME (transmission of 8-bit data as against 7-bit data), AUTH (user authentication by username and password), ATRN (authenticated TURN), ETRN, PIPELINING, HELP, DSN (delivery status notification), and many more features. It is a more widely adopted protocol for e-mail delivery.

**Internet Message Access Protocol (IMAP)**: IMAP is a widely used application layer protocol in an e-mail communication system and is used by e-mail clients to communicate with the recipients' e-mail server in order to manage mailboxes, incoming e-mail messages, and downloading e-mail messages for local storage. IMAP usually works on port 143 over an insecure connection. On a secure connection over SSL, IMAP is usually called IMAP over SSL (IMAPS) and works on port 993. IMAP allows more than one simultaneous connection to the server, support for message status flags, server-side searching of mailboxes, and the partial fetch of headers only, message or attachments separately. All modern e-mail servers support IMAP [4, 5].

**Post Office Protocol (POP3)**: POP3 is an e-mail communication protocol that works on the application layer and is used to download e-mail messages from the recipients' servers to e-mail clients. The protocol is designed to delete the e-mails on the server once they are downloaded and stored on the e-mail client. POP3 usually works on port 110 and port 995 in case it uses TLS or SSL. A large number of popular e-mail servers still support POP3 protocol even though it is old, and more advanced options like IMAP are available [4].

Table 1 shows the major e-mail transmission protocols used as on date in the e-mail communication system for submission, transport, and fetching e-mail messages. It further presents their main advantages and disadvantages.

**Table 1** E-mail communication protocols

| Protocol | Advantages | Disadvantages |
|---|---|---|
| SMTP | • Allows submitting e-mail messages for transmission and helps the transfer of e-mail messages between servers | • No security, encryption, authentication of senders<br>• Has a limited functionality |
| ESMTP | • Allows a secure transmission of e-mail messages<br>• Supports sender authentication<br>• UTF-8 encoding in mailbox names and headers<br>• Transparent exchange of 8-bit ASCII character set | • Same as SMTP unless the extensions are configured<br>• No default security unless STARTTLS extension is used |
| IMAP | • Allows mail clients to download e-mail messages from the server<br>• Supports for management of mailboxes, e-mails, etc.<br>• Supports for downloading individual components of headers, body | • No default security<br>• Enforces storage quotas<br>• Complex to maintain and unsupported by many hosts<br>• Connection required to read mail because of no default local storage |
| POP3 | • Allows downloading the full set of e-mail messages at one time from the e-mail server<br>• E-mails are entirely stored locally and not on the server once downloaded<br>• No size limit of e-mail messages | • Deletes the server copy immediately after the e-mail has been downloaded<br>• No support for fetching individual components of an e-mail message<br>• No default security |

## 1.2 Inherent Security Vulnerabilities in E-mail Communication System

**Insecure Communication Channels**: E-mail protocols such as SMTP, IMAP, and POP3 by default do not require or force the administrators to use security measures such as TLS or SSL for implementing a secure communication channel between the sending and the receiving servers. The communication by default occurs without encryption. E-mail messages are submitted from the client to the e-mail server, transferred between e-mail servers or relays, and downloaded by the recipient e-mail client without any form of data encryption. These e-mail messages can be easily accessed and even tampered with during transmission. Therefore, it is necessary that an e-mail server is configured to use SSL or TLS for secure transmission of e-mail messages.

**E-mails Stored on Servers during Hops**: The e-mail messages are transmitted from the sender through the sending server, possible hops, and the recipient server. These messages are stored on these servers in queues for delivery to the next hop. The e-mails are temporarily stored on all the hops except on the sending and receiving servers where they are permanently stored unless deleted by the sender or the receiver. In case of any security breach of any of the servers or hops, attackers can get access to the e-mail messages and even tamper with them. By default, the e-mail communication and protocols do not require e-mail messages to be encrypted. Even if the messages are transmitted over SSL and TLS securely, they are still stored on the servers in plaintext format and are vulnerable to unauthorized access and tampering.

**E-mail Stored on E-mail Client Systems**: Even if e-mail messages are transmitted over SSL and TLS and deleted from the e-mail servers and all intermediary nodes, they are still vulnerable to attacks because the messages when downloaded and stored locally on e-mail clients are vulnerable to unauthorized access as they are not encrypted. Therefore, e-mail messages should preferably be stored in encrypted form to safeguard it against this vulnerability.

## 2 Cryptographic Algorithms and E-mail Security

Cryptography is the practice of converting private and secret messages into a non-readable format and converting them back into a readable format only by the intended recipient of the messages in order to prevent unauthorized public or eavesdropping agents from reading the messages. It allows two parties to communicate safely in the public domain without anyone being able to understand their private communication. Cryptography is used extensively in electronic communication including e-mail communication. Cryptography is performed primarily using two main methodologies which are symmetric key cryptography and public key cryptography. The third type of encryption known as hashing, which is one-way encryption with no decryption, is also widely in use in information security.

**Symmetric Key Cryptography**: It is the most ancient system of encryption in which both the message sender and the recipient share a single secret key or closely related easily computed second secret key, used for encryption and decryption of the message. This system of encryption requires sharing the encryption key secretly between the two communicating parties before the actual communication. DES, AES, CAST-128/256, IDEA, Rivest cipher (Ron Code), Blowfish, Twofish [6], Camellia [7, 8], MISTY1, SAFER, KASUMI, SEED, ARIA, CLEFIA, etc., are some of the most common private key encryption algorithms in use today [9].

**Public Key Cryptography**: The public key or asymmetric key cryptography is a very modern system of message encryption which uses two different keys for encryption and decryption of the message. The recipient is required to generate a pair of private and public keys. The public key is shared publicly with the message sender, whereas the private key is kept secret by the recipient. The two keys are related to each other in a way that the public key is computed from the private key, but it is impossible to generate the private key from the public key. The encryption of the message is performed using the public key of the recipient, whereas the decryption can be performed only by using the private or secret key of the recipient. The advantage of this system of cryptography is that no key sharing in secret is required before the actual communication takes place. The key can be shared publicly without any danger of compromising the confidentiality of the message. RSA, Diffie–Hellman, DSA, ElGamal, and ECC are some of the most widely used public key encryption algorithms [10].

**Hashing**: Hashing [10] is one-way cryptography which converts data to a hash and does not use any keys for encryption, and there is no decryption of the hash, back to the original data, takes place. It is typically used to generate a digital fingerprint of message data or file contents. The fingerprint of the contents is usually sent to the recipient in encrypted form along with the original contents so that the recipient can recalculate the hash and compare with the one that came with the contents in order to make sure the contents have not changed during transmission. Some of the popular hashing algorithms include MD algorithms (MD3, MD4, MD5), SHA (SHA1, SHA2, SHA3), HAVEL [10, 11], and RIPEMED [12].

## 2.1 Basic Cryptographic Algorithms and their Implementation in E-mail Communication System

**DES**: Data Encryption Standard is a 56-bit symmetric/private key algorithm selected by the US National Bureau of Standards as a standard for the protection of sensitive information. It converts a 64-bit block-sized plaintext into ciphertext by a sequence of complex operations using a 56-bit private key. It is considered as a weak algorithm due to its small key size and can be broken by Brute force attack. This algorithm is more secure in its Tripple DES variation and the original DES algorithm is not a standard anymore.

**IDEA**: International Data Encryption Algorithm (IDEA) is also a symmetric key encryption technique which uses a key size of 128 bits to encrypt data of 64-bit blocks in eight transformation repetitions and a half round of output transformation. The decryption of the ciphertext is done similarly. This encryption algorithm has been broken using meet-in-the-middle and narrow-bicliques attacks, but the algorithm is still considered practically strong [13].

**AES**: Advanced Encryption Standard is a symmetric key encryption algorithm which uses 128-, 192-, and 256-bit keys for encryption in 12, 12, and 14 rounds, respectively. There are no computationally feasible attacks for this type of encryption [14].

**RSA**: Rivest–Shamir–Adleman is a public key encryption algorithm which 1024-4096 bits key size to encrypt data in a single round. It is extensively used on the Internet for secure data transmission. The encryption is done by the public key of the recipient, and the ciphertext can be decoded with the help of the private key of the recipient. RSA can also be used to encrypt data using the sender's private key, and the resulting ciphertext or signature can be decoded only using the public key of the sender. Hence, this algorithm is used both for data encryption and digital signature [15].

**ECC**: Elliptic curve cryptography is a system of data encryption which is based on the structure of plane elliptic curve over a finite field. ECC may use different key sizes for encryption including 128-bit key as well as a 3072-bit public key for encryption. This encryption system is highly secure and can be used for encryption, digital signatures, and more [16].

**Diffie–Hellman**: Diffie–Hellman (DH) secure key exchange method is used to exchange secret keys between two communicating parties over insecure public communications networks. Before this algorithm, encryption keys had to be shared secretly and physically between the communicating parties [17].

**MD5**: MD5 is a message digest algorithm used to convert 512-bit blocks into 128-bit hash. It is a very insecure hashing method because it is easy to find collisions.

**SHA1**: Secure Hashing Algorithm 1 converts a message input into 120-bit message digest. It is considered weak and replaced by more powerful SHA2 and SHA3 methods.

Table 2 shows the most popular cryptographic algorithms used in data encryption and communication security and their implementation in e-mail communication infrastructure. The strength regarding key length and block size of data has been mentioned along with the known attacks that can break the techniques. The attacks mentioned in the table can theoretically break the algorithms and compute the secret key or the original plaintext, but practically most of these attacks are harmless because of the limited computing power of individual computers or even distributed and grid computing systems.

**Table 2** Cryptographic algorithms

| Algorithm | Timeline | Key length (in bits) | Block size | Structure | Known attacks | Implementation in e-mail system |
|---|---|---|---|---|---|---|
| DES | 1976 | 56 | 64 | Feistel | Brute force attack | OpenPGP (as triple DES), PEM |
| IDEA | 1992 | 128 | 64 | Substitution-permutation | Differential timing attack, key-schedule attack | OpenPGP, SSL, TLS, PEM |
| AES | 2001 | 128,192 or 256 | 128 | Substitution-permutation | Side channel attack | OpenPGP (AES-128), SSL, TLS, S/MIME |
| RSA | 1977 | 512 to 4096 | Variant | Factorization | Factoring the public key | SSL, TLS, DKIM, PGP, PEM, S/MIME |
| ECC | | 128, 3072 | Variant | Algebraic | Quantum computing | OpenPGP (ECDSA and ECDH), SSL, TLS |
| DH | 1976 | 512, 1024, 2048 | – | – | Man-in-the-middle attack | PEM, OpenPGP (ECDH), TLS |
| DSS & DSA | 1994 | 512, 1024 | | Algebraic | Lattice attacks | SSL, TLS |
| MD5 | 1992 | – | 512 | Merkle–Damgård construction | Length extension | RSA |
| SHA1 | 1995 | 160 | 512 | Merkle–Damgård construction | Length extension | OpenPGP, SSL, TLS, PGP, S/MIME |

# 3 Cryptographic E-mail Algorithms

Cryptography is extensively used to secure e-mail communication in various ways which include the following:

**SSL**: Secured Socket Layer is a cryptographic protocol which helps to secure the communication in a computer network by implementing various security and encryption techniques. It allows secure and private communication over a public network by using public key cryptographic techniques for key exchange, data encryption, and integrity checks. SSL is used to secure the SMTP protocol in ESMTP in which all the communication between the e-mail client and e-mail servers is fully encrypted. SSL is deprecated and is intended to be replaced by TLS.

**TLS**: Transport Layer Security is also a cryptographic protocol used to secure the communication channels between communicating devices in public networks. It works in similar ways as SSL, but the two have significant differences. TLS is also used to secure the SMTP protocol to provide encryption to the data communication over public networks. It helps in prevention against eavesdropping and tampering with the e-mail messages.

**PEM**: Privacy-Enhanced Mail is a file format and is used to encrypt e-mail messages, keys, and certificates. This standard requires a pre-existing public key infrastructure in place requiring all users to trust the third party and therefore has been replaced by PGP and S/MIME.

**PGP**: Pretty Good Privacy offers privacy, authentication, data integrity, and encryption to e-mail communication. It makes use of message digest, message compression, private key encryption, and public key cryptography. It does not require the sharing of the public key of the recipient directly and instead uses a fingerprint of the e-mail address from which it can extract the public key. It protects e-mail messages from being accessed on the e-mail server by unauthorized personnel. OpenPGP, the most widely used e-mail encryption standard, is a derivate of PGP, and GnuPG is a free and complete implementation of OpenPGP.

**S/MIME**: Secure/Multipurpose Internet Mail Extensions provides complete end-to-end secure communication offering authentication, message integrity, non-repudiation of origin, privacy, and data security to e-mail messages with the help of digital signatures and data encryption. The digital signature proves the authorship of the e-mail message and prevents tampering [18].

**HTTPS**: Hypertext Transfer Protocol over SSL is used to transfer data between client and server on the internet securely. HTTPS is used to safeguard the downloading and management of e-mail messages using a web browser by a browser-based e-mail client. Without the use of SSL, unauthorized access can be obtained over the network [19].

**DKIM**: Domain Keys Identified Mail is a method for detecting sender spoofing or forgery in e-mail communication. E-mail sender forgery is often used for phishing and spamming attacks and is done by sending e-mail messages to victims pretending to be sent by the recipients known and trusted acquaintance or from a reputed company or organization. DKIM helps the recipients to verify whether or not the e-mail has come from the server of the domain name it claims to have come from or any server authorized by the domain owner. The authorized e-mail sending server attaches a digital signature to the e-mail created with the help a private key provided to it by the domain owner. The recipient server can verify the digital signature with the help of the public key published by the domain owner via DNS records which help both verify the sender of the e-mail and check the integrity of the e-mail message. DKIM makes use of the RSA algorithm for signing and verification of the e-mails [18].

Table 3 shows various cryptographic systems that have a role in securing e-mail communication. The type of security provided by these cryptographic systems and authentication mechanisms has been listed. The table outlines the significance of these cryptographic systems in e-mail communication, their advantages, and disadvantages. Some of the listed security mechanisms intended to make the data communication private and safe from network sniffing while others attempt to provide end-to-end security to e-mail which includes encrypting the e-mail message before sending via the network. DKIM, however, is used to prevent sender address forgery.

# 4 Cryptographic Systems for E-mail

Desktop and web-based e-mail clients offer a wide range of cryptographic security features to their users. Desktop e-mail clients like MS Outlook offer users the possibility to use SSL and TLS for SMTP, IMAP, and POP3 protocols if supported by the e-mail server. These features allow the desktop clients to always communicate with the e-mail server over secure channels to avoid unauthorized viewing or tampering of the e-mail message during transmission over the network. E-mail clients also offer the possibility to manage e-mail digital signature for digitally signing e-mail messages. Multiple e-mail digital certificates presented in [20] enable e-mail clients to sign e-mail messages sent from multiple e-mail accounts using a single digital signature certificate. Hence, there is no need to manage multiple digital certificates by a user for different e-mail accounts. E-mail clients help users encrypt the e-mail messages and attachments before sending them with the help of various encryption techniques, e.g., outlook helps users obtain a Digital ID and share the Digital ID with the recipients before sending encrypted e-mail. Only the recipients who possess the Digital ID of the sender can decrypt and read their encrypted e-mail messages. In this example, outlook manages the encryption keys for the users. Web-based e-mail clients offered by significant e-mail service providers by default use SSL to encrypt the communication between the mail server and the browser-based e-mail client.

The e-mail transmission protocols discussed in this paper have been implemented in various open-source and closed-source e-mail server software as well as e-mail clients. Table 4 compares popular open-source and proprietary e-mail server software packages implementing the e-mail communication protocols. The software packages have been categorized according to the e-mail communication protocols they implement, the availability of source code, SMTP authentication, and their built-in e-mail security protocols and other security features which include antivirus and antispam

**Table 3** E-mail cryptographic algorithms

| Cryptographic system, security type, and authentication method | Significance in e-mail communication system | Advantages | Disadvantages |
|---|---|---|---|
| **SSL** Transport layer RSA (challenge response) and DSA digital signature | • SSL is used by SMTP, IMAP and POP3 enable clients and server perform encrypted communication instead of plaintext communication over public networks | • Encrypts information<br>• Provides authentication<br>• Protects against phishing<br>• Provides secure transactions | • Uses many server resources<br>• Expensive and to be renewed periodically<br>• Slows down the connection as it encrypts and decrypts the data<br>• Mixed mode |
| **TLS** Transport layer RSA (challenge response) and DSA digital signature | • Similar to SSL with more advanced features enable secure communication between the servers and clients | • Does not require enhancement to each application<br>• NAT-friendly<br>• Firewall friendly | • Embedded in the application stack (some mis-implementation)<br>• Is protocol specific and needs to duplicate for each transport protocol<br>• Need to maintain context for connection (not currently implemented for UDP)<br>• Does not protect IP addresses and headers |
| **Bit message** End-to-end ECC (challenge response) and ECDSA digital signature | • Decentralized, completely anonymous and encrypted | • Completely anonymous and private e-mail communication<br>• Decentralized, end-to-end encrypted e-mail messages<br>• No trusted third party is required in this protocol<br>• Avoids sender spoofing | • Sender and recipient both must use bit message<br>• Messages are stored for short duration on the network<br>• IS experimental<br>• Unable to handle a large volume of e-mails |
| **PEM** End-to-end DES | • Provides end-to-end encryption of e-mail messages | • End-to-end privacy, encrypted e-mails | • PEM is outdated<br>• No support for MIME<br>• 7-bit encoding support for English language only<br>• Too short keys and weak support for cryptographic algorithms |
| **PGP** End-to-end RSA (challenge response) and DSA digital signature | • Provides end-to-end encryption, and authentication for data communication using public key cryptography<br>• Provides digital singing of e-mails and attachments | • End-to-end e-mail communication privacy<br>• The encrypted e-mail message storage | • Compatibility issues between sender and recipient<br>• Complexity in learning and Implementing<br>• Loss of messages due to lost unrecoverable passwords |

(continued)

**Table 3** (continued)

| Cryptographic system, security type, and authentication method | Significance in e-mail communication system | Advantages | Disadvantages |
|---|---|---|---|
| **S/MIME** End-to-end RSA (challenge response) and DSA digital signature | • Provides end-to-end e-mail encryption using asymmetric key cryptography<br>• Allows signing e-mails digitally | • End-to-end encryption of e-mail and attachments<br>• Authentication<br>• Message integrity<br>• Non-repudiation of origin (using digital signatures)<br>• Privacy and data security (using encryption) | • No proper support in web browser-based clients<br>• No virus scanning possible on e-mail server<br>• Administrative overhead of certificates |
| **DKIM** Sender address forgery RSA (challenge response) and DSA digital signature | • Allows verification of e-mail sender | • Verified e-mail sender<br>• No mail sender spoofing<br>• Provides data integrity by digital signature | • No verification of sender in the e-mail header<br>• Verification of authorized sending server only |

mechanisms. According to the data in the table, all of the e-mail server packages provide an SMTP authentication mechanism. Most of the packages come with built-in antispam and antivirus protection systems, while some of them support direct integration of third-party antispam and antivirus protection systems. Various supported protection mechanisms include SPF, DKIM, Bayesian SPAM filters, antivirus, and antispam modules.

Table 5 compares some popular e-mail clients and the e-mail communication protocols supported by each. It also shows whether or not the e-mail clients support LOGIN and PLAIN authentication. The table also shows which of the clients support the various protocols over SSL and TLS. Almost all the e-mail clients support SMTP, IMAP, and POP3 protocols. Almost all the clients also support PLAIN and LOGIN authentication. Except for a few of the e-mail clients, the e-mail clients support the underlying protocols over SSL and TLS. Hence, it can be safely deduced that almost all the clients support the e-mail communication protocols in plaintext as well as securely over SSL or TLS.

**Table 4** E-mail server systems

| Mail server | Source | Protocols | SMTP Auth. | Antispam features | | | | Antivirus | Antispam |
|---|---|---|---|---|---|---|---|---|---|
| | | | | Bayesian filter | SPF | DKIM | | | |
| Apache James | Open | SMTP/IMAP/POP3 | Yes | Yes | No | No | | Yes | Yes |
| Agorum core | Proprietary | SMTP/IMAP/SMTP over TLS/POP3 over TLS | Yes | No | No | No | | No | No |
| Axigen | Proprietary | SMTP/IMAP/POP3/SMTP over TLS/POP3 over TLS | Yes | Yes | Yes | Yes | | Yes | Yes |
| Citadel/UX | Open | SMTP/IMAP/POP3/SMTP over TLS/POP3 over TLS | Yes | Yes | Yes | No | | Yes | Yes |
| Courier Mail Server | Open | SMTP/IMAP/POP3/SMTP over TLS/POP3 over TLS | Yes | Yes | Yes | Yes | | Yes | Yes |
| IceWarp Mail Server | Proprietary | SMTP/IMAP/POP3/SMTP over TLS/POP3 over TLS | Yes | Yes | Yes | NA | | Yes | Yes |
| FirstClass | Proprietary | SMTP/IMAP/POP3/SMTP over TLS/POP3 over TLS | Yes | Yes | Yes | NA | | Yes | NA |
| Gordano Messaging Suite | Proprietary | SMTP/IMAP/POP3/SMTP over TLS/POP3 over TLS | Yes | Yes | Yes | NA | | Yes | Yes |
| Exim | Open | SMTP/SMTP over TLS | Yes | Yes | Yes | Yes | | Yes | Yes |
| hMailServer | Open | SMTP/IMAP/POP3/SMTP over TLS/POP3 over TLS | Yes | No | Yes | Yes | | Yes | Yes |
| Halon | Proprietary | SMTP | Yes | Yes | Yes | Yes | | Yes | Yes |
| Haraka | Proprietary | SMTP | Yes | Yes | Yes | Yes | | Yes | Yes |
| IBM Lotus Domino | Proprietary | SMTP/IMAP/POP3/SMTP over TLS/POP3 over TLS | Yes | NA | No | NA | | NA | NA |
| Kolab | Open | | Yes | Yes | Yes | NA | | Yes | Yes |

(continued)

**Table 4** (continued)

| Mail server | Source | Protocols | SMTP Auth. | Antispam features | | | Antivirus | Antispam |
|---|---|---|---|---|---|---|---|---|
| | | | | Bayesian filter | SPF | DKIM | | |
| OpenSMTPD | Open | SMTP/SMTP over TLS | Yes | NA | Yes | Yes | NA | NA |
| Oracle Communications Messaging Server | Proprietary | SMTP/IMAP/POP3/SMTP over TLS/POP3 over TLS | Yes | Yes | Yes | NA | Yes | Yes |
| Postfix | Open | SMTP/SMTP over TLS | Yes | Yes | Yes | Yes | Yes | Yes |
| Qmail | Open | SMTP/POP3 | Yes | No | NA | NA | No | No |
| Sendmail | Open | SMTP/SMTP over TLS | Yes | Yes | Yes | Yes | Yes | Yes |
| Synovel Collabsuite | Closed | SMTP/IMAP/POP3/SMTP over TLS/POP3 over TLS | Yes | Yes | Yes | NA | Yes | Yes |
| Zimbra | Open | SMTP/IMAP/POP3/SMTP over TLS/POP3 over TLS | Yes | Yes | Yes | Yes | Yes | Yes |
| Mailsite | Proprietary | SMTP/IMAP/POP3/SMTP over TLS/POP3 over TLS | Yes | Yes | Yes | NA | Yes | Yes |
| Microsoft Exchange Server | Proprietary | SMTP/IMAP/POP3/SMTP over TLS/POP3 over TLS | Yes | Yes | NA | NA | NA | NA |
| Mercury Mail | Proprietary | SMTP/IMAP/POP3/SMTP over TLS/POP3 over TLS | Yes | Yes | NA | NA | Yes | NA |

**Table 5** E-mail client systems

| Mail client | POP3 | IMAP | SMTP | Authentication | | Secure POP | | Secure IMAP | | Secure SMTP | | PGP support | S/MIME support |
|---|---|---|---|---|---|---|---|---|---|---|---|---|---|
| | | | | LOGIN | PLAIN | TLS | SSL | TLS | SSL | TLS | SSL | | |
| Alpine | Yes | Yes | Yes | Yes | Yes | Yes | Yes | Yes | Yes | Yes | Yes | Yes | Yes |
| Becky! Internet Mail | Yes | Yes | Yes | Yes | Yes | Yes | Yes | Yes | Yes | Yes | Yes | Yes | Yes |
| Claws Mail | Yes | Yes | Yes | Yes | Yes | Yes | Yes | Yes | Yes | Yes | Yes | Yes | Yes |
| Courier | Yes | Yes | Yes | Yes | Yes | No | No | No | No | No | No | Yes | No |
| eM Client | Yes | Yes | Yes | Yes | Yes | Yes | Yes | Yes | Yes | Yes | Yes | Yes | Yes |
| EmailTray | Yes | Yes | Yes | Yes | Yes | Yes | Yes | Yes | Yes | Yes | Yes | No | No |
| Evolution | Yes | Yes | Yes | Yes | Yes | Yes | Yes | Yes | Yes | Yes | Yes | Yes | Yes |
| IBM Notes | Yes | Yes | Yes | Yes | – | Yes | Yes | Yes | – | Yes | Yes | Yes | Yes |
| KMail | Yes | Yes | Yes | Yes | Yes | Yes | Yes | Yes | Yes | Yes | Yes | Yes | Yes |
| MS Entourage | Yes | Yes | Yes | Yes | No | Yes | – | Yes | – | Yes | – | Yes | Yes |
| MS Office Outlook | Yes | Yes | Yes | Yes | No | Yes | No | Yes | Yes | Yes | Yes | No | Yes |
| Mozilla Thunderbird | Yes | Yes | Yes | Yes | Yes | Yes | Yes | Yes | Yes | Yes | Yes | Yes | Yes |
| Mulberry | Yes | Yes | Yes | Yes | Yes | Yes | Yes | Yes | Yes | Yes | Yes | Yes | Yes |
| Outlook Express | Yes | Yes | Yes | Yes | No | Yes | Yes | Yes | Yes | Yes | Yes | No | No |
| Pegasus Mail | Yes | Yes | Yes | Yes | – | Yes | Yes | Yes | Yes | Yes | Yes | Yes | No |
| Roundcube | Yes | Yes | Yes | Yes | Yes | Yes | Yes | Yes | Yes | Yes | Yes | Yes | Yes |
| Squirrel Mail | Yes | Yes | Yes | Yes | Yes | Yes | Yes | Yes | Yes | Yes | Yes | Yes | Yes |
| Sylpheed | Yes | Yes | Yes | Yes | Yes | Yes | Yes | Yes | Yes | Yes | Yes | Yes | – |

(continued)

**Table 5** (continued)

| Mail client | POP3 | IMAP | SMTP | Authentication | | | Secure POP | | Secure IMAP | | Secure SMTP | | PGP support | S/MIME support |
|---|---|---|---|---|---|---|---|---|---|---|---|---|---|---|
| | | | | LOGIN | PLAIN | TLS | SSL | TLS | SSL | TLS | SSL | | |
| The Bat! | Yes | Yes | Yes | Yes | Yes | Yes | Yes | Yes | Yes | Yes | Yes | Yes | Yes |
| Windows Live Mail | Yes | Yes | Yes | Yes | No | Yes | No | Yes | No | Yes | No | – | Yes |
| Windows Mail | Yes | Yes | Yes | – | – | Yes | No | Yes | – | Yes | – | – | – |
| YAM | Yes | No | Yes | Yes | Yes | Yes | Yes | No | No | – | – | Yes | – |
| Zimbra | Yes | Yes | Yes | Yes | Yes | Yes | Yes | Yes | Yes | Yes | Yes | No | – |

# 5   Conclusion

This paper discusses and illustrates the mechanism of e-mail communication in brief, lists, and explains various e-mail communication protocols in practice today and draws a comparison between their features. The use and significance of the e-mail communication protocols have been discussed, highlighting the various security risks involved in, and vulnerable spots of the e-mail communication system. The paper highlights the existing vulnerabilities and weak points of e-mail communication which need the attention of researchers. Some of the e-mail communication protocols mentioned in this research attempt to provide solutions for these security issues, but they are not widely implemented for use because of their complexity in implementation, and requirements of infrastructure and trust in third-party intermediaries. The default behavior and implementation of the e-mail communication protocols, even over SSL or TLS, does not provide any security to the e-mail communication. The paper also discusses cryptography, various encryption mechanisms, and algorithms used in securing information communication. The work discusses two primary systems of encryption, namely public key cryptography and private key cryptography with an overview of their most popular algorithms and implementations. Hashing or message digest cryptography, which is a one-way encryption system with no decryption, has been discussed as well, highlighting its uses in communication security. A comparison of various parameters of the encryption techniques has been drawn. The implementation of cryptography in securing e-mail communication against unauthorized access, tampering during transmission and other types of risks, has been discussed as well which fall mainly the four categories of privacy, authentication, integrity, and non-repudiation. The paper explains the application of various encryption systems in the different e-mail communication protocols highlighting their significance in the protocols and the enhancements they provide.

**Acknowledgements**   This research work has been supported by Science and Engineering Research Board (SERB), Department of Science and Technology (DST), Government of India, under its file no. EMR/2016/006987.

# References

1. M.T. Banday, J.A. Qadri, N.A. Shah, A practical study of E-mail communication through SMTP. Working Papers on Information Systems, Sprouts. ISSN: 1535-6078, 10(20) (2010)
2. Klensin, Simple Mail Transfer Protocol, IETF RFC 2821 (2001)
3. J. Klensin, N. Freed, M. Rose, E. Stefferud, D. Crocker, SMTP service extensions (1995). 1017487RFC1869
4. P. Tzerefos, C. Smythe, I. Stergiou, S. Cvetkovic, A comparative study of simple mail transfer protocol (SMTP), post office protocol (POP) and X.400 electronic mail protocols, in *Proceedings of the 22nd Annual IEEE Conference on Local Computer Networks* (1997), pp. 545–554
5. M. Crispin, RFC 3501: Internet Message Access Protocol Version 4rev1. IETF (2003)

6. H. Zoran, G. Druga, Varaždin, Croatia, Comparative analysis of cryptographic algorithms. Int. J. Digit. Technol. Econ. **1**, 127–134 (2016)
7. X. Zhao, T. Wang, Y. Zheng, Cache timing attacks on camellia block cipher. IACR Cryptology ePrint (2009)
8. D. Lei, L. Chao, K. Feng, New observation on Camellia, in *Proceedings of SAC'05, Lecture Notes in Computer Science 3897* (Springer, Berlin, 2006), pp. 51–64
9. M. Rouvala, White paper: Security and WSN. New Nordic Engineering (2017)
10. B. Scheneir, *Applied Cryptography: Protocols, Algorithms, and Source Code in C* (Wiley, New York, 1996)
11. J. Kim, A. Biryukov, P. Bart, S. Hong, On the security of HMAC and NMAC based on AVAL, MD4, MD5, SHA-0 and SHA-1, in *International Conference on Security and Cryptography for Networks, SCN 2006: Security and Cryptography for Networks* (2006), pp 242–256
12. A.A. Milad, H.Z. Muda, H.Z. Muhamad Noh, M.A. Algaet, Comparative study of performance in cryptography algorithms (Blowfish and Skipjack). J. Comput. Sci. **7**, 1191–1197 (2012)
13. V. Prajwal, K. Prema, User defined encryption procedure for IDEA algorithm, in *International Conference on Advances in Computing, Communications and Informatics (ICACCI)* (2018)
14. FIPS, FIPS Pub 197: Advanced Encryption Standard (AES). US Nat'l Institute of Standards and Technology (NIST) (2001)
15. X. Zhou, X. Tang, Research and implementation of RSA algorithm for encryption and decryption, in *Proceedings of 2011 6th International Forum on Strategic Technology*. IEEE (2011)
16. Y. Zhang, C. Tianxi, T. Hong, A new secure E-mail scheme based on elliptic curve cryptography combined public key, in *IFIP International Conference on Network and Parallel Computing* (2008)
17. W. Diffie, M.E. Hellman, New directions in cryptography. IEEE Trans. Inf. Theory **22**(6), 644–654 (1976)
18. M.T. Banday, J.A. Qadri, A study of E-mail security protocols. eBritian, British Institute of Technology and E-commerce, UK, Issue 5, summer 2010 (2010), pp. 55–60. ISSN: 1755-9200
19. E. Rescorla, HTTP Over TLS. IETF RFC 2818 (2000)
20. M.T. Banday, S.A. Sheikh, Multiple E-mail address certificate, in *International Conference on Advances in Computing, Communications, and Informatics (ICACCI-2013)* (IEEE, 2013), pp. 1134–1139

# Current Trends in Cryptography, Steganography, and Metamorphic Cryptography: A Survey

Ranjana Joshi, Munesh C. Trivedi, Avadhesh Kumar Gupta
and Paritosh Tripathi

**Abstract** World without Internet is very difficult to recognize. Increasing dependency and popularity of Internet among its users make its user life easier and fast. Some of the information of its users can be public without taking much pain but, on the other hand side, some information cannot be made public. It needs not only privacy but security too. Security at the cost of asset. How much security is enough in terms of cost of asset? Several researcher advocates use cryptography to provide required security to sensitive information. Some popular algorithms which perform cryptography are RSA with AVK, ECC, Hybrid, CE, Blowfish, Twofish, Threefish, etc. Some researchers advocate the use of steganography. Some popular algorithms are LSB, modified LSB, etc. Some researchers advocate the use of both approach, i.e., cryptography and steganography. This approach is popularly known as metamorphic cryptography. Cryptography helps when caught by some unauthorized user and steganography helps in creating less suspect. In this paper, a survey is carried out over popular techniques used by researcher in performing cryptography and steganography. Paper tried to discuss the methodology these algorithms use. Paper also tries to discuss the strength and limitation of various proposed algorithm according to respective authors claims in their research work.

**Keywords** Steganography · Cryptography · Chaotic sequence · DNA sequence · Encryption · LSB · Blowfish · RSA etc.

R. Joshi
JRN Rajasthan Vidyapeeth University, Udaipur, Rajasthan, India
e-mail: rans.josh@gmail.com

M. C. Trivedi (✉)
Department of Computer Science and Engineering, National Institute of Technology Agartala,
Agartala, Tripura, India
e-mail: munesh.trivedi@gmail.com

A. K. Gupta
Information Technology Department, IMS, Ghaziabad, Uttar Pradesh, India
e-mail: sarthakcc@gmail.com

P. Tripathi
IT Department, IET, Dr. RML Avadh University, Faizabad, Uttar Pradesh, India
e-mail: paritoshtripathi@rmlau.ac.in

© Springer Nature Singapore Pte Ltd. 2021                                                237
X.-Z. Gao et al. (eds.), *Advances in Computational Intelligence and Communication
Technology*, Advances in Intelligent Systems and Computing 1086,
https://doi.org/10.1007/978-981-15-1275-9_20

# 1 Introduction

There are millions of activities taking place on world of Internet. And there could be many things which can go wrong, may be fall of sensitive information in the wrong handsRole of digital world cannot be ignored in twenty-first century. Several activities or tasks which are performed with the help of digital technologies not only makes life easier but also very fast, which in return saves time. Every activities of day-to-day life such as online shopping, ticket booking (which can of train/plane/water ships, etc.), online hotels booking, searching restaurants, food order, bill payments (ex-credit card, electricity, insurance, policy, etc.), recharges, online appointments (for ex-doctors), online feedback and complain, activities related to social sites, cab booking, and uploading videos on channels are performed with the help of digital world technologies. World without digital technologies realization seems to be very hard. There are millions of activities taking place on world of Internet. And there could be many things which can go wrong, may be fall of sensitive information in wrong hands. These all activities involve exchange of some messages which sometimes needs security. Security in terms of

1. Confidentiality
2. Integrity
3. Availability

*Confidentiality*: The term confidentiality means that information any how must be kept hidden or kept secret from unauthorized access from authorized or unauthorized users. To understand the term confidentiality, consider an example of military exchange of sensitive information. Military works on the principle of "need to know." They conceal their sensitive information. Only the authorized person or receiver can see the information. This is only because he only knows the keys in order to decrypt the information. This key is called "cryptographic key." This cryptographic key supports access control mechanism. Those who have the right cryptographic key will only able to unlock the information from sensitive data. But cryptographic key needs to be protected from unauthorized access.

Access control mechanisms support confidentiality. One access control mechanism for preserving confidentiality is cryptography, which scrambles data to make it incomprehensible.

A cryptographic key controls access to the unscrambled data, but then the cryptographic key itself becomes another datum to be protected.

*Integrity*: Authorized or unauthorized user unable to make unauthorized change in information.
*Availability*: Information should be available when needed by the authorized user.

The goal of the work presented here in this paper focusses on the confidentiality aspects of the security. Question is how to provide confidentiality to the end users while using the public channel of internet.

Snooping and traffic analysis are two different kinds of major attacks which impact the confidentiality of information.

*Snooping*: Unauthorized access of messages during communication between end users by the authorized or unauthorized users.

*Traffic analysis*: Locating some useful information, for example, who is communicating and at what timing? Duration of communication, communicating parties IP addresses etc.

Security mechanisms recommendation can be found in ITU-T (X.800) [1]. According to recommendations of X.800, data confidentiality can be achieved with the help of encipherment. Encipherment can be achieved through covering or hiding approach (cryptography/steganography). Some of the researcher's advocate that the message which end user wants to keep secret is to be encrypted using some procedure with the help of strong algorithm and keys to avoid snooping and traffic analysis kind of threat. This process is known as cryptography.

Cryptography word, with Greek origins, means out to be "secret writing." Similarly, steganography, with Greek origins "steganos," means covered writing.

Rest of the paper is organized as follows: Section 2 contains literature survey, in which discussions have been made about the existing technique of steganography, cryptography, and also about metamorphic cryptography. This section also contains comparisons of three famous algorithms (blowfish, twofish, threefish, etc.). Section 3 contains the conclusion of whole literature survey.

## 2 Literature Survey

Saumya Batham et al. in their research paper present an encryption algorithm which uses the concept of indexed based chaotic sequence [1]. Along with this concept, they have used hybrid video codec to compress video frames. Shilpi et al. in their research paper proposes an algorithm which works on the two tier of security: steganography and cryptography [2]. Researcher in their paper exploited the properties of audio as carrier medium to hide their message. For encrypting the text, they have used indexed based chaotic sequence. For hiding this encrypted text, LSB technique has been used. According to authors, simple LSB creates less distortion in its carrier. But to increase the strength of simple LSB, XOR concept has been introduced. XOR concepts help in hiding the secret encrypted text but creating less distortion in comparison to simple LSB.

Shivani et al. in their paper introduced the concept of zero distortion technique [3]. In their research, they talk about the limitation of LSB approach. They agree on this part that LSB is still popular and most utilized steganographic approach but still have some limitations. Altering the pixel value of LSB is least visible from naked eyes but if histogram analysis is performed, changes can be detected easily. Zero distortion technique proposed by them claims to create zero distortion in cover image as it does not alter the pixel value of carrier image [3, 4].

**Fig. 1** Security goals

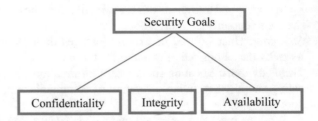

In other paper titled "Analysis of Different Text Steganography Techniques: A Survey," same author has conducted the survey of steganographic techniques which was applied on text [4, 5]. In their paper introduction part, they have mentioned that text steganography is difficult to perform if text is chosen to act as carrier. One strong reason for this approach is that text file lacks redundancy. On the other hand, if video, image, or audio is chosen as carrier medium, they are generally enriched with redundant data. They broadly classified the text steganographic technique in two major categories.

1. Changing the format of text.
2. Changing the meaning of text.

Figure 1 shows the semantic approach of text steganography. With the help of this approach which is also known as synonym substitution, a single or multiple bits of secret message can be made hidden.

Figure 2 shows the word spelling approach. In this approach, authors tried to exploit some approach which works on the basis that how the same meaning words spelled differently in British and American English language. Author claims that if this approach is utilized, it can increase the payload capacity of hiding secret message text.

Another approach which was discussed in this survey is used of white or open spaces for hiding purposes [4, 5]. If anyone wants to hide "0" bit in the cover file, then he has to put one blank space in cover text file. Similarly, if any one wants to hide "1" bit in cover file, then he has to put two blank spaces. Generally, in text file, white spaces are used between words in sentences or at the end of the sentences. Spaces between the HTML tags can also be utilized for hiding purposes. Spaces between HTML tags hardly affect the viewing aspect of HTML document.

Some more approaches such as line shifting and word shifting are proposed in this survey paper [4, 5].

Munesh Trivedi et al. in their research paper titled "Analysis of Several Image Steganography Techniques in Spatial Domain: A Survey" carried out survey which

**Fig. 2** Semantic approach [5] of text steganography

| WORD | SYNONYM |
|---|---|
| Lazy | Idle |
| Hard | Difficult |
| Unhappy | Sad |

**Table 1** Indicator values-based action [21]

| Indicator channel | Channel 1 | Channel 2 |
|---|---|---|
| 00 | No hidden data | No hidden data |
| 01 | No hidden data | 2 bits of hidden data |
| 10 | 2 bits of hidden data | No hidden data |
| 11 | 2 bits of hidden data | 2 bits of hidden data |

was based on image steganography technique [6, 7, 21]. In Sect. 2.1 of their survey work, they mentioned pixel indicator technique (PIT technique ). This technique works in two steps. In the first step, it exploits the properties of RGB color image. RGB color image consists of three planes: red, green, and blue. Any of these frames is selected and its two LSB bit is chosen to act as indicator. These two LSB bits of chosen frame tell that there exists secret data in other two components. According to the survey, many combinations of RGB frames are possible, for example, RGB, BGR, GBR, GRB, BRG, etc., and communicating parties are free to choose which channel is chosen to act as data carrier and which to choose to act as indicator. Table 1 shows indicator values-based action.

Authors [8, 22] proposed a concept which uses Modulus 3 function. A RGB image consists of three planes red, blue, and green. So, first step is to divide the color image into three components, i.e., red, blue, and green. Every pixel of the color image contains 24 bits, i.e., red-8, blue-8, and green-8. Authors have used all the three planes of RGB to embedding data. Pixel value modification (PVM ) is used for hiding secret data. These secret data are embedded by PVM technique in sequential manner, i.e., first three bits of secret data are hidden in first pixel RGB components sequentially. Same procedure is followed for rest of the bits. Function and cases used in the proposed approach are:

$$f = f(g_1, g_2, g_3, \ldots, g_n) = \sum_{gri} \mathrm{mod}\, 3 (\mathrm{for}\, i = 1\, \mathrm{to}\, n) \tag{1}$$

Three cases in this proposed algorithm are as follows:

Case 1: If $f$ is equal to $d$, then no modification is needed. The old value of $gri$ is assigned as a new pixel value.
Case 2: If $f$ is not equal to $d$ and $f < d$, then the value of $gri$ is increased by 1, i.e., $gri + 1$ is the new modified value.
Case 3: If $f$ is not equal to $d$ and $f > d$, then the value of $gri$ is decreased by 1, i.e., $gri - 1$ is the new modified value.

Some authors advocate the ROI type steganography. ROI stands for region of interest. The algorithms which were proposed on this basis target specific regions in cover image. They select only those regions that create less distortion in carrier image after embedding procedure. Different carrier images have different region of interest. So, payload hiding is random depending upon the ROI in carrier image. This approach has randomness creating difficulties for a steganalyst [3, 9, 23].

Some researchers advocate hiding data in edge pixel of the carrier image [10, 24]. Idea is that edge pixel can tolerate variation in comparison to smooth area in carrier image. Edge pixels have generally less similarity with nearby pixels (neighborhood). In the year 2007, a new approach has been proposed which uses the combination of both concepts: LSB and edge pixel. This approach is known as RELSB.

Some more approaches have been introduced which use the LSB substitution and PVD. PVD stands for pixel value differencing. According to authors, their proposed concept has high payload capacity and high imperceptibility [11, 25].

Sonka et al. in their work propose an algorithm that combines two popular algorithms: Canny edge detection algorithm and fuzzy edge detection algorithm [12, 26, 27]. Fusion of these two edge detection algorithm provides more number of carrier pixels for hiding data.

Many of the proposed image steganographic algorithms focus on the edge pixels of the carrier image or cover image. One of the algorithms which was proposed [13, 28] uses popular edge detection technique, i.e., Canny edge detection technique. Selected edge pixels were chosen for hiding purposes. To increase the security of payload, from few years chaotic maps have been used.

Shivani Sharma et al. in their paper title "Audio Steganography using ZDT: Encryption using Indexed Based Chaotic Sequence" proposed a concept that works on their proposed technique i.e. ZDT (Zero Distortion Technique). According to authors HVS, i.e., human visual system is less sensitive in comparison to HAS, i.e., human audio system. Proposed ZDT helps in achieving zero distortion in cover audio file. Authors along with ZDT approach have used the concept of Indexed based chaotic sequence to add one more tier in security [13, 29].

C. P. Sumathi in their research classified the steganographic methods in six categories: substitution method, transform domain, Sspread spectrum, statistical method, distortion technique, and cover generation [30]. In substitution method, redundant parts of carrier image is exploited. In transform domain procedure, transform space is exploited (transform space is utilized for hiding purposes). Spread spectrum uses the concept of spectrum (spread) communication. Statistical procedure as it names implies uses cover image statistical properties [14, 30].

Shilpi Mishra et al. in their research title "Audio steganography technique: a survey" carried out survey about the different types of audio steganography. Authors in their survey mentioned the different types of methods or procedures which are used by researchers nowadays [15, 31].

Pooja et al. in their research conducted a survey on encryption techniques [16]. Figure below shows the performance of different techniques: hybrid technique, double encryption, CE, and RSA with AVK and CBE.

Authors in their paper discusses about symmetric key and asymmetric key cryptography. Authors also put stress on the security of traditional cryptography schemes in reference of modern computing machines. In their survey paper, they also discuss about some modern techniques: RSA based on ECC with AVK, DES-RSA, 3d chaotic map technique, blowfish, JCE, etc. [17].

Mamta et al. in their paper titled "Designing of Robust Image steganography Technique based on LSB insertion and encryption" proposed a methodology in which they encrypt the secret text with the help of RSA algorithm (encryption process) and then hides their data using LSB approach (steganography). In the limitation section of their proposed paper, author clearly mentioned about payload capacity which is less. They also mentioned the image format issue which the user should consider before utilizing this proposed concept [5]. Kairullah in their research work proposed a technique which uses invisible ink concept for hiding the text data. Text medium is chosen as carrier medium because of smaller size than other media [7].

# 3 Conclusion

In the survey paper presented, an effort has been made to discuss the popular algorithms in the area of steganography, cryptography, and metamorphic cryptography. Paper also tries to discuss the procedure they actually follow during their implementation. Paper presented also tries to discuss the strength and limitation of various proposed algorithm in these area. Figure 3 shows the semantic approach of text steganography. Similarly, Fig. 4 contains word spelling approach. Figure 5 shows the performance of five secured algorithms: hybrid technique, double encryption, CE, and RSA with AVK and CBE. Table 2 contains the comparison of blowfish with some of other algorithms on the basis of different parameters. Nowadays, various algorithms have been proposed which are secured enough but for how long? Inventions in the area of computing world where modern computers are becoming more powerful not only in the area of computing but becoming smart too. In such a fast changing world and emerging technologies, there is always a need of security and to face the challenge in the way of security, algorithms must be kept updated or evolved with time.

**Fig. 3** Word spelling approach [5]

| American English | British English |
|---|---|
| Airplane | Aeroplane |
| Fiscal | Financial |
| Unalike | Unlike |

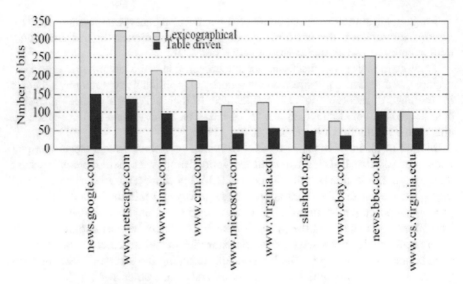

**Fig. 4** Lexicographical versus table driven HTML tags in different Web sites [5]

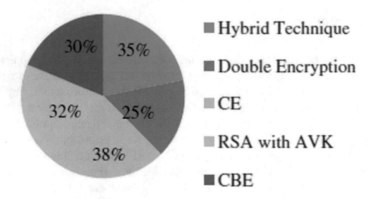

**Fig. 5** Performance of technique [16]

**Table 2** Comparison of various algorithms on the basis of different parameters [3, 16, 17]

| Parameter | Blowfish | Two fish | Three fish |
|---|---|---|---|
| Development | Bruce Schneier in 1993 | Bruce Schneier in 1998 | Bruce Schneier, Niels Ferguson, Stefan Lucks in 2008 |
| Key length (bits) | 32–448 | 128, 192, 256 | 256, 512, 1024 |
| Rounds | 16 | 16 | For 256, 512 key = 72 |
| | | | For 1024 key = 80 |
| Block sizes (bits) | 64 | 128 | 256, 512 and 1024 |
| Attack found | No attack is found to be successful against blowfish | Differential attack, related key attack | Improved related key |
| | | | Boomerang attack |
| Level of security | Highly secure | Secure | Secure |
| Possible keys | 232, 2448 | 2128, 2192, 2256 | 2256, 2512, 21024 |
| Time requires to check all possible keys | For a 448 bit $10^{116}$ year | | For 512 bit and 33 round |
| | | | Time complexity is 2355.5 |

| Parameter | Camellia | ECC | SAFER |
|---|---|---|---|
| Development | By Mitsubishi Electric and NTT in 2000 | Victor Miller from IBM and Neil Koblitz in 1985 | By Massey in 1993 |
| Key length (bits) | 128, 192 or 256 bits | Smaller but effective key (example 512 bit) | For SAFER K-64, 64 bit |
| | | | For SAFER K-128, 128 bit |
| Rounds | 18 or 24 | 1 | 4.75 |
| Block sizes (bits) | 128 bits | Stream size is variable | 64 |
| Attack found | In future, algebraic attack, such as Extended Sparse Linearization | Doubling attack | Linear cryptanalytic attack |
| Level of security | Secure | Highly secure | Secure |
| Possible keys | 2128, 2192, 2256 | 2512 | 264, 2128 |
| Time requires to check all possible keys | – | For 512 bit, $3 \times 10^4$ MIPS-years | – |

**Acknowledgements** I would like to thanks my guide, my parents and my friends without their help and support it was very difficult to carry out this survey work.

# References

1. S. Batham, A.K. Acharya, V.K. Yadav, R. Paul, A new video encryption algorithm based on indexed based chaotic sequence, in *Proceedings of Fourth International conference Confluence 2013: The Next Generation Information Technology Summit*, 27–28 Sept, pp. 139–143. (Available at IET and IEEE explorer)
2. M.C. Trivedi, S. Mishra, V.K. Yadav, Metamorphic cryptography using strength of chaotic sequence and XORing method. J. Intell. Fuzzy Syst. (SCI Indexed Journal) **32**(5), 3365–3375
3. R. Ganga Sagar, N. Ashok Kumar, Encryption based framework for cloud databases using AES algorithm. Int. J. Res. Stud. Comput. Sci. Eng. (IJRSCSE) **2**(6), 27–32 (2015)
4. M.C. Trivedi, S. Sharma, V.K. Yadav, Zero distortion technique: an approach to image steganography using strength of indexed based chaotic sequence, in *Symposium Proceedings Published by Springer in Communications in Computer and Information Science Series (CCIS), SSCC-2014*, vol. 467 (2014), pp. 407–416. ISSN: 1865:0929
5. M. Juneja, P.S. Sandhu, Designing of robust image steganography technique based on LSB insertion and encryption, in *Proceedings of 2009 International Conference on Advances in Recent Technologies in Communication and Computing* (available at IEEE Explore). https://doi.org/10.1109/artcom.2009.228
6. S. Sharma, A. Gupta, M.C. Trivedi, V.K. Yadav, Analysis of different text steganography techniques: a survey, in *Proceedings of Second International Conference on Computational Intelligence and Communication Technology*. IEEE CPS and made available on IEEE Explorer
7. M. Khairullah, A novel text steganography system using font color of the invisible characters in Microsoft Word Documents, in *Proceeding of 2009 Second International Conference on Computer and Electrical Engineering*. https://doi.org/10.1109/iccee.2009.127
8. M.C. Trivedi, S. Sharma, V.K. Yadav, Analysis of several image steganography techniques in spatial domain: a survey, in *Proceedings of Second International Conference on Information and Communication Technology for Competitive Strategies (ICTCS-2016)*. ACM-ICPS (Proceeding volume ISBN No 978-1-4503-3962-9). https://doi.org/10.1145/2905055.2905294
9. A.A.-A. Gutub, Pixel indicator technique for RGB image steganography. J. Emerg. Technol. Web Intell. **2**(1) (2010)
10. V. Nagaraj, et al., Color image steganography based on pixel value modification method using modulus function, in *International Conference on Electronic Engineering and Computer Science* (Elsevier, 2013)
11. R. Roy, et al., Chaos based edge adaptive image steganography, in *Elsevier International Conference on Computational Intelligence: Modeling Techniques and Applications (CIMTA)* (2013)
12. M. Singh, B. Singh, S.S. Singh, Hiding encrypted message in the features of images. IJCSNS **7**(4) (2007)
13. H.C. Wu, N.I. Wu, C.S. Tsai, M.S. Hwang, Image steganographic scheme based on pixel-value differencing and LSB replacement methods. Proc. Instrum. Electr. Eng. Vis. Images Signal Process. **152**(5), 611–615 (2005)
14. C.P. Sumathi, T. Santanam, G. Umamaheswari, A study of various steganographic techniques used for information hiding. Int. J. Comput. Sci. Eng. Surv. (IJCSES) **4**(6) (2013). https://doi.org/10.5121/ijcses.2013.46029
15. W.-J. Chen, C.-C. Chang, T. Hoang Ngan Le, High payload steganography mechanism using hybrid edge detector. Expert Syst. Appl. **37**, 3292–3301 (2010)

16. P. Dixit, A. Gupta, M.C. Trivedi, V.K. Yadav, Traditional and hybrid encryption techniques: a survey, in *Proceedings of International Conference on Recent Advancement in Computer, Communication and Computational Sciences (ICRACCCS-2016)*. Lecture Notes on Data Engineering and Communications Technologies (Springer, 2016)
17. R. Bhanot, R. Hans, A review and comparative analysis of various encryption algorithms. Int. J. Secur. Appl. **9**(4), 289–306 (2015)
18. M. Sonka, V. Hlavac, *Boyle, Image Processing, Analysis, and Machine Vision* (Thomson Brooks/Cole, 1999)
19. J. Canny, A computational approach to edge detection. IEEE Trans. Pattern Anal. Mach. Intell. **8**(6), 679–698 (1986)
20. S. Sharma, M.C. Trivedi, V.K. Yadav, A. Gupta, Audio steganography using ZDT: encryption using indexed based chaotic sequence, in *Proceedings of Second International Conference on Information and Communication Technology for Competitive Strategies (ICTCS-2016)*. ACM-ICPS (Proceeding volume ISBN No 978-1-4503-3962-9). http://dx.doi.org/10.1145/2905055.2905272

# Web and Informatics

# Enhancement in Algorithmic Design for Scalability Measurement of Web Application

**Pankaj N. Moharil, Sudarson Jena and V. M. Thakare**

**Abstract** Scaling the Web applications with more workload and the number of users becomes very crucial. Performance testing of a Web application specifies the applications' functionality, stability, and robustness. As the scalability testing is the extension of performance testing, it identifies the major workload and alleviates bottlenecks that can hamper the scaling of the application and it specifies the degree to which the test is developed. This paper proposes an algorithmic design for measuring scalability of Web applications to get the increased performance with few steps to do a test. This design allows adding additional users and working load to the available environment and encountering the linear changes with software and hardware which affect the server performance. An application development process requires careful planning for making scalable application, and it is important to test scalability problems regularly and rigorously.

**Keywords** Requirement analysis · Performance · Scalability design · Scalability testing

## 1 Introduction

The development process of an application requires careful planning for making scalable applications. Performance testing of an application verifies that the system meets the specification stated by the client or end user, whereas the scalability of a system indicates its ability to either administrate an increasing amount of work to quickly expand as demands increase. Scalability planning should be done at the time of project development. To determine the scalability of a Web application, one should go through the performance testing factors such as response time, load balancing, mean time to failure and perform tuning. Scalability is not unprocessed pace about operating system or particular software technologies, hardware technologies and

P. N. Moharil (✉) · V. M. Thakare
SGBAU, Amaravati, India
e-mail: pmohril@gmail.com

S. Jena
SUIIT, Sambalpur, Odisha, India

© Springer Nature Singapore Pte Ltd. 2021
X.-Z. Gao et al. (eds.), *Advances in Computational Intelligence and Communication Technology*, Advances in Intelligent Systems and Computing 1086,
https://doi.org/10.1007/978-981-15-1275-9_21

storage technologies. It does not get a scale if it is not designed to scale. Now the question arises on how to test Web application and what are the factors that need to be considered to check the scalability of the application [1]. Scalability in Web and other distributed applications is a complicated problem; thus, it is compulsory to amend the behavior of the application or system after making changes.

Scalability is the adequacy to boost the ability to produce continuous access in service amplitude. The key normal for scalable application is that extra load just requires extra assets instead of broad adjustment of the application itself [2]. Scalability of an application should have equivalent assistance among the two clear domains: software and hardware [3]. While designing the scalability of an application, it is necessary to clarify the efficiency of resource management. The designer should know that the scalability is not limited to any component of an application. An application designer should consider the scalability at all levels. The designer should consider some points at the time of designing scalability of a system [4].

- Use of short-time interval.
- Clear resource management.
- Arrangement for changeability.
- Arrangement for replaceability.
- Distribution of services and tasks.

In clear resource management, one thing should be concentrated on dispute resources which is the mean reason for the scalability problem. The application cannot be scaled when there is inadequate memory, processor, bandwidth, and cycles. One should order the use of resources from large to shot. Round robin technique should be used while acquiring resources. The application has a finite amount of resources while getting short resources to obtain the resource as early as possible. Shorter the amount of time that a process is using a resource, the sooner the resource is will be available to another process [5]. In the arrangement of changeability, wherever the resources are derived you can make it change. Resource amalgamation takes advantage of the change in resources. Com+ and ODBC amalgamation are the best examples of resource amalgamation. The Universal Scalability Law (USL) is an analytical model used to quantify system application scaling. It is universal because it includes Amdahl's law and Gustafson linear scaling as special cases. USL model helps to predict how a system will perform under huge load beyond what you can observe. Unacceptable workload boundaries are highly nonlinear. Capacity limits are scalability limits [6]. The USL point of maximum predicts is the system's maximum throughput. It is a way to access a system's capacity and helps to get a better idea of how close the system is to its maximum capacity. Some performance presentation techniques like parallelization, pooling, caching, and collocation are used for getting better scalability, but these are only qualitative representations. Quantitative representation explicates the performance of the system. Performance models are necessary not only for forecasting but also for explaining scalability measurements. Many people use the term scalability without intelligibly defining it. System scalability is quantifiable, and it must be quantified [7]. The universal law of computational scaling contributes that quantification. USL parametric model has the ability to analytically model the reverse throughput. Parametric model of scalability using USL

$$x(N) = \frac{\gamma(N)}{1 + \alpha(N - 1) + \beta(N - 1)} \tag{1}$$

Explaining the system throughput $x(N)$ at given load $N$, fundamental scaling effects contained in the USL can be represented by measuring system throughput and load on the system as

When ($\alpha = 0$, $\beta = 0$): Equal crash for resist.
When ($\alpha = 0$, $\beta = 0$): Cost of sharing resources.
When ($\alpha = 0$, $\beta = 0$): Decreasing returns from contention.
When ($\alpha = 0$, $\beta = 0$): Negative returns from incoherency.

Each of these contributions can be integrated into an equation model, Eq. (1), to compute the relative capacity. Relative capacity $C(N)$ is the normalized throughput at each successive load $N$, and thus, any throughput can be expressed as $x(N) = C(N) x(1)$.

The load might be induced either by $N$ virtual users generated by load testing tools or real user on a production platform. The three coefficients $\alpha$, $\beta$, $\gamma$ in Eq. (1) can be identified, respectively, with

- Concurrency—This can be interpreted as slope associated with linear rising scalability, when $\alpha = \beta = 0$.
- Contention—(with proportion $\alpha$) Queuing for shared resources or due to serialization. When $\alpha > 0$, $\beta = 0$.
- Coherency—(with proportion $\beta$) Because of deferral for data to wind up predictable by ideals point to point trade of data between resources that are distributed, when $\alpha$, $\beta > 0$.

The independent variable $N$ can speak to either software scalability or hardware scalability.

Software scalability—The number of clients or load generators ($N$) is increased on the settled hardware setup. Hardware scalability—The number of physical processors ($N$) is augmented in the hardware setup while keeping the user load per executing per processor is thought to be preserving the equivalent for each additional processor. Equation (1) expresses that the hardware and software scalability are two sides of the same coin [7]. A nonzero value of $\beta$ is associated with measured throughput that goes reverse. The fitted coefficient in Eq. (1) provides an estimate of the user at maximum scalability will occur with this formula $N \max = \sqrt{((1 - \alpha)/\beta)}$. The two theorems provide the justification for applying the same USL equation to both hardware and software scalability measures.

**Theorem 1** *Amdahl's law for parallel speedup is equivalent to the synchronous queuing bound on throughput in the machine repairman model of processor.*

**Theorem 2** *The USL is equivalent to the synchronous queuing bound on throughput for linear load-dependent machine repairman model of the multiprocessor.*

In this paper, an algorithmic design is introduced to measure the scalability of Web applications. The paper is organized as follows. The proposed algorithmic design is defined in Sect. 2. The implementation of algorithmic design using automated tool NeoLoad is described in Sect. 3. Section 4 describes the results, Sect. 5 describes the conclusion, and Sect. 6 is defined with the future scope.

## 2   Algorithmic Design

This section is proposed with the study of an algorithmic design for measuring the scalability of Web applications to increase the performance of Web applications under huge load.

**Algorithm**

Step 1:   Initialization of requirement.
Step 2:   Hands on work area by user.
Step 3:   Functioning of user.
Step 4:   Performance checking of software, hardware, and network with different load.
Step 5:   Calculation of response time, throughput, threshold load, memory and CPU usage, network, and Web server request by user.
Step 6:   Performance identification of different classes of Web sites.
Step 7:   Plot performance table.
Step 8:   Stop.

A logical representation of scalability measure proposed under algorithmic design is as follows.

Scalability will not scale if it is not designed to scale. Before going to scale the application, first resource collection should be there. Through initial step, requirement gathering is done. The entire software and hardware requirements get gathered. Necessities can be useful and non-useful or non down to earth [8]. The application framework will manufacture scale bunch; the improvement individuals or engineers need to bring versatility into thought through the whole framework advancement cycle. The prerequisite social event will determine the limit of versatility which too is accomplished. Instatement step indicates a few which ought to be viewed as that are

• Collection of software and hardware of requirements.
• Analyzing the goals, objectives, and scope of the project.
• Creating a checklist.

Requirements can be useful or reasonable and non-utilitarian or non-pragmatic. While gathering, the requirement essential objective is the remaining task at hand test. Load ought to get determined with the number of clients, throughput or output, and response time. It is essential to know the association between the number of clients, throughput or output, and response time. System traffic load and discharge times ought to be approved. The workload estimation incorporates the number of functionalities of the programming framework with the assessed response time. Try to determine the no users of Web applications with maximum no parallel or concurrent users where the system is maintaining stability and providing accurate response time to these users. One thing should be assured that the user limit is more than that the required simultaneous users where the application must help when it is conveyed. The user should get the application in a timely manner, and users should be able to do transactions within an assigned time period there should be the identification of failure rates those real user observers under huge load and with the number of transactions. Using automated tools.

Actuate server side strength and abasement. Developers and testers have to create a checklist for scalability testing as follows:

- When a server is at heavy load.
- Check an application server crashes under huge load.
- Check for other middle tier server slow down within the huge load.
- Check for the database server slow down under a huge load.
- Make a check for load balancing of the system and its functionality.
- Need to make a change with the hardware for greater performance.
- Check for deadlock or deadly embrace conditions of assigned resources.

Greater designs of an application have a greater impact on the scalability of an application. There should be the best adjustment with the hardware and software product with qualitative and quantitative analysis and coding of the product. While designing the scalability of an application, make a check for the operating system, higher RAM, processor speed, and automated load testing tools. While selecting the automated tools to guarantee about the outstanding burden where all parts of a web application including the hardware, load adjusting segments and all product segments which are incorporated and the scalability are approved [9, 10]. There should be monitoring on actual users or real-life usages, real load generation, and refinement on scalability tests for accurate emulation of the load. Nowadays, peoples are demanding Internet of things and technical breakthroughs, where developers and tester have to put the real-time user experiences. Challenge is to make such an application which will work as expected. For the greater result of scalability, the tester has to use automated load test tools, like Web load, WAPT, Apache JMeter, and LoadStrom. Utilization of these testing instruments will repeat a practical load which will relate with genuine use of an application. While creating the scalability test plan for performance measure, some points should be considered which are as follows:

- Try to generate load which will stress all n-tier applications and permits matching of the earthy blend of various gatherings accomplishing distinctive kinds of business exercises on the site amid pinnacle workload.
- Validation of response time from the server to ensure the number of parallel users requesting for the pages. Make a check for the correct pages to get served or not by the Web application under huge load and greater stress.
- Allow for the simple safeguarding of visual contents as the application changes. With the goal that the scalability can be re-certify each time when the framework is changed.
- Introduce centralized control for distributed load tests, and try to measure response times of entire transactions.
- Introduce real-time monitoring of real-time users at any point when the load test is in advancement in appointing time frame.
- Introduce a real-time graph which helps in understanding the scalability characteristics.

The scalability test plan should specify the business analysis of an application, the number of users and test situations that will be required for load testing.

To calculate the scalability, attributes create a planned process which should include test plan/scenarios with increased workload as per the scaling target and hitting the server, collection of data, data analyzing, investigation of the problem, and updating the test documents. The use of automated load testing tools will generate the visual script which gives an idea about functionality as well as user actions; time bounding, number of tests to perform and data sources. Single visual scripts should be checked with multiple virtual users to execute load tests' scenarios. Before executing the full load test with an immense number of virtual users of each gathering sort, the following two scenarios should be executed to check for any defects:

- First, execute the test with the small number of virtual users to verify that the system scales up accurately. Every one of business exchange starts with 10 virtual users scaling up to 30–50 virtual users.
- Second, execute the blend of various business exchange contents which starts with 5 virtual users scaling up to 30 virtual users.

Create the documentation by generating different reports from full load test, update the test documents with the past implications, and verify the scalability. Reports can obtain with the virtual user, throughput, response time, threshold load transaction done, and Web server request versus response.

- Virtual Users—Report specifies the elapsed time for each user in time format hh:mm:ss.
- Response Time—This identifies the response time of the server under various types of load and breakpoint load of the user request.
- Throughput—It is the rate at which the number of demand is served by the Web server effectively.
- Threshold Load—It is the number of users requests that Web server could handle with the desired throughput.

- Hit Per Second—This report indicates the number of requests made by the user per second.
- Web Server asks for versus Response—A proportion of how productively the Web servers are grouped to do the required load adjusting.

For doing performance testing of Web application, multiple Performance testing tools are available like load runner, WAPT, and NeoLoad. Scalability test of the Web application should be done using all the classes of a Web application like social, government, organizational, educational, e-commerce, and individual sites to represent the scalability test with its all main characteristics. For doing this, the automated tool should be which is available. By assigning huge load as virtual users with a ramp-up scenario, this can include one or more profiles. This scenario is mostly used for performance, stress, and capacity test. After executing the performance test, results should be presented in a statistical manner with server response, bandwidth, response time, average time, throughput, number of users, response time per second, average request response time, transactions, and error count.

## 3  Implementation of Proposed Algorithmic Design

While executing the performance and versatility of Web applications, this paper utilizes different classes of Web application like social, government, instructive, online business, and amusement. Out of these sites, some sites are separated by utilization and load as nearby and worldwide sites. These local and global sites are contrasted and immense load. The performance is performed utilizing NeoLoad 6.6 performance testing tool. The instrument permits to gauge Web application performance and versatility measure under tremendous load. The heap of 10–50 virtual users is doled out to check the conduct and scalability of the applications. Results are appeared as tables and charts.

### 3.1  Tables Shown Below Provide the Basic Details of the Components and Values Used in the Project

Table 1 shows general statistics for all users, pages, request, throughput, response time, average time, and alerts.

Table 2 shows statistic of controller performance count of all the components with average time—avg., minimum time—min, maximum time—max, truncated mean—avg. %, and standard deviation—std. dev. (Time is calculated in seconds.)

Table 3 shows local host performance count of all the components with average time—avg., minimum time—min, maximum time—max, truncated mean—avg. %, and standard deviation—std. dev. (Time is calculated in seconds.)

**Table 1** Total transaction summary

| Total pages | 65 | Average pages/s | 0.4 |
|---|---|---|---|
| Total requests | 94 | Average requests/s | 0.5 |
| Total users launched | 10 | Average request response time | 10.99 s |
| Total iterations completed | 0 | Average page response time | 5.5 s |
| Total throughput | 1.13 MB | Average throughput | 0.05 Mb/s |
| Total request errors | 0 | Error rate | 0% |
| Total action errors | 0 | Alerts total duration | 0% |

**Table 2** Controller performance counter

| Controller | Min | Avg. | Max | Med | Avg. 90% | Std. dev. |
|---|---|---|---|---|---|---|
| User load | 0 | 7.79 | 10 | 10 | 8.03 | 3.62 |
| Throughput | 0 | 0.061 | 0.273 | 0.05 | 0.056 | 0.064 |
| Alerts | 0 | 0 | 0 | 0 | 0 | – |
| Critical alerts | 0 | 0 | 0 | 0 | 0 | – |
| Warning alerts | 0 | 0 | 0 | 0 | 0 | – |
| CPU load | 1 | 5.81 | 58 | 4 | 4.62 | 7.23 |
| Memory used | 32 | 43.8 | 58 | 42 | 43.6 | 7.83 |
| Thread count | 561 | 566.2 | 574 | 563 | 566 | 5.17 |
| Connected | 1 | 1 | 1 | 1 | 1 | 0 |
| Total disconnections | 0 | 0 | 0 | 0 | 0 | 0 |
| Total reconnections | 0 | 0 | 0 | 0 | 0 | 0 |
| Network/segments sent/sec | 0 | 1138 | 2615 | 1342 | 1126 | 753.4 |
| Network/wlan0 (Intel(R) Wi-Fi Link 1000 BGN)/Mb/r/s | 0 | 0.077 | 0.224 | 0.093 | 0.073 | 0.067 |
| Network/wlan0 (Intel(R) Wi-Fi Link 1000 BGN)/Megabits sent/sec | 0 | 0.01 | 0.093 | 0.006 | 0.007 | 0.013 |
| Default zone/user load | 0 | 7.79 | 10 | 10 | 8.03 | 3.62 |
| Default zone/population1/user load | 0 | 7.79 | 10 | 10 | 8.03 | 3.62 |
| Default zone/population1/user path/user load | 0 | 7.79 | 10 | 10 | 8.03 | 3.62 |

## *3.2 Graphical Representation of Performance*

See Figs. 1, 2, 3, and 4.

**Table 3** Local host performance counter

| Local host | Min | Avg. | Max | Med | Avg. 90% | Std. dev. |
|---|---|---|---|---|---|---|
| CPU load | 1 | 5.63 | 58 | 4 | 4.59 | 6.68 |
| Memory used | 15 | 25.7 | 29 | 27 | 25.8 | 3.21 |
| Throughput | 0 | 0.061 | 0.273 | 0.05 | 0.056 | 0.064 |
| User load | 0 | 7.79 | 10 | 10 | 8.03 | 3.62 |
| Thread count | 42 | 70.2 | 75 | 70 | 70.5 | 4.29 |
| Thread pool used | 0.1 | 0.101 | 0.11 | 0.1 | 0.101 | 0.003 |
| Task pool used | 0 | 0 | 0 | 0 | 0 | 0 |
| Connection/disconnections | 0 | 0 | 0 | 0 | 0 | – |
| Connection/reconnections | 0 | 0 | 0 | 0 | 0 | – |
| Connection/ping duration (ms) | 3 | 3.79 | 4 | 4 | 3.82 | 0.406 |
| Network/segments sent/sec | 0 | 16.7 | 790.4 | 0 | 0.831 | 83 |
| Network/segments retransmitted/sec | 0 | 0 | 0 | 0 | 0 | 0 |
| Net./wlan0/Wi-Fi Link 1000.gb received/sec | 0 | 0.077 | 0.23 | 0.094 | 0.073 | 0.068 |
| Net./wlan0/Wi-Fi L 1000 BGN)/Mb sent/sec | 0 | 0.01 | 0.096 | 0.006 | 0.007 | 0.014 |

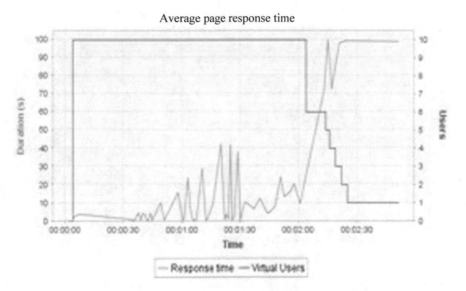

Average page response time

**Fig. 1** Average response time in seconds of all pages during test. Time (Min—0.155, Avg.—5.49, Max—41.932)

Average request response time

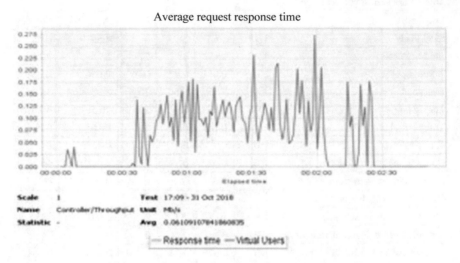

**Fig. 2** Average request response time in seconds of all user requests during test. Time (Min—0.155, Avg.—10.986, Max—99.483)

Request per second

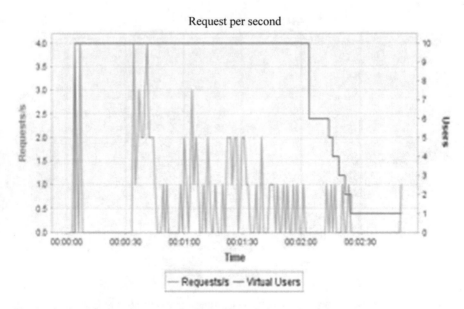

**Fig. 3** Number of request per seconds. Time (Min—0, Avg.—0.5, Max—4)

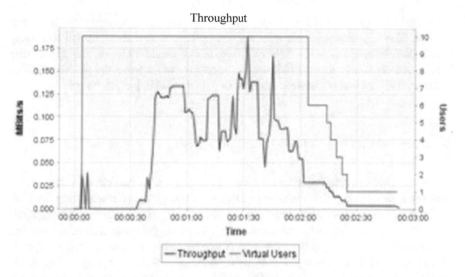

**Fig. 4** Number of megabits of information every second returned by the server. Time (Min—0, Avg.—0.05, Max—0.19)

# 4 Results

Scalability will not scale if it is not designed to scale. It is the adequacy to boost the ability to produce continuous access in service amplitude. This paper uses the NeoLoad automated performance tool to measure the scalability attributes with the heavy load, where the load of 10–50 is assigned as the virtual user in while loading the request to the server. The tool provides means of identifying potential performance bottlenecks of the applications. Here, applications of different domains with different loads are used in ramp scenario to generate the test. Analysis of test results is done by comparing the results from different runs of the same scenario. Results show the numerous statistics on Web applications mainly focuses on the number of pages requested, page response time, and average response time. It is also useful and instructive to combine and check the number of figures against statistics coming from the component in applications' architecture such as application server and database server. Table 1 shows the results statistics with the total transaction count. The tool provides the controller and load generator where the controller launches the scenario with user load and counts the performance. Performance counters which put on the server automatically count number request per second sent to the server to calculate the pages requested, response time, throughput, average response time, and threshold values as shown in Tables 2 and 3. Scalability attributes are defined with graphical representation shown in Figs. 1, 2, 3, and 4, where average page response time, average request response time, average response time of all transactions, throughput, error counts, response time against load, overall throughput of transactions, and load generator CPU usage are shown with time boundaries' minimum time, maximum time, and average time per seconds required by the server to response the users' request. The key characteristic of scalable application is that additional load only requires additional resources rather than extensive modification of the application itself. Add on Hardware and Network is necessary to improve the performance of web applications under different loads with a different kind of hardware.

## 5 Conclusion

Automated load testing tools are utilized for scaling the normal client–server architectures which require the recreation of thousands of virtual users. Scaling the Web application and performance streamlining is the need of the present Web applications. While presenting the adaptability one imperative point ought to be considered is that the financial aspects of scalability. Now and again, the cost of versatility turns into an issue while expanding scalability of the applications. The proposed algorithmic structure clears the scaling of Web application performance under colossal load. Scalability is quantifiable. Hardware and software scalability measures are critical to know while working with the number of resources, virtual users, constant client experiences, and their inputs. Scalability testing is essential and spreads utilitarian and non-practical components of Web applications.

## 6 Future Scope

This paper concentrated on scalability measurement of Web applications and its services. Recognizing and addressing performance of Web application with more noteworthy scalability enhance the nature of Web applications. Scalability measure expands the capacity to deliver nonstop access in administration measurements under enormous load and time required by the executives. An algorithmic design introduced in this paper helps to improve load balancing. It attempts to decide the number of users of the Web application with the most extreme number of parallel and simultaneous users with a variable load, where the framework keeps up dependability and gives exact response time and throughput. Scaling client–server architecture with the automated performance testing tool improves the quality of Web applications.

## References

1. S. Moore, *Centralized Performance Testing. Software Test Professional Spring* (Springer, 2012)
2. E. Proko, I. Ninka, Analyzing & testing web application performance. Int. J. Eng. Sci. **3**(10) (2013)
3. I. Enesi, E. Zanaj, S. Kokonozi, B. Zanaj, Performance evaluation of state full load balancing in predicted time intervals and CPU load, in *IEEE Eurocon*, 6–8 July 2017, Orchid, R. Macedonia

4. P. Moura, F. Kon, Automated scalability testing of software as service, in *IEEE 8th International Workshop on Automation of Software*, May 2013
5. L.G. Williams, C.U. Smith, Web application scalability: a model based approach. Software Research Performance Engineering Services, May 2004
6. J. Hotman, N.J. Gunther, Getting in the zone for successful scalability, arxiv:0809:2541v/[CS.PF], 15 Sept 2008
7. N.J. Gunther, A general theory of computational scalability based on rational functions, arxiv:0808:1431v2/[CS.PF], 25 Aug 2008
8. N.J. Gunther, A new interpretation of Amdahl's geometric scalability, arxiv.org/abs/cs/02, Oct 2007
9. W.T. Tsai, Y. Huang, X. Bai, J. Gao, Scalable architectures for SaaS, in *IEEE 15th International Symposium on Object/Component/Service-Oriented Real-Time Distributed Computing Workshops*, Apr 2012, pp. 112–117
10. S. Sharmila, E. Ramadevi, Analysis of performance testing on web applications. International J. Adv. Res. Comput. Commun. Eng. **3** (2014)

# A Survey on Internet of Things (IoT): Layer-Specific, Domain-Specific and Industry-Defined Architectures

Harmanjot Kaur and Rohit Kumar

**Abstract** Labelled as "The next Industrial Revolution", Internet of Things (IoT) is the technology giant which is rapidly gaining ground in the scenario of modern wireless sensor networks (WSNs) where physical objects (or "things") are connected through the common platform, i.e. Internet. The Internet of Things paradigm is paving the way towards success, providing services in different domains. Single architecture is not enough to provide essential services in all domains. The reference model can be made to provide a starting point for developers looking further for developing strong IoT solutions. In order to facilitate future research and to help the product builder to choose from the different architectures, this paper does an analysis of layer-specific, domain-specific and industry-defined IoT architectures. Intel, IBM and CISCO have also made a push towards developing new models that aims to standardize the complex and fragmented IoT industry by releasing IoT platform reference architectures. This paper contributes the IoT, basic elements of IoT, systematic classification of different IoT architectures and comparison of industry-defined architectures.

**Keywords** Internet of Things · IoT architecture layers · Domain-specific architecture · Industry-defined architecture · CISCO architecture · IBM architecture · Intel architecture

## 1 Introduction

Internet of Things (IoT) term was introduced for the first time in the year 1999 by *Kevin Ashton*, which is an intelligent network infrastructure that allows numerous individually identifiable objects (sensors, actuators, wireless devices, etc.) to interconnect with each other to perform intricate tasks in a supportive manner [1]. According to 2020 conceptual framework, the term Internet of Things (IoT) can be represented using the following general statement [2]:

H. Kaur · R. Kumar (✉)
Chandigarh University, Gharuan, Mohali, Punjab 140413, India
e-mail: rohitbhullar@gmail.com

H. Kaur
e-mail: hammu153@gmail.com

© Springer Nature Singapore Pte Ltd. 2021
X.-Z. Gao et al. (eds.), *Advances in Computational Intelligence and Communication Technology*, Advances in Intelligent Systems and Computing 1086,
https://doi.org/10.1007/978-981-15-1275-9_22

$$IoT = Sensors + Data \text{ (or information)} + Network + Services$$

The International Telecommunication Union (ITU) described about IoT in a report in 2005 that: "*IoT will interconnect objects from the world together in an intelligent and sensory manner*" [3]. In the IoT foresight, the term "things" is a pervasive term that embraces various physical objects, moveable personal items such as smartphones, digital cameras and tablets [4]. As the Internet-connected devices are rapidly growing in number, the traffic generated by them is likely to augment considerably. For instance, Cisco guesstimates about the generation of Internet traffic by IoT devices will climb to 70% in year 2019 from 40% in 2014 [5]. One more anticipation made by Cisco is that the number of device-to-device (D2D) connections will augment from 24% (in year 2014) to 43% (in year 2019) of all connected IoT devices [6].

IoT is a parasol term that comprises diverse categories [7]: Wireless sensors or actuators, RFID tags, wearables connected to Internet, devices with Internet connectivity facility, Bluetooth-enabled smartphones, smart homes, connected cars, etc. So, single architecture is not enough that goes well with all these applications of IoT for their diverse needs. Vendors like CISCO, INTEL and IBM have come up with their own reference architectures. When the enthusiasts and the researchers are developing any IoT product, the scarcity of comprehensive knowledge regarding these IoT architectures puts a ceiling on them to prefer any one particular architecture [8].

The rest of the paper is arranged as Sect. 2 presents the basic elements of Internet of Things in tabular form, in Sect. 3, architectural considerations are summarized, and Sect. 4 demonstrates the comparisons of different IoT-based industry-defined architectures. Section 5 describes the conclusion of the paper that shows the usefulness of the work presented in this paper.

# 2   Basic Elements of IoT

To better understand the concept of the emerging technology—Internet of Things, one should have the general knowledge of the basic building blocks of this technology. IoT implementation includes the following basic elements as shown in Table 1 [9].

# 3   Architectures of IoT

There are numerous architectures that have been proposed by various researchers in the recent years of emergence of IoT technology. They can be classified in three different ways:

i.   Layer-specific architectures
ii.  Domain-specific architectures

**Table 1** Basic elements of IoT

| Elements of IoT | Description |
| --- | --- |
| Sensors | Sensors perceive information from the environment |
| Actuators | Actuators perform the actions |
| Things | Responsible for communication and gathering the data without any human involvement |
| Gateways | Acts as a middleware between the devices and network, enables connectivity among devices and other infrastructures, and during the information flow, it provides security and management facilities |
| Network infrastructure (NI) | NI IoT devices include routers, aggregators, gateways, repeaters, it provides data flow control from devices or things to the cloud infrastructure, and during flow of data, it also enables security |
| Cloud infrastructure (CI) | CI IoT devices are virtualized server (VS) and data storage unit (DSU), and it allows advanced computing functionalities and the storage facilities |
| RFID tags | Physical objects are equipped with radio frequency identification (RFID) tags or other identification barcodes. Smart sensor devices can sense these codes [10] |

iii. Industry-defined architectures

These are further classified as shown in Fig. 1.

## 3.1 Layer-Specific Architectures

On the basis of number of layers, IoT architectures can be classified as shown in Fig. 2.

**Three-Layer Architecture**: According to the work performed in paper [11], it comprises three layers: (1) *Perception Layer* (or sensor layer/recognition layer) [17] is the lower most layer that interrelates the physical things and components via smart devices (sensors, actuators, RFID tags, etc.) (2) *Network Layer*, well-known as gateway layer or transmission layer, is imposed as the central layer of IoT architecture [18]. It allows the integration of various devices (switches, hub, gateway, etc.) and various communication technologies (like LTE, Bluetooth, WiFi, etc.) (3) *Application Layer*, or business layer, is the upmost layer in IoT architecture [19]. It receives the information transmitted from the network layer and uses this information to allocate required services [11]. *Advantages*: simplicity, effortless problem identification and flexibility. *Limitations*: absence of application layer security [20].

**Four-Layer Architecture**: P. P. Ray, in his work [12], describes a service layer, that acts as a middleware between network and interface (or application) layer, that is responsible for providing the data services in IoT. Following this concept, a service-oriented architecture (SoA) has recently been introduced by the researchers to sustain

**Fig. 1** Classification of IoT architecture

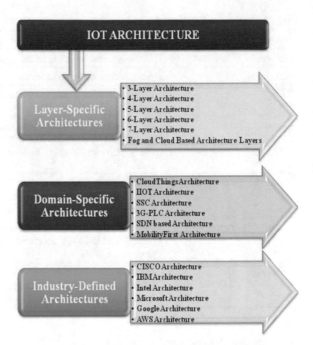

**Fig. 2** Classification of layer-specific architecture [8, 9, 11–16]

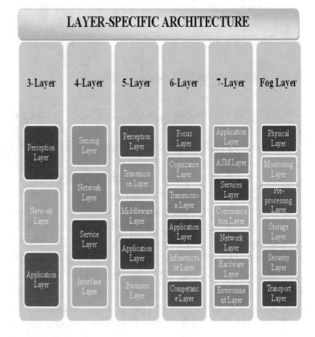

IoT [19, 21]. Its main functions are the SOA-based IoT helps in reduction of time of product development, it helps in designing the workflow in simple way, and it makes the process of marketing the commercial products easier.

**Five-Layer Architecture**: Rafiullah Khan in [13] illustrates the IoT architecture as arrangement of five layers as shown in Fig. 2. Main functions of layers of five-layer model are (1) *Perception layer* perceives and gathers the information about physical objects and passes it to the next layer (2) *Transmission layer* helps in secure transmission of data to the next layer. It encompasses wireless networks like LTE, Wifi, WiMax, Zigbee, etc (3) *Middleware layer* is associated with data storage (4) *Application layer* performs the function of application management (5) *Business layer*, the upmost layer, helps in data analytics, and based on that, it takes the further decisions [22].

**Six-Layer Architecture**: In paper [14], the author introduced a six-layer architecture in which two new layers are augmented into the previous IoT model. These are (1) *MAC layer* that helps to monitor and control the devices (2) *Processing and Storage layer* that helps to process the queries, analyse and store the data [22].

In paper [9], six-layer architecture was proposed for IoT as shown in Fig. 2. The different functions of these layers are (1) *Focus layer* is mainly concerned with identification of devices (2) *Cognizance layer* determines the sensing adequacy (3) *Transmission layer* allows transmission of data (4) *Application layer* determines the collection and categorization of information according to the necessity in the application region (5) *Infrastructure layer* provides facilities that are concerned with SOA (6) *Competence business layer* allows to analyse the business networks of the IoT systems.

**Seven-Layer Architecture**: Researchers proposed a seven-layer architecture [15], which consists of seven-layers as shown in Fig. 2: (1) *Application layer* gathers data about a particular task according to client's needs (2) *Application support and management layer* determines the security and is responsible for management of IoT devices and their operations (3) *Services layer* determines various activities performed by the developers in order to provide essential services to the customers (4) *Communication layer* provides a connection between the sensing layer and service layer to transmit the data (5) *Network layer* helps devices in transmission and processing of data through Internet (6) *Hardware layer* allows the integration of hardware components that are needed for deploying the IoT system (7) *Environment layer* determines the possibility of detection of physical objects [9].

**Fog- and cloud-based Architecture Layers**: In paper [8], a fog layered architecture [16] is presented which comprises six layers as demonstrated in Fig. 2: (1) *Physical Layer* is concerned with the analysis of things. (2) *Monitoring layer* monitors the factors like resources and services by the customers and diverse responses. (3) *Pre-processing layer* performs the functions like processing, filtering and analysing the data. (4) *Storage Layer* performs the function of storing the information and distributing the data when needed. (5) *Security layer* ensures the security and privacy of the data. (6) *Transport Layer* is the uppermost layer that allows transmission of data.

## 3.2  Domain-Specific Architectures

**CloudThings Architecture**: In his work [23], J. Zhou described smart home-based scenario for IoT. In this perspective, a CloudThings architecture was introduced that is a cloud-centric IoT platform which makes use of three services: Platform as a Service (PaaS), Infrastructure as a Service (IaaS) and Software as a Service (SaaS). Collaboration of IoT with Cloud presents a feasible approach that smooths the progress of development of an application [20].

**Industrial IoT Architecture**: The architecture for green industrial IoT is demonstrated in Fig. 3. It comprises: (1) *Sense entities domain* that involves sense nodes (SNs), gateway nodes (GNs) and control nodes (CNs). (2) *Constrained RESTful Network* (3) *Cloud Server* (4) *User applications* [24].

**Smart and Connected Communities (SCC) IoT Architecture**: In their work [26], the authors introduced an IoT-based SSC architecture for smart cities that comprises four different layers, as illustrated in Fig. 4: (1) *Responding layer (or Sensing Layer)* is the outer most layer which contains smart devices that interact with the

**Fig. 3**  Energy-efficient IIoT architecture [25]

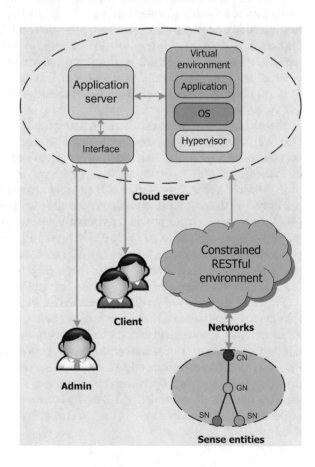

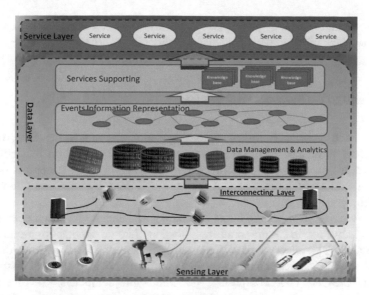

**Fig. 4** IoT architecture for smart and connected communities [26]

IoT system (2) *Interconnecting layer* allows information exchange and data transfer among different devices and different domains (3) *Data layer* performs the tasks like storing huge amount of heterogeneous and trivial data, extracting necessary knowledge from perceived data (4) *Service layer* provides the interface between IoT system and users in the town.

**3G-PLC (Power Line Communication) Architecture**: In paper [27], the authors described 3G-PLC-based IoT architecture, which collaborates two complex communication networks: Third generation (3G) network and Power Line Communication (PLC) network. Scalability factor is the main motive after using the 3G and PLC networks. The IoT framework layers such as perception layer, network layer, aggregation layer and application layer are integrated with the purpose of discovering a new 3G-PLC IoT-based architecture. *Advantages*: Enhanced services as compared to backhaul network opponents, reduced cost of network construction. *Limitation*: Deficiency of incorporation of network heterogeneity factor [20].

**Software-Defined Network (SDN)-Based Architectures**: The author of the paper [10], Z. Qin et al. designed a SDN-based IoT architecture. The motive behind implementation of this architecture was to enable various IoT tasks in heterogeneous wireless network environments with high-level QoS. *Advantages*: Flexibility, effectiveness in terms of data flow and efficient resource management, improved scalability, context-aware semantic information retrieval and quick and easy deployment of resources. *Limitation*: lacks in the management of heterogeneous IoT networks.

**MobilityFirst Architecture**: In paper [28], the authors introduced a name-specific Future Internet Architecture (FIA) termed as MobilityFirst that helps in addressing various challenges related to mobile phones when they act as spontaneous gateways of WSNs in IoT systems. The capacity of the system is analysed and compared to the sensor data rate at a given hotspot. *Advantages*: Ad hoc services, high security. *Limitation*: lack of incentive mechanisms for mobile contributors [20].

## 3.3   Industry-Defined Architectures

**CISCO Architecture**: It follows an edge model, the processing layer of five layer model is divided into three layers, where data aggregation like collecting data from group of devices is done by data abstraction layer, storage is done by data communication layer, and real processing is done by edge computing layer [8, 29, 30].

**IBM Architecture**: IBM's cloud computing architecture has been converted to model suitable for IoT with some modifications like device handling and device management [31]. Most of the IBM architecture for IoT deals with middleware rather than the complete architecture. Bringing the power of IBM's Watsons IoT [32] with CISCO's edge computing, the architecture components revolve around the four major components: connect, information management, analytics and risk management [8].

**INTEL Architecture**: The INTEL System Architecture Specification (SAS) architecture comprises 8-layers: the business layer, application layer, control layer, management layer, data layer, communications and connectivity layer, security layer and developer enabling layer [33].

**Microsoft Architecture**: The architecture recommended by Microsoft in document [34] for IoT applications is cloud native, micro-service and serverless-based. Microsoft recommended sub-systems communicate over HTTPS/REST using JSON (as it is human readable) though binary protocols should be used for high performance needs. The architecture also supports a hybrid cloud and edge compute strategy.

**Google Architecture**: Google reference architecture for IoT introduced by Google Cloud platform comprises three basic components: (1) *Device*: includes hardware/software that has direct contact with the world and establishes connection to a network for communication among each other, (2) *Gateway*: helps devices to connect to the cloud services, (3) *Cloud*: device data is sent to the cloud for processing and combining it with the data generated from other devices [35].

**AWS IoT Architecture**: Amazon Web Services (AWS), reference architecture for IoT released in 2015, is a managed cloud platform within Amazon IoT solutions. It comprises device gateway, message broker, rules engine, security and identity service, thing registry, device shadows and device shadow service [36].

| Company | Searches on Google | Tweets on Twitter | Newspaper & Blogs | LinkedIn survey |
|---------|--------------------|--------------------|--------------------|-----------------|
| Intel | 1K | 2.6K | 4K | 616 |
| Microsoft | 480 | 1.6K | 26K | 545 |
| Cisco | 1K | 1.4K | 5K | 719 |
| Google | 390 | 3.1K | 21K | 99 |
| IBM | 720 | 1.5K | 7K | 504 |

**Fig. 5** Analysis of top five IoT companies on the basis of survey performed in 2015 [37]

**Fig. 6** Top five IoT companies ranking [37]

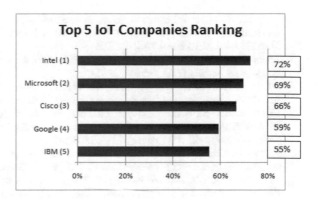

## 4   Comparison of Industry-Defined Architectures for IoT

Lueth [37] performed a survey to know about the top most Internet of Things companies. The author collected the data from different sources like Google, LinkedIn, Company Websites, Twitter, IoT Analytics as described in Fig. 5.

Based on the scores gathered from different resources, it was clear that the Intel company has overtaken Google in 2015 in terms of searches on Google and LinkedIn survey of the number of employees that carry "Internet of Things" as their tag. Therefore, on this basis, the growth of the companies can be represented in terms of a bar chart as shown in Fig. 6 which represents the top five IoT companies rankings.

## 5   Conclusion

As IoT is rapidly gaining popularity in the technological environment, there is also a need of specific standard architecture. However, there is no one particular architecture standardization. Therefore, in order to gain knowledge about the IoT architectures, this paper summarized the current state-of-art of Internet of Things architectures that are classified on three different bases, i.e. layer-specific architectures, domain-specific architectures and industry-defined architectures along with the comparison of the industry-defined architectures. However, IoT architecture in some application domains like in agriculture, in noise detection systems, etc., still needs to be explored.

# References

1. T.D. Nguyen, J.Y. Khan, D.T. Ngo, Energy harvested roadside IEEE 802.15.4 wireless sensor networks for IoT applications. Ad Hoc Netw. (2016). http://dx.doi.org/10.1016/j.adhoc.2016. 12.003
2. M. Bilal, A review of internet of things architecture, technologies and analysis smartphone-based attacks against 3D printers (2017). arXiv, arXiv:1708.04560. Available: https://arxiv.org/abs/1708.04560. Accessed 22 Sept 2018
3. International Telecommunications Union, ITU Internet Reports 2005: The Internet of Things. Executive Summary (ITU, Geneva, 2005)
4. I. Ungurean, N. Gaitan, V.G. Gaitan, An IoT architecture for things from industrial environment, in *2014 10th International Conference on Communications (COMM)*, Bucharest, 2014, pp. 1–4
5. Cisco Visual Networking Index: Forecast and Methodology, 2014–2019. Cisco, 27 May 2015. http://www.cisco.com/c/en/us/solutions/collateral/service-provider/ip-ngn-ip-next-generation-network/white_paper_c11-481360.pdf
6. K. Rose, S. Eldridge, L. Chapin, The Internet of Things: an overview—understanding the issues and challenges of a more connected world. The Internet Society (ISOC), Oct 2015
7. M. Swan, Sensor Mania! The Internet of Things, wearable computing, objective metrics, and the quantified self 2.0. J. Sens. Actuator Netw. (2012). https://doi.org/10.3390/jsan1030217
8. N.M. Masoodhu Banu, C. Sujatha, IoT Architecture a Comparative Study. Int. J. Pure Appl. Math. **117**(8), 45–49 (2017). https://doi.org/10.12732/ijpam.v117i8.10
9. N.M. Kumar, P.K. Mallick, The Internet of Things: insights into the building blocks, component interactions, and architecture layers, in *International Conference on Computational Intelligence and Data Science (ICCIDS)*, 2018
10. Z. Qin, et al., A software defined networking architecture for the Internet-of-Things, in *2014 IEEE Network Operations and Management Symposium*, May 2014, pp. 1–9
11. J. Lin, W. Yu, N. Zhang, X. Yang, H. Zhang, W. Zhao, A survey on Internet of Things: architecture, enabling technologies, security and privacy, and applications. IEEE Internet Things J. **4**(5), 1125–1142 (2017)
12. P.P. Ray, A survey on Internet of Things architectures. J. King Saud Univ. Compu. Inf. Sci. (2016). http://dx.doi.org/10.1016/j.jksuci.2016.10.003
13. R. Khan, S.U. Khan, R. Zaheer, S. Khan, Future internet: the internet of things architecture, possible applications and key challenges, in *10th International Conference on Frontiers of Information Technology*, 2012
14. I.B. Thingom, Internet of Things: design of a new layered architecture and study of some existing issues. IOSR J. Comput. Eng. (IOSR-JCE) (2015)
15. D. Darwish, Improved layered architecture for Internet of Things. Int. J. Comput. Acad. Res. (IJCAR) **4**, 214–223 (2015)
16. P. Sethi, S.R. Sarangi, Internet of Things: architectures, protocols, and applications. J. Electr. Comput. Eng. https://doi.org/10.1155/2017/9324035
17. H. Suo, J. Wan, C. Zou, J. Liu, Security in the internet of things: a review, in *2012 international conference on Computer Science and Electronics Engineering (ICCSEE)*, 2012, pp. 648–651
18. M. Leo, F. Battisti, M. Carli, A. Neri, A federated architecture approach for internet of things security, in *Proceedings of 2014 Euro Med Telco Conference (EMTC)*, Nov 2014

19. M. Al-Fuqaha, M. Guizani, M.Aledhari Mohammadi, M. Ayyash, Internet of Things: a survey on enabling technologies, protocols, and applications. IEEE Commun. Surv. Tutor. **17**(4), 2347–2376 (2015)
20. I. Yaqoob et al., Internet of Things architecture: recent advances, taxonomy, requirements, and open challenges. IEEE Wirel. Commun. **24**(3), 10–16 (2017)
21. L.D. Xu, W. He, S. Li, Internet of Things in industries: a survey. IEEE Trans. Ind. Inf. **10**(4), 2233–2243 (2014)
22. T. Ara, P.G. Shah, M. Prabhakar, Internet of Things architecture and applications: a survey. Indian J. Sci. Technol. **9**(45), 0974–5645 (2016). https://doi.org/10.17485/ijst/2016/v9i45/106507
23. J. Zhou, et al., Cloudthings: a common architecture for integrating the Internet of Things with cloud computing, in *IEEE 17th International Conference on Computer Supported Cooperative Work in Design*, 2013, pp. 651–657
24. K. Wang, Y. Wang, Y. Sun, S. Guo, J. Wu, Green industrial Internet of Things architecture: an energy-efficient perspective. IEEE Commun. Mag. **54**(12), 48–54 (2016). https://doi.org/10.1109/MCOM.2016.1600399CM
25. Green Industrial Internet of Things Architecture: An Energy-Efficient Perspective—Scientific Figure on Research Gate. Available from: https://www.researchgate.net/figure/Energy-efficient-IIoT-architecture_fig1_311750469. Accessed 23 Sept 2018
26. Y. Sun, H. Song, A.J. Jara, R. Bie, Internet of Things and big data analytics for smart and connected communities. IEEE Access **4**, 766–773 (2016). https://doi.org/10.1109/ACCESS.2016.2529723
27. H.-C. Hsieh, C.-H. Lai, Internet of things architecture based on integrated PLC and 3G communication networks, in *IEEE 17th International Conference on Parallel and Distribution Systems*, 2011, pp. 853–56
28. J. Li et al., A mobile phone based WSN infrastructure for IoT over future internet architecture, in *IEEE International Conference on Cyber, Physical and Social Computing and Green Computing and Communications, IEEE and Internet of Things*, pp. 426–33, 2013
29. The Internet of Things Reference Model, Whitepaper from Cisco, 2014
30. Z. Javeed, Edge Analytics, the pros and cons of immediate, local insight, May 2018. Available: https://www.talend.com/resources/edge-analytics-pros-cons-immediate-local-insight/. Accessed 22 Sept 2018
31. Gerber, Simplify the development of your IoT solutions with IoT architectures, IBM Document, 7 Aug 2017
32. Watson IoT Platform Feature Overview, IBM Cloud, 2018. Available: https://console.bluemix.net/docs/services/IoT/feature_overview.html#feature_overview
33. The Intel IoT platform, Architecture Specification White Paper
34. Microsoft Azure IoT Reference Architecture, Microsoft (2018) https://azure.microsoft.com. https://azure.microsoft.com/en-us/blog/getting-started-with-the-new-azure-iot-suite-remote-monitoring-preconfigured-solution/
35. Overview of Internet of Things (2018) https://cloud.google.com/. https://cloud.google.com/solutions/iot-overview. Accessed 29 Sept 2018
36. 9 platforms that have revolutionised the world of things connectivity — part 1. https://itnext.io/9-platforms-that-have-revolutionised-the-world-of-things-connectivity-part-1-2664624d6f65. Accessed 29 Sept 2018
37. K.L. Lueth, The top 20 Internet of Things companies right now: Intel overtakes Google, IoT Analytics, 24 Feb 2015. Available: https://iot-analytics.com/20-internet-of-things-companies/

# A Survey on Social Networking Using Concept of Evolutionary Algorithms and Big Data Analysis

Rutuja Nawghare, Sarsij Tripathi and Manu Vardhan

**Abstract** The primary motive of this survey is to impart complete analysis of social connectivity issues using concepts of big data analysis. Issues related to social networking include community detection, cluster analysis, personalized search, anomaly detection, searching cut space, friendship selection, load balancing, structural balancing, etc. These issues can be improvised using machine learning techniques including particle swarm optimization, evolutionary algorithms, genetic algorithm, properties of graphs such as embeddedness and triadic closure. These properties and algorithms enhance the behavior of social networking issues when applied to datasets of various social networking platforms.

**Keywords** Evolutionary algorithms · Community detection · Social connectivity · Genetic algorithm · Particle swarm optimizer

## 1 Introduction

Social network manages the proportions of collaboration among the general population. It incorporates subjective and quantitative associations and collaborations with other individuals inside a group of friends or family, companions and colleagues. The network can be expanded through different routes inside social destinations. This can be accomplished by means of direct or indirect network existing inside a network. Probability of two networks expanding their availability by trading thoughts of their own region important to one another is also there. Thus, lots of suggestions and thoughts get broadcast inside a social networking site.

R. Nawghare (✉) · S. Tripathi · M. Vardhan
National Institute of Technology Raipur, Raipur, India
e-mail: rnawghare.mtech2017.it@nitrr.ac.in

S. Tripathi
e-mail: tripathi.cs@nitrr.ac.in

M. Vardhan
e-mail: mvardhan.cs@nitrr.ac.in

© Springer Nature Singapore Pte Ltd. 2021          277
X.-Z. Gao et al. (eds.), *Advances in Computational Intelligence and Communication Technology*, Advances in Intelligent Systems and Computing 1086,
https://doi.org/10.1007/978-981-15-1275-9_23

In parallel to the massive growth of social connectivity, interconnection within data is also increasing. This diversity has influenced people to interpret and process new knowledge. As most of these data are derived and resides within the Internet, there is an open challenge is deciding how to gather, dissect and follow up on generated enormous information.

Enormous information is regularly portrayed utilizing five Vs: velocity, value, volume, veracity and variety. Volume refers to the measure of the information created, velocity refers to the rate at which information is produced and moved around, variety refers to various kinds of information, veracity refers to the governance and reliability of information, and value intends to derive some valuable information [1].

Ongoing long range informal communication sites like Twitter, Facebook, LinkedIn, YouTube and Wikipedia have caught exabytes of data coordinated with day by day cooperation of individuals alongside associating huge client populace. Social networking is crafted by social researchers in the brace to human interpersonal organizations, mathematicians and physicists in the subject of complex system hypothesis, and most recently, PC researchers in the examination of data or Internet-empowered social networks [1].

As shown in Fig. 1, data from social sites like Twitter, Facebook, LinkedIn, Instagram, etc., web services, mobiles, and tablets are collected from various platforms. Processing based on various attributes is done on saved dat7a. These processed data are then further analyzed based on various constraints such as efficiency, connectivity and number of nodes.

To deal with the increasing data, evolutionary algorithm is the better choice. Also, the evolutionary algorithm has various real-time applications. It includes biometric inventions, cluster detection, gene expression profiling, encryption and code breaking, real parameter optimization, fuzzy cognitive map learning, automotive learning, etc.

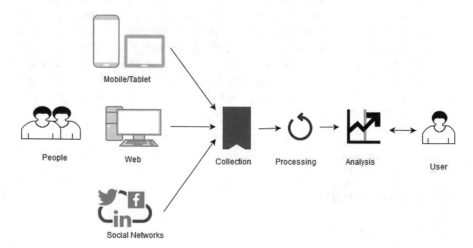

**Fig. 1** Process model of big data for social media

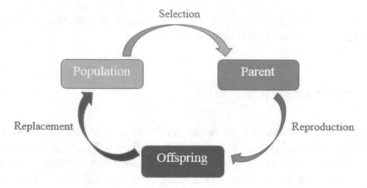

**Fig. 2** Flowchart of evolutionary algorithms

Evolutionary algorithms are the ones whose processing is being inspired by biological evolution. Evolutionary algorithms can be classified into various types. First is the bio-inspired evolutionary algorithms which are inspired from living creature and human life. Second is the nature-inspired evolutionary algorithm which is inspired from the natural phenomenon and physical rule. The third is the socio-inspired evolutionary algorithms inspired by the political rules that govern humans [2].

Figure 2 shows the flow of the evolutionary algorithm. The first step in the evolutionary algorithm is selection where the best-fitted individual is selected for the further processing and remaining are discarded or used for the next iteration. Selected individuals are survived till the occurrence of next individual who is more fit than the current one comes into the picture. This process is continued until the selection criterion is met.

Basically, in order to solve complex problems, evolutionary algorithms imitate biological processes in the form of computer application. Concepts of biology such as selection, mutation, crossover and reproduction are used in the evolutionary algorithm. In divergent to traditional algorithms, evolutionary algorithms use random sampling. In the traditional classical system, they basically try to maintain single best solution, whereas the population of the candidate solution is present for an evolutionary algorithm. Evolutionary algorithms are considered as the best solution for various social networking issues.

Evolutionary algorithms have different advantages. One of the greatest favorable circumstances is its adaptability gains, as most evolutionary algorithms ideas are adaptable to even complex issues. Most evolutionary algorithms are fit to meet the target focus also. Better improvement is conceivable with evolutionary algorithms, as the number of inhabitants in arrangements keeps the calculation from getting secured a specific arrangement.

In this paper, issues of social networking are listed such as cluster formation, community detection, anomaly detection, seed selection and dynamic as well as hybrid community detection, cut space problem, personalized search, friendship selection and structural balance using evolutionary algorithms as well as various machine learning and artificial intelligence concepts. The issue of clustering is discussed in detail along with various methods to handle it.

## 2  Issues of Social Networking

In today's world, social media has turned out to be most advanced and extensively used platform for establishing connection, gathering and transfer of information. Various platforms such as Facebook, Gmail, LinkedIn and YouTube provide these functionalities. With the increase in the use of the Internet, lots of population are engrossed in these platforms for their respective purpose. They are turning out to be good for their work, but there are various issues which need to be considered while using such platforms.

Given are some issues related to social networking.

### 2.1  Cluster/Community Detection

Community detection is an optimization within the social network for finding partitions with maximum modular density within a network. Modular density refers to the strength of the division formed within a network. High modularity means the presence of dense connection within nodes. Fundamentally, it is an NP-hard issue which can be illuminated utilizing developmental calculation [3]. Cluster recognition depends on the property that solid network has thick intra-association joins and scanty between associations join.

As shown in Fig. 3, total four clusters are formed based on the selection feature. Thus, based on the nature of selection feature used, the number of the clusters varies. Also, all clusters are not necessarily of same size.

**Fig. 3**  Cluster detection [4]

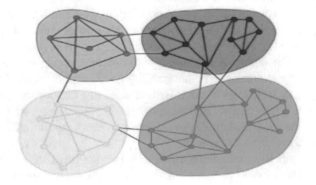

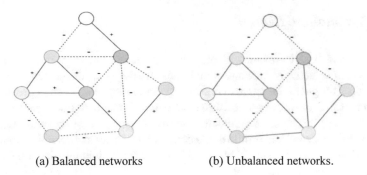

(a) Balanced networks                    (b) Unbalanced networks.

**Fig. 4** Balanced and unbalanced network

## 2.2 Cluster/Community Detection

For the better understanding of social network, signed networks are an effective tool. Here, the graph is represented in the form of a triangle. The triangle is considered as balanced if the product of signs which is present there is positive else it is considered as imbalanced. This was proposed by Heidre [5].

Heider proposed the following balancing principles:

- Friend of a companion is companion.
- Enemy of a companion is adversary.
- Friend of an enemy is adversary.
- Enemy of an enemy is adversary.

From Fig. 4a, the first graph is structurally balanced as all the formed triangles within a network follow the properties of structural balancing. Thus, it is a structurally balanced network.

From Fig. 4b, the given graph does not follow all properties of structural balancing. This graph exhibits the property of friend of enemy is friend which makes the given graph unbalanced. Blue, pink and white colored nodes forms triangle which violates the balancing property of graph. Thus, it is unbalanced network.

## 2.3 Seed Selection

From the group of vertex present, selection of specific vertex increases the chances of increasing the connectivity within the social network. Thus, it is a problem of settling the nodes that increase or expand the spread of enthusiasm for social networks. Because of NP-hard nature, various heuristics can be applied including greedy algorithm and evolutionary algorithm [6].

## 2.4   Personalized Search

It is a method based on human–computer interaction where preference is given to the user to effectively evaluate individuals. It speeds up the search process and reduces user fatigue. Interactive evolutionary computing (IEC) which provides user involvement in evolutionary algorithm proves to be better than traditional evolutionary computing [7]. The main goal is to acquire a customized query item required by the client, by making client profile dependent on interpersonal interaction action. The client profile is really developed by pages liked by the individual client on Facebook on their particular client account.

## 2.5   Anomaly Detection

It is a self-organizing inconsistency detection system which detects inconsistency or anomaly based on nature-inspired techniques such as insect colony optimization. It detects the anomaly based on the collaborative effort of nodes by using local information and to identify and misuse basic assets in condition, and distributed decision-making algorithms are used [8].

## 2.6   Security Analysis

Security checkup of data is a crucial point. Users communicate with the webs as well as the communities which also contain enemies. Because of this, there are different types of security attacks which may cause the stealing of data. Existing tools such as Apache spark can be used for security analysis as huge amount of data is gathering on regular basis. It also provides machine learning algorithms.

## 2.7   Routing

Social networks involve vast communication among users. Thus, routing is an important issue related to social networks as one user can communicate with another user via intermediate users. Successive forwarding of the message can be considered in geographical routing and social routing. Social routing includes the mechanism which provides free network sharing between Internet communities and groups. Geographical routing is based on the geographical position of the information residing within the network.

# 3  Related Work

As we are discussing issues of social networking issues, there is a lot of work done in this area. Given below are list of related work based on the issues of social networking issues.

## 3.1  Related Work Based on Community Detection

The issue of clustering was seriously examined in the previous years, and a variety of arrangements toward tackling this issue were displayed and executed. Liu et al. [9] proposed the paper on community detection in the year 2014. This paper includes the community detection algorithm which handles large-scale signed networks. This is done by multi-objective evolutionary algorithm for signed networks. Multi-objective evolutionary algorithms are those where multiple decision making is done including mathematical optimization problems that contain more than one optimization problem to be optimized. Indirect and direct merged representation of communities is designed. Hence, because of this, multi-objective evolutionary algorithm for signed networks can shift between various portrayals during the evolutionary process.

Another work includes the wellness works in an evolutionary methodology toward network identification in the complex system given by Jora and Chira [3] in the year 2016. Here, detection of clusters is done via study of various fitness functions which are based on evolutionary search process. Fitness function included here is the community score.

Another work includes the community detection based on evolutionary algorithm given by Jami and Reddy [10] in the year of 2016. Here, community detection is done with the use of cat swarm optimization (CSO), particle swarm optimization (PSO), genetic algorithm using simulated annealing (GA-SA) and genetic algorithm (GA). Fitness function used is modularity density function. Genetic algorithm with simulated annealing performs better than the other listed methods as simulated annealing takes population and applies a gradually reducing random variation to each member of the population. It is based on the physical process of annealing which does exactly that.

Balegh and Farzi [11] proposed the algorithm for recognition of community in social networks in the year of 2017 by overcoming the disadvantages of k-means clustering and also by utilizing k-means++ and k-rank-D algorithm. Here, adjacency graph representation is converted to Euclidean space. Accuracy and normalized mutual information (NMI) were the evaluation criterion. To determine initial centers, it chooses the node which has the highest ingress to their neighboring vertex and is not surrounding to another vertex which is called as the center of the clusters.

Lighari and Hussain [12] introduced the composite model for clustering analysis and rule-based for big data security using Apache Spark in 2017. It includes the normalization of the dataset using mean and standard deviation from where the type of attacks is detected. Apache Spark provides an open-source framework for cluster computing.

Apache Spark is utilized for the security examination which offers great machine learning library. To tackle the clustering issue with attributes, multi-objective evolutionary algorithm in view of property likenesses is been proposed by Li et al. [9] in the year of 2018. Multi-objective evolutionary algorithm builds on attribute and structural similarities (MEA-SN) is used here. The great harmony between vertex properties and topological structure is accomplished by MOEA-SA.

## 3.2 Related Work Based on Other Issues of Social Connectivity

Evolutionary algorithm is not only used in clustering issue of social networking but also in other issues of social networking. Lots of work in done on social connectivity issues using evolutionary algorithms.

**Work based on structural balancing**. The equalization shows that a system is fundamentally adjusted if a hub can be partitioned into two subdivisions, inside every subdivision, there are just positive edges, and in another subdivision, there are just negative edges. It gives data with respect to the relations that keep arranging from being adjusted. Yan et al. [5] introduced a novel-based bi-target display for structural balancing and a multi-objective discrete PSO to advance the bi-target demonstrate. For structural balance, a novel bi-objective model for social networks is used, and a multi-objective discrete PSO is used to enhance the bi-objective model. After each run of the algorithm, it produces a set of Pareto solutions, every one of which yields definite system partitions that separate the network into certain signed networks.

**Work based on seed selection**. Seed choice is the issue of finding a littlest subset of hubs, i.e., seed hubs that could amplify the spread of impact. Weskida and Michalski [6] expressed developmental calculation for seed choice in a social network. This is executed utilizing general reason registering on illustrations preparing unit (GPGPU). As compared to greedy algorithm, GPGPU scales up well. Seed node determination should be possible by greedy algorithms, while combined nodes can beat an outcome collected by two dissimilar nodes. This disadvantage can be rectified by utilizing developmental calculation. This is executed utilizing general reason registering on designs handling unit (GPGPU). This scales up well when contrasted with the greedy algorithms.

**Work based on personalized search**. Personalized search is basically the search based on individual interest. Personalized search is been accomplished by an interactive evolutionary algorithm. Choi et al. [7] have utilized domain knowledge of personalized search for quick interactive estimation of distribution algorithm (IEDA). EDA merges genetic learning and probabilistic learning models. Steps involved are space reduction with domain knowledge, interactive-based preference surrogate model and interactive (EDA).

**Work based on anomaly detection**. Anomaly detection in social networks is concerned about the potential for self-sorting out irregularity discovery on a system. In [8], there is a HONIED: hive oversight for network intrusion early warning using DIAMoND—a honey bee enhanced strategy for completely dispersed digital protection. DIAMoND allows protection of potentially sensitive information of individual participating groups. Neighborhood strategy used is based on previous knowledge and assumptions. Performance is based on sensitivity which indicates number of malicious packets, specificity which is number of legitimate packets and overall system accuracy which means correctly identified malicious and legitimate.

**Work based on searching cut space**. Generally, vertex-based encoding was utilized in developmental calculations for taking care of graph issue where every quality relates to a vertex. For addressing the outstanding maximum cut issue, Seo et al. [13] presented an edge-set encoding in light of the spreading over the tree, sort of edge-based encoding. There is utilization of edge-set encoding depending on spreading over the tree for tending to the outstanding maximum cut issue rather than vertex-based encoding. Edge-set encoding performs well for certifiable issues than vertex-set encoding. Change in encoding plan can improve execution. Vertex-based encoding is ideal for dense network.

# 4   Related Papers

This section includes the list of papers studies related to the various issues of social networking. There are two tables listed below. As an issue of community detection is been more focused, the first table contains the paper related to only community detection issue, whereas the second table contains papers related to issues of social connectivity excluding community detection.

From Table 1, it can be seen that community detection issue of social networking can be resolved with various types of evolutionary algorithms. There are various algorithms such as PSO, GA, GA-SA, MEA-SN listed in the above table which can be used to deal with the issue of community detection. K-means algorithm is also used to resolve the issue of community detection.

From Table 2, it can be seen that other issue of social networking such as personalized search, seed selection, anomaly detection, cut space problem and structural balancing can be resolved with various types of evolutionary algorithms. There are various algorithms such as GA, EDA, multi-objective discrete particle swarm optimizer, DIAMoND listed in the above table which can be used to deal with the issues. Embeddedness and triadic closure property of the graph are also used.

Thus, comparing both the tables, evolutionary algorithms act as a medium which resolves almost all the issues of social networking issues. Also, results obtained by applying evolutionary algorithms are more efficient than results obtained by traditional methods.

**Table 1** Related papers on community detection

| Paper name | Description | Dataset | Technology used | Outcome |
|---|---|---|---|---|
| Evolutionary Community Detection in Complex and Dynamic Networks [3] | Investigate a few wellness works in an evolutionary methodology toward network identification in complex system | Football network for 2000 season Zachary's karate club network, Dolphins network, Krebs network for political books | Evolutionary algorithm | This model can be altered in an extremely normal manner to dynamic systems |
| Community Detection in Social Networks using a Novel Algorithm without Parameters [11] | Algorithm is proposed for detection of community in social networks | Karate club, Soccer team and Dolphin dataset | K-means++, K-means and k-rank-D algorithms | Calculation to distinguish bunch inside chart and sharp decision of starting focal point of the group is been done |
| A Hybrid Community Detection based on Evolutionary Algorithms in Social Networks [10] | Community detection is done using evolutionary algorithms | American college football, Dolphin social network dataset and Zachary's karate club dataset | PSO, CSO, GA, SA-GA | Compared to CSO, PSO, GA-SA and GA, GA-SA performs better |
| A Multi-objective Evolutionary Algorithm Based on Similarity for Community Detection from Signed Social Networks [9] | Includes community detection algorithm which handles large-scale signed networks | American football network, Bottlenose Dolphin network and Zachary's karate club | Multi-objective evolutionary algorithm for signed networks (MEA-SN) | Examination between MEAs-SN and other three existing calculations recorded in this paper indicates better execution of MEAs-SN over different calculations |

(continued)

**Table 1** (continued)

| Paper name | Description | Dataset | Technology used | Outcome |
|---|---|---|---|---|
| Hybrid model of rule-based and clustering analysis for big data security [12] | Big data tool, Apache Spark is used for the security analysis which offers good machine learning library | Kddcup99 | K-means using Apache Spark | In this procedure, it is been perceived that the grouping goes about as the normalizer for the anomalies in dataset, in light of the fact that subsequent to clustering, less time is required to work the rest of the dataset |
| A Multi-objective Evolutionary Algorithm Based on Structural and Attribute Similarities for Community Detection in Attributed Networks [14] | Apache Spark is utilized for the security examination which offers great machine learning library. To tackle the ascribed diagram issue, Multi-objective transformative calculation in view of property likenesses (MOEA-SA) is been proposed | Facebook ego-networks with multi-attribute and attributed graphs | Multi-objective evolutionary algorithm derived from attribute and structural similarity (MEA-SN) | Through all the experiments, it is been proved that great harmony between topological structure and vertex properties is accomplished by MOEA-SA |
| Community Detection in Graphs based on Surprise Maximization using Firefly Heuristics [15]. | To achieve well-performing universal tool for graph clustering grounded on surprise maximization, a novel heuristic community detection approach is proposed | Zachary's karate club (ZKC), Relaxed Caveman Model (RCM), Power Law Cluster Graph Model (PLCGM) [16] and Erdos-Renyi Model (ERM) graph | Evolutionary algorithm(Naive Firefly Algorithm) | Simulation results have statistically verified that the optimality of the proposed method is broader when connected over particular graph |
| CAP: Community Activity Prediction Based on Big Data Analysis [17] | Network movement designs are separated by examining the huge information gathered from virtual social space and physical world | Data collected from Huazhong University of Science and Technology [18] | Singular value decomposition (SVD), clustering and tensors | CAP (community activity prediction) concept achieves good augury performance. It can likewise be adequately used to decrease the unpredictability of information |

**Table 2** Related papers on other issues of social networking

| Paper name | Description | Dataset | Technology used | Outcome |
|---|---|---|---|---|
| A survey on platforms for big data analytics [19] | It gives top to bottom investigation of various stages accessible for big data analysis | – | K-means clustering | The broadly utilized k-implies bunching calculation was picked as a contextual analysis to exhibit the qualities and shortcomings of various stages |
| Expansion of Social Connectivity: a concept of Big Data Analysis and Genetic Algorithm Modeling [20] | Concept of big data and genetic algorithm used in social connectivity | Research gate dataset | Genetic algorithm, embeddedness and triadic closure | By using genetic algorithm, the resources have been optimized |
| Evolutionary Algorithm for Seed Selection in Social Influence Process [6] | Seed selection is done using concept of evolutionary algorithm. Seed selection can be done by greedy algorithm, but combined nodes can beat an outcome created by two unique nodes. This drawback can be improvised by using evolutionary algorithm. The general-purpose computing on graphical processing unit (GPGPU) is used for computing | UC Irvine message network, Digg network [21] and Facebook wall posts network [22] | Genetic algorithm and social influential model | Evolutionary algorithm scales up well as compared to greedy algorithm |

(continued)

**Table 2** (continued)

| Paper name | Description | Dataset | Technology used | Outcome |
|---|---|---|---|---|
| Personalized Search Inspired Fast Interactive Estimation of Distribution Algorithm and Its Application [7] | There is a utilization of quick intuitive estimation of distribution algorithm (EDA) by utilizing space learning. EDA merges genetic learning and probabilistic learning models | Laptop data from JD.com | Interactive genetic algorithm and estimation of distributed algorithm (EDA) | The execution of IEDA in diminishing client weakness and accelerating the PC, the inquiry is tentatively shown in laptop search. PS-IEDA-DK is a substitute for refining the customized seek in E-business |
| Hive Oversight for Network Intrusion Early Warning Using DIAMoND: A Bee-Inspired Method for Fully Distributed Cyber Defense [8] | It is concerned about the capability for self-organizing anomaly revelation system inspired from Honey bee | – | DIAMoND: A Bee-inspired method for fully distributed cyber defense | DIAMoND has shown 20 percent enhancement in sensitivity without relinquishing specificity |
| An Edge-Set Representation Based on a Spanning Tree for Searching Cut Space [13] | There is a use of edge-set encoding in light of spanning tree for coordinating the well-known maximum cut problem instead of vertex-based encoding | – | Genetic algorithm, edge-based encoding | The proposed edge-set encoding in directing spanning tree has solid benefits for coordinating real worlds parse graph. Likewise, changes in encoding strategy could be good to enhance execution |

(continued)

**Table 2** (continued)

| Paper name | Description | Dataset | Technology used | Outcome |
|---|---|---|---|---|
| A Novel Bi-objective Model with Particle Swarm Optimizer for Structural Balance Analytics in Social Networks [5] | It introduces a novel bi-objective display for social organizations auxiliary equalization, and a multi-objective discrete PSO agent is utilized to upgrade the bi-objective model | Slovene Parliamentary network (SPP), Epidermal growth factor receptor signaling pathway network (EGFR), Yeast network and Election network. Macrophage network and Gama subtribes network (GGS) | Multi-objective discrete particle swarm optimizer | Broad tests contrasted and a few different models and calculations have been done to approve the adequacy of given model and calculation |
| Genetic Algorithm for friendship selection in Social IoT [23] | SIoT is a concatenation between IoT and social networks where every object will be able to explore the required service using friendship connections | Network Brightkite | Genetic algorithm | Discussion regarding the connection determination system that can be utilized by genetic calculation to locate the close ideal arrangement is done here |

## 5 Conclusion and Future Work

In this paper, from various issues of social networking, clustering is discussed in detail. This paper surveys various platforms through which these issues can be resolved and enhanced. An evolutionary algorithm has proven to be very effective in solving the issue of clustering and other issues of social networking. Evolutionary algorithms that are been used are GA, GA-SA, CSO, PSO, multi-objective evolutionary algorithm. All the above evolutionary algorithms were applied on various real-time dataset to detect the behavior of the algorithm.

For seed selection in social influence process, as graphical processing unit is used, it has a limitation of memory size. By use of compressed sparse row technique, this limitation can be overcome and can be implemented for large-scale network. In personalized search, dynamically employing group intelligence to interactive evolutionary computing (IEC) for getting the more accurate preference model and reducing evaluation uncertainties in personalized search can be achieved. For edge-set representation in spanning tree, encoding scheme can be further refined and can be applied on various graph.

Clustering is crucial in data mining applications and data analysis. The task of clustering is to differentiate between two groups where one group contains objects with similar properties than another group of objects.

Strategies for cluster detection in systems have been proposed which enable clusters to override. Thus, it can be further be improvised to detect the overlapping communities present within the detected communities. This can be achieved through multi-objective evolutionary algorithms. Cluster detection techniques can also be applied for dynamic community structure. Multi-objective evolutionary algorithms for sign networks are applied on undirected graph only. This can be further extended for directed graphs as well.

## References

1. W. Tan, M.B. Blake, I. Saleh, S. Dustdar, Social-network-sourced big data analytics. IEEE Internet Comput. **17**(5), 62–69 (2013)
2. A. Shefaei, B. Mohammadi-Ivatloo, Wild goats algorithm: an evolutionary algorithm to solve the real-world optimization problems. IEEE Trans. Ind. Inform. **14**(7), 2951–2961 (2018)
3. C. Jora, C. Chira, Evolutionary community detection in complex and dynamic networks, in *Proceedings of 2016 IEEE 12th International Conference on Intelligent Computer Communication and Processing, ICCP 2016*, 2016, pp. 127–134
4. M. Eslami, A. Aleyasen, R.Z. Moghaddam, Friend grouping algorithms for online social networks: preference, bias, and implications, in *International Conference on Social Informatics*, 2014, pp. 34–49
5. J. Yan, S. Ruan, Q. Cai, J. Shi, Z. Wang, M. Gong, A novel bi-objective model with particle swarm optimizer for structural balance analytics in social networks, in *2016 IEEE Congress on Evolutionary Computation (CEC) 2016*, 2016, pp. 728–735

6. M. Weskida, R. Michalski, Evolutionary algorithm for seed selection in social influence process, in *Proceedings of 2016 IEEE/ACM International Conference on Advances in Social Networks Analysis and Mining, ASONAM 2016*, 2016, pp. 1189–1196

7. Y. Chen, X. Sun, D. Gong, Y. Zhang, J. Choi, S. Klasky, Personalized search inspired fast interactive estimation of distribution algorithm and its application. IEEE Trans. Evol. Comput. **21**(4), 588–600 (2017)

8. M. Korczy et al., Hive oversight for network intrusion early warning using DIAMoND: a bee-inspired method for fully distributed cyber defense. IEEE Commun. Mag. **54**(6), 60–67 (2016)

9. Chenlong Liu, Jing Liu, Zhongzhou Jiang, A multiobjective evolutionary algorithm based on similarity for community detection from signed social networks. IEEE Trans. Cybern. **44**(12), 2274–2287 (2014)

10. V. Jami, G.R.M. Reddy, A hybrid community detection based on evolutionary algorithms in social networks, in *2016 IEEE Students' Conference on Electrical, Electronics and Computer Science, SCEECS 2016*, 2016

11. B. Balegh, S. Farzi, Community detection in social networks using a novel algorithm without parameters, 2017, pp. 355–359

12. S.N. Lighari, D.M.A. Hussain, Hybrid model of rule based and clustering analysis for big data security, in *2017 1st International Conference on Latest trends in Electrical Engineering and Computing Technologies, INTELLECT 2017*, vol. 2018, Jan 2018, pp. 1–5

13. K. Seo, S. Hyun, Y.H. Kim, An edge-set representation based on a spanning tree for searching cut space. IEEE Trans. Evol. Comput. **19**(4), 465–473 (2015)

14. Z. Li, J. Liu, K. Wu, A multiobjective evolutionary algorithm based on structural and attribute similarities for community detection in attributed networks. IEEE Trans. Cybern. **48**(7), 1963–1976 (2018)

15. J. Del Ser, J.L. Lobo, E. Villar-Rodriguez, M.N. Bilbao, C. Perfecto, Community detection in graphs based on surprise maximization using firefly heuristics, in *2016 IEEE Congress on Evolutionary Computation, CEC 2016*, 2016, pp. 2233–2239

16. P. Holme, B.J. Kim, Growing scale-free networks with tunable clustering. Phys. Rev. E Stat. Phys. Plasmas Fluids Relat. Interdiscip. Top. **65**(2), 2–5 (2002)

17. Y. Zhang, M. Chen, S. Mao, L. Hu, V. Leung, CAP: community activity prediction based on big data analysis. IEEE Netw. **28**(4), 52–57 (2014)

18. ICDAR 2017 Page Object Detection Competition. http://www.icst.pku.edu.cn/cpdp/ICDAR2017_PODCompetition/dataset.html

19. D. Singh, C.K. Reddy, A survey on platforms for big data analytics. J. Big Data **2**(1), 1–20 (2015)

20. G. Ghosh, N. Kasturi, Expansion of social connectivity: a concept of big data analysis and genetic algorithm modeling, in *2017 Conference on Information and Communication Technology, CICT 2017*, vol. 2018, Apr 2018, pp. 1–6

21. M. De Choudhury, H. Sundaram, A. John, D.D. Seligmann, Social synchrony: predicting mimicry of user actions in online social media, in *Proceedings of 12th IEEE International Conference on Computational Science and Engineering, CSE 2009*, vol. 4, 2009, pp. 151–158

22. B. Viswanath, A. Mislove, M. Cha, K.P. Gummadi, On the evolution of user interaction in Facebook, in *Proceedings of 2nd ACM Workshop Online Social Networks—WOSN'09*, 2009, p. 37

23. W. Mardini, Y. Khamayseh, M.H. Khatatbeh, Genetic algorithm for friendship selection in social IoT, in *Proceedings of 2017 International Conference on Engineering & MIS, ICEMIS 2017*, vol. 2018, Jan 2018, pp. 1–4

# A Comparative Study of Deployment Tools from IoT Perspective

Raghav Maheshwari and K. Rajalakshmi

**Abstract**  Nowadays, Internet of Things has become one of the most trending technologies in the field of computer science. Many businesses are accepting the various practices of Internet of Things to bring reliability, stability, and security in their development and production phases of products or services. Deploying their application to a large scale is the key challenge faced by many users. Choosing the right tool according to the needs is a decision one should make taking various parameters into consideration. This paper studies various deployment tools, based on various researches and practices, on platforms like Raspberry pi board, Linux devices, and so on and tried to give a uniformed comparative study among them on various useful and important parameters.

**Keywords**  Dev-ops · Internet of Things · IoT · Deployment tools · Deployment · Development · Production · Docker · Mender · AWS Greengrass · Kaa · Android Things · Raspberry Pi · Comparison

## 1   Introduction

The Internet of Things (IOT) is considered to be a boom for computer science and technology. Heterogeneous devices, like mobile phones, laptops, cameras, sensors, embedded devices, and others, become part of a mesh network, thus enabling them to interact and communicate among them and use the shared resources [1]. Inspired by this, many started practices, like continuous development, continuous testing, continuous integration, and continuous deployment, are commonly known as Dev-Ops [2].

The practices involved users to maintain a stream of incessant flow of data, from the development phase till the deployment phase. Though there are some individual problems and errors in each of the phases, but it is the phase of continuous deployment

R. Maheshwari (✉) · K. Rajalakshmi
Jaypee Institute of Information Technology, Noida, India
e-mail: raghavddn16@gmai.com

K. Rajalakshmi
e-mail: k.rajalakshmi@jiit.ac.in

© Springer Nature Singapore Pte Ltd. 2021
X.-Z. Gao et al. (eds.), *Advances in Computational Intelligence and Communication Technology*, Advances in Intelligent Systems and Computing 1086,
https://doi.org/10.1007/978-981-15-1275-9_24

**Fig. 1** Phases in the
deployment cycle

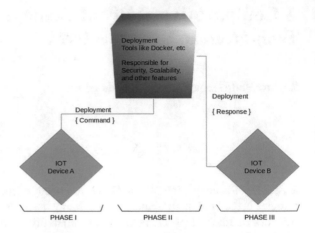

which involves a deep research to counter the problem. Continuous deployment cycle
incorporates automated flow of data from one device to another, through some tools,
as shown in Fig. 1.

Various companies have business which requires them to choose some tools for
the deployment of their application to remote devices. There is not only one tool,
and all the tools do not possess the same features. To calculate their needs and then
choose a deployment tool suitable for them, much man working hours are required.
The idea of lessening such problems and having a comparative study of some such
major tools paved way to the research and practices explained in later sections.

Deployment tools, namely Docker, Mender, AWS Greengrass, Kaa, and Android
Things, were chosen and were studied to present a comparison among them, to
increase the users' convenience.

## 2 Objectives

To present a comparison among the deployment tools, it is a necessary step to properly
install them on some device and generate some application which could be deployed
to other remote devices. Thus, the main objectives of the paper are:

A. Installation of tools—First and foremost step of the research is the proper installation of the deployment tools. These tools are Docker, Mender, AWS Greengrass, Kaa, and Android Things. The proper installation is essential as it forms
   the basis of the research. The installation of the tools is explained in later sections,
   respectively.
B. Application development—For capturing the real-time data, this paper focused
   on developing an Android application which uses the camera functionality of
   the mobile device to capture the image and send it to the system. The application

was developed keeping in mind the deployment process, thus all the tools used the same.

C.  Identification of the component—The components or measures of the comparative study among different tools of deployment could be divided broadly into two categories: functional and non-functional. Functional issues could include security, privacy, internet layer, languages used, and others. Non-functional measures include scalability, elasticity, and Raspberry Pi support, and so on.

D.  Comparative study—The final step is to list down a detailed comparative study among different tools of deployment, on different functional and non-functional measures working on similar kind of data. This gave us a clear view of pros and cons of the tools in comparison with one another.

# 3 Layers of Internet of Things

It is very important to understand the seed of deployment tools. These tools work for the enhanced experience of Internet of Things. IOT is not just devices, but consists of different layers. [3] As shown in Fig. 2, they are:

A.  *End systems*—At the very bottom layer comes the end systems. These are the equipment or the hardware used in any IoT-based project. It could be sensors like temperature, heat, humidity, or could be cameras, mobile devices, capturing equipment, RFID tags, and other wired or wireless devices, or a combination of both.

   Its key feature is to collect information from its surroundings and pass on the information. For this paper, mobile devices and cameras have been used to capture the information. Being responsive in nature, they are able to get live data as per the need.

B.  *Connections*—The connections used by the end systems to pass on the information further could be a normal router Wi-Fi, or a subnet of the net. It could use

**Fig. 2** IoT architecture

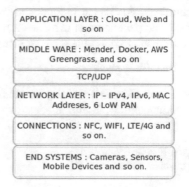

LTE/4G, WAN, or maybe LANs too, depending on the need. Its key function is to provide a connection among the end systems and the network layer.

C. *Network layer*—Next comes the network layer. Network layer contains the IPv4/IPv6, and so on, which acts as a unique identity for different components of the system, to ease the process of the communication. The sending and receiving of the data is done through these addresses, like the MAC addresses or 6LoW-PAN and others. It is responsible for providing addresses to different components for making the communication possible.

  This paper focuses more on IPv4 for addressing in the project. In other words, it could be said that doorway helps in to and fro movement of the data. It provides network connectivity to the data, which is essential for any IoT system.

D. *TCP/UDP*—The connections are established either by *TCP or UDP* protocol. TCP or Transmission Control Protocol is connection oriented—once a connection is established, data can be sent bidirectional. UDP or User Datagram Protocol is a simpler, connectionless Internet protocol.

E. *Middleware*—For deployments and over the air transmission, comes the middleware. Tools like Mender, Docker, AWS Greengrass, Jenkins, Apache, Kaa, and so on are built solely for deployments of data from system to cloud or from one system to another.

  Its main function is to securely deploy the data to its destination. The destination may be cloud of any other system. The various deployment tools falls under this layer of the IoT architecture.

F. *Application layer*—The final layer is the application layer, where the data rests on reaching. It could be cloud servers, Web hosts, Android platform, and so on. Data analytics or visualization can be under this layer. Applications are controlled by users and are the delivery point of a particular service.

## 4 Getting Started with Docker

Docker is a platform to develop, deploy, and run applications with containerization technology. An image is launched by running a container [4]. A container is an executable package that provides a ready-to-use environment on any machine [5]. The following shows the steps to install Docker successfully for efficient working.

A. **Installing Docker**

(1) Install Docker from the terminal using the convenience script, as a standard command, as in Fig. 3.

**Fig. 3** Standard command for the convenience script [5]

```
curl -fsSL get.docker.com -o get-docker.sh
sudo sh get-docker.sh
```

**Fig. 4** Local registry script
standard command [5]

```
$ docker run -d -e
http://192.168.1.33:5001
--restart=always --name registry
```

(2)   To use as a non-root user [5]

$$\text{sudo usermod -aG docker your-user} \tag{1}$$

B.  Building Images of Application

Here, take the application ready to be deployed. After creating the script for the application in the conventional way, one needs to build it.

(1)   Build the image by getting into the directory of the image [5].

$$\text{\$ docker build -t⟨image name⟩.} \tag{2}$$

(2)   To run the image [5]:

$$\text{docker run -d⟨image name⟩} \tag{3}$$

(3)   To save the image in.tar format [5]:

$$\text{\$ docker save -o⟨file name⟩.tar⟨image name⟩} \tag{4}$$

C.  **Pushing and Pulling the Image**

(1)   Create a local registry, the standard command is (Fig. 4):
(2)   Tag the image [5]:

$$\text{\$ docker tag⟨image name⟩ 192.168.5.45:5010/new name} \tag{5}$$

(3)   Pushing the image into the registry [5]:

$$\text{\$ docker push 192.168.5.45:5010/new name} \tag{6}$$

(4)   Pulling the image on from the registry [5]:

$$\text{\$ docker pull 192.168.5.45:5010/new name} \tag{7}$$

To felicitate remote pulling of the image, push the image to the Docker Hub instead of local registry.

Thus, the Docker is ready and up for working.

## 5   Getting Started with Mender

Mender is an open source, over the air deployment tool for Linux devices. As a prerequisite Docker Engine is required for Mender to work [6], for the correct installation of Mende, the following steps can be followed.

### A.   Installing Mender

(1)   Mender integration environment is initially downloaded (Fig. 5):
(2)   Create the initial user and allow the secure connection afterward [6]:

$$\text{sudo ./demo exec mender-useradm /usr/bin/useradm--create-user}$$

$$\text{user name} = \text{newname@emailaddress.com--password} = \text{newpassword} \quad (8)$$

### B.   Deploying to Devices

Open the Mender UI by navigating to https://localhost/ in the same browser as you accepted the certificate. One can also download the demo images for testing purposes.
   There are two types of images [6]:

– Disk images (*.sdimg): they are used to establish the device memory for devices without Mender running already.
– Mender Artifacts (*.mender): they are used to push images to the Mender server so that new root file systems could be deployed to devices in which Mender is already running.

   The images could be deployed to devices like Raspberry pi and Beagle Bone, and its progress can be tracked on the UI itself.

### C.   Maintaining the Environment

(1)   Stopping the Mender Services [6]

$$./\text{stop} \quad (9)$$

(2)   Starting the Mender Services

$$./\text{start} \quad (10)$$

(3)   Wrapper script to bring up the environment

$$./\text{up} \quad (11)$$

   Thus, the Mender is ready and up for running.

**Fig. 5** Mender initial script
standard command [6]

```
git clone –b 1.6.0
https://github.com/mendersoftware/inte
gration.git integration-1.6.0
cd integration-1.6.0
```

# 6  Getting Started with AWS Greengrass

AWS Greengrass comes under the umbrella of AWS IOT Console, to spread the capabilities to local devices, making a secure network and medium for information exchange [7].

To run Greengrass on a Raspberry Pi board, firstly, the P2 SSH option should be enabled. After creating a group and a user, one need to make sure that the Linux Kernel is up to date and all the dependencies are met. To check them, the standard commands are (Fig. 6):

## A.  Installing Greengrass Core Daemon

(1)  After signing into the AWS Management Console, browse to Greengrass and select Create an Greengrass Group.
(2)  Follow the steps, by giving the Group name, Core name, Easy Group Creation and save the Certificates on the system. Also, download the Software Configuration for Raspbian on Linux.
(3)  Now, send the zipped downloaded files to the Raspberry Pi board (preferred). To transfer the compressed files from your computer to a Raspberry Pi AWS Greengrass core device, open a terminal window on your computer and run the following standard commands (Fig. 7).

   On the Raspberry Pi board (Fig. 8).
(4)  Install the root—CA certificate and run the daemon through the standard command [7]:

cd/greengrass/ggc/core/

**Fig. 6** Greengrass dependency script standard command [7]

```
cd /home/pi/Downloads
git clone https://github.com/aws-
samples/aws-greengrass-samples.git
cd aws-greengrass-samples
cd greengrass-dependency-checker-
GGCv1.6.0
sudo modprobe configs
sudo ./check_ggc_dependencies | more
```

**Fig. 7** System script standard command [7]

```
cd path-to-downloaded-files
sudo scp greengrass-OS-architecture-
1.6.0.tar.gz  pi@IP-address:/home/pi
sudo scp GUID-setup.tar.gz pi@IP-
address:/home/pi
```

**Fig. 8** Raspberry Pi board system script standard command [7]

```
cd path-to-compressed-files
sudo tar -xzvf greengrass-OS-architecture-
1.6.0.tar.gz -C /
sudo tar -xzvf GUID-setup.tar.gz -C
/greengrass
```

**Fig. 9** Publisher script standard command [7]

```
python basicDiscovery.py --endpoint
AWS_IOT_ENDPOINT --rootCA root-ca-
cert.pem --cert publisher.cert.pem --key
publisher.private.key --thingName
HelloWorld_Publisher --topic
'hello/world/pubsub' --mode publish --
message 'Hello, World! Sent from
HelloWorld_Publisher'
```

**Fig. 10** Subscriber script standard command [7]

```
python basicDiscovery.py --endpoint
AWS_IOT_ENDPOINT --rootCA root-ca-
cert.pem --cert subscriber.cert.pem --key
subscriber.private.key --thingName
HelloWorld_Subscriber --topic
'hello/world/pubsub' --mode subscribe
```

$$\text{sudo/greengrassd start} \tag{12}$$

## B. Interacting with Devices

(1) After choosing Add Device on the Group Configuration page, give the name, like Publisher or Subscriber, and also download the security certificates. The devices are added to the Greengrass Group.

(2) On the group configuration page, select the source after Add Subscription. Choose Select, Devices, and choose Publisher. Similarly, for a target, choose Subscriber. Finally, give a topic name for the deployment.

(3) Choose Deploy option to deploy the updated sample group configuration. The working could be seen by opening two terminal windows as Publisher and Subscriber, respectively, and giving the respective standard commands, as shown in Figs. 9 and 10.

Lambda functions could also be created for further functionality and triggered for more accurate communications and data processing.

Thus, the AWS Greengrass is set up and ready for working.

## 7 Getting Started with KAA

Kaa is an open source IoT platform. Its architecture consists of a Kaa server, Kaa extensions, and endpoint SDK, along with Apache zookeeper. To set up Kaa, this paper chose virtual sandbox, which could run on any system. [8] Kaa installation is done on a virtual machine following the steps given below.

A. **Installing Kaa Sandbox**

(1) Download Oracle Virtual Box (preferred) on the system, and then download the Sandbox.ova image.
(2) Once installed, adjust the memory and processors to be distributed to the virtual box.
(3) Once all done, it opens an URL, directing us to the Kaa UI page.
(4) Go to Sandbox Management Page and specify the real IP of the system.

B. **Launching Kaa Application**

(1) After installation, Kaa provides sample applications for test running. Using the sandbox, one can even download the source codes of several applications or make a new one (as made earlier). To work with user defined instances, download an SDK Library and deploy it to the required endpoint.
(2) On the Administrative UI, login and go to the Add Application and enter the required information. Create file data-scheme.json and configuration-scheme.json to give the required information.
(3) Following this, upload the file, choose the tenant developer, and finally, click on Add Scheme from the options. After giving the appropriate description, add the scheme.
(4) To generate the SDK, on the UI, go to Generate SDK option, and create the profile. Fill out the required details and choose the corresponding option to generate one.
(5) For the Client Side Application, write the code either in C/C++/Java after installing the respective dependencies. For the Java application.

Save the application in the demo_app directory and then build the java application [8]:

$$java -cp * jar * .java \qquad (13)$$

Launch the application to get started it working [8]:

$$java\ java -cp\ '. : ./'\ Filename \qquad (14)$$

Other configuration changes could also be done, like endpoints and so on as per our need.

Thus Kaa is ready and up for working.

# 8 Getting Started with Android Things

Android Things is an IoT platform based on Android, from Google. As the name suggests, application developed in Android can be deployed and Android Studio can be used for debugging mechanisms [9].

## A. **Things SDK**

With additional APIs from Support Library, Android Things encompasses the utility and functionality of the core Android frameworks. This lets users to make possible the communication and message passing among different types of hardware devices. Some features of such apps are:

1. It has increased access to hardware resources and drivers.
2. The starting up and memory requirements are not optimized by system apps.
3. Automatic launching of apps on startup of the device.

## B. **Console**

For user convenience and higher working experience, Android Things presents an UI, known as Console. It enables users to install or update, push or pull images on connected devices. It increases the user functionality the ease at which one can deploy applications.

## C. **Launching**

To get started, create the application to be deployed using the new project wizard:

1. Choose Android Things as the form factor.
2. Select Android Oreo (8.1) or higher. This enables user to incorporate other APIs and dependencies.
3. Give a name to the empty new activity, like New Activity or Home Activity.
4. Android Things add the dependency to the Things Support Library to the application build gradle file: [9]

$$dependencies\{$$

$$...$$

$$compileOnly$$

$$'com.google.android.things:androidthings: + '$$

$$\}                                                                          (15)$$

5. Connect to Raspberry Pi or other hardware through TTL Cable.

    Android Things is set up and ready for working.

# 9  Application

An Android application is written that uses the camera property of the mobile devices to capture the image of the surrounding (preferably a person). The image is then transmitted by a local node js server to the local host, thus showing the image on the browser. Figures 11 and 12 show the snippet of the Android application and node js server, respectively.

```
@app.route('/',methods=['GET','POST'])

def index(): #storing mechanism
        request_data = request.get_json()
        nm = request_data['name']
        #print len(nm)
        image_64_decode=base64.b64decode(nm)

        filepath = os.path.join(path_to_folder,"a"+".PNG")

        if not os.path.exists(path_to_folder):
                os.makedirs(path_to_folder)
        file1 = open(filepath, "wb")

        file1.write(image_64_decode)

        file1.close()
        return "File created"


    if(__name__=="__main__"): #local host address
            app.run(debug=True,host='192.168.43.26',port=5021)
```

**Fig. 11** Snippet for the Android application

```
    <configuration>
      <option name="BUILD_FOLDER_PATH" value="$MODULE_DIR$/build" />
      <option name="BUILDABLE" value="false" />
    </configuration>
  </facet>
</component>
<component name="NewModuleRootManager" LANGUAGE_LEVEL="JDK_1_7" inherit-compiler-outp
  <exclude-output />
  <content url="file://$MODULE_DIR$">
    <excludeFolder url="file://$MODULE_DIR$/.gradle" />
  </content>
  <orderEntry type="jdk" jdkName="1.8" jdkType="JavaSDK" />
  <orderEntry type="sourceFolder" forTests="false" />
</component>
```

**Fig. 12** Snippet for the node.js server

# 10   Comparative Study

Table 1 shows the comparative study among four deployment tools on various functional and non-functional parameters, based on researches and practical applications.

### A.   Comparison Table

The comparison table for the five tools could be summarized as follows.

### B.   Analysis

Docker gives us the easiest way to transfer application to remote systems security. It uses the concept of containerization, which is the application packed with the

**Table 1** Comparative study

| | Docker | Mender | AWS green grass | KAA | Android things |
|---|---|---|---|---|---|
| Base | Lightweight as above Host OS | Need docker engine as a prerequisite | Use of lambda functions and triggers | Use of sandbox or virtual machine | Android platform |
| Concept | Containerization | Uses docker theory of containerization | Uses pub-sub and OPC-UA protocol | Kaa clusters | Google's developers kit (things SDK) |
| Security | Good | Good | Good | Good | Good |
| Scalability | High | High | High | High | High |
| RPI support | Yes | Yes (better in later versions) | Yes | Yes | Yes (just for development) |
| Development level | Good | Good (needs errors resolving) | Good (a bit confusing) | Good | Good |
| Production level | Better | Good | Good | Good (need changes) | Good |
| Paid version available | Yes | Yes | Yes | Yes | Yes |
| OTA updates | Yes | Yes | Yes | Yes | Yes |
| Documentation | Available | Available | Available | Available | Available |
| Java | Yes | Yes | Yes | Yes | Yes (Android) |
| Python | Yes | Yes | Yes | No | No |
| C/C++ | Yes | Yes | No | Yes | No |
| Internet | Could run below or above the internet | Internet connection is required for the setup and updates | Devices should be sharing the same network | Needs an Internet connection to connect to the cloud, gateways otherwise | Wired connection for initial testing, Internet otherwise |
| Cloud support | Docker hub | Mender server | AWS IOT console | Kaa cloud | Google store |
| Requirements | Docker engine | Docker engine and internet | AWS core and device SDKs | Kaa sandbox | Google SDK and android studio |

dependencies, just like a ship container. One can make multiple images of the same container and distribute them to various locations. One can create a local registry, or make an account on the Docker Hub for pushing and pulling images. The building process is simple and easy to handle, but some concerns occur over the security [10] automation of the deployments.

Mender incorporates Docker theory of containerization and Docker Engine, the latter being a prerequisite for working with Mender. It provides sample images and user friendly UI for better working experience. Deploying to Raspberry Pi boards has errors in earlier versions, with its control measures released with the subsequent versions.

AWS Greengrass, generally comes under the shade of AWS IOT Platform, gives a wide range of functionalities, but tends to take a confusing road for new users. With the concept of device shadows, it helps interacting within a local network. Lambda functions, user modified programming modules, increase functionality. It is more onto the side of messages communication via OPC-UA, with machine learning models to work with.

Kaa provides a platform for device communication with the cloud via protocols like MQTT. It also provides gateway architecture in cases of Internet failure. It helps doing data analytics and data visualization over the data collected. Though it gives a good UI interaction, it sometimes get confusing working with it. It provides Link Encryption type security with real-time computations.

Android Things, by Google, brings on Android Platform as its basis. One should have a good knowledge of Android and Android Studio, before starting with it. It supports Raspberry Pi board, but only for development level, as of now. Debugging of errors becomes simpler with Android Things. It also provides its own kit, for initial understanding of its working.

# 11   Conclusion

The Dev-Ops and IoT are nowadays becoming very closely related fields. Deployment tools are quite responsible for the smooth relationship between them. Based on researches and practical applications, this paper tries to find out various pros and cons of different tools and bring them together in a comparative study.

Docker shows very promising results for the deployment of various applications. The brand name of Amazon Web Services and Google behind tools Greengrass and Android Things, respectively, gives an edged value to them. Mender and Kaa also showed good results and scope of improvement. Hence, one needs to select the tool depending on the needs and necessity of the products or services. This paper also paves way for further researches on similar topics.

# References

1. S. Haller, S. Karnouskos, C. Schroth, The Internet of Things in an enterprise context. Future Internet – FIS 2008 Lecture Notes in Computer Science, vol. 5468 (2009), pp 14–28
2. M. Callanan, A. Spillane, DevOps: making it easy to do the right thing. IEEE Softw. **33**(3), 53–59 (2016). H. Xuyong, Basic *Research of Wireless Sensor Networks and Applications in Power System* (Huazhong University of Science & Technology, Wuhan, 2008)
3. M. Weyrich, C. Ebert, Reference architectures for the internet of things. IEEE Softw. **33**(1), 112–116 (2016). D. Liu, L. Zhao, The research and implementation of cloud computing platform based on docker. ACM J. (2015). International Center for Wavelet Analysis and Its Applications, School of Information and Software Engineering
4. C. Pahl, Containerization and the paas cloud. IEEE Cloud Comput. **2**(3), 24–31 (2015)
5. Documentation on Docker. https://www.docker.com/
6. Documentation on Mender. https://mender.io/
7. Documentation on AWS GreenGrass. https://aws.amazon.com/greengrass/
8. Documentation on Kaa. https://www.kaaproject.org/
9. Documentation on Android Things. https://developer.android.com/things/get-started/
10. T. Combe, A. Martin, R. Di Pietro, To docker or not to docker: a security perspective. IEEE Cloud Comput. **3**(5), 54–62 (2016)

# A Perspective of Missing Value Imputation Approaches

Wajeeha Rashid and Manoj Kumar Gupta

**Abstract** A massive amount of data is emerging continuously in the era of big data; the missing value is a common yet challenging problem. Missing data or missing value can be described as the data whose values in any given dataset are unknown. It is a recurrent phenomenon and can give biased results from the data to be observed and affect the value of the learning process. Therefore, it is mandatory to use missing value imputation techniques. Missing value imputation provides (MVI) optimal solutions for missing values in a dataset. There are diverse missing value imputation techniques such as statistical method and machine learning methods proposed till date, each having their own significances and flaws. Reasonably, good imputation results may be produced by machine learning MVI approaches but they take greater imputation times than statistical approaches usually. In this paper, we have reviewed some of the missing value imputation approaches proposed. In order to show the efficiency of their proposed approaches, they had compared their proposed methods with some baseline imputation algorithms like mode/median, k-nearest neighbor, Naive Bayes, support vector machine. The outcome of each proposed method is analyzed and their extension scopes from the perspectives of research focus. The main idea of this literature is to give a general review of missing value imputation approaches.

**Keywords** Big data · Missing data · Missing data imputation · Data mining · Machine learning · MCAR · MAR · MNAR

## 1 Introduction

Big data refers to all the data that is being generated all over the world at a tremendous rate. It is so voluminous and complex that traditional databases and software are insufficient for them. The data can be structured, semi-structured, or unstructured. Structured data is the one that can be categorized and analyzed. It can be stored in a

W. Rashid · M. K. Gupta (✉)
School of Computer Science and Engineering, Shri Mata Vaishno Devi University, Katra, India
e-mail: manoj.gupta@smvdu.ac.in

W. Rashid
e-mail: wajeeha1592@gmail.com

© Springer Nature Singapore Pte Ltd. 2021
X.-Z. Gao et al. (eds.), *Advances in Computational Intelligence and Communication Technology*, Advances in Intelligent Systems and Computing 1086,
https://doi.org/10.1007/978-981-15-1275-9_25

relational database in the form of rows and columns. Examples include name, gender, age, phone number, data generated from forms, sensors, etc. Semi-structured data is unorganized structured data. As unorganized, it is difficult to analyze, process, recover compared to structured data. Examples include XML file, CSV documents, and weblogs. Unstructured data as the name indicates is the data that is unstructured or unorganized. Examples include audio, video, images, PowerPoint presentations, etc. Doug Laney first introduces 3 V's associated with big data:

- Velocity: It deals with the speed at which the huge amounts of data are generated. Data is arriving as a stream of continuous data and at an immense rate. Every day number of emails, Twitter posts, Instagram photos, video clips, etc. are increasing with speed of light. With every passing second, data is increasing. Such a plethora of data needs to be stored, analyzed, and processed.
- Variety: It is defined as the type of data. Today's data is far different than the data in the past. Nowadays, all the data is unstructured. In fact, almost 80% of world data is unstructured including text, sensor data, audio, video, clickstreams, social media updates, etc.
- Volume: It refers to the massive amount of the data generated every second from social media, cell phones, cars, sensors, photos, videos, etc. The amount of data in the world is expected to double every two years. By 2020, we will have 50 times the amount of data that we had in 2011.

Now, two more V's are added:

- Veracity: It refers to the quality or trustworthiness of the data. It refers to the imprecision of data. Examples include Twitter posts with the hashtag, GPS data.
- Value: It refers to the capability of ours to change data into value. Having an endless amount of data is useless until it can be inverted into some value. There are various issues related to mining big data, one of the major issue is missing values in a dataset. Missing data or missing value can be described as the data whose values in any given dataset are unknown. It is a recurrent phenomenon and can give biased results from the data to be observed. There are the various rationales for missing data, some arise unintentionally where some knowingly. Three types of missingness are there (i) Missing completely at random (MCAR) (ii) Missing at random (MAR) (iii) Not missing at random (NMAR). (i) In missing completely at random (MCAR), there is no correlation between how the data is missing and the data we have and we do not have. When data is MCAR, the fact that the data is missing is independent of the observed and unobserved data. In other terms, missingness does not rely on input values. (ii) In missing at random (MAR), sequence of missingness has a correlation with observed data. In MAR, missing is systematically related to the observed but not the unobserved data. In another way, missingness depends only on observed input data. (iii) Missing not at random (MNAR) in theory is a sequence of missingness that has a correlation to both the observed data and the missing data. Missingness depends on missing variables. This type of missingness is problematic as it depends on observed and unobserved data. Till date, directly handling missing dataset is perverse. There are

various methods for handling missing data such as deletion methods and imputation methods. Deletion methods are useful only when the missing rate is below 5%. If the missing rate is high, deleting missing values can affect the quality of the results.

On the other hand, imputation of missing values can give optimal solutions to missing values in a dataset. Two types of imputation techniques are there: statistical and machine learning. The statistical imputation methods are mean imputation, regression imputation, and multiple imputations. Machine learning imputation methods for handling missing data are k-nearest neighbor imputation, fuzzy methods, decision trees, support vector. The objective of this survey is to guide the researchers to choose the efficient method for imputation of missing data.

## 2    Various Missing Value Imputation Approaches

Some of the imputation techniques discussed are as follows:

(a)  **Evidence chain technique**

This algorithm searches all the related information of missing values in each dataset and then merges the related information to form a chain of information for additional estimation of missing values [1]. This algorithm is divided into five steps. The first step marks the missing datasets. The second step is subdivided into four parallel sections:

In the first section, after the missing values are marked, it is used to determine a set of complete data present in an incomplete dataset which is used as relevant information in determining missing data values. Section second computes the feasible values of missing data from complete data tuples. Complete dataset for each data tuple is computed in section third which is further used in finding the probability of missing values. Section fourth calculates missing datasets and complete datasets in an incomplete data tuples.

The third step combines the related information determined from the first section with the feasible values calculated in Sect. 2. Step fourth evaluates possible values of missing values. Finally, the values computed in the fourth step replace the missing values in the given dataset.

The performance of this algorithm is better in comparison with various baseline algorithms such as K-nearest neighbor [2, 3], Naive Bayes and mode. Also, the efficiency of this algorithm increases with an increase in missing rates.

(b)  **Estimation of missing values in mixed-attribute datasets**

This technique deals with the imputation of missing values in a dataset containing heterogeneous attributes (attributes are of various types, i.e., a combination of discrete and continuous) which are mentioned as mixed-attribute imputation [4]. It uses a mixed kernel (combination of two kernel functions [5])-based iterative estimator for the imputation of mixed-attribute datasets. This technique is significantly better than other kernel-based algorithms with respect to classification accuracy and root mean square error (RMSE) at various missing ratios. The comparison is made with some existing methods, such as the nonparametric imputation method with a single kernel, the nonparametric method for continuous attributes, and frequency estimator (FE) and traditional nonparametric kernel-based imputation of missing values from [6]. Imputation of missing values takes place first with the mode or mean for discrete and continuous variables. Then afterward, previous imputed missing values are termed as known values for next iteration of imputation. Observed information is utilized with this method in incomplete instances.

Extended method of nonparametric estimation of regression functions for continuous data and categorical (2004) [7] is used for mixed attribute missing value imputation. All the observed information is taken under consideration in this method including observed information in a dataset containing missing values.

### (c) A class center-based approach

The class center is calculated in this approach for replacing missing values. The threshold is also set for further imputation. The threshold is set by calculating the median of the distances between the center of each class and the observed data [8]. In this approach, center value is substituted for missing value and then the distance is measured between imputed value and class center value by some distance function (Euclidean distance) for comparison with a threshold. The imputation process terminates for distances less than the threshold and for distances greater than threshold imputed values are added and subtracted with calculated standard deviation. This approach outperforms other approaches like K-nearest neighbor imputation (KNNI) [9], mean and support vector machine (SVM) [10] in terms of numerical datasets with high missing rates. Even the computational efficiency of this algorithm is far better than SVM and KNNI.

### (d) Decision tree and decision forests

In addition to decision tree, this approach makes use of decision forests to locate horizontal segments of the dataset where we have maximum resembles and attribute correlation and using similarities and attribute correlation imputation of missing values takes place [11]. Decision tree divides the given dataset into a number of leaves and each leaf consists of mutually exclusive records. In decision tree-based missing value imputation (DMI), categorical missing values are imputed within leaves, whereas for numerical imputation, expectation maximization imputation (EMI) algorithm is applied within leaves. This method uses a combination of the decision tree and expectation maximization algorithm [12]. Extension of DMI technique is SiMI that uses decision forest algorithm and EMI plus splitting and merging technique. Decision forest consists of a number of decision trees. Similarities and correlation of records are higher in the intersection of leaves of the different decision trees. The small-sized intersection is merged in such a way that they produce better similarities and correlations. With the use of segments, one can increase the imputation accuracy for missing values. Also, this technique is better than EMI [13] as it uses only a group of records for imputing missing values, whereas EMI uses all the records in a dataset for missing value imputation. These two techniques are also better than existing imputation technique based on similar records called Iterative bicluster-based least square (IBLLS) [14]. IBLLS uses only those attributes having a correlation greater than the threshold value while DMI and SiMI select all attributes based on the power of correlation.

### (e) Combining instance selection

This approach improves the imputation process by filtering some of the noisy data from a dataset [15]. For better classification, performance imputation and instance selection are combined. Four different strategies are used for a combination of instance selection and imputation of missing values. In the first strategy, imputation is conducted first and then instance selection is employed to strain out noisy data. In the second, first, instance selection is accomplished than the imputation takes place on that minimized dataset. In the third and fourth strategy, instance selection is again employed on the first two strategies. These methods outperform final classification performance than the baseline methods. The second strategy improves the classification performance of KNN and SVM when imputation is done on the categorical and numerical datasets. With regard to classification accuracy and the reduction rate, the fourth strategy is best suitable.

### (f) Fuzzy-Rough method

This is one of the effective approaches for missing value imputation as these methods are brilliant methods for dealing with uncertainty with robust and noise-tolerant property [16]. Three imputation techniques are employed on the basis of fuzzy-rough method, viz. vaguely quantified rough sets, average ordered weighted-based rough sets and implicator-based fuzzy rough sets. This method combines fuzzy rough with the nearest neighbor algorithm to acquire easiness and precision of nearest neighbor plus robust and noise-tolerant property of the fuzzy rough set. The extension of fuzzy-rough nearest neighbor algorithm (FRNN) [17] is used to estimate missing values in a dataset. This extended approach is named as fuzzy-rough nearest neighbor imputation (FRNNI). For an occurrence of the dataset say $x$, comprising of minimum single missing value, the algorithm after searching for its k-nearest neighbors will place them in a set, $M$. The algorithm will then estimate the missing value by $x$'s nearest neighbors. This algorithm then calculates lower and upper approximations of $x$ with respect to some instance say $z$ to produce final membership $M$. This approximation is conducted for every instance present in set M and algorithm will return a value based on the above calculations.

### (g) Stochastic Semi-Parametric regression model

This model overcomes some limitations of linear and nonparametric models. This method is far better as compared to the previous deterministic semi-parametric imputation in terms of effectiveness and efficiency. Kernel-based semi-parametric regression method is used for imputing missing values [18]. This model works in two steps: one is the nonparametric model and other is a parametric model. In a dataset, if the missing values occur, they are being upgraded by the mean of the familiar values in that dataset and a random value. This method is better when there is the least information regarding observed and missing attribute variables. In terms of accuracy of prediction and different missing rates, stochastic semi-parametric model outperforms the deterministic semi-parametric regression model.

# 3   Comparison of Various Approach

Table 1 gives a summary of the different approaches mentioned above and a comparative analysis is performed depicting the effectiveness of various techniques at the different missing rates.

In Table 1, different researchers compare their proposed approach with various existing baseline imputation methods on the bases of RMSE, classification accuracy, and speed up. Xu et al. [1] use evidence chain to improve the imputation accuracy. The imputation accuracy of this algorithm is high and steady with the change in missing rate compared to other missing data imputation algorithms such as Naive Bayes, mode imputation and KNN. However, this algorithm is well suited for discrete missing data only and is time-consuming. Zhu et al. [4] propose two estimators for imputation of missing values with heterogeneous datasets as no estimators were there for mixed-attributed datasets. This method exhibits better interpolation and extrapolation, as well as exponential time, is reduced to almost polynomial time. This technique outperforms existing approaches for imputation of both discrete and continuous missing value on bases of RMSE and classification accuracy. In Tsai et al. [8], a class center-based approach is used for better imputation, classification accuracy and time required for imputation is least compared to machine learning imputation techniques. For numerical dataset, the classification accuracy of this algorithm is remarkably better than the baseline algorithm when the missing rates are above 10%. For categorical dataset, the accuracy rate is less than mode nearly 1%. For mixed dataset classification, accuracy is similar to the numerical dataset. RMSE and imputation time of this algorithm are far better than baseline approaches. Computational cost is also less than SVM and KNNI. Rahman and Islam [11] to find similar records use decision trees and decision forests as correlation and similarities among attributes are high using trees and forests. More the correlation and similarities, higher is the imputation accuracy. Although this approach improves imputation accuracy, the time complexity is high. Tsai and Chang [15] perform instance selection to filter some of the noisy data from the dataset for improving classification performance. Amiri and Jensen [16] combine fuzzy and rough to deal with uncertainty. The performance of this method is outstanding in general but with some complexity which can be enhanced using optimization. Qin et al. [18] use stochastic semi-parametric regression imputation for better effectiveness and efficiency as compared to the existing deterministic semi-parametric regression imputation method. This method is better when there is least priori knowledge regarding observed dataset.

**Table 1** Comparison of various missing value imputation approaches

| Author and year | Imputation | Type of missingness | Compared algorithm | Percentage of missing value (%) | Dataset type | Remarks |
|---|---|---|---|---|---|---|
| Xu et al. [1] | Evidence chain | Not mentioned | Naive Bayes, Mode and K-Nearest Neighbor (KNN) | 10, 20, 30 | UCI adult dataset with 5 attributes | Time consumption increases rapidly |
| Zhu et al. [4] | Estimation of mixed-attribute datasets | MCAR | Poly[a], Normal[b] and FE[c] | 10, 20, 30, 50, 80 | 6 UCI | Exponential time decreases to polynomial one |
| Tsai et al. [8] | A class center based | MCAR | Mean, Mode, KNNI, and SVM[d] | 10, 20, 30, 40, 50 | 34 UCI (15 numerical, 11 categorical and 8 mixed) | Less imputation time than Machine Learning MVI method |
| Rahman and Islam [11] | Decision trees and decision Forests | MAR | EMI[e], IBLLS[f] | 1, 3, 5, 10 | 9 UCI (numerical, categorical) | Improved imputation accuracy. High time complexity |
| Tsai and Chang [15] | Combining instance selection | MCAR | KNN, SVM | 10, 20, 30, 40, 50 | 29 UCI (11 numerical, 9 categorical and 9 mixed) | Better classification accuracy |
| Amiri and Jensen [16] | Fuzzy Rough | MAR | BPCAI[g], CMCI[h], FKMI[i], EMI, KNNI[j] | 5, 10, 20, 30 | 27 datasets from KEEL dataset repository | High time complexity |
| Qin et al. [18] | Stochastic semi-parametric | MAR, MCAR | Nonparametric Linear regression | 10, 20, 40 | DELL Workstation | Well suited when having less priori knowledge |

[a]Poly: nonparametric with single-kernel imputation
[b]Normal: Existing nonparametric kernel-based imputation method
[c]FE: conventional frequency estimator
[d]SVM: Support Vector Machine
[e]EMI: Expectation Maximization Imputation
[f]IBLLS: Iterative Bicluster-based Least square
[g]BPCAI: Bayesian PCA Imputation
[h]CMCI: Concept Most Common Imputation
[i]FKMI: Fuzzy K-Means Imputation
[j]KNNI: K-Nearest Neighbor Imputation

# 4  Conclusions

In this paper, various novel imputation techniques have been reviewed such as evidence chain technique, estimation of missing values in mixed-attribute datasets, class center-based approach, decision tree, and decision forests, combining instance selection, fuzzy rough set, and stochastic semi-parametric model. In Table 1, we compute performance parameters, the percentage of missing value and concluded that the class center-based approach is best than other approaches as it takes meager time than machine learning method and has better imputation accuracy.

# References

1. X. Xu et al., MIAEC: Missing data imputation based on the evidence chain. *IEEE Access* (2018)
2. J. Han, M. Kamber, *Data Mining Concept and Techniques*, 3rd edn. (Multiscience Press, USA), pp. 226 (2012)
3. P. Keerin, W. Kurutach, T. Boongoen, Cluster-based KNN missing value imputation for DNA microarray data, in *2012 IEEE International Conference on Systems, Man, and Cybernetics (SMC)*, Seoul (2012), pp. 445–450
4. X. Zhu et al., Missing value estimation for mixed-attribute data sets. IEEE Trans. Knowl. Data Eng. **23**(1), 110–121 (2011)
5. Y.S. Qin et al., POP algorithm: kernel-based imputation to treat missing values in knowledge discovery from database
6. S.C. Zhang, Parimputation: from imputation and null-imputation to partially imputation. IEEE Intell. Inform. Bull. **9**(1), 32–38 (2008)
7. J. Racine, Q. Li, Nonparametric estimation of regression functions with both categorical and continuous data. J. Econometrics **119**(1), 99–130 (2004)
8. C.-F. Tsai, M.-L. Li, W.-C. Lin, A class center-based approach for missing value imputation. Knowl. Based Syst. **151**, 124–135 (2018)
9. R. Pan, T. Yang, J. Cao, K. Lu, Z. Zhang, Missing data imputation by K nearest neighbors based on the grey relational structure and mutual information. Appl. Intell. **43**, 614–632 (2015)
10. K. Pelckmans, J.D. Brabanter, J.A.K. Suykens, B.D. Moor, Handling missing values in support vector machine classifiers. Neural Netw. **18**, 684–692 (2005)
11. M.G. Rahman, M.Z. Islam, Missing value imputation using decision trees and decision forests by splitting and merging records: two novel techniques. Knowl. Based Syst. **53**, 51–65 (2013)
12. M.G. Rahman, M.Z. Islam, iDMI: a novel technique for missing value imputation using a decision tree and expectation-maximization algorithm, in *2013 16th International Conference on Computer and Information Technology (ICCIT)*, Khulna (2014), pp. 496–501
13. T. Schneider, Analysis of incomplete climate data: estimation of mean values and covariance matrices and imputation of missing values. J. Clim. **14**(5), 853–871 (2001)
14. K. Cheng, N. Law, W. Siu, An Iterative bicluster based least square framework for estimation of missing values in microarray gene expression data. Pattern Recogn. **45**(4), 1281–1289 (2012)
15. C.-F. Tsai, F.-Y. Chang, Combining instance selection for better missing value imputation. J. Syst. Softw. **122**, 63–71 (2016)
16. M. Amiri, R. Jensen, Missing data imputation using fuzzy-rough methods. Neurocomputing **205**, 152–164 (2016)

17. R. Jensen, C. Cornelis, Fuzzy-rough nearest neighbor classification, in *Transactions on Rough Sets XIII* (Springer, Berlin, 2011), pp. 56–72
18. Y. Qin et al., Semi-parametric optimization for missing data imputation. Appl. Intell. **27**(1), 79–88 (2007)

# Advanced Deadline-Sensitive Scheduling Approaches in Cloud Computing

Duraksha Ali and Manoj Kumar Gupta

**Abstract** Cloud computing technology in the field of high-performance distributed computing has become a milestone as it provides shared computing and storage resources as a service upon request of user. Also, with the advent of the computers, scheduling problem got good attention in innumerous fields and arose in industries and technology as well. Service provider's main requirement is the productive utilization of available resources because it is altogether demanding to supply on-demand resources to the users in a best possible manner for improving performance and faster computation time. In this paper, we present a review to act as an aid to the newcomer researchers to understand various advanced techniques proposed by researchers on deadline-sensitive task scheduling in cloud computing. The schemes reviewed in this paper include grouped task scheduling, deadline-sensitive lease scheduling, scientific workflow scheduling, resource provisioning, pair-based task scheduling, workflow in multi-tenant cloud environment, etc. We also performed a comparative analysis of these approaches and their parameters. Our goal of reviewing the scheduling literature is to scrutinize the interest in different problems among the proposed techniques and algorithms.

**Keywords** Deadline sensitive · Backfilling · Decision maker · Resource allocation · Task scheduling · Cloud computing

## 1 Introduction

Cloud computing has turned out to be a buzzword due to the ever-growing data and computing requirements. As stated by service-level agreement (SLA), service provider requires to attain the on-demand request of cloud user to execute user applications. User requests can arrive at any time [1].

D. Ali · M. K. Gupta (✉)
School of Computer Science and Engineering, Shri Mata Vaishno Devi University, Katra, India
e-mail: manoj.gupta@smvdu.ac.in

D. Ali
e-mail: darakhshan558@gmail.com

© Springer Nature Singapore Pte Ltd. 2021
X.-Z. Gao et al. (eds.), *Advances in Computational Intelligence and Communication Technology*, Advances in Intelligent Systems and Computing 1086,
https://doi.org/10.1007/978-981-15-1275-9_26

In cloud computing SaaS, there are three key roles, viz. cloud providers, application providers, and users. Cloud providers as suppliers offer the virtual machines (VM's) as a service and manage the infrastructure of data center. Application providers rent the VM from cloud providers and offer different applications as a service to users. Users as consumers request for the particular application and get the desired results from application providers [2]. Resources in bulk are to be issued to the millions of distributed users in cloud dynamically, reasonably and in the in cost-effective manner. There are some common constraints within which the scheduling has to be accomplished which are: priority constraint, dependency constraint, deadline constraint, and budget constraint [3].

## 2 Peculiarities and Challenges

The prime issues often pertinent to cloud computing are resource management, virtualization, multi-tenancy, network infrastructure management, security, data management, application programming interfaces (APIs), etc. [4]. The plethora of virtualized resources in the cloud and the prodigious different virtualized assets being used per task in the workflow, are needed to be managed and executed within deadlines. As the manual scheduling is not a reasonable solution, the main motive of the cloud providers is better deadline-sensitive scheduling in cloud environment [5]. Various algorithms have been polished up to attain the demands of cloud computing, and one among them is task scheduling in which we assign priorities to different tasks on the basis of their attributes and criterion and the tasks with high priority are scheduled first [6].

## 3 Deadline-Sensitive Scheduling Approaches in Cloud Computing

Various recent scheduling techniques and algorithms studied in this paper are as:

### 3.1 Scientific Workflows with Task Replication

There is sanguine notion about the cloud environment performance. There are performance irregularities up to 35% in terms of execution time and 65% in terms of data transfer time. These variations occur because of the shared character of virtualized environment and the time an application uses up in communication [7]. Various limiting factors that become hurdles for scientific workflow execution are budget constraints, workflow structure because of dependencies, charge for integer time units, historical data analysis for proper estimation of execution time to meet deadlines, performance variations, and failure of one or more VM's during execution. Dependencies among the tasks and adjusts employed by algorithm are two main reasons for the idle slots. VM with smallest cost in the descendingly sorted list is chosen as a new VM to be used and thus provides the selection process of choosing the type of VM. All the idle time slots including paid and unpaid slots are listed; tentative order of task replication to available slot is specified based on execution time and number of children. A task is checked if it can fit in a particular time slot depending on the execution time of task, size of time slot, violation of dependencies, etc. Lastly, if task fits, replica is created.

To render more contingency strategies for turning down the delays due to the performance fluctuations of the leased cloud resources, replication of tasks in the scientific workflows can be performed [8]. EIPR [Enhanced IC-PCP (IaaS Cloud-Partial Critical Path) algorithm] for scientific workflows is specified in three steps: (a) Provisioning and scheduling: determine type and number of VM's to be provisioned; ordering and placement of tasks on those resources. (b) Data transfer-aware provisioning: besides specifying the start and stop time of VM's, we need to consider the data transfer time as well because data are transferred from parent task to VM before task starts to execute. (c) Task replication: the motive of task replication is to hike the performance rather than fault tolerance so the tasks are replicated only on different VM's.

## 3.2 Resource Provisioning for Data-Intensive Applications

Additional resources are added from public cloud to the pool of a private cloud organization with limited resources especially when QoS is required within deadline restrictions. There is a need of hybrid cloud middleware to deal with cloud bursting model in which, on requirement of more resources in private cloud, it bursts into public cloud.

Middleware like Aneka can be used which is a PaaS used to deliver resource scheduling and provisioning. Aneka is employed for resource provisioning from clusters, grids, private clouds, public clouds, and hybrid clouds [9]. Aneka components are middleware, container, and platform abstraction layer (PAL). Further, middleware is subdivided into two components: Aneka Daemon and Aneka container. Location of data, network latency, and bandwidth constraints are taken into consideration while dealing with resource provisioning for data-intensive applications in hybrid cloud environment [10]. If an entire dataset is to be moved to the cloud, it would not be efficient because of data transfer time owing to bandwidth and data size [11].

Figure 1 shows the working of this resource provisioning algorithm in steps.

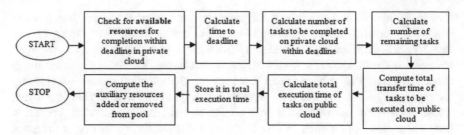

**Fig. 1** Resource provisioning for data-intensive applications

## 3.3 Grouped Task Scheduling

Tasks are distributed into five categories in grouped task scheduling algorithm (GTS) as: CategoryUrgentUser&Task, CategoryUrgentUser, CategoryUrgentTask, CategoryLongTask, and CategoryNormalTask (with priority in the descending order). Categorization is based on four attributes such as task size, task type, user type, and task latency. After setting the tasks into appropriate groups, they are scheduled onto available resources. The algorithm works in two phases: first phase is deciding which class to be scheduled primarily and second phase is selecting a task in the particular chosen class to be scheduled primarily. The class having elevated importance of attributes will be primarily scheduled because former phase depends on the weights of attributes of all the tasks inside a class. The tasks having smaller execution time will be primarily scheduled because the later phase depends on the execution time of tasks [4].

GTS algorithm is based on three basic algorithms: task scheduling algorithm, Min-Min algorithm, and improved cost-based algorithm. GTS takes the advantage of low latency of task scheduling algorithm and low execution time of Min-Min algorithm so results in minimal latency with minimal execution time [6].

## 3.4 Deadline-Sensitive Lease Scheduling Using AHP

Most of the leases in real-time environment are deadline-aware leases and backfilling algorithm is used to schedule them. But backfilling algorithm cannot work well sometimes when similar kind of leases occur. A decision maker like AHP can be incorporated with backfilling algorithm to schedule greater number of leases and prevent lease rejection [1]. Besides AHP, there are some other decision makers, like PROMETHEE, TOPSIS, WSM, OWA, etc., that can be used. References [12–14] describe how to select and how to rank projects using PROMETHEE.

A lease is chosen from best effort queue to issue unused resources of free slot to deadline-aware lease in backfilling algorithm. Backfilling algorithm combined with a decision maker can be applied when there are similar leases. Then, it checks for the

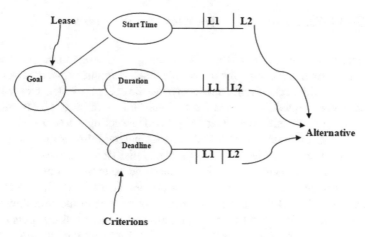

**Fig. 2** Decision hierarchy tree

lease if deadline sensitive or not, (determined by slack value) which depicts whether a lease can be scheduled or not. Decision hierarchy tree (DHT) of the problem has goal, criterions, and alternatives. The goal of AHP (Analytical Hierarchy Process) is to schedule a lease and criterions for AHP are three parameters: deadline, duration and start time. Also, similar leases L1 and L2 are alternatives as shown in Fig. 2.

### 3.5 Pair-Based Task Scheduling Algorithm

A collection of tasks are divided into a group of categories, to be scheduled on a set of clouds with a motive to lower the overall layover time. Before the scheduling, tasks are paired and those paired tasks should hail from two different categories. Categories are disjoint but contain equal number of tasks. Tasks are mapped to various clouds initially but finally allocated to one convenient cloud. Two tasks, $Ti$ and $Tj$ (where $i, j$ is task number and $j = i+1$) can be executed on the same cloud only if $Ti$ successfully completes its execution, however, $Tj$ can be processed on the same cloud if the end time of $Ti$ [et($Ti$)] and the transfer time ($\alpha$) between the $Ti$ and $Tj$ is less than start time of $Tj$ [st($Tj$)] i.e., et($Ti$) + $\alpha$ < st($Tj$).

Hungarian algorithm [15] can be used in paired tasks to lessen the total layover time, within deadlines. It divides the tasks into even number of categories (usually two C1 and C2) and computes converse lease time and lease time among tasks of categories C1 and C2 [16]. It computes column opportunity matrix (COM) and row opportunity matrix (ROW) and with the help of lease time schedules the tasks.

## 3.6 CEDA Scheduling for Workflow Applications

Cost-effective deadline-aware (CEDA) scheduling focuses on minimizing total economic cost and total execution time for workflows. It computes upward rank of tasks,
selects the task with highest rank, and assigns to the cheapest instance of VM keeping in view VM acquisition time and additional price overhead. CEDA algorithm
reduces cost while meeting deadlines [17]. While selecting a VM instance, CEDA
wishes to choose an already active VM instance having leftover time of its charge
time interval suffice to run the task prior its latest finish time because of no extra time
for VM acquisition to launch new VM instance and no extra cost overhead of VM.
CEDA works in two phases. First is task prioritizing phase in which after initialization, latest finish time (LFT) and minimum execution time of the workflow (MTW)
are to be computed followed by computing upward rank of tasks. It puts the tasks
into task list and selects the first task from the list if it is not a pseudo-task. Second
is planning phase in which CEDA finds the most optimized schedule by finding the
cheapest VM instance. CEDA assigns the task to such VM instance and updates the
pool.

## 3.7 Resource Provisioning and Scheduling for Algorithm for Scientific Workflows

There are two main phases of workflow execution in cloud computing. First one
is resource provisioning in which resources are selected and provisioned. Second
one is scheduling in which tasks are organized into a schedule S = (R, M, TET,
TEC) in terms of set of **R**esources $R = \{r_1, r_2, \ldots, r_n\}$, **M**apping of tasks to
resources, **T**otal **E**xecution **T**ime, and **T**otal **E**xecution **C**ost) are mapped onto best-
suited resource. Deadline-based resource provisioning and scheduling algorithm for
scientific workflows is based on particle swarm optimization (PSO) [18].

PSO is based on the social conduct of bird flocks [19]. Swarm of particles move
in the defined problem space trying to optimize a problem [20] until the stopping
criterion is met. PSO works in two phases. First one is encoding of problem or solution
representation and second is defining fitness function. Encoding is to determine
meaning and dimension of a particle. Here, particle is a workflow along with its
tasks. Dimension of a particle is the coordinate system determined by number of
tasks in the workflow. The number of available resources for the execution of tasks
where a particle can move is the range here. Based on the objective of the problem,
fitness function is minimized.

Whenever two solutions are compared, if both are feasible then choose the one
with greater fitness. If only one is feasible, then choose the feasible one. If both are
infeasible, then choose the one with less deadline violation.

## 3.8 Workflow Scheduling in Multi-tenant Cloud Computing Environment

Multi-tenancy is a feature of cloud computing which allows an architecture for cloud users to share the same infrastructure with computing resources in isolation and invisible to other tenants. Multi-tenancy is a feature of cloud computing which allows a platform for cloud users to share the same infrastructure with computing resources in isolation and invisible to other tenants. Resource management in multi-tenancy nowadays is an issue. Cloud-based workflow scheduling (CWSA) algorithm caters this issue thus reduces the execution time of workflow, tardiness, and cost of workflow execution and maximizes resource utilization. CWSA makes use of idle gaps between scheduled tasks to schedule more tasks thus minimizing makespan [21].

In multi-tenancy, a new layer is introduced above SaaS or IaaS called workflow as a service (WaaS). By using scheduler and central submission interface, it automates the process thus reducing total turn-around time of submitted tasks. Broker chooses the largest possible backfill job to be allocated to idle VM thus guaranteeing deadline [2]. In workflow scheduling structure, tenants present their workflow applications to the cloud system via tenant layer. Between tenant layer and infrastructure layer, there is middleware layer which acts as a bridge between them and is meant to minimize the complexity such as organizing and managing.

When tasks are submitted to the system, scheduler checks for the dependencies to determine the order in which tasks should be executed. Tasks are prioritized by their deadlines in the service queue. An idle schedule gap or idle CPU time between tasks is computed and used to schedule other tasks to minimize makespan within deadlines.

The comparison analysis of all the above approaches and their different features is performed in Table 1.

## 4 Analysis and Discussions

Variant parameters used by the above approaches are tabulated in Table 2. From Table 2, we can analyze that the most functional parameters of deadline-sensitive scheduling in cloud computing are execution time, cost, and deadline that are taken into consideration by several researchers while latency, resource utilization, makespan, load balancing, and wastage of time slots are considered by some researchers in the relevant field. Though very less number of researchers had studied the metrics like number of leases scheduled, performance fluctuations, number of virtual machine instances launched, tardiness, layover time, and time slots allocated, these metrics require more to be focused by upcoming researchers in future.

As EIPR algorithm [8] outperforms other algorithms because apart from reducing execution time drastically, it also deals with the performance fluctuations rendered in cloud computing environment while keeping in view the deadlines too. It minimizes

**Table 1** Comparison of various deadline-sensitive scheduling approaches in cloud computing

| Improvement strategy | Performance metrics | Nature of tasks | Environment | Datasets used | Baseline Algorithm | Future scope |
|---|---|---|---|---|---|---|
| GTS algorithm [4] | Latency of tasks, load balancing, execution time span | Independent | Hybrid cloud | Not mentioned | Min-Min Algo, TS Algo | Increment in no of attributes, independent tasks |
| Resource provisioning for data-intensive applications [10] | Execution time, total cost, total number of VM instances launched | Independent | Aneka | Geospatial datasets provided by (PSMA) Australia | Default, enhanced | For workflows |
| Deadline sensitive lease scheduling using AHP [1] | Number of leases scheduled, number of time slots assigned, wastage of time slots | Independent | MATLAB R2010a | Not mentioned | Backfilling algorithm | Implementation in OpenNebula, use other decision maker |
| Scientific Workflows with Tasks Replication [8] | Total execution time, performance fluctuations | Workflow | CloudSim toolkit | MONTAGE, Cybershake, LIGO, SIPHT | IC-PCP algorithm | Replication across multiple clouds |
| Pair-based task scheduling algorithm [16] | Total layover time | Independent | MATLAB R2014a, Haizea | 22 different datasets generated by MATLAB using Monte Carlo method | FCFS, HA-LT, HA-CLT | Test with large no. of tasks (>500), increase in no. of groups |
| CEDA scheduling for workflow Applications [17] | Total execution time (TET), Total price of execution (TPE). | Workflow | MATLAB | MONTAGE, Cybershake, LIGO, SIPHT | ICPC, ICPCPD2 | Termination delay of VMs, VMs in different regions, communication cost |
| Resource Pro-visioning and Scheduling for Scientific Workflows [18] | Overall workflow execution cost | Workflow | CloudSim framework | MONTAGE, Cybershake, LIGO, SIPHT | IC-PCP, SCS | Data transfer cost between DC's |
| Workflow Scheduling in Multi-Tenant Cloud Environments [21] | Makespan, tardiness, execution time, cloud resource utilization, cost of workflow execution | Workflow | CloudSim framework | Cybershake, SIPHT | FCFS, EASY, backfilling, minimum completion time | Resource failures, application in context of mobile cloud computing |

**Table 2** Scheduling parameters considered by recent deadline-sensitive scheduling approaches

| Paper | Execution Time | Cost | Load Balancing | Latency | Number of VM Instances Launched | Makespan | Time Slots Allocated | Tardiness |
|---|---|---|---|---|---|---|---|---|
| GTS algorithm [4] | ✓ | ✗ | ✓ | ✓ | ✗ | ✗ | ✗ | ✗ |
| Resource provisioning for data-intensive applications using Aneka [10] | ✓ | ✓ | ✗ | ✗ | ✓ | ✗ | ✗ | ✗ |
| Deadline sensitive lease scheduling using AHP [1] | ✗ | ✗ | ✗ | ✗ | ✗ | ✗ | ✓ | ✗ |
| Scientific Workflows with Tasks Replication [8] | ✓ | ✗ | ✗ | ✗ | ✗ | ✗ | ✗ | ✗ |
| Pair-based task scheduling algorithm [16] | ✗ | ✗ | ✗ | ✗ | ✗ | ✗ | ✗ | ✗ |
| CEDA scheduling for workflow Applications [17] | ✓ | ✗ | ✗ | ✗ | ✗ | ✗ | ✗ | ✗ |
| Resource provisioning and scheduling algorithm for scientific workflows [18] | ✗ | ✗ | ✗ | ✗ | ✗ | ✗ | ✗ | ✗ |

(continued)

**Table 2** (continued)

| Paper | Execution Time | Cost | Load Balancing | Latency | Number of VM Instances Launched | Makespan | Time Slots Allocated | Tardiness |
|---|---|---|---|---|---|---|---|---|
| Workflow scheduling in multi-tenant cloud environment [21] | ✓ | ✗ | ✗ | ✗ | ✗ | ✓ | ✗ | ✓ |

| Paper | Layover Time | Price of Execution | Workflow Execution Cost | Wastage of Time Slots | Number of Leases Scheduled | Performance Fluctuations | Resource Utilization | Deadline Guarantee |
|---|---|---|---|---|---|---|---|---|
| GTS algorithm [4] | ✗ | ✗ | ✗ | ✗ | ✗ | ✓ | ✗ | ✓ |
| Resource provisioning for data-intensive applications using Aneka [10] | ✗ | ✗ | ✗ | ✗ | ✗ | ✗ | ✗ | ✓ |
| Deadline sensitive lease scheduling using AHP [1] | ✗ | ✗ | ✗ | ✓ | ✓ | ✗ | ✗ | ✓ |
| Scientific Workflows with Tasks Replication [8] | ✗ | ✗ | ✗ | ✗ | ✗ | ✗ | ✗ | ✓ |

(continued)

**Table 2** (continued)

| Paper | Layover Time | Price of Execution | Workflow Execution Cost | Wastage of Time Slots | Number of Leases Scheduled | Performance Fluctuations | Resource Utilization | Deadline Guarantee |
|---|---|---|---|---|---|---|---|---|
| Pair-based task scheduling algorithm [16] | ✓ | ✗ | ✗ | ✗ | ✗ | ✗ | ✗ | ✓ |
| CEDA scheduling for workflow Applications [17] | ✗ | ✓ | ✗ | ✗ | ✗ | ✗ | ✗ | ✓ |
| Resource provisioning and scheduling algorithm for scientific workflows [18] | ✗ | ✗ | ✓ | ✗ | ✗ | ✗ | ✗ | ✓ |
| Workflow scheduling in multi-tenant cloud environment [21] | ✗ | ✗ | ✓ | ✗ | ✗ | ✗ | ✓ | ✓ |

the execution time up to 59% compared to baseline algorithm. Thus, based on this analysis, EIPR is a better algorithm in terms of execution time and performance fluctuations comparatively. GTS algorithm [4] attains less latency for long tasks than baseline algorithm by 54 and 57% and it turns down the execution time by 9% and 12% for 100 and 50 services, respectively. Thus, we can analyze that GTS algorithm is a better approach in terms of latency and execution time comparatively.

## 5 Conclusions

In cloud computing, scheduling is one of the most critical problems. So, an opportunity always lies there for the enhancement of previously completed work. This paper studies various recent approaches in deadline sensitive scheduling in cloud computing environment. In this literature survey, most of the authors have concentrated on minimization of execution time whereas others have taken under consideration, cost, latency, resource utilization, tardiness, layover time, wastage of time slots, number of leases scheduled, etc. Comparative analysis of these approaches is based on nature of tasks, datasets used for simulation, platforms, baseline algorithms, performance metrics, and future scope. This gives the clear idea of the several potential directions for future work that can be explored and investigated by newcomer researchers.

## References

1. S.C. Nayak, C. Tripathy, Deadline sensitive lease scheduling in cloud computing environment using AHP. J. King Saud Univ. Comput. Inf. Sci. **30**(2), 152–163 (2018)
2. S.M. Shin, Y. Kim, S.K. Lee, Deadline-guaranteed scheduling algorithm with improved resource utilization for cloud computing, in *2015 12th Annual IEEE Consumer Communications and Networking Conference (CCNC)* (IEEE, 2015)
3. M. Kalra, S. Singh, A review of metaheuristic scheduling techniques in cloud computing. Egypt. Inform. J. **16**(3), 275–295 (2015)
4. H.G.E.D.H. Ali, I.A. Saroit, A.M. Kotb, Grouped tasks scheduling algorithm based on QoS in cloud computing network. Egypt. Inform. J. **18**(1), 11–19 (2017)
5. A.R. Arunarani, D. Manjula, V. Sugumaran, Task scheduling techniques in cloud computing: a literature survey. Future Gener. Comput. Syst. **91**, 407–415 (2019)
6. X. Wu et al., A task scheduling algorithm based on QoS-driven in cloud computing. Procedia Comput. Sci. **17**, 1162–1169 (2013)
7. K.R. Jackson et al., Performance analysis of high performance computing applications on the amazon web services cloud, in *2nd IEEE International Conference on Cloud Computing Technology and Science* (IEEE, 2010)
8. R.N. Calheiros, R. Buyya, Meeting deadlines of scientific workflows in public clouds with tasks replication. IEEE Trans. Parallel Distrib. Syst. **25**(7), 1787–1796 (2014)
9. C. Vecchiola et al., Deadline-driven provisioning of resources for scientific applications in hybrid clouds with Aneka. Future Gener. Comput. Syst. **28**(1), 58–65 (2012)
10. A.N. Toosi, R.O. Sinnott, R. Buyya, Resource provisioning for data-intensive applications with deadline constraints on hybrid clouds using Aneka. Future Gener. Comput. Syst. **79**, 765–775 (2018)

11. X. Xu, X. Zhao, A framework for privacy-aware computing on hybrid clouds with mixed-sensitivity data, in *2015 IEEE 7th International Symposium on High Performance Computing and Communications (HPCC), 2015 IEEE 12th International Conference on Cyberspace Safety and Security (CSS), 2015 IEEE 17th International Conference on Embedded Software and Systems (ICESS)* (IEEE, 2015)
12. Z. Zhao, Y. Jiang, X. Zhao, SLA_oriented service selection in cloud environment: a PROMETHEE_based approach, in *2015 4th International Conference on Computer Science and Network Technology (ICCSNT)*, vol. 1 (IEEE, 2015)
13. K. Kaur, H. Singh, PROMETHEE based component evaluation and selection for Component Based Software Engineering, in *2014 IEEE International Conference on Advanced Communications, Control and Computing Technologies* (IEEE, 2014)
14. J.-P. Brans, P. Vincke, B. Mareschal, How to select and how to rank projects: the PROMETHEE method. Eur. J. Oper. Res. **24**(2), 228–238 (1986)
15. A. Frank, On Kuhn's Hungarian method—a tribute from Hungary. Nav. Res. Logistics (NRL) **52**(1), 2–5 (2005)
16. S.K. Panda, S.S. Nanda, S.K. Bhoi, A pair-based task scheduling algorithm for cloud computing environment. J. King Saud Univ. Comput. Inf. Sci. (2018)
17. R.A. Haidri, C.P. Katti, P.C. Saxena, Cost effective deadline aware scheduling strategy for workflow applications on virtual machines in cloud computing. J. King Saud Univ. Comput. Inf. Sci. (2017)
18. M.A. Rodriguez, R. Buyya, Deadline based resource provisioning and scheduling algorithm for scientific workflows on clouds. IEEE Trans. Cloud Comput. **2**(2), 222–235 (2014)
19. J. Kennedy, *Particle Swarm Optimization. Encyclopedia of Machine Learning* (Springer, Boston, MA, 2011), pp. 760–766
20. A. Lazinica (ed.), *Particle Swarm Optimization* (InTech, Kirchengasse, 2009)
21. B.P. Rimal, M. Maier, Workflow scheduling in multi-tenant cloud computing environments. IEEE Trans. Parallel Distrib. Syst. **28**(1), 290–304 (2017)

# A Survey of Text Mining Approaches, Techniques, and Tools on Discharge Summaries

Priyanka C. Nair, Deepa Gupta and Bhagavatula Indira Devi

**Abstract** The discharge summary contains voluminous information regarding the patient like history, symptoms, investigations, treatment, medication, etc. Though the discharge summary has a general structured way of representation, it is still not structured in a way that clinical systems can process. Different natural language processing (NLP) and machine learning techniques have been explored on the discharge summaries to extract various interesting information. Text mining techniques have been carried out in public and private discharge summaries. This survey discusses different tasks performed on discharge summaries and the existing tools which have been explored. The major dataset which has been used in existing research is also discussed. A common outline of system architectures on discharge summaries across various researches is explored. Major challenges in extracting information from discharge summaries are also detailed.

**Keywords** Natural language processing · Discharge summary · Information extraction · Text classification · Relation extraction · NLP tools

## 1 Introduction

Most of the healthcare organizations are increasingly moving toward maintenance of data in the form of Electronic Health Record (EHR). The data gathered from healthcare organizations globally would lead to practices that can improve the diagnosis and treatment of a plethora of health conditions [1, 2]. Data mining techniques,

P. C. Nair · D. Gupta (✉)
Department of Computer Science and Engineering,
Amrita School of Engineering, Amrita Vishwa Vidyapeetham, Bengaluru, India
e-mail: g_deepa@blr.amrita.edu

P. C. Nair
e-mail: v_priyanka@blr.amrita.edu

B. I. Devi
Department of Neurosurgery, National Institute of Mental Health and Neurosciences,
Bengaluru, India
e-mail: bidevidr@gmail.com

© Springer Nature Singapore Pte Ltd. 2021
X.-Z. Gao et al. (eds.), *Advances in Computational Intelligence and Communication Technology*, Advances in Intelligent Systems and Computing 1086,
https://doi.org/10.1007/978-981-15-1275-9_27

vis-a-vis information extraction, are critical in handling such large and diverse set of clinical data. Medical records of patients are maintained in the health center in forms like narrative text, numerical measurements, recorded signals, and images. Feature extraction is applied across major, broadly classified categories of medical imaging data [3, 4], electrophysiological monitoring data, and clinical–textual data to gather various insights. Computational interpretational challenges are numerous including, but not limited to, color variation during slide preparation and inability to generate beyond a 2D planar image [5]. Although it might be subjective to a degree, clinical–textual data are ideal for meaningful extraction using machine learning techniques, hence able to provide better insights. Aforementioned textual data appear to be the most critical and information rich. Most useful data here have been observed to be from radiology, pathology, and transcription reports [6] which include Papanicolaou (Pap) reports for cervical cancer patients [7], oncology narratives [8], and clinical notes from internal emergency unit.

Instead of mining individual clinical reports, mining discharge summary appears to be much more efficient as it is the most critical since it conveniently covers diverse set of information. Discharge summary provides condensed and relevant information like history of the patient, diagnosis, treatment, medication information, condition of the patient at the time of discharge, and advice on discharge. This can give great insight in medical field if studied and analyzed. The aims could be identifying the diagnosis, medical information, or medication from the discharge summary [9, 10] and identifying the relationship between history of the patient, the treatment provided, and its outcome. The discharge summary is not optimally structured for the clinical application systems to process. It is difficult to have a clinician or any person to interpret the record for every case for large studies. Hence, various natural language processing (NLP) techniques are explored in several research works on discharge summary of different kinds of diseases. Named entity recognition (NER), tagging, parsing, rule-based extraction, text-based classification and various other NLP approaches have been applied to extract relevant information from the discharge summary. There are different NLP-based tools [11–14] available which are used for extraction of information from clinical documents that are discussed in detail in remaining sections of the paper.

Section 2 describes various publicly available and private data sources. A generally followed system architecture in various set of studies has been detailed in Sect. 3. Different approaches in NLP that are applied on discharge summaries are described in Sect. 4. The existing tools which are available for processing and extracting information from the discharge summary are explained in Sect. 5. The discussion and challenges of applying text mining in discharge summary are explained in Sects. 6 and 7.

## 2 Data Source

The discharge summaries are either private data collected from a hospital for a team specifically for their use or a publicly available data. This section discusses various publicly available and private discharge summaries for research work. The public data are mainly provided by Informatics for Integrating Biology and the Bedside (i2b2) challenge since 2006. Several kinds of discharge summary are provided in each challenge with a different objective. Various discharge summaries, the count, task, and techniques explored on them and NLP tools applied on them are summarized in Table 1. The initial subsection of Table 1 shows the works done on publicly available discharge summary, obtained from various challenges conducted by i2b2. Each challenge has a different task as described in Table 1. 2006 challenge was for deidentification of patient details and identifying the smoking status, while the 2008 Obesity challenge was to identify the obesity status and related comorbidities from the discharge summaries. The challenges in 2010 and 2012 aimed at exploring the relations in the discharge summary. The tasks for 2014 challenge were deidentification and heart diseases risk factor annotation in discharge summary. Different teams participated in these challenges have applied different techniques for performing the task on the respective dataset as discussed in Table 1. The later subsection represents the works done on private data. Conditional Random Field (CRF), rule-based method, and Support Vector Machine (SVM) are the techniques mainly used in most of the works as can be seen in the table. The table also discusses various tools explored on the data for different tasks. Some of the tools are open source and some are commercial. Few tools like LingPipe are NLP tools that have been used for information extraction. Majority of the tools, like Metamap, GENIA tagger, cTAKES, and MedEx that have been applied on the data, are NLP tools specifically used for medical texts. The detailed description of various tools for discharge summaries is explained in detail in Sect. 5.

The general system architecture for automating a discharge summary to a structured form is explained in Sect. 3. The tasks and techniques mentioned here are discussed in detail in the following sections of the paper.

## 3 System Architecture

Before discussing the system architecture, we need to understand a general structure of a discharge summary which is discussed in subsection of this section, and the later subsection discusses the general work flow for a system that applies various natural language processing techniques on discharge summary.

**Table 1** Dataset for discharge summaries

| Dataset | | Techniques | Tools applied | Count | Publication |
|---|---|---|---|---|---|
| *Publicly available Data from i2b2 challenge* | | | | | |
| 2006 Deidentification and Smoking Challenge | Concept Identification Text Classification | Conditional random field, rule-based, SVM, hybrid methods kNN (k nearest neighbor), Naïve Bayes, SVM, decision tree, AdaBoost, rule-based system, latent semantic analysis, artificial neural network | LingPipe and Carafe toolkit-CRF | 889 | [21, 37–39] |
| 2008 Obesity Challenge | Text Classification | SVM classifier, context aware rule-based classifier, automated isolation of hotspot passages, hybrid approach | LingPipe, GENIA tagger | 507 | [40–44] |
| 2009 Medication Challenge | Information Extraction | Rule-based, ensemble of classifiers, rule-based, CRF, and SVM | | 1243 | [10, 20, 24, 45] |
| 2010 Relations Challenge | Concept Extraction, Relation classification, Assertion classification | CRF, SVM, rule-based, boosting, logistic regression | SharpNLP, GENIA part-of-speech tagger, WordNet, MetaMap, SRL | 1703 | [26, 46] |
| 2012 Temporal relations challenge | Concept, Event and Temporal expression and Temporal Relation Extraction | CRF, name entity extraction, SVM, rule-based methods, Brown clustering, maximum entropy | GENIA tagger TARSQI, Metamap, cTAKES, HeidelTime, SUTIME GUTIME | 310 | [33, 47, 48] |
| *Private data from various health care centers* | | | | | |

(continued)

**Table 1** (continued)

| Dataset | | | Techniques | Tools applied | Count | Publication |
|---|---|---|---|---|---|---|
| EMU (Epilepsy Monitoring Unit) | Named Entity Recognition | | Regular expression-based named entity recognition, rule-based method | Metamap | 662 | [19] |
| Discharge summary | Information and Relation extraction | | CRF, SVM, and pattern-based method | | 3012 | [17] |
| The South London and Maudsley NHS Foundation Trust (SLaM) | Information Extraction, sentence classification | | Regular expression, SVM | TextHunter, Con Text | 23,128 | [31] |
| Discharge summary of asthma and COPD patients | Concept extraction Sentence classification | | Regular expression-based Bayes, SVM, and decision trees | HITEx | 150 | [28] |
| Patients with history of ischemic heart disease | Text classification | | Rule-based classifier, SVM, nearest neighbor classifier, repeated incremental pruning to produce error reduction (RIPPER), decision tree | MTERMS | 1200 | [29] |

## 3.1   Discharge Summary

Discharge summary is made publicly available to carry out different research by i2b2 challenge. Different challenge focuses on different task as discussed in Sect. 2. 2006 Deidentification and Smoking Challenge discharge summary consists of admission and discharge date and sections like admission diagnosis, history of present illness, previous medical history, medications on admission, physical examination, laboratory data, hospital course and treatment. Few discharge summaries included additional sections like condition at discharge. 2008 i2b2 challenge provides discharge summary whose sections are similar to smoking challenge discharge summary. Additional sections include social history, allergies, discharge medications, and follow-up date. 2009 medications challenge also provide similar discharge summary, but the task mainly focuses on extracting medication, dosages, frequencies, and their effects. The discharge summary provided in 2014 i2b2 challenge is to identify the presence of heart disease risk factor in diabetic patients. This document includes date of visit and sections like current condition of the patient, problems of the patients, medications advised, flow sheets that discuss the measurements (like BP, height, and weight) and plan that indicates the treatment or the change in treatment. Though there are minor variations in the structure of different discharge summaries provided, the general structure is common. All sections are not present in all documents.

One sample discharge summary is Fig. 1. Discharge summary of a patient has a semi-structured form with sections including History, On-Examination, Investigation, Medical Reports, Treatment, and Condition at Discharge as highlighted in sample discharge summary (Fig. 1). While not a universal format, this follows a similar general structure. Record of past events that might be relevant to a patient's current health condition, which could be an account of past diseases, injuries, treatments and other medical facts, is noted down in the History section. Post-collection of history, the On-Examination section that describes the patient condition at the time of admission is noted down. Various investigations are conducted on the patient, and

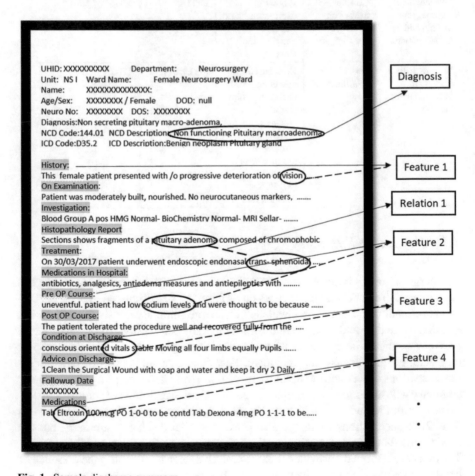

**Fig. 1** Sample discharge summary

the results go in the Investigation section. Treatment and medications provided in the hospital for the patient are recorded. These are the major commonly occurring sections in all kinds of discharge summary. It thus contains diverse set of information from history to advise on discharge, thus making it important to extract relevant information. The features that could be extracted from the discharge summary are shown in Fig. 1 and explained in Sect. 3.2.

## 3.2 Generic System Architecture

Though a structure is followed, the discharge summary is not structured enough for clinical decision support system to make predictions using the data in it. Hence, it needs to be processed to obtain information from it. For information extraction from these discharge summaries, it has to go through various stages that are depicted in the general system architecture as shown in Fig. 2. The discharge summaries are semi-structured and hence require a lot of preprocessing. To split the sections in discharge summary, regular expressions and rule-based approach have been used after which the sentences under each section need to be separated. Text normalization plays a major role in text mining specially in clinical text. Similar abbreviation may be used in different ways like short form. For instance, the term "lle" can be used to represent *left lower extremity* as well as *Edema of lower limbs*. Such cases need to be handled using text normalization techniques. Term matching is the step where

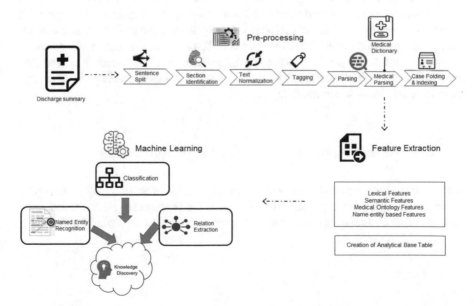

**Fig. 2** Generic System Architecture for works related to discharge summary

morphologically similar words need to be identified as same. Irrelevant sentences if any need to be filtered out. Parsing is performed, and relevant medical terms are extracted with the help of medical dictionaries like Systematized Nomenclature of Medicine (SNOMED). Case folding needs to be employed to manage the usage of words in different cases (upper/lower case) in the discharge summary. For instance, headache can be written as Headache, HeadAche, headache, or HEADACHE in a discharge summary. Hence, such cases need to be handled using case folding where all the characters are converted to same case.

Features are extracted from each sections of the discharge summary separately to form a feature after various preprocessing steps. Figure 1 shows the sample features that could be extracted from the discharge summary. Feature 1 corresponds to a clinical entity *vision* which is present under history section. Similarly, *sodium levels* under PreOP course are considered to be another feature. Thus, different medical entities under each section of the discharge summary are considered to be the different features obtained. Various kinds of features include Lexical Features, Semantic features, Pragmatic features, and name entity-based features using various methods as mentioned in following sections. Lexical features (at the level of word) include n-gram, lemma, prefixes, and suffixes. Section to which the sentence belongs to is considered to be pragmatic (at the level of context) feature. Semantic feature (at the level of meaning) could be the semantic classes in UMLS, DrugBank, and MedDRA. A sign or symptom is a kind of semantic type, and hence, the clinical entities recognized as sign or symptom can be considered to be semantic feature. For instance, feature 1 in Fig. 1 is vision which is identified as sign or symptom semantic type can be categorized as a semantic feature. The kind of features to be extracted is highly dependent on the kind of task the work focuses on. The analytical base table which represents each patient's value for each of the feature can be created. Performance is evaluated generally using Precision, Recall, and F-measure. Some works attempt to assess the performance using micro-averaged and macro-averaged Precision, Recall, and F-measure. The different text mining approaches in discharge summary are discussed in Sect. 4.

## 4   Text Mining Approaches in Discharge Summary

Different information can be extracted from discharge summaries using various NLP approaches. The main approaches in NLP are named entity recognition (NER), text classification and relation extraction which are discussed in following sections.

## 4.1 Information Extraction

One critical component in clinical text document processing is information extraction (IE) task that extracts medical information from clinical documents automatically. Named entity recognition is one of the main techniques used in IE [15]. The approaches generally fall into four categories: lexicon-based, rule-based, machine learning-based, and hybrid [16]. The lexicon-based NER techniques work with creating a dictionary of all terms which match a specific named entity. Unstructured text is mapped with phrases and words in the dictionary, and matched terms are extracted using a string-matching algorithm. Rule-base techniques consist of a combination of pattern matching methods, linguistic features, and heuristics for extraction of named entity from the text. Machine learning-based NER uses mainly supervised techniques especially Conditional Random Field (CRF) and Support Vector Machine (SVM). Different techniques have been used in various works on discharge summary.

Drug and symptom expressions are extracted from 3012 discharge summaries collected at the University of Tokyo Hospital [17, 18]. CRF-based NER method is used for the extraction of terms that include Remedy, Medical Test, Disease, Medication and Patient action and then converted to a table structure called TEXT2TABLE. Complex epilepsy phenotype has been extracted from discharge summaries using rule-based NER [19]. The system known as Phenotype Extraction in Epilepsy (PEEP) made use of a corpus of 662 discharge summaries from Epilepsy Monitoring Unit (EMU) at UH CMC in which 400 were selected randomly as training set and 262 were selected as test set with Epilepsy and Seizure Ontology (EpSO) as the knowledge base. The phenotype and anatomical location correlation candidate generation is performed using MetaMapRENER. This is formed by embedding RENER (rule-based epilepsy named entity recognition) into MetaMap. Regular expression-based epilepsy named entity recognition has been performed in this work. Medication information is extracted from discharge summaries provided by i2b2 2009 challenge using lexicons and rules, where drug names are identified and then drug-related information (mode, dosage, duration) is mined by exploring the context of these drug names [20]. A drug lexicon with 183,941 drug names has been constructed from Food and Drug Administration and RxList websites and Unified Medical Language System (UMLS) Metathesaurus. As it is understood from various studies that medication-related information is found mostly after the drug name in the text, rules are generated to extract this information from those subparts following the drug name. Extraction of medication information from discharge summaries is carried out using ensemble classifiers that combine outputs from different classifiers which are rule-based, Support Vector Machine (SVM) and Conditional Random Field (CRF)-based system [10]. The data used are the datasets available from the i2b2 challenge 2009 in which 268 discharge summaries out which 17 forms the train set and 251 forms the test set. Medication, dosage, mode, frequency, duration, and reason are the different fields that have been identified from the discharge summary using the classifiers. The work applies TinySVM for extracting the medication information and uses tenfold cross-validation on all 268 datasets to evaluate the performance of the model. CRF which is a probabilistic graphical model is used for extracting the information. Simple majority voting, local CRF-based voting, and local SVM-based voting are the strategies used in combining outputs of different classifiers. Results show the ensemble classifier has better performance compared to individual classifiers. The best system for tenfold cross-validation is found to be CRF-based voting system in this work. Deidentification of discharge summaries (provided by 2006 i2b2 challenge), where private health information (PHI) is automatically removed, is considered to be similar to a NER problem [21]. CRF is employed to learn the features that are important to identify PHI and marks all tokens using BIO tagging where B shows that the word is beginning of the phrase, I represents that the word is inside or it is at the end of the phrase, and O represents a word is outside a phrase [22]. A rule-based system has been used to identify PHI from the discharge summary of i2b2 2006 challenge [23]. Medication information has been extracted from discharge summary using a hybrid NER [24] which combines machine learning and rule-based techniques. The discharge summaries available from i2b2 challenge in 2009 have been used, out of which 110 annotated discharge summaries have been used as training set and 35 summaries have been used as the development set. A total of 251 summaries were provided as test set. In the data, the medication field is supposed to represent the name field, and all other fields are treated as non-name fields. NER is used as a sequence labeling task with BIO tagging scheme to identify the name and non-name fields. An ontology-based approach has been implemented to extract medical terms of 125 breast cancer patients as an alternative to the typical NER method used for term extraction in MedIE [25]. Part-of-speech patterns have been used to generate term candidates and then checked for existence of those terms in ontology. Grouping of medical terms has been performed using lookup of synonyms in ontology as doctors are interested in predefined diseases like hypertension.

The rule-based approach is usually domain specific and requires domain experts to generate most relevant rules for NE extraction. The main issue while using supervised techniques in machine learning-based NER is that training requires annotated data which is not available easily. Annotating the data is a time-consuming task in addition to the need of domain experts. Hence, hybrid model, which is a combination of the other models, is more suitable technique in NER as it utilizes advantage and avoids the constraints of various models. Although ontology-based information extraction has higher performance than general NER approaches, it requires intensive searching. Also, it fails sometimes to retrieve all terms as the ontology is incomplete.

## 4.2 Text Classification

Classification algorithms are used in information extraction of clinical entities of discharge summaries which are already discussed in the subsection—information extraction. Sentence-level classification is another major task carried out on discharge summaries using various classifiers. Classification algorithms are also used to classify the discharge summaries to different classes after the feature extraction. Assertions are classified from discharge summaries provided by 2010 i2b2 challenge on concepts, assertions, and relations [26]. Assertion classification is the classification of medical problem into one of the specified categories (not_associated_with_the_patient, conditional, hypothetical, possible, absent, and present in this work). A hybrid approach where one rule-based classifier has been combined with four statistical classifiers by voting has been implemented for assertion classification. Dictionary has been created manually for the rule-based classifier. 3-g words and 2-g LGP syntax, verb and section name are also used, in addition to the features of rule-based classifier, for the statistical classifiers. The classifiers used are SVM-binary, boosting, logistic-regression-binary, and SVM-multiclass, and the final class is decided by voting of all these classifiers. A different kind of assertions classifications is assigning a specific assertion category (present, possible, and absent) to each medical concept (pneumonia and chest pain) in discharge summaries of 2010 i2b2 challenge using SVM classifier [27]. The smoking status, from discharge summaries of around 97,000 asthma and COPD patients from Partners' Health care Research Patient Data Repository, has been classified into past smoker, current smoker, non-smoker, patient denies smoking, and insufficient data [28]. The classification of smoking status is performed using SVM classifier using single words as features. Patients with depression are identified using classification on the features extracted from the discharge summary of patients from Partners Health care, with a history of ischemic heart disease [29]. Relevant sections of discharge summary have been identified using MTERMS NLP system [30], and then, specialized lexicon consisting of depression-related terms has been created. The classification algorithms like SVM, nearest neighbor classifier, RIPPER, and decision tree were applied on the features created and classified the cases into three classes (depression with high, intermediate, and low confidence). User-defined decision tree which had been created in the work had outperformed the other classifiers. The kinds of symptoms have been classified from the discharge summaries of patients who had received a severe mental illness (SMI) diagnosis using SVM classifier [31]. The symptoms have been classified into one of the following five categories—positive symptoms, negative symptoms, disorganization symptoms, manic symptoms, and catatonic symptoms by the classifier using the features provided by TextHunter. The breast cancer patients are classified into non-smoker, former smoker, and current smoker using ID3-based decision tree with fivefold cross-validation in MedIE system [25]. Link-grammar parser has been used to extract features from discharge summary of 125 breast cancer patients that have been provided to the decision tree for classification.

## 4.3 Other Approaches

This section discusses other different works carried out on discharge summary. One of the main works is extraction of relation between various clinical concepts. Relation extraction is the task of finding pairs of two terms in a sentence or a set of multiple sentences which have either syntactic or semantic relationship between them. A system has been developed that translates discharge summary which is the form of free text into structured representation [26]. Relations that have been identified in this work are problem and a concept. The relations to be identified are problem-treatment class (TrIP, TrWP, TrCP, TrAP, TrNAP), problem-test class (TeRP, TeCP, Non-Tep), and problem-problem (PIP and Non-PIP). Features used for this work have been generated using concept co-occurrence, verb-based, $N$-gram sequential patterns. The classification of relation into the three types has been performed using voting of the different classifiers which are SVM-binary, boosting, logistic-regression-binary, and SVM-multiclass. MedIE extracts patient information from free-text medical records of patients with breast complaints [25]. Relations between symptoms or disease with persons, or parts of the body are kinds of relations that are extracted using a graph-based approach which make use of parsing result of link-grammar parser. Initially parsed sentence is transformed to form a graph representation, and then, concept association is done based on the graph generated. In the graph, a word is represented as the node and a link would be represented as an edge. Edge could be given weights depending on the type of link. In MedLEE (Medical Language Extraction and Encoding), the system has a parser that is driven by a semantic grammar [12]. The grammar contains rules that specify semantic patterns and the underlying target structures to which it should be mapped to. When a well-formed semantic pattern is identified, it shows the underlying semantic relations between the concepts. The adverse effect relation between a drug and its effect is extracted from 3012 discharge summaries gathered at University of Tokyo Hospital using both pattern-based method and SVM, which is a machine learning method [17]. The pattern-based method judges whether each drug symptom pair has an adverse effect relation or not using heuristic rule-based patterns. SVM-based relation identifier utilizes symptom lexicon, drug lexicon, word chain, and distance as features. Relation between the Problem-Action (medical problem of the patient and the doctor's action for that problem) has been extracted from 150 discharge summaries from patients with systemic lupus erythematosus in a Korean Hospital [32]. A segmentation unit and a clinical semantical unit are defined that defines relevant events in a causal relationship and within a relevant time. SVM has been used for classification of a clinical sematic unit. Problem-Action relations within the unit have been then identified by event causality pattern. An end-to-end system has been developed to identify the temporal relations in discharge summaries provided by 2012 i2b2 challenge [33]. Temporal relation is the relation between an event and temporal expression (timex) or two timexes. For relation identification, ten multi-SVM classifiers are used.

The existing tools which have been applied on the discharge summaries to achieve particular tasks are summarized in Sect. 5.

## 5  Existing Tools

Numerous NLP tools are available in the open-source domain as listed in Table 2 particularly since a decade. Most of these tools are initially developed for a specific kind of clinical text and later adapted for various other types of clinical texts like discharge summary. LifeCode [11] is an NLP system which extracts clinical and demographic information from the medical records that are in the form of free text. Though initially it had been developed for the emergency medicine clinical specialty, later it was updated to work with diagnostic radiology and other clinical texts. It seeks assistance from a human expert when it encounters information which is beyond its competence. MedLEE [12], an NLP system, has been originally developed for processing radiological reports of the chest to identify patients who are suspicious for tuberculosis. It has been later applied to discharge summary too. It has preprocessor, parser, phrase regularization, and encoding modules. The MetaMap System [14], provided by the National Library of Medicine (NLM), performs the mapping of natural language text to concepts of the UMLS Metathesaurus. It has been developed for extracting information from MEDLINE abstracts, but later applied to clinical documents like pathology reports and discharge summaries as well [34]. cTAKES [13], which is a generic open-source NLP system developed at the Mayo Clinic, has been developed to extract information from the narrative found in EMRs with the help of NLP modules like sentence boundary detector, normalizer, tokenizer, POS tagger, NER annotator, and shallow parser. Other tools that have been applied on various discharge summaries are also listed in Table 2. Named entity recognition and parsing are the main NLP tasks performed by all kinds of tools discussed. Few have certain characteristics specific to the type of medical text for which it was developed for. The tools discussed in Table 2 are majorly publicly available making it easy for applying on various kinds of discharge summaries, yet do not extract all relevant clinical entities from the discharge summaries, thus leading to scope of developing new tools for ontology-based information extraction and a decision support system. The challenges faced while applying text mining techniques on the different categories of discharge summaries are summarized in Sect. 6.

## 6  Summary

This paper focuses on dataset available, tools, and techniques applied on discharge summary which is of the semi-structured form. Discharge summary consists of large variety of information under each section and hence provides a great scope for applying text mining. Various kinds of data source can be publicly available or private. These are discussed with the works done on each kind of data. A generic design model for any system that tries to extract information from discharge summary and use it for prediction is discussed. Different works may follow variations of this model according to the task, but the main work flow is common. Information extraction,

**Table 2** Existing NLP tools for clinical texts in the form of discharge summary

| Tool | Description | Task | Availability | Year |
|---|---|---|---|---|
| Lifecode [11] | Clinical and demographic information | Document segmenter, lexical analyzer, phrase parser, and concept matcher. | Commercial | 1996 |
| MEDLEE [12, 49] | Initially developed for radiology reports, then applied to mammography report and later to discharge summary too. It captures most frequent clinical information in discharge summaries. | Preprocessor, parser, phrase regularization, encoding to controlled vocabulary | Commercial | 1994 |
| MetaMap [14] | to map biomedical text to the Metathesaurus | Parsing, variant generation, candidate retrieval and evaluation, mapping constructions. | Open source | 2000 |
| MedIE [25] | Mines information from clinical text records | Ontology-based term extraction, graph-based relation extraction, decision tree-based text classification | Commercial | 2006 |
| cTAKES [13] | Extract information from clinical text | Sentence boundary detector, normalizer, tokenizer part-of-speech (POS) tagger, named entity recognition (NER) annotator and shallow parser | Open source | 2006 |
| HITEx [28] | Use GATE (General Architecture for Text Engineering) modules—extracts smoking status, principal diagnoses, comorbidities from discharge summaries of asthma patients | Parser, POS tagger, UMLS concept mapper, n-gram tool, noun phrase finder, classifier, regular expression-based concept finder | Open source | 2006 |
| MERKI parser [50] | Extracts patient medication information | Concept definition, parsing | Open source | 2008 |

relation extraction, and text classification are the major tasks using techniques like rule-based, ML, or hybrid techniques applied on discharge summary. Various kinds of NLP tools are available for extracting information from clinical texts. The main systems which have been developed for discharge summary or later adapted to work for discharge summary are discussed in this paper. There are lots of challenges while applying NLP techniques on discharge summary which are discussed in the Sect. 7.

# 7 Challenges

Preprocessing is one of the biggest challenges in the discharge summary as the style followed by different doctors may vary. Some clinical texts that do not follow the rules of grammar are short and consist of telegraphic notes. Discharge summary is full of abbreviations, acronyms, and other short-hand phrases. These short-hand phrases may have different interpretations in different context, making it ambiguous for NLP technique. As discharge summary is often created without any spelling support, mis-spellings are frequent. It usually consists of pasted sets of vital signs or laboratory results with embedded non-text strings complicating otherwise straightforward tasks like sentence splitting. All these would make preprocessing of discharge summary a challenge. Recognizing medical terms in the text helps for further extraction of the relationship and patterns from them. This is very challenging in healthcare domain. One of the several reasons is that there is no complete dictionary for the clinical named entities, thus making the simple text matching algorithms insufficient for these tasks. Relation extraction has not been explored extensively in discharge summaries. If the systems developed that do not identify the syntactic or semantic relations between words, serious misinterpretations might occur if there is partial matching with the patterns. Text mining is not explored on all kinds of discharge summary yet. Employing any supervised machine learning algorithm in classifying the documents or in NER requires annotated discharge summaries. This would require manual annotation of large number of discharge summary which is a tedious task. Few new techniques like Active learning [35] need to be explored to resolve this. Although there are clinical NLP tools which are open source, a generic tool which works for any kind of discharge summary is not available. One of the widely used tools is MetaMap, the output of which is multiple mapping. Appropriate mapping needs to be selected for final decision. It also does not extract all concepts owing to restriction to some selected semantic type [36]. Other tools available have been developed for a specific kind of clinical texts, thereby making it insufficient to work with all kinds of data. There exists no decision support system that is capable of making predictions and deciding based on the discharge summary given.

# References

1. S.S. Shastri, P.C. Nair, D. Gupta, R.C. Nayar, R. Rao, A. Ram, Breast cancer diagnosis and prognosis using machine learning techniques, in *The International Symposium on Intelligent Systems Technologies and Applications* (Springer, Cham, 2017)
2. S. Khare, D. Gupta, K. Prabhavathi, M.G. Deepika, A. Jyotishi, Health and nutritional status of children: survey, challenges and directions, in *International Conference on Cognitive Computing and Information Processing* (Springer, Singapore, 2017)
3. D.P. Pragna, S. Dandu, M. Meenakzshi, C. Jyotsna, J. Amudha, Health alert system to detect oral cancer, in *Inventive Communication and Computational Technologies (ICICCT)* (2017)
4. T. Babu, T. Singh, D. Gupta, S. Hameed, Colon cancer detection in biopsy images for Indian population at different magnification factors using texture features, in *2017 Ninth International Conference on Advanced Computing (ICoAC)* (IEEE, 2017)
5. A. Madabhushi, G. Lee, Image analysis and machine learning in digital pathology: challenges and opportunities 170–175 (2016)
6. S.V. Iyer, R. Harpaz, P. LePendu, A. Bauer-Mehren, N.H. Shah, Mining clinical text for signals of adverse drug-drug interactions. J. Am. Med. Inform. Assoc. **21**(2), 353–362 (2014)
7. K.B. Wagholikar, K.L. MacLaughlin, M.R. Henry, R.A. Greenes, R.A. Hankey, H. Liu, R. Chaudhry, Clinical decision support with automated text processing for cervical cancer screening. J. Am. Med. Inform. Assoc. **19**(5), 833–839 (2012)
8. R. Angus, R. Gaizauska, M. Hepple, Extracting clinical relationships from patient narratives, in *Proceedings of the Workshop on Current Trends in Biomedical Natural Language Processing* (2008)
9. W. Long, Extracting diagnoses from discharge summaries, in *AMIA Annual Symposium Proceedings* (2005)
10. S. Doan, N. Collier, H. Xu, P.H. Duy, T.M. Phuong, Recognition of medication information from discharge summaries using ensembles of classifiers. BMC Med. Inform. Dec. Mak. **12**(1), 36 (2012)
11. D.T. Heinze, M.L. Morsch, R.E. Sheffer Jr, M.A. Jimmink, M.A. Jennings, W.C. Morris, A.E. Morsch, LifeCode™—a natural language processing system for medical coding and data mining, in *AAAI/IAAI* (2000)
12. C. Friedman, P.O. Alderson, J. Austin, J. Cimino, S. Johnson, A general natural-language text processor for clinical radiology. J. Am. Med. Inform. Assoc. **1**(2), 161–174 (1994)
13. G.K. Savova, J.J. Masanz, P.V. Ogren J. Zheng, Mayo clinical Text Analysis and Knowledge Extraction System (cTAKES): architecture, component evaluation and applications. J. Am. Med. Inform. Assoc. **17**(5), 507–513 (2010)
14. A.R. Aronson, Effective mapping of biomedical text to the UMLS Metathesaurus: the MetaMap program, in *Proceedings of the AMIA Symposium* (2001)
15. X. Zhou, H. Han, I. Chankai, A.A. Prestrud, A.D. Brooks, Converting semi-structured clinical medical records into information and knowledge, in *21st International Conference on Data Engineering Workshops* (2005)
16. S. Keretna, C.P. Lim, D. Creighton, A hybrid model for named entity recognition using unstructured medical text. in *2014 9th International Conference on System of Systems Engineering (SOSE)* (IEEE, 2014)
17. E. Aramaki, Y. Miura, M. Tonoike, T. Ohkuma, H. Masuichi, K. Waki, K. Ohe, Extraction of adverse drug effects from clinical records, in *MedInfo* (2010)
18. E. Aramaki et al., Text2table: medical text summarization system based on named entity recognition and modality identification, in *Proceedings of the Workshop on Current Trends in Biomedical Natural Language Processing. Association for Computational Linguistics* (2009)
19. L. Cui, S.S. Sahoo, S.D. Lhatoo, G. Garg, P. Rai, A. Bozorgi, G.-Q. Zhang, Complex epilepsy phenotype extraction from narrative clinical discharge summaries. J. Biomed. Inform. **51**, 272–279 (2014)

20. L. Deléger, C. Grouin, P. Zweigenbaum, Extracting medical information from narrative patient records: the case of medication-related information. J. Am. Med. Inform. Assoc. **17**(5), 555–558 (2010)
21. Ö. Uzuner, Y. Luo, P. Szolovits, Evaluating the state-of-the-art in automatic de-identification. J. Am. Med. Inform. Assoc. **14**(5), 550–563 (2007)
22. E. Aramaki et al., Automatic deidentification by using sentence features and label consistency, in *i2b2 Workshop on Challenges in Natural Language Processing for Clinical Data*, vol. 2006 (2006)
23. R. Guillen, Automated de-identification and categorization of medical records, in *i2b2 Workshop on Challenges in Natural Language Processing for Clinical Data* (2006)
24. H. Scott, F. Xia, I. Solti, E. Cadag, Ö. Uzuner, Extracting medication information from discharge summaries, in *Proceedings of the NAACL HLT Second Louhi Workshop on Text and Data Mining of Health Documents. Association for Computational Linguistics* (2010)
25. X. Zhou, H. Han, I. Chankai, A. Prestrud, A. Brooks, Approaches to text mining for clinical medical records, in *Proceedings of the 2006 ACM Symposium on Applied Computing* (2006)
26. Y. Xu, K. Hong, J. Tsujii, E.I.-C. Chang, Feature engineering combined with machine learning and rule-based methods for structured information extraction from narrative clinical discharge summaries. J. Am. Med. Inform. Assoc. **19**(5), 824–832 (2012)
27. C.A. Bejan, L. Vanderwende, F. Xia, M. Yetisgen-Yildiz, Assertion modeling and its role in clinical phenotype identification. J. Biomed. Inform. **46**(1), 68–74 (2013)
28. Q.T. Zeng, S. Goryachev, S. Weiss, M. Sordo, S.N. Murphy, R. Lazarus, Extracting principal diagnosis, co-morbidity and smoking status for asthma research. BMC Med. Inform. Decis. Mak. **6**(1), 30 (2006)
29. M. Sordoa, M. Topazb, F. Zhongb, M. Murralid, S., Navathed, R.A. Rochaa, Identifying patients with depression using free-text clinical documents, in *MEDINFO* (2015)
30. L. Zhou, J.M. Plasek, L.M. Mahoney, N. Karipineni, F. Chang, X. Yan, F. Chang, D. Dimaggio, D.S. Goldman, R.A. Rocha, Using Medical Text Extraction, Reasoning and Mapping System (MTERMS) to process medication information in outpatient clinical notes, in *AMIA Annual Symposium Proceedings*, vol. 2011
31. R.G. Jackson, R. Patel, N. Jayatilleke, A. Kolliakou, M. Ball, G. Gorrell, A. Roberts, R.J. Dobson, R. Stewart, Symptoms of severe mental illness from clinical text: the Clinical Record Interactive Search Comprehensive Data Extraction (CRIS-CODE) project. BMJ Open **7**(1), e012012 (2017)
32. J.-W. Seol, W. Yi, J. Choi, K.S. Lee, Causality patterns and machine learning for the extraction of problem-action relations in discharge summaries. Int. J. Med. Inform. **98**, 1–12 (2017)
33. Y. Xu, Y. Wang, L. Tianren, J. Tsujii, E.I.-C. Chang, An end-to-end system to identify temporal relation in discharge summaries: 2012 i2b2 challenge. J. Am. Med. Inform. Assoc. **20**(5), 849–858 (2013)
34. A.R. Aronson, F.-M. Lang, An overview of MetaMap: historical perspective and recent advances. J. Am. Med. Inform. Assoc. **17**(3), 229–236 (2010)
35. M. Kholghi, L. Sitbon, G. Zuccon, A. Nguyen, Active learning: a step towards automating medical concept extraction. J. Am. Med. Inform. Assoc. **23**(2), 289–296 (2015)
36. K. Denecke, Extracting medical concepts from medical social media with clinical NLP tools: a qualitative study, in *Proceedings of the Fourth Workshop on Building and Evaluation Resources for Health and Biomedical Text Processing* (2014)
37. B. Wellner, M. Huyck, S. Mardis, J. Aberdeen, A. Morgan, L. Peshkin, A. Yeh, J. Hitzeman, L. Hirschman, Rapidly retargetable approaches to de-identification in medical records. J. Am. Med. Inform. Assoc. **14**(5), 564–573 (2007)
38. A.M. Cohen, Five-way smoking status classification using text hot-spot identification and error-correcting output codes. J. Am. Med. Inform. Assoc. **15**(1), 32–35 (2008)
39. Ö. Uzuner, I. Goldstein, Y. Luo, I. Kohane, Identifying patient smoking status from medical discharge records. J. Am. Med. Inform. Assoc. **15**(1), 14–24 (2008)
40. H. Yang, I. Spasic, J.A. Keane, G. Nenadic, A text mining approach to the prediction of disease status from clinical discharge summaries. J. Am. Med. Inform. Assoc. **16**(4), 596–600 (2009)

41. Ö. Uzuner, Recognizing obesity and co-morbidities in sparse data. J. Am. Med. Inform. Assoc. **16**(4), 561–570 (2009)
42. I. Solt, D. Tikk, V. Gál, Z.T. Kardkovács, Semantic classification of diseases in discharge summaries using a context-aware rule-based classifier. J. Am. Med. Inform. Assoc. **16**(4), 580–584 (2009)
43. V.N. Garla, C. Brandt, Ontology-guided feature engineering for clinical text classification. J. Biomed. Inform. **45**(5), 992–998 (2012)
44. K.H. Ambert, A.M. Cohen, A system for classifying disease comorbidity status from medical discharge summaries using automated hotspot and negated concept detection. J. Am. Med. Inform. Assoc. **16**(4), 590–595 (2009)
45. Ö. Uzuner, I. Solti, E. Cadag, Extracting medication information from clinical text. J. Am. Med. Inform. Assoc. **17**(5), 514–518 (2010)
46. Ö. Uzuner, B.R. South, S. Shen, S.L. DuVall, 2010 i2b2/VA challenge on concepts, assertions, and relations in clinical text. J. Am. Med. Inform. Assoc. **18**(5), 552–556 (2011)
47. K. Roberts, B. Rink, S.M. Harabagiu, A flexible framework for recognizing events, temporal expressions, and temporal relations in clinical text. J. Am. Med. Inform. Assoc. **20**(5), 867–875 (2013)
48. W. Sun, A. Rumshisky, O. Uzuner, Evaluating temporal relations in clinical text: 2012 i2b2 challenge. J. Am. Med. Inform. Assoc. **20**(5), 806–813 (2013)
49. C. Friedman, Towards a comprehensive medical language processing system: methods and issues, in *Proceedings of the AMIA Annual Fall Symposium* (American Medical Informatics Association, 1997)
50. S. Gold, N. Elhadad, X. Zhu, J.J. Cimino, G. Hripcsak, Extracting structured medication event information from discharge summaries, in *AMIA Annual Symposium Proceedings* (2008)

# Smart Environmental Monitoring Based on IoT: Architecture, Issues, and Challenges

**Meenakshi Srivastava and Rakesh Kumar**

**Abstract** With the advent of time, the fast progress in the communication technologies, and development of Internet of Things (IoT), the modern world is able to invisibly intertwine with sensors and different computational devices while sustaining constant network connectivity. The constant connected modern world with computational devices and actuators develops a smart environment. Smart environmental monitoring is a systematic sampling that helps to understand the natural environment such as air, water, biota, and land. In general form, environmental monitoring is the application of IoT. It normally uses the sensor to help in environmental protection by monitoring parameters such as air quality, water quality, etc., and it can include areas like monitoring the condition of wildlife. This paper explores IoT architecture in the smart environment as well as a comparison between technologies used in environmental monitoring. It basically emphasizes on various issues and challenges of smart environmental monitoring. A layered framework of the smart environment is also presented. It is an informative paper for the researchers that suggest effective research in this domain.

**Keywords** Internet of Things (IoT) · Smart environment · Network connectivity

## 1 Introduction

Environmental monitoring plays an important specific role in eco-systematic health and the parameter such as air quality, water quality, etc. verifies our surrounding environment [1]. The tracing of environment parameters is necessary to find out the quality of the environment. The collected data include the mandatory detail for various organizations and agencies. With this result, the government can form knowledgeable decision regarding the environment, how society will be affected by it and

M. Srivastava (✉) · R. Kumar
Department of CSE, M.M.M. University of Technology, Gorakhpur, UP 273010, India
e-mail: srivastavameenakshi05@gmail.com

R. Kumar
e-mail: rkiitr@gmail.com

© Springer Nature Singapore Pte Ltd. 2021
X.-Z. Gao et al. (eds.), *Advances in Computational Intelligence and Communication Technology*, Advances in Intelligent Systems and Computing 1086,
https://doi.org/10.1007/978-981-15-1275-9_28

how it will affect society. Outside the organization and other agencies, the knowledgeable information taken by multiple people, since of the climate effect on a huge range of human condition. The knowledgeable information is taken by public health specialist to create successful policies. The main aim of environmental monitoring is to detect human risk and wildlife risk. Environment change and climate condition have an effect on human health condition, for example, high air pollution damages the human health condition such as asthma. Standard environmental monitoring technique has developed significantly in the past few eras, the challenges imposing by site-related problems such as rough terrains and the wide number of the method need to deploy and gathered data from a sensor at a consistent intervening time. The word smart about the capacity to autonomously acquire and apply apprehension and the sentence environment refers that the circumstances object or condition by which one is surrounded. Consequently, a smart environment while as a very small world where the various type of smart object is constantly working to form resident lives extra pleasant.

IoT is the main technology for wide industries and smart environmental monitoring. IoT is an abstract idea and prototype that being examine present prevalent in the environment of the various object that connected through wired or wireless connection and individual addressing plan are interact with everyone and collaborate with another device to develop new service and approaches.

IoT connects various types of devices like smart internet TVs, smart ACs, sensors, etc. through the internet to cloud so that all devices and sensors are connected together and form a relationship between people and things. In this paper, some latest concepts related to IoT architecture in the smart environment and layered framework for IoT related smart environment and challenging security-related issues and also discuss in brief. Comparative studies of latest technology and analysis are also explained. It helps for a developer for the latest application development and future work.

Rest of the paper is organized as follows: Sect. 2 explains the related work and comparison between technologies used in environmental monitoring in detail. IoT architecture in smart environment is given in Sect. 3. Section 4 explains the sensor network architecture in detail. Environmental monitoring related challenges and issues are given in Sect. 5. Finally, Sect. 6 contains the conclusion of this paper.

## 2 Related Work

Ibrahim et al. [1] suggest an approach to create a cost effectual systematize environmental monitoring device using the Raspberry Pi (RPi) single-board computer. It gives detail about the environment through a sensor and sends it instantly to the internet. Gaur et al. [2] proposed smart city architecture and its application based on IoT. They aim to address few services in a smart city environment by using morphological modeling and then we aim to address few services for the smart city by using Dempster–Shafer theory. It is very hard to cover the various situation of the smart city through our planning. We direct focus on the very important region of the smart city. In this paper, a smart appliances control system based on IoT technology has been given to solve the problem that how to control these various devices efficiently so that to achieve extra pleasant security area home [3]. In this paper, the author describes adopting smart sensor alliances for industry WSN in IoT smart environment. The system can gather data knowledgeable. It was planning by IEEE1451 protocol by merging with CPLD and wireless communication application. Jaladi et al. [4] proposed short-range wireless technology in which she works on sensor logging application in a smart environment.

It is very acceptable for actual time [5]. Abraham and Beard [6] give an IoT solution for the remote environment. It suggests an approach to create and use smart attach sensor by using IoT technology so that it provides instant monitoring of air quality, soil quality and gathered data will be shown on GUI that provides actual time detail. It may be used to explain the actual condition of the region being monitored. In 2016, Wang et al. [3] work on smart control system in which he defines how to control smart home with different technologies. Mois et al. [7] proposed three IoT-based wireless sensors in which he suggested an approach of smart communication environmental monitoring. In 2017, Ramesh et al. [8] proposed remote temperature monitoring and thermal protection system. Sastra et al. [9] worked on smart environmental monitoring as IoT application and smart campus and smart city are the main application. Prathibha et al. [10] worked on agriculture monitoring system in which smart farming is the main application. In Table 1, there is a brief comparison among different IoT technologies on the basis of techniques and applications. In this table, different authors give various technology and application. Remote monitoring technology widely used in the industry of IoT. Most of the researchers mainly emphasize on how to enable the device to see, hear, touch, and smell the physical world from themselves and communicate to people [11]. IoT has also solved lots of security-related issues.

**Table 1** Comparison between technologies used in environmental monitoring

| Technology | Author | Year | Application |
|---|---|---|---|
| Short-range wireless technology | Jaladi et al. [4] | 2017 | Sensor logging application in a smart environment |
| Remote Monitoring | Abraham and Beard [6] | 2017 | Real-time monitoring |
| A Smart control system for smart home | Wang et al. [3] | 2013 | Appliance control system for smart home |
| Three IoT based wireless sensor | Mois et al. [7] | 2017 | Smart communication, environmental monitoring |
| Remote temperature monitoring | Ramesh et al. [8] | 2017 | Environmental control systems, Thermal protection |
| Environmental monitoring as IoT application | Sastra et al. [9] | 2016 | Smart campus, smart city |
| Agricultural monitoring | Prathibha et al. [10] | 2017 | Smart farming, smart agriculture |

## 3   IoT Architecture in Smart Environment

Smart environment plays an important role as an unhealthy environmental condition affects human, animal, birds, and plant health [12]. Efforts had been made by many analysts to solve the issues of environmental pollution. Due to the industries and transportation wastes, the development of good environment is not easy and the environment is damaged because of the daily factor of the thoughtless human task. For monitoring the environment, it requires a smart new technology and smart management.

Figure 1 shows that monitoring the environment condition is very significant in order to evaluate the real-time limitation of the monitoring environment to take right life conclusion according to gather data for monitoring environment system, and it is used in inclusion to reduce the factories and vehicles squander. The smart environment is main technology in our day-to-day life which gives me numerous facilities and gives me many solutions for smart environments such as smart healthcare, smart grid, air pollution, weather monitoring, etc. and they are connected to every people

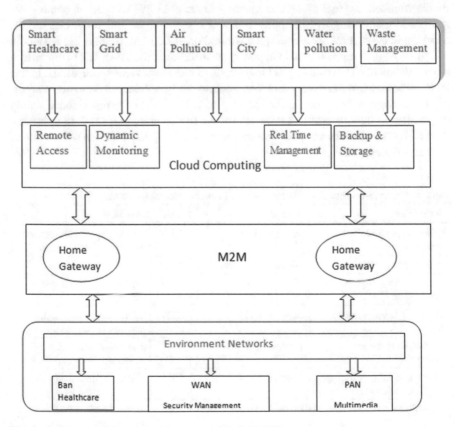

**Fig. 1** Architecture in a smart environment taken from [12]

between home area network and then smart environment can use various type of network such as BAN, WAN, PAN for different applications. Smart environment tool combines with IoT technology. It is developed for tracking the smart environment object which gives us the potential advantage to successfully achieve a renewable life. The Internet of Things (IoT) technology has the ability to monitor the quality of air by gathering the environment data from remote sensor beyond the area. IoT technology provides a geological area to achieve a process of best managing traffic in main cities. IoT is used to calculate the water pollution level in sequence to notify decision on water consumption and therapy. Another smart environment feature is to weather forecasting and weather monitoring. It can give a high decision and it can provide precision for weather forecasting by monitored data interchange and sharing environmental information. It also helps in gathering weather data which is stored in a cloud platform to examine.

## 4  Sensor Network Architecture

Figure 2 shows the sensor network architecture of environmental monitoring. Sensor nodes are the basic part of WSN and it produced the following basic process [13].

(i)   Data collected by different sensors.
(ii)  Collected data are stored as a temporary form.
(iii) Processing the sensor data using a microcontroller.
(iv)  Send the monitored data to the cloud using node mcu.
(v)   Study the processed data for alert generation.

**Arduino**: It is a microcontroller board based on ATMEGA 328p microcontroller. It consists of six analog pin (A0-A5) and 14 digital output pin (D0-D13) [14]. It has

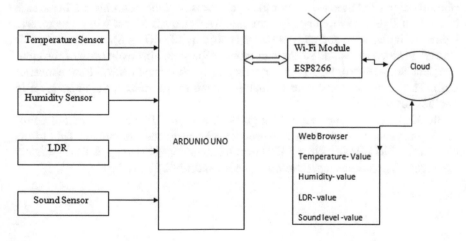

**Fig. 2** Sensor network architecture

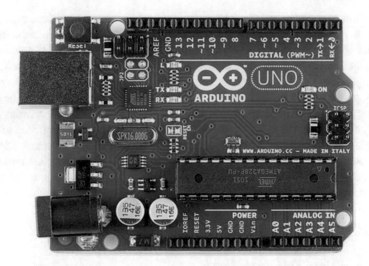

**Fig. 3** Arduino Board taken from [14]

input ADC. It simply powered it to a computer with USB cable or powered it with a 9-volt battery. It is programmable with Arduino IDE. For USB to serial converter, it takes the ATmega 16U2 programmed. Arduino IDE helps to write code in Arduino board. These codes are written in c or c++ language. It provides a software library (Fig. 3).

**Temperature Sensor**: Temperature sensors sense the temperature data of neighboring environment. It consists of four pins. Pin1 should be connected to 3.3 volts; P2 is digital output pin so that it provides input to node MCU. Pin3 is empty pin and Pin4 should be connected to ground (Fig. 4).

**Light Dependent Resistor (LDR)**: Light-dependent resistor (LDR) also known as a photoresistor. LDR used to monitor the light intensity of the environment. In the case of LDR, if light intensity level is come to lower than LED light will be increased. Cameras shutter open and close speed is control by LDR (Fig. 5).

**Sound Sensor**: The sound sensor can monitor the sound intensity level of the environment. Sound sensor board has a dynamic microphone and LM393 level converter chip. This sound sensor output is analog. Sound sensor detects the noise level in decibel (dB) (Fig. 6).

**Node MCU**: It is an Internet of Things (IoT) platform. It is a Wi-Fi module. Processing the sensors data using a microcontroller and send the monitor data to the cloud using Node MCU. Node MCU development board made up of Wi-Fi module ESP8266. Espressif system developed the low-cost ESP8266 (Fig. 7).

**Fig. 4** DHT11 Sensor taken
from [15]

**Fig. 5** LDR taken from [16]

**Fig. 6** Sound Sensor taken
from [17]

**Fig. 7** Node MCU taken
from [18]

## 5 Issues and Challenges

There are many issues and challenges. Security issues are interconnected to the
threats. In our environment, monitoring security-related issues can damage the monitoring system, violation of the transmission channel connecting many parts of the
system, unauthorized access.

**Damage The Monitoring System Architecture**: The aim of destroying the environmental monitoring system can be placed at threats to the confidentiality, probity, and
availability of both the environment-related data. Assume that the Mumbai municipality desires to construct a new playing field for children and it decides it is the
safest area, and it also examines the collected data to discard air polluted location of
the city. Alice damages the sensor node to A, analysis result shows that the Mumbai
municipality location where playfield construct is close to Alice. When every sensor
node working appropriately than the same risk occurs. These risks can effect some
of the three phases of environment lifecycle. Some problems occur when opponent
replaces in compromising the nodes gather data and the data where environment
gather data are a store.

**Violation of The Transmission Channels**: Every transmission is interconnected
to the different objects of a sensor network. It can represent a feasible object for
an opponent. In a certain, the opponent may be a passive opponent, i.e., she only
observed to monitor the transmission channel to observe knowledgeful information
that she is not able to access. Two situations configure two traditional security risks,
which can be instinctive violate the data of confidentiality and integrity of data.

**Unauthorized Access**: It can possibly require the database after the assembly and
detailed examination of environmental data, it is stored. Under the rule of external
owner, another party storage server, the storage server can be a local server. In the first
case, the server can be examined trusted and privacy control should only be imposed
against opposed to users requesting ingress to the stored data. In the second case, the
exterior storage server is not considered trusted and accordingly access restriction

should also take into vision the reality, after that the external server should not be allowed to access the external data.

**Precision**: In numerous Internet of Things (IoT) based smart environment, for example, transport, health, industries, and vehicular network, where appliance and organization are worlds widely connected. Precision is very important challenges that require to be labeled. The two important factors that effect the precision of IoT (Internet of Things (IoT)-based smart environment are network latency and bandwidth. Accordingly, these two factors require to be considered when utilizing distributed Internet of Things (IoT) in a smart environment.

**Compatibility**: It is the biggest challenge in Internet of Things (IoT)-based smart environment, where each and every device, product are connected with everyone. Due to the deficiency of general language, some of the devices and products are not able to connect with everyone.

# 6 Conclusions

In this paper, we mainly focused on the domain of smart environment by considering its architecture, related issues and challenges. The knowledge of smart environment architecture and security issues is essential for the developer. Comparison between technology and application of smart environment is also discussed. This paper is informative support for the researcher as it is a comparative study of smart monitoring using IoT that will help for their future work in the smart environment.

# References

1. M. Ibrahim, A. Elgamri, S. Babiker, A. Mohamed, *Internet of things based smart environmental monitoring using the Raspberry-Pi computer* (IEEE, 2015)
2. A. Gaura, B. Scotneya, G. Parra, S. McClean, Smart city architecture and its applications based on IoT, in *The 5th International Symposium on Internet of Ubiquitous and Pervasive Things*, vol. 52 (Elsevier, 2015), pp. 1089–1094
3. M. Wang, G. Zhang, C. Zhang, J. Zhang, C. Li, An IoT-based appliance control system for smart homes, in *Fourth International Conference on Intelligent Control and Information Processing (ICICIP)*, June 2013
4. A.R. Jaladi, K. Khithani, P. Pawar, K. Malvi, G. Sahoo, Environmental monitoring using wireless sensor networks (WSN) based on IOT. Int. Res. J. Eng. Technol. **4**(1) (2017)
5. Q. Chi, H. Yan, C. Zhang, Z. Pang, L. Da Xu, A reconfigurable smart sensor interface for industrial WSN in IoT environment. IEEE Trans. Ind. Inform. **10**(2), 1417–1425 (2014)
6. S. Abraham, J. Beard, Remote environmental monitoring using internet of things (IoT) (IEEE, 2017)
7. G. Mois, Analysis of three IoT-based wireless Sensors for environmental monitoring. IEEE Trans. Instrum. Measur. **66**(8), 2056–2064 (2017)
8. V. Ramesh, M. Sankaramahalingam, M.S. Divya Bharathy, R. Aksha, Remote temperature monitoring and control using IoT, in *IEEE International Conference on Computing Methodologies and Communication* (2017)

9. N.P. Sastra, D.M. Wiharta, Environment monitoring as on IoT application in building smart campus of University Udyana, in *ICSGTEIS*, October 2016
10. S.R. Prathibha et al., IoT based monitoring system in smart agriculture, in *International Conference on Recent Advances in Electronics and Communication Technology* (2017)
11. J. Shah, B. Mishra, IoT enabled environmental monitoring system for smart cities, in *International Conference on Internet of Things and Applications*, January 2016
12. K.A.M. Zeinab, S.A.A. Elmustafa, Internet of Things applications, challenges and related future technologies. *Research Gate*, January 2017
13. L.M.L. Oliveira et al., Wireless sensor network: a survey on environment monitoring. J. Commun. **6**, 143–151 (2017)
14. R.K. Kodali et al., IoT based weather station, in *International Conference on Control, Instrumentation, Communication and Computational Technologies (ICCICCT)* (2016)
15. https://components101.com/dht11-temperature-sensor
16. https://www.aam.com.pk/shop/light-dependent-resistor-ldr-sensor-5mm-gl5528/
17. Nerokas engineering solutions. Retrieved from https://store.nerokas.co.ke/index.php?route=product/product&product_id=1423
18. Pimoroni Ltd (GB). Retrieved from https://shop.pimoroni.com/products/nodemcu-v2-lua-based-esp8266-development-kit

# Wideband Band-stop Filter for C-Band and S-Band Applications Using Finite Integration in Time-Domain Technique

Ajay Gupta, Anil Rajput, Monika Chauhan and Biswajeet Mukherjee

**Abstract** A band-stop filter (BSF), which shows the features of large bandwidth, is present in this paper. Changing the arrangement of quarter wavelength associating shapes of ordinary open-stub band-stop filter with the coupled line section is utilized in this paper. The filter characteristics are controlled by the gap between the parallel lines. This band-stop filter was design using single $\lambda/4$ resonator with one section of counter-coupled lines with open stub. The 20 dB rejection band of the proposed filter is from 2.7 to 11 GHz, and insertion loss of passband is below 1.2 dB. There are four transmission zeros in the stopband. The circuit size of the proposed filter is 13.2 mm $\times$ 26.25 mm ($0.17\lambda_g \times 0.33\lambda_g$). The application of these filter in the field of communication, biomedical, public address system etc.

**Keywords** Band-stop filter (BSF) · Microstrip · Transmission zeros · Coupled line · Quarter-wavelength resonator ($\lambda/4$) · Ultra-wideband (UWB) · Electrocardiogram (EGC)

## 1 Introduction

Band rejection filters are key components in microwave for suppressing unwanted signals. Active device like mixer and oscillator is frequently pursued by BSF to eliminate the excessive order harmonics and added undesirable signal separately. A few literary works similar to open-stub filters [1], coupled end filters [2], and parallel-coupled filters [3] had been investigated. The above filters give the detail at resonant frequency band. Notwithstanding, because of conveyed qualities of transmission media, the above filter experiences ill effects for the issue of false passbands. For wide stopband in BSFs, periodic structure for harmonics tuning [4] and double plane arrangement [5] is effective arrangements. Parallel coupled line connected with short circuit stub create multiple transmission zero in the stop band and frequency selecting, coupled line structure have been explored in [4, 5]. A spur line has been

A. Gupta (✉) · A. Rajput · M. Chauhan · B. Mukherjee
PDPM Indian Institute of Information Technology, Design and Manufacturing,
Jabalpur, Jabalpur, M.P., India
e-mail: gupta.ajay73@ymail.com

© Springer Nature Singapore Pte Ltd. 2021
X.-Z. Gao et al. (eds.), *Advances in Computational Intelligence and Communication Technology*, Advances in Intelligent Systems and Computing 1086,
https://doi.org/10.1007/978-981-15-1275-9_29

utilized to enhance the stopband suppression for a band-stop filter [6]. Additionally, various microwave components including diplexers and switches are likewise included in BSF. For suppressing the pole, λ/4 open stub is used and its effort as K inverter. Numerous helpful strategies [7–14] have demonstrated promising outcomes managing the symphonious issues.

The parallel-coupled microstrip filter [7] and endless disruption of the width of the coupled line [9] filters have high-order harmonic rejection enhancement of more than 30 dB. Filter utilizing a stepped impedance resonator travels the second passband greater than two times of center frequency [8]. Input and output selecting line has a additional conduction zero in the stop band has been impacted [10]. The electromagnetic-band gap (EBG)-based filter offers more than 25 dB elimination at the high-order harmonic [11]. Cascading extra band-stop filter in [12] and [14] has been demonstrated as a clear strategy to eliminate the harmonics, but this will also rise the overall circuit size. The shunt λ/4 open stubs were found to form conduction zeros in rejection band [13]. UWB characteristics have been proposed to operate from 2.7 to 11 GHz. This range of bandwidth is also used in C-band and S-band applications.

## 2 Mathematical Analysis of Wideband Filter

The center frequency selection, coupled line structure, and its equivalent circuit [2] can be investigated based on even and odd modes of transmission line. Transmission coefficient from port one to port two $S_{21}$ is given by [1]:

$$s_{21} = \frac{\Gamma_e - \Gamma_o}{2} = \frac{Y_{\text{ino}}}{Y_o + Y_{\text{ino}}} - \frac{Y_{\text{ine}}}{Y_o + Y_{\text{ine}}} \tag{1}$$

The transmission zero condition of coupled line structure is given by:

$$y_{0e} \cot \theta = y_{0o} \cot \theta \tag{2}$$

$Y_0$ = characteristic admittance for matching the line
$Y_{0e}$ = even mode characteristic admittances.
$Y_{0o}$ = odd mode characteristic admittances.
$\Gamma_e$ = even mode reflection coefficients.
$\Gamma_o$ = odd mode reflection coefficients.
$S_{21}$ = Transmission of the signal from port one to port two.

Single transmission zero is obtained after $\theta = 90°$ on $f_r$ where $f_r$ is the center frequency by which coupled line is λ/4 long. In this design, equation two is obtained by doing even and odd mode analysis of the proposed filter by crating even mode acts as open circuit and odd mode acts as short circuit. In this work, changing the coupling distance marginally longer than λ/4 resonator has been utilized to move

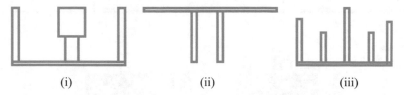

| (i) | (ii) | (iii) |

**Fig. 1** (i) Double-mode resonator. (ii) Three-mode resonator. (iii) Four-mode resonator

the transmission zero toward center frequency. The gap between two lines regulates the bandwidth of stopband. A less gap between parallel lines indicates the strong connection, and the high gap between the parallel lines indicates the weak coupling. Due to strong coupling, bandwidth is high as compared to weak coupling.

## 3  Multimode Resonator

For obtaining wideband filter, multimode resonator is used in the place of single-mode resonator because it makes filter compact. Multimode resonator increases the number of transmission zeros because of which the bandwidth of the filter increases. Figure 1(i) shows double-mode resonator which gives two transmission zeros in stopband. Figure 1(ii) shows three-mode resonator which gives three transmission zeros in stopband. Figure 1(iii) shows four-mode resonator which gives four transmission zeros in stopband.

## 4  Design Detail of the Wideband Band-stop Filter

RO4003 substrate with a relative permittivity of $\varepsilon_r = 3.38$, loss tangent of 0.0027, and thickness of substrate $h = 0.51$ mm is used in design for simulation purpose. The input and output lines are 50 $\Omega$ because ideally waveguide port is available for 50 $\Omega$ line. To reduce the reflection at input and output port 50 $\Omega$ lines is use for matching purpose. The physical dimension of proposed BSF is $L_1 = 4$ mm, $W_1 = 2$ mm, $L_2 = 8.75$ mm, $W_2 = 0.6$ mm, $L_3 = 4$ mm, $W_3 = 0.6$ mm, $L_4 = 5$ mm, $W_4 = 0.4$ mm, $L_5 = 7$ mm, $W_5 = 0.2$ mm, $L_6 = 6.8$ mm, $W_6 = 2$ mm, $L_7 = 2$ mm, $W_7 = 2$ mm, $G_1 = 0.25$ mm, $G_2 = 0.3$ mm, $G_3 = 0.25$ mm. Figure 2 shows the wideband band-stop filter, which is symmetric about $Y$-axis. This structure consists of coupled line which is connected with three open stub line, which is responsible for center frequency and together this structure acts as a frequency selecting coupling structure. $L_5$ stub with coupled line shows the characteristic of four-mode resonator, which produce four transmissions zeros in stopband. $L_1$ stub controls the reflection coefficient as well as harmonics.

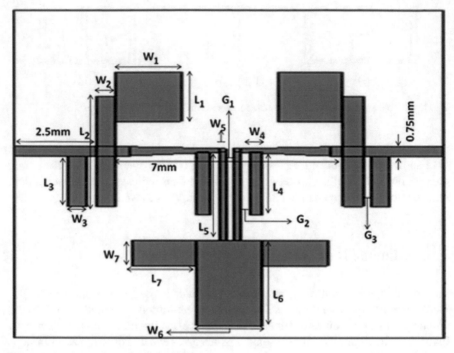

**Fig. 2** Proposed wideband band-stop filter

Figure 3a shows the simulation result without $L_1$ stub. Because of the absence of $L_1$ stub, reflection coefficient of BSF is very poor or above 20 dB. Figure 3b shows the variation in bandwidth with $G_2$, which is the gap between $L_4$ and $L_5$ stubs. $G_2 = 1.9$ mm shows wide bandwidth as compared to $G_2 = 0.15$ mm. In Fig. 3c, it can be seen that reflection coefficient and bandwidth are also controlled by $L_5$ stub. $L_5 = 5$ mm offers good reflection coefficient but poor bandwidth; $L_5 = 5.5$ mm gives good bandwidth but poor reflection coefficient. So the bandwidth and reflection coefficient can be controlled by $L_5$ stub. Waveguide port for transmission of the signal has been employed with horizontal dimension of port as three times of 50 $\Omega$ line width and vertical dimension as five times of height of substrate.

## 5 Simulated Results and Discussion

Based on the discussion of parametric value analysis, the optimum dimension of the BSF is proposed as discussed in Sect. 3. The Computer Simulation Technology (CST) microwave studio has been employed to perform all simulations. The simulations are based on finite integration in time-domain (FIT) technique. Simulation model includes refined meshing of 10 cells per unit wavelength and 10 cells per maximum

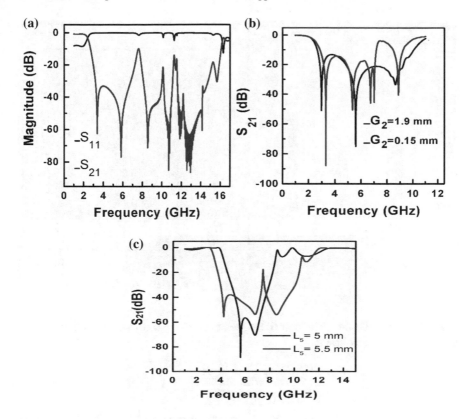

**Fig. 3** **a** Simulated result without $L_1$ stub. **b** Variation in bandwidth with $G_2$. **c** Variation in return loss and bandwidth with $G_2$ stub

modal box edge. Due to this setting, the whole design is divided into 24,120 cells. Simulation is done for smallest cell 0.15 mm to largest cell 1.05 mm. The open boundary condition is preferred for simulation. The open boundary condition is the closest approximation of the filter in practical application, and hence, it has been employed for all the simulation. The transient domain or time-domain solver has been used to perform the FIT operation in CST.

Figure 1 represents the physical dimension of wideband BSF. Coupled line with three open stubs is used as a frequency selecting coupling structure, which gives the center frequency and three transmission zeros in the stopband. When $L_4$ stub is added in the structure, it gives four transmission zeros in the stopband and fractional bandwidth is increased up to 127% (from 2.7 to 11 GHz). $L_4$ stub acts as quarter-wave resonator. $L_1$ stub is used for improved reflection coefficient.

Figure 4 shows the simulated result of the wideband BSF. The reflection coefficient is below 20, and 20 dB rejection bandwidth is up to 8.25 GHz from 2.7 to 11 GHz. Simulated insertion loss is 1.2 dB for complete bandwidth of operation. The second harmonic is suppressed up to two times of $f_r$. There are four transmission zeros at

**Fig. 4** Simulated result of
wideband band rejection
filter

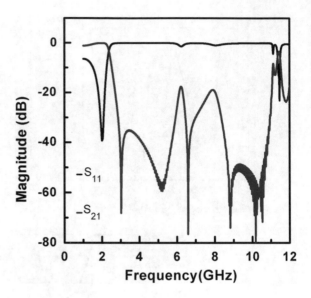

3.5, 6.5, 9, and 10 GHz. The center frequency and 20 dB rejection bandwidth have
been calculated as follows:

$$f_r = \frac{f_h + f_l}{2} \tag{3}$$

$$20 \text{ dB Fractional Bandwidth} = \frac{f_h - f_l}{f_r} \tag{4}$$

$f_r$ = Center frequency
$f_h$ = Upper frequency in bandwidth
$f_l$ = Lower frequency in bandwidth

Table 1 shows the 20 dB fractional bandwidth of the previous research work.
From the table, it is evident that proposed BSF is novel and has better characteristics
then existing design.

## 6 Conclusion

A BSF with wide stopband is proposed in this paper. The improvement of bandwidth
is because of the creation of additional transmission zeros in the stopband by means
of combining one shunt open-circuited stub with parallel-coupled line stub, offering
a good stopband performance. The center frequency of proposed BSF is close to
6.5 GHz, and 20 dB rejection bandwidth is 127%. The 20 dB rejection band is from

**Table 1** Evaluation of the proposed work with the previous works

| References | Center frequency (in GHZ) | 20 dB rejection fractional bandwidth (%) | Bandwidth (GHz) |
|---|---|---|---|
| [15] | 06 | 120 | 2.3–9.5 |
| [16] | 02 | 100 | 1–3 |
| [17] | 03 | 94 | 1.43–4.57 |
| [18] | 2.05 | 90 | 1.5–2.6 |
| [19] | 04 | 80 | 3–5 |
| [20] | 1.5 | 100 | 0.5–2.5 |
| This work | 6.5 | 127 | 2.7–11 |

2.7 to 11 GHz, and insertion loss of passband is below 1.2 dB. In communication system, original signal can be distorted due to some harmonics or noise; this filter is used to remove these unwanted noises, e.g., signal and image processing. The other example of this BSF in medical field application like biomedical instrument eliminates line noise in EGC. The proposed BSF finds applications in short-distance wideband communication system setup for higher-frequency range of operation.

# References

1. J.-S. Hong, M.J. Lancaster, *Microstrip Filters for RF/Microwave Applications* (Wiley, New York, 2000)
2. G.L. Matthaei, L. Young, E.M.T. Jones, *Microwave Filters, Impedance-Matching Networks, and Coupling Structures* (McGraw-Hill, New York, 1980)
3. S. Cohn, Parallel-coupled transmission-line resonator filters. IRE Trans. Microw. Theory Tech. **6**(4), 223–231 (1958)
4. C.Y. Hang, W.R. Deal, Y. Qian, T. Itoh, High efficiency transmitter front-ends integrated with planar an PBG, in *Asia-Pacific Microwave Conference* (2000), pp. 888–894
5. J.-Y. Kim, H.-Y. Lee, Wideband and compact bandstop filter structure using double-plane superposition. IEEE Microw. Wirel. Compon. Lett. **13**(7), 279–280 (2003)
6. W.-H. Tu, K. Chang, Compact microstrip bandstop filter using open stub and spurline. IEEE Microw. Wirel. Compon. Lett. **15**(4), 268–270 (2005)
7. J.-T. Kuo, W.-H. Hsu, W.T. Huang, Parallel coupled microstrip filters with suppression of harmonic response. IEEE Microw. Wirel. Compon. Lett. **12**(10), 383–385 (2002)
8. M. Makimoto, S. Yamashita, Band pass filters using parallel coupled stripline stepped impedance resonators. IEEE Trans. Microw. Theory Tech. **28**(12), 1413–1417 (1980)
9. T. Lopetegi, M.A.G. Laso, J. Hernandez, M. Bacaicoa, D. Benito, M.J. Grade, M. Sorolla, M. Guglielmi, New microstrip 'wigglyline' filters with spurious passband suppression. IEEE Trans. Microw. Theory Tech. **49**(9), 1593–1598 (2001)
10. J.-T. Kuo, E. Shih, Microstrip stepped impedance resonator bandpass filter with an extended optimal rejection bandwidth. IEEE Trans. Microw. Theory Tech. **51**(5), 1554–1559 (2003)
11. Y.W. Kong, S.T. Chew, EBG-based dual mode resonator filter. IEEE Microw. Wirel. Compon. Lett. **14**(3), 124–126 (2004)
12. L.-H. Hsieh, K. Chang, Piezoelectric transducer tuned bandstop filter. Electron. Lett. **38**(17), 970–971 (2002)

13. L. Zhu, W. Menzel, Compact microstrip bandpass filter with two transmission zeros using a stub-tapped half-wavelength line resonator. IEEE Microw. Wirel. Compon. Lett. **13**(1), 16–18 (2003)

14. J. Garcia, F. Martin, F. Falcone, J. Bonache, I. Gil, T. Lopetegi, M.A.G. Laso, M. Sorolla, R. Marques, Spurious passband suppression in microstrip coupled line bandpass filters by means of split ring resonators. IEEE Microw. Wirel. Compon. Lett. **14**(9), 416–418 (2004)

15. M.-Y. Hsieh, S.-M. Wang, Compact and wideband microstrip bandstop filter. IEEE Microw. Wirel. Compon. Lett. **15**(7), 472–474 (2005)

16. M.Á. Sánchez-Soriano, G. Torregrosa-Penalva, E. Bronchalo, Compact wideband bandstop filter with four transmission zeros. IEEE Microw. Wirel. Compon. Lett. **20**(6), 313–315 (2010)

17. L. Liang, Y. Liu, J. Li, S. Li, C. Yu, Y. Wu, M. Su, A novel wide-stopband bandstop filter with sharp-rejection characteristic and analytical theory. Prog. Electromagn. Res. C **40**, 143–158 (2013)

18. A. Görü, C. Karpuz, Uniplanar compact wideband bandstop filter. IEEE Microw. Wirel. Compon. Lett. **13**(3), 114–116 (2003)

19. H. Shaman, J.-S. Hong, Wideband bandstop filter with cross-coupling. IEEE Trans. Microw. Theory Tech. **55**(8), 1780–1785 (2007)

20. K. Divyabramham, M.K. Mandal, S. Sanyal, Sharp-rejection wideband bandstop filters. IEEE Microw. Wirel. Compon. Lett. **18**(10), 662–664 (2008)

# A Two-Step Technique for Effective Scheduling in Cloud–Fog Computing Paradigm

Ashish Mohan Yadav, S. C. Sharma and Kuldeep N. Tripathi

**Abstract**  As we know the Internet of Things applications are emerging as a help-ing hand for the ease of mankind in day-to-day life but when it clubs with cloud computing comes up with the limitation of far distance among Internet of Things gadgets and cloud computing infrastructure which gives an idea to work with a new distributed computing environment with the combination of "cloud computing" and fog computing. "Fog computing" majorly can be used to minimize the transmission delay (latency) and the cost for use of cloud assets as cloud computing helps us to use the complex, large, and heavy tasks to be offloaded on cloud. Here, with this article, we are showing a study for the trade-off between cloud cost and makespan whenever we are scheduling applications in such a kind of environment. We give an algorithm called BAS to sequence applications with the balance between performance and cost of cloud usage. With the simulated results, we have shown that our proposed method is working better compared to some peer methods.

**Keywords**  Cloud computing · Fog computing · Internet of Things (IoT) · Task scheduling

## 1  Introduction

These days, the smart gadgets work as Internet of Things is the area developing revolutionarily in data and communication innovation, the scholarly community, and industry have been creating capable structures to guarantee the saturating availability of smart gadgets. Clients are enabled by these smart gadgets by giving them a passage to a scope of extraordinary performance and cost-effective savvy offices in various kinds of situations including smart urban communities, smart metering associated vehicles, smart homes, large-scale wireless sensor systems, and so on.

Be that as it may, various angles, for example, treating, constrained comput-ing, and networking capacities of IoT gadgets make them inappropriate to execute diverse, processor, or memory concentrated applications. To beat these limitations,

A. M. Yadav (✉) · S. C. Sharma · K. N. Tripathi
Indian Institute of Technology, Roorkee, Roorkee, India
e-mail: ayadav@pp.iitr.ac.in

© Springer Nature Singapore Pte Ltd. 2021
X.-Z. Gao et al. (eds.), *Advances in Computational Intelligence and Communication Technology*, Advances in Intelligent Systems and Computing 1086,
https://doi.org/10.1007/978-981-15-1275-9_30

cloud computing which gives powerfully ascendable and regularly virtualized assets is given as a service by communication web may be a strong add on pillar to IoT. The cloud defeats these restrictions of such IoT gadgets known as savvy gadgets (e.g., low powered battery, low in computational power, low with memory, and low accessibility of system resources) can be helped by offloading high computational intensive, broad computational asset required tasks up to an incredible computationally capable environment in the cloud, pushing just straightforward jobs to the less efficient tech-savvy gadgets.

With this exponential ascent in the quantity of associated gadgets, conventional cloud-based frameworks, in which storage and computation are done in a few large data centers, would not have the capacity to deal with the IoT's data. This comes as a result of the separation between smart gadgets and the cloud. The trading of tremendous information or demand of services between the gadgets and the cloud causes overwhelming traffic and weight on network data bandwidth results transmission delay with greater latency.

To have the capacity to give such sort of design, another computational architype called fog computing recently been proposed. Fog computing is based on idea of providing computing capabilities specifically at the verge of the cloud as opposed to being conveyed profoundly of the cloud. Fog computing is the thought for distributed computing setup that spreads computational facilities introduced by the distributed computing to the edge of the network as appeared in Fig. 1.

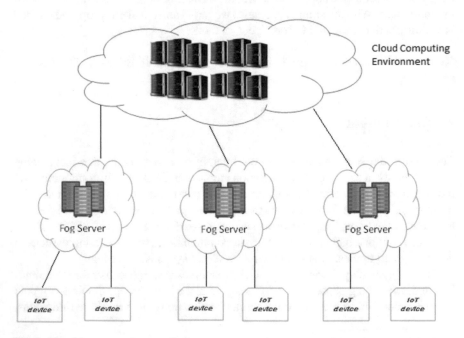

**Fig. 1** Cloud-fog computing paradigm

Practically speaking, many offloading applications are made out of numerous units carrying out diverse jobs. These jobs can be either autonomous of one another or commonly dependent [1]. In the case of autonomy, all jobs can be offloaded all the while and computed in parallel but in the case of dependency, notwithstanding, the application is comprised of jobs that require a contribution from some other different jobs as input then that execution in parallel may not be as easy. The last application is typically demonstrated as a work process, that is demonstrated by a directed acyclic graph (DAG) [2], which tells about the arrangement of nodes speaks about various jobs and the arrangement of links indicates their priority constraints [3, 4]. Whenever this streamed data travels from source (e.g., sensors) to cloud, every point with whom it touches from the system bottom to the system roof (e.g., cloud) everyone is likely to be a processing potential. So, the major objective in sequencing of these jobs is to achieve a suitable processing hub in this paradigm of cloud-fog.

Here, in the distributed computing environment, a new system has emerged with the collaboration of cloud and fog for executing extensive scale computing DAG-based applications [5]. We at that point propose a new application planning heuristic, budget-aware scheduling (BAS) algorithm to take care of the scheduling issues in this stage. In addition, we present a task reschedule technique which is in charge of used to refine the output sequences of the BAS technique to fulfill the client characterized deadline requirements, consequently enhance the QoS of the framework.

## 2 Related Work

There have been different examinations that endeavor to take care of job sequencing issue in heterogeneous computing frameworks, where the arrangement of jobs (workflow) is effectively shown by a DAG [2, 6–8]. Because DAG job sequencing is a non-deterministic polynomial-time finish problem [2], it is required to be understood by heuristic calculations for finding estimated ideal sequences. The heterogeneous earliest finish time (HEFT) [6] calculation is one of the well-known and broadly utilized techniques. The HEFT incorporates two fundamental stages: a job organizing stage for registering the priorities of all jobs dependent on upward rank esteem and a processor choice stage for choosing the job with the most elevated upward rank (priority) at each progression and doling out the chosen job to the processor which reduces the job completion time. Wang et al. [8] present techniques named entry task duplication selection policy and ITS to enhance the productivity of the sequencing technique. But as they are meant to limit makespan, do not take a look at money related expense for large tasks.

Fog processing, described by stretching out distributed computing to the edge of system, has as of late gotten impressive consideration. The seed commitment on fog processing is the one exhibited by Bonomi et al. [9], in which the creators give bits of knowledge identified with the principle prerequisites and qualities of fog system and feature the significance of cloud–fog exchange and the job of fog system with regards to IoT.

Souza [10, 11] location QoS-efficient service distribution issue in a consolidated cloud–fog design comprising a double layer fog intending to decrease use of cloud access defer in IoT situations. Alsaffar et al. [5] present a design of IoT service assignment and asset allotment, in which client solicitations can be proficiently overseen and appointed to suitable cloud or fog dependent on linearized choice tree which considers tree conditions (benefit estimate, completion time, and VM limit). Nan et al. [12] consider a three-level cloud of things (CoT) architecture and a technique called unit-spot enhancement for appropriating the incoming information to the relating levels that can give practical processing while at the same time ensuring saving response time.

Our work fills in as a beginning stage to decide an ideal sequence for huge -scale process serious applications dependent on the cooperation among cloud and fog systems.

## 3 Problem Model

Job sequencing on a target framework is characterized as an issue of dispensing the jobs, form as application, to an arrangement of different working capacities processors in order to reduce the makespan of the sequence. In this way, the contribution of job sequencing takes a task diagram as an input and also takes a processor diagram as an input, and the output is a sequence which tells that which job will take which processor to execute.

Task diagram: A task diagram which is shown by a directed acyclic graph, TD, in which the arrangement of vertices shows arrangement of parallel assignments, and each edge $e_{ij} = (t_i, t_j) \in E$ tells about the correspondence or priority limitation between two tasks $t_i$ and $t_j$. Each job has positive work amount $w_i$ that shows the measure of processing works (e.g., the quantity of instructions). The arrangement of every immediate forerunner and successors of $t_i$ are signified as pred $(t_i)$ and succ$(t_i)$, respectively. A task with no ancestors is known as an entry task and with no successors is known as an exit task.

Processor diagram: A processor diagram, PD, shows a model of a cloud–fog environment, where the arrangement of nodes indicates the arrangement of processors, every one of each is either a cloud or fog hub, and a connection $d_{ij} \in D$ means an edge between two processors. The arrangement of $N_{cloud}$ and $N_{fog}$ signifies about cloud or fog hub. Every processor $P_i$ has the capacity $p_i$ and $bw_i$ to demonstrate data transfer on the connection associating it to different processors.

### 3.1 Proposed Method

Our method works in two noteworthy stages; the first one which takes a task diagram and processor diagram as input allots each job to a proper processing hub (cloud or fog) for the accomplishment of ideal estimation of utility function which decides a bargaining between sequence time (makespan) and expense for cloud usage.

After that, task rescheduling stage methodology is applied which takes the output of first stage as an input works to fulfill the client characterized deadline.

Node choice stage: A task $t_i$ just starts its execution after every preceding assignment of job i is finished. Let $FT(t_i)$ be the completion time of the latest former $i$th job. Now, it is the time for execution of $t_i$ at the node which is selected. In this way, it is said the information exchange time of $i$th job and characterized by

$$FT(t_i) = \max_{t_j \in \text{pred}(t_i), \ p_m \in N} \left[ ft(t_j, P_m) \right] \tag{1}$$

where $ft(t_j, P_m)$ is called the finish time of job $t_j$ on processor $P_m$.

To calculate communication time between node $P_m$ and $P_n$ a function $ct(e_i^{mn})$ is used, describe below

$$ct(e_i^{mn}) = \begin{cases} \left( d_i^m + \sum_{\substack{t_j \in \text{pred}(t_i) \\ t_j \in \text{exec}(P_m)}}^{} ct_{ji} \right) * \left( \frac{1}{bw_m} + \frac{1}{bw_n} \right) \\ \qquad \text{if } m \neq n \\ 0 \qquad \text{if } m = n \end{cases} \tag{2}$$

where $d_i^m$ is the quantity of data used for processing the task $t_i$ and is saved on processor $P_m$.

Below equation shows the preparation time of task i which is used to calculate as when all the information is transferred from storage to any selected hub, is characterized by

$$rt(t_i, P_n) = FT(t_i) + \max_{P_m \in N} \left( ct(e_i^{mn}) \right) \tag{3}$$

Let $E_{-\text{ST}}(t_i, P_n)$ and $E_{-\text{FT}}(t_i, P_n)$ are the time of starting and finishing the job i at processor $P_n$ and $c(t_i, P_n)$ is the total time taken to execute the job.

Let the time interval $[t_A, t_B]$ is the time when processor $P_n$ is idle then a task i is able to run on that processor if

$$\max\{t_A, rt(t_i, P_n)\} + c(t_i, P_n) \cdot t_B \tag{4}$$

For the found interval $[t_A, t_B]$, the $E_{-\text{ST}}(t_i, P_n)$ and $E_{-\text{FT}}(t_i, P_n)$ are computed as follows

$$E_{-\text{ST}}(t_i, P_n) = \max\{t_A, rt(t_i, P_n)\} \tag{5}$$

$$E_{-\text{FT}}(t_i, P_n) = E_{-\text{ST}}(t_i, P_n) + c(t_i, P_n) \tag{6}$$

Consequently, let $\text{cost}(t_i, P_n)$ be the money related expense for executing task $i$ on the node $P_n$. In the event that $P_n$ is a cloud hub, the financial expense incorporates different costs such as preparation, storage, and memory for job i on hub $P_n$, and correspondence cost for measuring active information which flows from other hubs to the objective hub $P_n$ to process task $i$. Conversely, if $P_n$ in that case a fog hub, the customers are just billed for exchanging the active information from cloud hubs to the objective fog hub in the framework. In this way, the $\text{cost}(t_i, P_n)$ is characterized as

$$\text{cost}(t_i, P_n) = \begin{cases} \text{cp}(t_i, P_n) + \text{cs}(t_i, P_n) + \text{cm}(t_i, P_n) + \sum_{P_m \in N_{\text{cloud}}} \text{ccom}(t_i, P_n) \text{ if } P_n \in N_{\text{cloud}} \\ \sum_{P_m \in N_{\text{cloud}}} \text{ccom}(t_i, P_n) \qquad\qquad\qquad\qquad\qquad \text{if } P_n \in N_{\text{fog}} \end{cases} \tag{7}$$

In Eq. (7), additional terms are expressed as

$$\text{cp}(t_i, P_n) = c_1 * c(t_i, P_n) \tag{8}$$

where $c_1$ is the cost for processing used to calculate processing cost cp at node $P_n$. where $c_2$ is the cost for storage used to store data per unit. Let $st_i$ is the size of storage used for task i so cost for storage cs is calculated as follows.

$$\text{cs}(t_i, P_n) = c_2 * st_i \tag{9}$$

then cost for memory used is calculated as follows where $c_3$ is the cost per data unit charged $s_m$ is the total memory used for task i.

$$\text{cm}(t_i, P_n) = c_3 * s_m \tag{10}$$

Let $c_4$ be the measure of charge per information unit for exchanging outgoing information from hub $P_m$, and after that, the correspondence measure is determined as pursues

$$\text{ccom}(t_i, P_n) = c_4 * \left( d_i^m + \sum_{\substack{t_j \in \text{pred}(t_i) \\ t_j \in \text{exec}(P_m)}} ct_{ji} \right) \tag{11}$$

From this $\text{cost}(t_i, P_n)$, we can draw a function which calculates a balance between the performance and expense as describes

$$U(t_i, P_n) = \frac{\min_{P_k \in N}[\text{cost}(t_i, P_k)]}{\text{cost}(t_i, P_n)} * \frac{\min_{P_k \in N}\left[E_{-\text{FT}}(t_i, P_k)\right]}{E_{-\text{FT}}(t_i, P_n)} \tag{12}$$

At that point, the task $t_i$ is doled out to the hub $P_n$, which gives the maximal estimation of the trade-off $U(t_i, P_n)$. After assignment $t_i$ is booked on hub $P_n$, the real complete time of task $t_i$, AFT $(t_i)$, is equivalent to $E_{-FT}(t_i, P_n)$. After all assignments are sequenced, the resulting time would be the genuine complete time of the $i$th task. Node choice stage: For any application, the client-defined due time is vital data for QoS for any framework. This is a must for any application scheduler that job sequencing is done in a way that it does not complete later than a predefined deadline time. To confine the due time infringement, it is needed for the fog broker in the fog environment to develop the timeline plans (schedule) after the node choice stage so the job can be executed while meeting the client characterized due time with the potentially most reduced measure of the money related expense for cloud assets. In this manner, we present a procedure in Algorithm 1, node choice stage, to enhance the finishing time of job execution as due time requirements. The real thought of the procedure is that the makespan of a sequencing is altogether affected by the aggregate processing time of tasks on the critical path of an application graph [13]. The critical path alludes to the longest execution way between the entry task and exit task of the application and the assignments in this way are called critical tasks.

In the next technique, the enhancement for the makespan of sequencing can be found by rescheduling the critical tasks to better handling hubs which can further diminish the finishing time of each critical task. The processing hubs which ensure the EFT estimation of a critical task is not more prominent than the task sub-deadline are considered as the candidate hubs to which this task thought to be allocated. In view of the rundown of the candidate hubs, the hub determination still depends on the most extreme estimation of the function $U(t_i, P_n)$ referenced in the previous paragraph. Consequently, all the critical tasks are reassigned, whatever is left, are non-critical tasks, may likewise be rescheduled based on the reports developed after the selection on the critical tasks.

In fact, it is extremely troublesome for any scheduling algorithm to create such kind of timelines which can meet every single tight due time. In particular, the task reschedule stage will be carried forward until the point when the timeline is fulfilled (current_makespan ≤ deadline) or there is no timeline giving preferable makespan over the most recent timeline that has quite recently been found in a similar stage (current_makespan ≥ old_makespan).

# 4 Implementation and Analysis

First, to show the performance of proposed algorithm (BAS), we have compared our algorithm in terms of makespan length, cloud cost, and compromise level. We have matched greedy for cost that is known for minimal workflow cost of execution; HEFT [6], which tries to timeline interdependent tasks at heterogeneous environment for minimal execution time and CCSH [14], which works both on efficient finish time and cost for customers.

## 4.1   Experimental Setup

The setup used for our work is appeared in Table 1. The simulated environment is created utilizing iFogSim with Eclipse 4.4 which is a structure to create fog processing infrastructures and services. Our first examination covers an arbitrary task diagram TD with the different dimensions ranging from 25 to 100 and a heterogeneous processor diagram PD which is a blend of total 40 hubs (25 cloud + 15 fog) with distinctive capacities. It is expected that the handling rate of fog hubs is set littler compared to cloud hubs as appear in Table 2. Furthermore, the expense of utilizing cloud assets is determined in Table 3.

**Table 1** Simulation characteristics

| Processor | Intel Xenon 2.5 GHz |
|---|---|
| Memory | 16 GB |
| Simulator | iFogSim/CloudSim 3.0.3 |
| Operating system | Windows 10 student edition |
| Topology model | Fully connected |

**Table 2** Statistics of the cloud and fog systems

| Parameters | Fog architecture | Cloud architecture |
|---|---|---|
| No of processors | 15 | 25 |
| Processing rate | [20, 400] MIPS | [300, 1800] MIPS |
| Bandwidth | 512 Mpbs | 10, 200, 512, 1024 Mbps |

**Table 3** Cost for usage of cloud assets

| Parameters | Value |
|---|---|
| "Cost for processing per unit time" | (0.1, 0.5) |
| Cost for communication per unit data | (0.3, 0.7) |
| Memory cost per unit used | (0.01, 0.1) |
| Storage cost per unit used | (0.05, 0.2) |

Algorithm 1. Budget-aware scheduling (BAS) algorithm

> $Input: TD\ and\ PD, TD: Task\ diagram, PD: Processor\ diagram$
> $Output: A\ task\ timeline\ sequence$

$Set\ priority\ pr(t_i)\ for\ every\ task\ by\ upward\ traversal\ starting\ from\ t_{exit}.$

> $Set\ all\ tasks\ T\ into\ Q\ by\ sorting\ in\ nonincreasing\ order\ of\ priority.$
> $Repeat\ for\ each\ t_i \in Q$
> > $Repeat\ for\ each\ P_n \in N$
> > > $Calculate\ E_{-ST}(t_i,\ P_n), E_{-FT}(t_i,\ P_n) and\ cost(t_i,\ P_n)$
> > > $Calculate\ function\ U((t_i,\ P_n)$
> > $End$
> > $Select\ processor\ P_n\ to\ process\ t_i\ for\ maximum\ value\ of\ U(t_i,\ P_n)$
> $End$

Algorithm 2. Task reschedule algorithm

> $Input: TD, PD\ and\ application\ schedule\ S;\ TD: Task\ diagram, PD: Processor\ diagram$
> $Output: A\ task\ timeline\ sequence$
> $Temp \leftarrow S$
> $Repeat$
> > $old\_makespan \leftarrow\ Temp.makespan$
> > $Do\ for\ each\ t_i \in TD$
> > $t_{i_{reassigned}} \leftarrow false$
> > $End$
> > $CT \leftarrow critical\ path\ of\ task\ diagram$
> > $distance \leftarrow \frac{old\_makespan - deadline}{length\ of\ CT}$
> > $Do\ for\ each\ t_i \in CT$
> > > $subdeadline(t_i) \leftarrow AFT(t_i) - distance$
> > > $if\ subdeadline(t_i) \geq 0\ then$
> > > > $EN \leftarrow list\ of\ eligible\ processing\ nodes$
> > > > $if\ EN \neq \emptyset\ then$
> > > > > $BN(t_i) \leftarrow max[U(t_i,\ P_n)]$
> > > > $else$
> > > > > $BN(t_i) \leftarrow min[E_{-FT}(t_i,\ P_n)]$
> > > > $end\ if$
> > > > $AFT(t_i) \leftarrow E_{FT}(t_i,\ BN(t_i))$
> > > > $t_{i_{reassigned}} \leftarrow true$
> > > $end\ if$
> > $end\ for$
> > $Temp \leftarrow new\ schedule$
> $until\ Temp.makespan \leq deadline\ or\ Temp.makespan \geq$
> $old_{makespan}$

**Fig. 2** Comparison of
schedule length

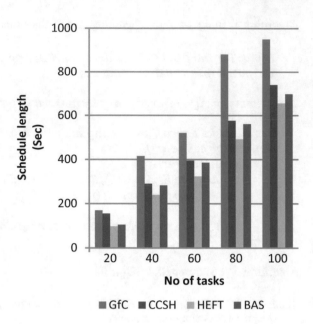

## 4.2 Evaluation of Algorithms

Figure 2 demonstrates that as far as timeline length, GfC calculation gets the most pessimistic scenario; HEFT calculation acquires the best outcome while CCSH and BAS are in the center.

In particular, proposed strategy is 5.61% superior to CCSH. What is more, contrasted and GfC, our technique even accomplishes a far superior execution, about 31.94%. In any case, in regards to the money related expense for cloud assets (showed in Fig. 3), it is clearly visible that HEFT [6] gives optimal execution but with highest elevation in expenses while the inverse is valid for GfC calculation. Contrasted with HEFT, our strategy can spare 31.08% of cloud cost while performance decrease is not over 22%. Likewise, contrasted with CCSH [14] calculation, our strategy can spare 5.07% of cloud cost, which implies that our technique has a little financial favorable position together with its adequacy.

In the next investigation, we assess the effect on cloud related usage expenses and also on timeline length with the expansion of total sum of cloud hubs using BAS technique with a settled figure of tasks. As in the case of previous algorithm, we had fixed number of processing hubs but there now we changed this from 5 to 25 to produce diverse recreated models of the computational environment. The investigation outcomes appeared in Figs. 4 and 5 show that the presence of added hubs, which give improved handling limit and in addition the bigger measure of input information for job execution than fog hubs, results in improved framework execution, however, the higher money related expense for cloud assets. It is very detectable that the timeline length decreases 40.98%; however, the cloud usage expenses rise 60.13% as

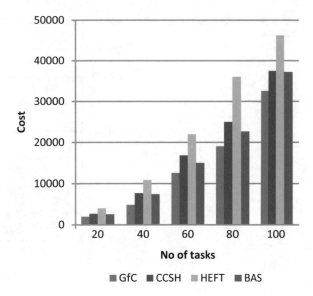

**Fig. 3** Comparison of costs

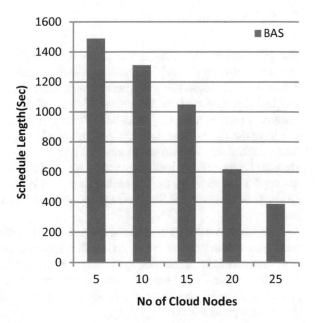

**Fig. 4** Effect of cloud nodes in schedule length

the quantity of cloud hubs increments from 15 to 20. So by these results, the consequences of the examinations can help us to decide that what could be a combination of both cloud and fog hubs in computing framework in order to fulfill sequence line time (makespan) requirement or spending constraints of the application scheduling issue.

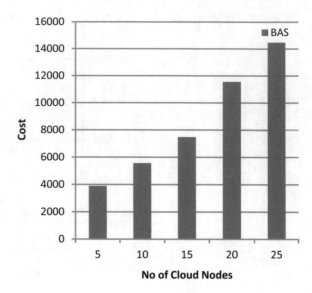

**Fig. 5** Effect of cloud nodes with cost

## 5 Conclusion

The IoT gadgets alongside the requests for administrations and applications are expanding quickly in terms of amount as well as scale. To deal with this challenge, this consolidated cloud–fog engineering environment is looking as a promising prototype that if all-round use can give proficient information handling different applications or administrations, particularly those which are complex in computing. For receiving the most out of such a paradigm, one has to distribute computing assignments deliberately at each computational hub (cloud or fog). We propose the BAS algorithm considering a balance among computation and expense saving to assemble the application timeline plan, which ensures job execution inside due time imperatives as well as diminishes the required expenditure for the utilization of cloud assets.

And furthermore, green processing is currently ending up imperative. With the enormous volume and regularly expanding application requirements, it is desired with cloud and fog processing systems to be more utilized in terms of power consumption. Accordingly, we can broaden the given algorithm for huge-scale computing requirements by likewise considering energy proficiency with a challenge of achieving a parameter QoS.

# References

1. P. Mach, Z. Becvar, Mobile edge computing: a survey on architecture and computation offloading. IEEE Commun. Surv. Tutorials **19**, 1628–1656 (2017). https://doi.org/10.1109/comst.2017.2682318
2. J.D. Ullman, NP-complete scheduling problems. J. Comput. Syst. Sci. **10**, 384–393 (1975). https://doi.org/10.1016/S0022-0000(75)80008-0
3. W.A. Higashino, M.A.M. Capretz, L.F. Bittencourt, CEPSim: modelling and simulation of Complex Event Processing systems in cloud environments. Future Gener. Comput. Syst. **65**, 122–139 (2016). https://doi.org/10.1016/j.future.2015.10.023
4. J.-H. Choi, J. Park, H.D. Park, O. Min, DART: fast and efficient distributed stream processing framework for Internet of Things. ETRI J. **39**, 202–212 (2017). https://doi.org/10.4218/etrij.17.2816.0109
5. A.A. Alsaffar, H.P. Pham, C.-S. Hong et al., An architecture of IoT service delegation and resource allocation based on collaboration between fog and cloud computing. Mob. Inf. Syst. **2016**, 1–15 (2016). https://doi.org/10.1155/2016/6123234
6. H. Topcuoglu, S. Hariri, Wu Min-You, Performance-effective and low-complexity task scheduling for heterogeneous computing. IEEE Trans. Parallel Distrib. Syst. **13**, 260–274 (2002). https://doi.org/10.1109/71.993206
7. H. Arabnejad, J.G. Barbosa, List scheduling algorithm for heterogeneous systems by an optimistic cost table. IEEE Trans. Parallel Distrib. Syst. **25**, 682–694 (2014). https://doi.org/10.1109/TPDS.2013.57
8. Z. Wang, Z. Ji, X. Wang et al., A new parallel DNA algorithm to solve the task scheduling problem based on inspired computational model. Biosystems **162**, 59–65 (2017). https://doi.org/10.1016/J.BIOSYSTEMS.2017.09.001
9. F. Bonomi, R. Milito, J. Zhu, S. Addepalli, Fog computing and its role in the internet of things, in *Proceedings of the first edition of the MCC workshop on Mobile cloud computing—MCC '12* (ACM Press, New York, NY, USA, 2012), p. 13
10. V.B. Souza, X. Masip-Bruin, E. Marin-Tordera et al., Towards distributed service allocation in fog-to-cloud (F2C) scenarios, in *2016 IEEE Global Communications Conference (GLOBECOM)* (IEEE, 2016), pp. 1–6
11. V.B.C. Souza, W. Ramirez, X. Masip-Bruin et al., Handling service allocation in combined Fog-cloud scenarios, in *2016 IEEE International Conference on Communications (ICC)* (IEEE, 2016), pp. 1–5
12. Y. Nan, W. Li, W. Bao et al. Cost-effective processing for delay-sensitive applications in Cloud of Things systems, in *2016 IEEE 15th International Symposium on Network Computing and Applications (NCA)* (IEEE, 2016), pp. 162–169
13. S. Gotoda, M. Ito, N. Shibata, Task scheduling algorithm for multicore processor system for minimizing recovery time in case of single node fault, in *2012 12th IEEE/ACM International Symposium on Cluster, Cloud and Grid Computing (CCGrid 2012)* (IEEE, 2012), pp. 260–267
14. J. Li, S. Su, X. Cheng et al. Cost-conscious scheduling for large graph processing in the cloud, in *2011 IEEE International Conference on High Performance Computing and Communications* (IEEE, 2011), pp. 808–813

# Design of a Drag and Touch Multilingual Universal CAPTCHA Challenge

Abdul Rouf Shah, M. Tariq Banday and Shafiya Afzal Sheikh

**Abstract** In order to protect sensitive data from bots, service providers protect their Web sites through some of the CAPTCHA tests. This test ensures that protected resources are accessible to only legitimate human users and no computer program (bot) gets their access. With a growing number of Web sites and Web portals offering more and more sensitive data for access by its legitimate human users coupled with the advances in techniques used to break CAPTCHA challenges, more robust and complex CAPTCHA challenges were developed. Though some of these tests offer the desired level of security, however, they often are inaccessible to a large population of the Web sites and also suffer from other usability issues. This paper presents a new CAPTCHA challenge based on the mouse motion event and is also usable on touch screens. This test can be made highly secure, lightweight, multilingual, and universal and does not suffer from any usability issues. The test can be coded for English or any other regional language, therefore, overcoming the language barrier of the CAPTCHA challenges. Experimentation with the proposed model implementation of the CAPTCHA challenge has proved its robustness, diversity, lightweight, and usability.

**Keywords** CAPTCHA · HIP · Protected resources · Website security · Regional CAPTCHA

## 1 Introduction

With the vast and rapid growth of the Internet and its expanding services, the requirement for improving the security of its offered services and the resources has increased multifold. Various security algorithms which provide robust security to various resources which a Web service provides to the user have been developed, however, as efforts were made to make the web service secure, equal efforts were made to make such type of techniques which can breach their security and hack the data or other types of resources which these websites or Web services are handling. Many

A. R. Shah · M. T. Banday (✉) · S. A. Sheikh
PG Department of Electronics and Instrumentation Technology, University of Kashmir, Srinagar, Jammu and Kashmir 190006, India

© Springer Nature Singapore Pte Ltd. 2021
X.-Z. Gao et al. (eds.), *Advances in Computational Intelligence and Communication Technology*, Advances in Intelligent Systems and Computing 1086,
https://doi.org/10.1007/978-981-15-1275-9_31

techniques were developed to interact with Web services to gain access to protected resources and services through automated programs which otherwise require human interaction, thus, making the Web resource vulnerable to various types of attacks such as data leak/theft, network blocking, or server storage blockage. To mitigate such type of attacks, a test called CAPTCHA challenge was designed by Luis Von Ahn in the year 2000 with Manuel Blum, Nicholas Hopper, and John Langford [1]. A CAPTCHA challenge is a test in which a computer program is made to test if the user of the Web service is a human or a Web bot. The challenge has been designed in such a way that it can only be solvable by a human and not by an automated computer program which pretends to be a human. The user is presented with a challenge to prove that it is a human which is interacting with the Web resource. Once the CAPTCHA server receives the attempted data from the user and if the attempt is correct, then only the user is allowed to access the protected Web resource. The main characteristics of a CAPTCHA test are:

- It should easily be solvable by a human user.
- It should be difficult or impossible for a computer/Web bot to solve.
- It should not impose much burden on the Web resource to render and incorporate the CAPTCHA test.
- It should be easy to generate.

Although the design of existing CAPTCHA techniques considers the above characteristics, however, the main problem which most of the modern CAPTCHA challenges is the lack of desired user-friendliness and universality. Various devised CAPTCHA techniques based on distorted image character recognition are hard for humans to solve. CAPTCHA tests like Fig. 1 shows an example of distorted image character CAPTCHA challenge.

Along with user-friendliness, another challenging issue with the current modern CAPTCHA tests is their non-universality. A CAPTCHA test should not depend on the user language, physical location, or its regional language. Practically speaking, it is challenging to develop a CAPTCHA challenge which is entirely universal, but taking certain assumptions into consideration, a generalized CAPTCHA test can be created. Another aspect of universality is to develop a CAPTCHA test which when attempted on any electronic devices can be attempted with equal ease and understanding. Various devised CAPTCHA challenges are accessible and user-friendly to attempt on a computer or on a laptop which has a physical keyboard attached to them; however, when it comes to devices which have a virtual keyboard or have a small

Fig. 1 An example of distorted image character CAPTCHA challenge

touch screen like smartphones, smart tabs, etc., they become challenging. This paper presents a CAPTCHA technique which promises to mitigate the above-highlighted challenges of CAPTCHA tests.

## 2 Related Work

Various techniques were developed to stop Web bots or computer programs from unauthorized accessing Web resources. Since the introduction of CAPTCHA tests in the year 2000, various new CAPTCHA techniques based on text and distorted images, such as PARC and AC Berkeley [2–4], were introduced with an aim to provide improved security. Apart from these, audio CAPTCHA techniques wherein noise is added to a clear voice in such a manner that only humans can recognize the actual voice and no speech recognition tools can recognize the actual speech have also been developed [5, 6]. In the process of providing new techniques and improved security against computer bot attacks, many CAPTCHA image, text, and voice-based techniques were designed. As new secure CAPTCHA techniques are developed, and more and more efforts are being made to break/crack such techniques. Thus, a race between developing and cracking CAPTCHA challenges has already begun. Many CAPTCHA techniques considered secure previously have been broken thus leading to the development of more difficult, robust, and complex CAPTCHA. As more complex CAPTCHA challenges are developed, it often became tough for human users to pass them. On the contrary, some more straightforward and user-friendly techniques [7–10] have also been developed. However, most of them lack diversity across different electronic devices. They are more accessible and straightforward for human users to solve on one type of devices but very difficult and often inaccessible on other devices leaving them unusable for such devices. Apart from a few, most of the CAPTCHA challenges require user interactions in the English language and thus rendering it unsuitable for non-English language Web sites. The current authors in [11, 12] have proposed multilingual CAPTCHA for non-native users and suggested a framework [13] for different organizations for its effective implementation. Further, an image-based CAPTCHA technique was proposed in [14] wherein a user is presented with a CAPTCHA image composed of several sub-images. In [15], a drag and drop CAPTCHA has been proposed in which a user has to identify two small images within a composite image and has to drag a source image to target image.

This paper presents a CAPTCHA technique which besides being user-friendly is universal and can be accessed across various electronic devices with equal ease of operation.

## 3 Proposed CAPTCHA Challenges

This paper presents a CAPTCHA technique which is based on a mouse drag/touch event. In this type of mouse drag event CAPTCHA challenge, a user is presented with a choice of selecting the language in which s/he wants the CAPTCHA challenge to be presented as shown in Fig. 2, and then a randomly generated character set which belongs to the selected language is generated. The individual characters of the generated character set are presented in individual HTML divisions; the user has to identify the natural writing flow of the language in which the CAPTCHA character set is being generated and just drag the mouse in the same order over the HTML divisions. The position of HTML divisions, the generation of characters, and the order or the position of the generated characters in the challenge character string are generated randomly at the runtime. On small touch screen devices, the user has to merely touch the individual divisions in the same order as presented in the CAPTCHA challenge string. Since the user has not to type any characters, the CAPTCHA challenge is universal across various smart devices.

Figures 3, 4, and 5, respectively, show the screenshots of the devised mouse drag/touch event-based multilingual universal CAPTCHA challenge.

The advantages of mouse drag/touch event-based CAPTCHA challenge are its universality and user-friendliness. Firstly, since a user can select a regional or an international language in which s/he wants to attempt the CAPTCHA challenge, the user is not bound to have any knowledge to any specific regional or international language to attempt the challenge, thus making the CAPTCHA challenge more universal. Secondly, as the proposed technique of attempting the CAPTCHA challenge does not require the user to type any alphabets in the textbox, the mouse drag/touch event-based CAPTCHA challenge can be attempted with equal ease on any small touch screen device as it can be attempted on any other computing device having a larger screen.

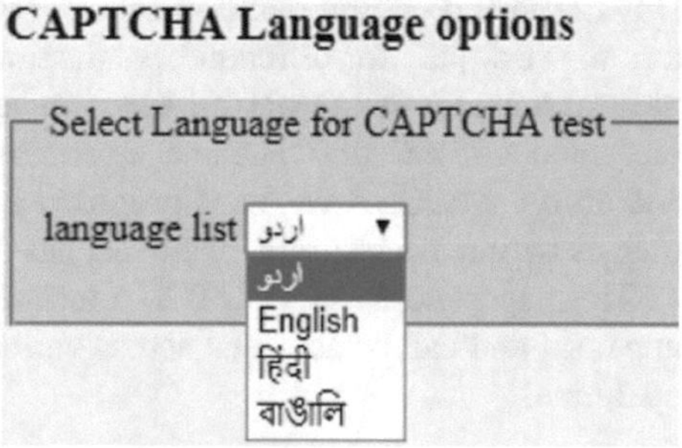

**Fig. 2** Language selection in the proposed CAPTCHA challenge

**CAPTCHA Language options**

Select Language for CAPTCHA test

language list اردو ▼

اردو
English
हिंदी
বাঙালি

**Fig. 3** A screenshot of the CAPTCHA challenge in the English language

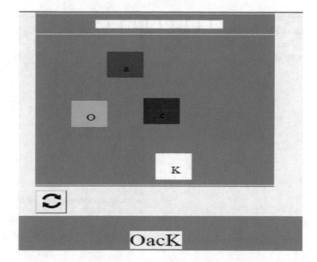

**Fig. 4** A screenshot of CAPTCHA challenge in the Urdu language along with mouse trails

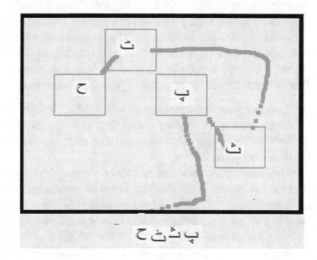

## 4   Implementation

In [1], the authors propose that a random number should be seen as a trusted public service. Various online as well as programming language resources provide support for generation of random numbers. The proposed CAPTCHA test has been developed in Java programming language and uses Java class library called Secure Random [3] for random number generation which is considered more secure than others among prominent programming languages. This class provides a cryptographically secure random number generator (RNG) with which it is tough for a bot to predict the pattern or the next random number. Various types of data can be created, saved, or maintained at runtime and can be used only once in a program or in a session. The

**Fig. 5** A screenshot of
CAPTCHA challenge in the
Hindi language along with
mouse trails

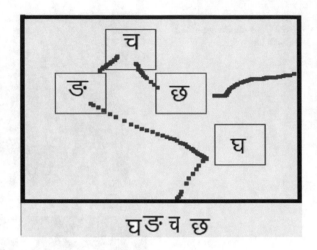

random coordinate of a division tag or a label is rendered on an HTML page. The user attempts the test by a mouse drag/touch over the HTML divisions. However, the mouse cursor events are not saved. In the proposed system, the validation, generation process of a CAPTCHA attempt is performed at the CAPTCHA server side. As such, there is limited scope for a bot to retrieve any attempted data at client side. The implementation of the proposed system is elaborated through the following steps:

- *The user selects a language in which s/he wants to attempt the CAPTCHA challenge, then a set of four characters are randomly generated/selected from the selected language character set the with the help of secure random number generator function.*
- *The generated/selected characters are then placed into separate HTML Divisions position coordinates of which are also generated randomly with the help of the same secure random number generation function with a different instance.*
- *The four generated characters are now placed into a string of fours character with random order of their position in the string.*
- *The order of character in which they are positioned in the string is stored in the database with an index of their corresponding random coordinates along with their requested server ID.*
- *The whole data of the randomly generated character, CAPTCHA challenge string along with the HTML Division position coordinates are sent to the requested server as raw data.*
- *The server on receiving the raw data renders the data on to the web page and presents it as a CAPTCHA challenge.*
- *The user attempts the CAPTCHA challenge by merely identifying the natural writing flow of the language in which the CAPTCHA string is being presented, and then the user drags the mouse cursor/touches over the HTML Division.*
- *The user data of the CAPTCHA challenge along with the data the user has submitted in the Web form is then transferred to the web server as raw data.*

- *The Web server on receiving the whole data retrieves the form data and transfer the CAPTCHA challenge data to the CAPTCHA server for validation.*
- *The CAPTCHA server on receiving the CAPTCHA attempt data retrieves the server ID from which the attempted data has come and then compares the received data with the answer data in the database with same server ID. In case, the data is validated successfully the user is identified as human. Otherwise, CAPTCHA validation fails.*

The proposed systems implementation is based on the internal working of the CAPTCHA server. The internal working of the proposed CAPTCHA server comprises of two phases namely CAPTCHA generation process and CAPTCHA validation process.

**CAPTCHA Generation Process**

Whenever a user from the client side tries to access a protected Web resource, the Web server sends a CAPTCHA challenge request to the registered CAPTCHA server. On receiving a request, the CAPTCHA server invokes the CAPTCHA generation process function. The generation function of the CAPTCHA server generates the random CAPTCHA challenge character set along with random coordinates for the divisions were the individual characters are placed on the Web form and sends the whole data to the requested Web server as raw data. Figure 6 illustrates the workflow of the CAPTCHA generation process.

**CAPTCHA Validation Process**

In this process after the user attempts the CAPTCHA challenge and submits the form data, the Web server on receiving the whole data retrieves the form data and sends the CAPTCHA attempted data to the CAPTCHA server for validation. This is illustrated in a workflow diagram in Fig. 7.

# 5   Security and Usability

**Security Analysis**

Whenever a user from the client side tries to access a protected Web resource, the Web server sends a CAPTCHA challenge. Most text-based image CAPTCHA breaking techniques accomplish their goal in three steps, viz. preprocessing, segmentation, and recognition. In the preprocessing, the image undergoes some transformations that help remove distractions, noise, backgrounds, etc. The transformations may include image binarization, image thinning, de-noising, etc. In the second step, the segmentation of different characters occurs based on various characteristics of the CAPTCHA including individual characters, character projection, connected components, character width, character features, character contour, etc. The segmentation process may be performed based on character structure and by applying a Gabor

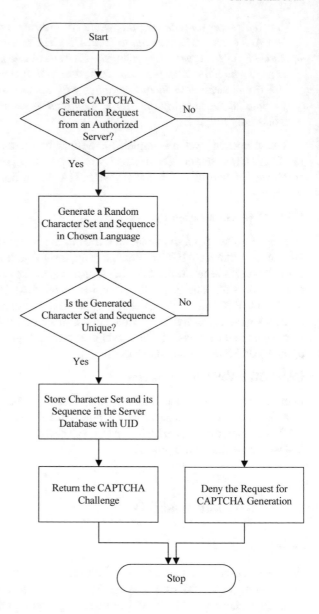

**Fig. 6** Proposed CAPTCHA generation process

filter. In the final step, the segmented characters are recognized using optical character recognition (OCR) techniques. Once the characters are recognized, the result is entered into the input field of the CAPTCHA challenge and submitted. In the CAPTCHA technique presented in this study, there is no such input field for entering the results, and the characters need not to be recognized and typed in anywhere. The CAPTCHA is verified based on the mouse pointer movements of the user. Once the user moves the pointer in a predefined pattern, the system recognizes the user as a

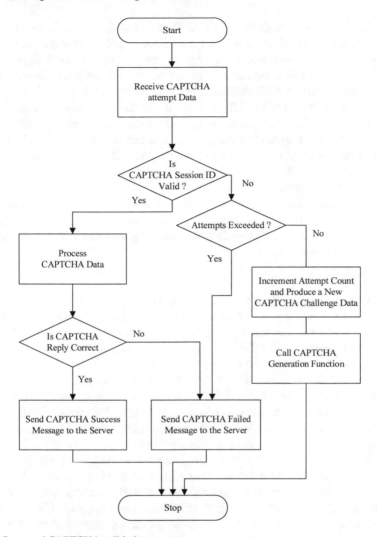

**Fig. 7** Proposed CAPTCHA validation process

legitimate human. The text-based image CAPTCHA breaking techniques are of no use against this type of CAPTCHA.

Another type of attack on CAPTCHA is farming out attack in which the bot copies the image presented in the CAPTCHA challenge and presents it to a human user somewhere on an entirely different Web site, who is usually paid for recognizing the contents of the image and solving the CAPTCHA challenges. The user input is sent back to the bot and the bot fills-in that answer into the input field of the original CAPTCHA challenge. This process is performed in real time and automatically by the attacking system. These types of attacks are prevalent and have high success rates because it involves cheap human labor to recognize the CAPTCHA challenges.

This type of attack can be used in the CAPTCHA challenge presented in this paper. The user is not required to recognize the characters and fill in somewhere, so the farming out attack cannot present an image to the human user for recognition. The CAPTCHA is required to be solved right on the website implementing it because there is a close connection between the client and the server for listening to the mouse pointer moves on the screen and the verification of the CAPTCHA challenge.

The presented CAPTCHA is also secured against brute force and hit and trial attack by limiting the number of maximum failed attempts from a given IP address. After a given threshold of failed attempts, the CAPTCHA server always returns a failure response for a chosen time duration.

In [7, 8, 10], the authors argue that for a computer bot or program, it is a significant problem to collect any data related to any mouse movements like a drag, drop, click, hover, etc., which an authentic human user performs at the client side. This characteristic of mouse usage by human users provides strength to the proposed CAPTCHA methodology.

**Usability Analysis**

The usability test was performed in three languages, i.e., English, Hindi, and Urdu. The test was performed with the help of 20 professional computer users, 10 users with intermediary computer skills and 10 non-professional computer users. The users were asked to solve the CAPTCHA challenge on the computer, tablet as well as smartphones. The results of the test are presented in Table 1.

The usability test results indicate that professional users took a minimum of 1.0 and maximum 2.1 s to solve the CAPTCHA, the intermediate users took minimum 1.9 and maximum 3.9 s to solve the CAPTCHA challenge. The non-professional users took a long time to solve the CAPTCHA which is minimum of 2.9 s and a maximum of 5.9 s. Professional users found it slightly easier to solve the CAPTCHA on a computer than handheld devices whereas intermediate and non-professional users considerably more time to solve the CAPTCHA on handheld devices than on the computer. The test also revealed that English CAPTCHA was the easiest and solved in the least time than Hindi and Urdu CAPTCHAs. Urdu language CAPTCHA was the most time-consuming CAPTCHA challenge for the test subjects. The overall average time to solve the CAPTCHA was found to be 2.72 s.

# 6 Discussions

CAPTCHA, also known as the reverse Turing test, is intended to determine whether a user trying to access a Web resource is a human and or a computer program. As various techniques were developed to save Web resources from various Web bot attacks, equal efforts were made to break such CAPTCHA security techniques. The CAPTCHA techniques which are present today do provide security from various Web bot attacks, but they lack user-friendliness and universality. This paper proposes a novel mouse drag and touch event-based CAPTCHA challenge, which a user can solve by a simple

**Table 1** Functionalities in Indian regional websites

| Language | Devices | Type of user | | | | | | | | | Average time per language per device |
|---|---|---|---|---|---|---|---|---|---|---|---|
| | | Professional (20) | | | Intermediate (10) | | | Non-professional (10) | | | |
| | | $T\_min$ | $T\_max$ | $T\_Avg$ | $T\_min$ | $T\_max$ | $T\_Avg$ | $T\_min$ | $T\_max$ | $T\_Avg$ | |
| English | Computer | 1.20 | 1.90 | 1.55 | 1.90 | 2.50 | 2.20 | 2.90 | 3.80 | 3.35 | 2.37 |
| | Tablet | 1.30 | 1.60 | 1.45 | 2.00 | 2.90 | 2.45 | 3.00 | 3.90 | 3.45 | 2.45 |
| | Smartphone | 1.00 | 1.30 | 1.15 | 1.90 | 2.90 | 2.40 | 2.90 | 3.90 | 3.40 | 2.32 |
| Hindi | Computer | 1.00 | 1.30 | 1.15 | 2.00 | 2.90 | 2.45 | 2.90 | 3.90 | 3.40 | 2.33 |
| | Tablet | 1.20 | 1.70 | 1.45 | 2.90 | 3.10 | 3.00 | 3.80 | 4.90 | 4.35 | 2.93 |
| | Smartphone | 2.10 | 1.30 | 1.70 | 3.00 | 3.60 | 3.30 | 3.90 | 5.00 | 4.45 | 3.15 |
| Urdu | Computer | 1.10 | 1.70 | 1.40 | 2.10 | 2.20 | 2.15 | 3.90 | 4.80 | 4.35 | 2.63 |
| | Tablet | 1.20 | 1.30 | 1.25 | 2.20 | 2.90 | 2.55 | 4.10 | 4.90 | 4.50 | 2.77 |
| | Smartphone | 1.50 | 1.90 | 1.70 | 3.00 | 3.90 | 3.45 | 5.10 | 5.90 | 5.50 | 3.55 |
| Average time/user type | | 1.29 | 1.56 | 1.42 | 2.33 | 2.99 | 2.66 | 3.61 | 4.56 | 4.08 | 2.72 |

mouse drag or by a touch gesture on a touch screen. The working of the CAPTCHA has been illustrated and explained in detail, and a model implementation has been verified which can further be improved regarding presentation, usability, and security. The paper also presents the security features of the proposed CAPTCHA challenge and how it is safe from the recent CAPTCHA breaking attacks. A comprehensive usability study of the CAPTCHA has also been provided demonstrating its ease of use in multiple languages across multiple devices.

Solving the CAPTCHA proposed in this paper needs human intervention using mouse movements which can make it difficult for users who might not be able to make use of a mouse or a touch screen. Solving the CAPTCHA using keyboard alone is not possible which can limit its usability in those areas—audio support for visually impaired users' needs to be included in the proposed CAPTCHA. The CAPTCHA requires the use of client-side scripting languages like JavaScript for properly handling mouse movement and related events. Disabling JavaScript in the browser might render the CAPTCHA non-functional, and thus users will not be able to access the protected content. Conventional text-based image CAPTCHAs do not usually require client-side scripting support and can usually be solved with the keyboard alone. The CAPTCHA needs to work in all kinds of browsers, and cross-browser testing is required. Browser updates might occasionally require upgrading and retesting of the CAPTCHA because browsers sometimes come with significant changes in their architecture which affect Web application rendering and functionality. In the proposed CAPTCHA, the server generates a random string and locations of the individual characters and sends the characters and locations to the client in the form of HTML. The client/browser renders the CAPTCHA challenge. When the user moves the mouse over the characters, the coordinates of the characters are noted and sent to the server for comparison with the location sequence from its database. If the sequence matches, it considers the answer as correct. Some bot can be programmed to obtain the locations of the characters from the HTML and send them in the sequence of the string to the server without the need of any OCR recognition and try to break the challenger. This issue can be reduced by replacing the characters at the bottom of the CAPTCHA challenge with a regular distorted text-based image and by replacing the HTML divisions with images. The proposed CAPTCHA verifies the user input by comparing the sequence of the coordinates of characters based on user mouse movements with a coordinate sequence of CAPTCHA as per database entry on the server. The CAPTCHA can be further improved to make it more robust and reliable by recording the entire mouse movement trail inside the CAPTCHA challenge and sending it to the server which can analyze the mouse trail to determine if the trail was left by a human using a mouse or generated automatically by a bot.

The presented limitations of the model implementation of the proposed CAPTCHA challenge can be overcome by improving the presented CAPTCHA as explained in this section. In the proposed CAPTCHA technique, the user does not require any knowledge of any regional or international language to attempt the CAPTCHA challenge successfully. Instead, the user can quickly solve the CAPTCHA across various electronic devices with equal ease and comfort, without compromising the security that a CAPTCHA test is expected to have.

**Acknowledgements** This research work has been supported by Science and Engineering Research Board (SERB), Department of Science and Technology (DST), Government of India under its file no. EMR/2016/006987.

# References

1. L. von Ahn, M. Blum, N.J. Hopper, J. Langford, *CAPTCHA: Using Hard AI Problems for Security*. Computer Science Dept., Carnegie Mellon University, Pittsburgh PA 15213, USA. IBM T.J. WRC, Yorktown Heights NY 10598, USA
2. H.S. Baird, K. Popat, Human interactive proofs and document image analysis, in *Proceeding of the 5th IAPR Workshop Document Analysis Systems*, Princeton, NJ (2002)
3. A. Coates, H. Baird, R. Fateman, Pessimal print: a Reverse Turing test, in *Proceeding of the IAPR 6th Int'l Conference on Document Analysis and Recognition,* Seattle, WA, 2001, pp. 1154–1158
4. M.T. Banday, N.A. Shah, A study of CAPTCHAs for securing web services. IJSDIA Int. J. Secure Digital Inf. Age **1**(2), 66–74 (2009). ISSN: 0975-1823
5. CAPTCHA website, Carnegie Mellon University, http://www.captcha.net
6. G. Kochanski, et al., A reverse turing test using speech, in *Proceedings of the Seventh International Conference on Spoken Language Processing (ICSLP2002 -INTERSPEECH 2002)*, Denver, Colorado, USA, 16–20 September 2002, pp. 1357–1360
7. D CAPTCHA—A Next Generation of the CAPTCHA Advanced Virtual and Intelligent Computing (AVIC) Research Center Department of Mathematics, Faculty of Science, Chulalongkorn University Patumwan, Bangkok, Thailand
8. A. Desai, P. Patadia, *Drag and Drop: A Better Approach to CAPTCHA*. Sardar Patel University New VidhyaNagar, Gujarat, India
9. R. Gossweiler, M. Kamvar, S. Baluja, What's Up CAPTCHA? A CAPTCHA based on image orientation, in *International World Wide Web Conference Paper at WWW 2009*
10. V.A. Thomas, K. Kaur, cursor CAPTCHA—implementing CAPTCHA using mouse cursor. IEEE Explore 3 Oct 2013
11. M.T. Banday, S.A. Sheikh, Design of secure multilingual CAPTCHA script. Int. J. Web Publishing IGI Global **7**(4), 1–27 (2015)
12. M.T. Banday, S.A. Sheikh, Design of CAPTCHA script for indian regional websites, in *Communications in Computer and Information Science*, ed. by S.M. Thampi, P.K. Atrey, C.I. Fan, G. Martinez Perez, vol. 377, pp 98–109, Print ISBN: 978-3-642-40575-4, Published by Springer Berlin Heidelberg. https://doi.org/10.1007/978-3-642-40576-1_11
13. M.T. Banday, S.A. Sheikh, Service framework for dynamic multilingual CAPTCHA challenges: INCAPTCHA, in *2014 International Conference on Advances in Electronics, Computers, and Communications (ICAECC-2014)*, Reva Institute of Technology and Management, Bangalore, India, published by IEEE, 10–11 October 2014. ISBN: 978147995497
14. M.T. Banday, N.A. Shah, Image flip CAPTCHA. ISeCure, The ISC Int. J. Inf. Secur. Iranian Society of Cryptology, Tehran, Iran **1**(2), 103–121 (2009). ISSN 2008-2045 and 2008-3076
15. N.A. Shah, M.T. Banday, Drag and drop image CAPTCHA. All Sprouts Content. 250 (2010). https://aisel.aisnet.org/sprouts_all/250

# Using Gaussian Mixtures on Triphone Acoustic Modelling-Based Punjabi Continuous Speech Recognition

Wiqas Ghai, Suresh Kumar and Vijay Anant Athavale

**Abstract** Continuous speech recognition for a particular language is always an area which relies, for its performance, on these major aspects: acoustic modelling and language modelling. Gaussian mixture model-hidden Markov model (GMM–HMM) is a part of acoustic modelling. These components are applied at the back end of ASR design to accurately and efficiently convert continuous speech signal to corresponding text. Triphone-based acoustic modelling makes use of two different context-dependent triphone models: word-internal and cross-word models. In spite of active research in the field of automatic speech recognition for a number of Indian and foreign languages, only few attempts have been made for Punjabi language, specially, in the area of continuous speech recognition. This research paper is aimed at analysing the impact of GMM–HMM-based acoustic model on the Punjabi speaker-independent continuous speech recognition. Recognition accuracy has been determined at word and sentence levels, respectively, with PLP and MFCC features by varying Gaussian mixtures from 2 to 32.

**Keywords** Acoustic model · Language model · Perplexity · n-grams · Probability estimate · HTK · Bigrams · Cross-word · Biphones · Triphones · Gaussian · Trigrams · GMM · HMM

W. Ghai (✉)
RIMT University, Punjab, India
e-mail: ghaialpha@gmail.com

S. Kumar
Sanskriti University, Mathura, UP, India
e-mail: sureshkaswan@gmail.com

V. A. Athavale
PIET, Panipat, Haryana, India
e-mail: vijay.athavale@gmail.com

© Springer Nature Singapore Pte Ltd. 2021                     395
X.-Z. Gao et al. (eds.), *Advances in Computational Intelligence and Communication Technology*, Advances in Intelligent Systems and Computing 1086,
https://doi.org/10.1007/978-981-15-1275-9_32

# 1   Introduction

Research in the area of automatic speech recognition (ASR) for Punjabi language is a multidisciplinary effort which covers the phonetics of Punjabi language [1, 2], speech processing techniques and pattern recognition. Figure 1 provides the architecture of ASR which is clearly describing the importance of three components: acoustic model, lexicon and language model [3–7] in the functioning of the decoder. Continuous speech [8, 9] differs from other mode of speech in many ways: Firstly, it contains words connected together but not separated by any pause. Secondly, it is less clearly articulated. Thirdly, phonetic context and co-articulation always have a great impact on speech from within the words and across the word boundaries. In this paper, we have explored language modelling for Punjabi language and impact of GMM–HMM [10–13]-based acoustic model on Punjabi speaker-independent continuous speech recognition [14]. A comparative study of CSR performance has also been made on Sinhala and Punjabi languages.

## 1.1   Triphone Modelling

In triphone-based modelling approach [3, 15], every phone is associated with its left and right neighbourhoods. As a result, state parameters for the same phone become different due to the presence of different neighbouring phones around it. Secondly, triphone-based modelling takes care of the context information. Thirdly, the number of triphones is more than the number of phones involved in the training data. There are two types of context-dependent triphone models which are dealt in this work:

- *Word-internal models*: In case of word-internal triphones, word boundaries in the training transcriptions are considered and marked. A word-internal triphone model makes use of biphones for word boundaries and triphones for word-internal triples. For a word ਪੜ੍ਹਦਾ:

IPA          pəɽd̪ɑː
English      PARHDAA

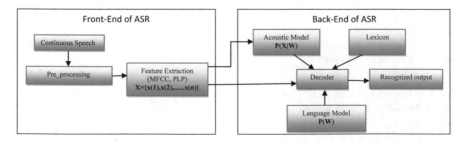

**Fig. 1**  ASR architecture

Phones　　　P a rh d aa
Triphones　　sil(p), p(sil, a), a(p, rh), rh(a, d), d(rh, aa), aa(d, sil), sil(aa)

In case of word-internal triphones, the phones at word endings are found to be context-independent, and language HMM complexity is same as that when all phones are context-independent.

- Cross-word models: A cross-word triphone captures co-articulation effects across word boundaries and as a result becomes important for continuous speech phenomenon. For constructing cross-word triphone models, word boundaries in the training data are to be ignored. Triphones are used at word boundaries and as a result, HMMs used to compose the word, become dependent on adjacent words. For a word sequence ਪੜ੍ਹਦਾ:

IPA　　　　　pəɽd̪ɑː
English　　　PARHDAA
Phones　　　p a rh d aa
Triphones　　a(p, rh), rh(a, d), d(rh, aa), aa(d, s), s(aa, ii)

Word-internal models are found to be comparatively less accurate for continuous speech modelling, but the models formed in this modelling are small in number. Therefore, word-internal models can be more accurately trained on any given set of data. In our work, both these types of triphone models have been dealt with.

## 1.2　Gaussian Mixtures HMM

A Gaussian mixture model (GMM) is defined as parametric probability density function. Gaussian mixtures [10, 12, 13] are used to model a complex speech signal as emission probability distribution functions in HMM. The number of Gaussian mixtures is one of the key issues among the representation of phonetic information and HMM topology which are to be dealt for the implementation of statistical techniques such as continuous density HMM [11, 16] and multivariate Gaussian mixtures. For a given observation vector $X_m$, output likelihood $P(X_m|S)$ for a HMM state $S$ is computed as weighted sum of probabilities $P(X_m)$ as:

$$\sum_{i=1}^{C} w_i \cdot P_i(X_m) \tag{1}$$

where $P_i(X_m)$ is $n$-dimensional multivariate Gaussian probability distribution function. State parameters are:

- Number of mixtures $C$
- Variance–covariance matrix of $i$th Gaussian mixture
- Weighing factor $w_i$
- Mean vector

Among number of techniques for estimating the parameters of GMM, maximum likelihood estimation (MLE) is most popular and effective. Parameters estimated by MLE are effective in maximizing the likelihood of GMM given the training data.

## 2 Language Model for Punjabi Language

A language model, as shown in Fig. 2, is designed to help an automatic speech recognizer to determine a probable word sequence-independent of the acoustics. A language model assists the speech recognizer to make the right decision about recognition when two different sentences sound the same. A language model is comprised of two components: vocabulary and grammar. Vocabulary provides set of words to be recognized, whereas grammar provides a set of rules for regulating the order in which the words may be arranged to form sentences. Grammar may be a stochastic model or may be comprised of formal linguistic rules. Stochastic model approach is better than linguistic model approach for speech recognition task because of two disadvantages in linguistic model approach: computationally demanding nature and problem of not allowing the appearance of grammatically incorrect sentences. Stochastic models have robustness and simplicity since they are based on probabilities for sequence of words. These features of stochastic models approach can be achieved only in the presence of large amount of training data. Mathematically, a statistical language model [5] estimates the probability $P(w_1, w_2, \ldots, w_n)$ of a word sequence. It is defined as follows:

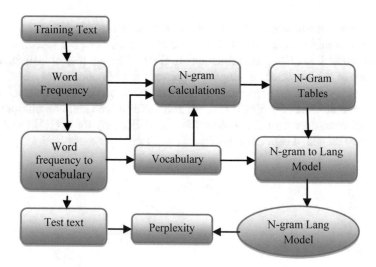

**Fig. 2** LM architecture

$$P(w_1, w_2, \ldots, w_n) = \prod_{i=1}^{n} P(w_i | w_1, w_2, \ldots, w_{i-1}) \qquad (2)$$

Here $w_1^n = (w_1, w_2, \ldots, w_n)$ is a sequence of words which forms a sentence. $P(w_i | w_1, w_2, \ldots, w_{i-1})$ denote conditional probability of producing word $w_i$ given the history $w_1, w_2, \ldots, w_{i-1}$. A language model, being an information source, utilizes all possible features of a language in predicting words and gives rise to a parameter per-word entropy $H$. Estimated per-word entropy of a language model determines the average difficulty or uncertainty experienced by an automatic speech recognition system in searching a word from the same source. Perplexity [3] is a measure related to $H$ which is used to determine the performance of a language model. In other words, perplexity $P$ measures the goodness of a language model by testing n-gram language model [4, 5, 7, 17–20] on test text as shown in the Fig. 2. Its relation with estimated per-word entropy $\hat{H}$ is defined as

$$P = 2^{\hat{H}} \qquad (3)$$

where

$$\hat{H} = -\frac{1}{n} \log_2 P(w_1, w_2, \ldots, w_n) \qquad (4)$$

Perplexity also acts as a measure of "how many different equally most probable words on an average follow a given word". Lesser the perplexity, better the language model is. Accordingly, such language model always models the language better. Both perplexity and entropy do not take into the acoustic difficulty of recognizing a word. Sometimes a language model, which differentiates acoustically similar words or sentences, may end up in lower word error rate in comparison to a model having lower perplexity. In addition to perplexity, variance and out of vocabulary (OOV) rate are also computed to evaluate the performance of designed language model. For Punjabi language CSR, a text corpus of 525 K Punjabi words has been used for training, and a test corpus of 40 K unseen text has been used as a testing data. This text corpus has been collected from online newspaper sites.

Conditioning of text corpus has been given due consideration by removing punctuation marks and replacing numbers in figures with numbers in words. While building a standard language model, text corpus has been gradually increased with a step size of 75 K. The number of bigrams and trigrams which has been generated at each step, as shown in Graph 1, clearly depicts the almost increase in the number of bigrams, whereas the increase in number of trigrams has been more pronounced from 300 K corpus. As shown in Table 1, perplexity has increased initially till 225 K and thereafter there is a decrease in perplexity in the range of 300–450 K which can be attributed to the steep rise in the number of trigrams shown in Graph 1.

**Graph 1** N-grams versus corpus size

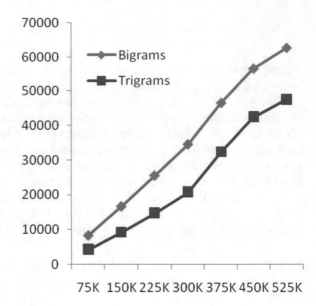

**Table 1** Evaluation of language model

| Corpus size (K) | Perplexity | OOV rate |
|---|---|---|
| 75 | 438.45 | 18.32 |
| 150 | 550.58 | 12.45 |
| 225 | 581.27 | 9.53 |
| 300 | 568.66 | 7.68 |
| 375 | 567.22 | 7.25 |
| 450 | 526.87 | 6.23 |
| 525 | 565.25 | 4.79 |

## 3  Performance Analysis

A data set of 408 different sentences of Punjabi language containing 6–7 words on an average has been created in-house. A speech corpus of 803 wav files has been created from a total 16 speakers (9 males and 7 females) using Audacity 2.0.2 at 16-bit sample format, 44,100 Hz sample rate and mono channel. HTK 3.4.1 [21] has been used to design this work. ASR has been trained for 891 Punjabi words. CSR results have been obtained from monophone-based acoustic model and triphone-based acoustic model. Basic difference between monophones and triphones is that the latter carries context information in them. Monophones behave as context-independent sub-word units. But context information available in Triphones always proves to be beneficial for continuous speech which has greater frequencies of successor and predecessor words through predefined contexts. In this paper, performance of triphone-based CSR has been compared with monophone-based CSR to depict the benefit of context information in CSR.

## 3.1 Word Recognition Accuracy

Using PLP features,

- Monophone-based acoustic model has provided word recognition accuracy of 91.23%, whereas triphone-based acoustic model has provided word recognition accuracy 97.66%.
- With the increase in number of Gaussian mixtures from 2 to 32, the range of word recognition accuracy is 9.49% for monophone-based acoustic model, whereas it is 3.47% for triphone-based acoustic model. This shows that word recognition accuracy has remained uniform for triphone-based acoustic model (Graph 2, 3).

Using MFCC features,

- Monophone-based acoustic model has provided word recognition accuracy of 88.36%, whereas triphone-based acoustic model has provided word recognition accuracy 97.90%.
- With the increase in number of Gaussian mixtures from 2 to 32, the range of word recognition accuracy is 18.79% for monophone-based acoustic model, whereas it

**Graph 2** Word recognition accuracy with PLP features for (a) monophones and (b) triphones

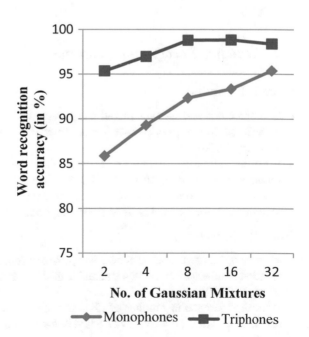

**Graph 3** Word recognition accuracy with MFCC features for (a) monophones and (b) triphones

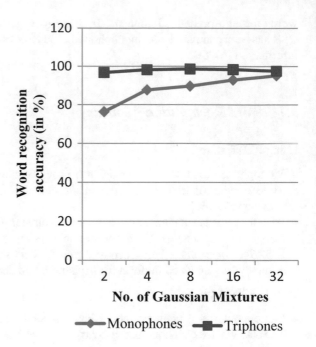

is 1.82% for triphone-based acoustic model. This shows that word recognition has remained uniform for triphone-based acoustic model.

## 3.2 Sentence Recognition Accuracy

Using PLP features,

- Monophone-based acoustic model has provided sentence recognition accuracy of 60.89%, whereas triphone-based acoustic model has provided sentence recognition accuracy 79.56%.
- With the increase in number of Gaussian mixtures from 2 to 32, the range of sentence recognition accuracy is 11.11% for monophone-based acoustic model, whereas it is 26.67% for triphone-based acoustic model. This shows that Sentence recognition has remained uniform for monophone-based acoustic model (Graph 4).

Using MFCC features,

- Monophone-based acoustic model has provided sentence recognition accuracy of 52.89%, whereas triphone-based acoustic model has provided sentence recognition accuracy 81.78%.
- With the increase in number of Gaussian mixtures from 2 to 32, the range of sentence recognition accuracy is 53.34% for monophone-based acoustic model,

**Graph 4** Sentence recognition accuracy with PLP features for (a) monophones and (b) triphones

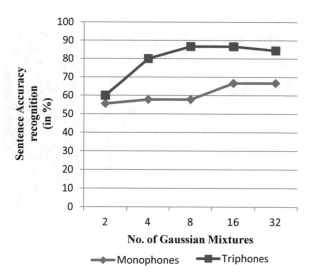

whereas it is 17.78% for triphone-based acoustic model. This shows that sentence recognition has remained uniform for triphone-based acoustic model (Graph 5).

**Graph 5** Sentence recognition accuracy with MFCC features for (a) monophones and (b) triphones

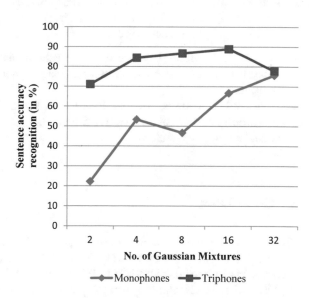

**Graph 6** Comparison of
ASR performance for
Sinhala and Punjabi
languages

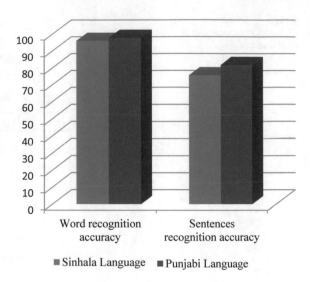

# 4 Comparison with Sinhala Language CSR

Reason behind this comparison is that both Punjabi [22] and Sinhala [23] are members
of same Indo-Aryan family of languages. The comparison of the CSR designed for
these two languages has been discussed below:

- Using MFCC features, triphone-based acoustic modelling has given word recognition accuracy of 96.14% for Sinhala language, whereas it is 97.90% for Punjabi
  language.
- Using MFCC features, triphone-based acoustic modelling has given sentence
  recognition accuracy of 75.74% for Sinhala language, whereas it is 81.78% for
  Punjabi language (Graph 6).

# 5 Conclusion

This paper has explored the impact of Gaussian mixtures on triphone-based acoustic model with two different types of features: MFCC and PLP. Word recognition
accuracy has improved from 91.23 to 97.66% with PLP features and from 88.36 to
97.90% with MFCC features by using triphone-based acoustic model. It is found that
word recognition accuracy has been better with MFCC for lower Gaussian mixtures
and better with PLP features for higher Gaussian mixtures. Sentence recognition
accuracy has improved from 60.89 to 79.56% with PLP features and from 52.89
to 81.78% with MFCC features by using triphone-based acoustic model. Sentence

recognition accuracy has been achieved better with MFCC features till 16 Gaussian mixtures. However, it has been found better with PLP features for 32 Gaussian mixtures.

**Acknowledgements** Our study aimed at investigating the impact of Gaussian mixtures on triphone-based acoustic model with two different types of features: MFCC and PLP. In spite of active research in the field of automatic speech recognition for number of Indian and foreign languages, only few attempts have been made for Punjabi language, specially, in the area of continuous speech recognition. All participants (speakers) involved are authors of the paper and given their consent for the study done. It is not important to increase the number of speakers with reference to presented work.

# References

1. R.K. Aggarwal, M. Dave, Using Gaussian mixtures for Hindi speech recognition system. Int. J. Signal Process. Image Process. Pattern Recogn. **4**(4) (2011)
2. Audacity 2.0.0, retrieved June 15, 2012 from http://download.cnet.com/Audacity/
3. S. Lata, Challenges for design of pronunciation lexicon specification (PLS) for Punjabi language (2011). http://hnk.ffzg.hr/bibl/ltc2011/book/papers/MPLRL-4.pdf
4. HTK Book, Retrieved on Mar 18, 2012 from http://htk.eng.cam.ac.uk
5. L. Rabiner, et al., *Fundamentals of Speech Recognition* (Pearson Publishers, 2010)
6. N. Souto, et al., Building language models for continuous speech recognition systems. $L^2$ F—Spoken Language Systems Laboratory, Portugal, 2001. http://l2f.inesc-id.pt/
7. B.J. Hsu, Generalized linear interpolation of language models, in *ASRU* (2007). ISBN: 978-1-4244-1746-9/07
8. M. Sanda et al., Acoustic modelling for croatian speech recognition and synthesis. INFORMATICA **19**(2), 227–254 (2008)
9. H. Ney et al., On structuring probabilistic dependences in stochastic language modeling. Comput. Speech Lang. **8**(1), 38 (1994)
10. M.N. Stuttle, *A Gaussian Mixture Model Spectral Representation for Speech Recognition* (University Engineering Department, Hughes Hall and Cambridge, 2003)
11. W. Ghai, N. Singh, Continuous speech recognition for Punjabi language. Int. J. Comput. Appl. **72**(14), 422–431 (2013)
12. S. Sinha, et al., Continuous density hidden markov model for hindi speech recognition. GSTF Int. J. Comput. (JoC), **3**(2) (2013). https://doi.org/10.7603/s40601-013-0015-z
13. M. Vyas, A gaussian mixture model based speech recognition system using MATLAB. Signal Image Process. Int. J. **4**(4) (2013)
14. G.S. Sharma et al., Development of application specific continuous speech recognition system in Hindi. J. Sign. Inf. Process. **3**, 394–401 (2012)
15. M. Dua et al., Punjabi automatic speech recognition using HTK. Int. J. Comput. Sci. Issues (IJCSI) **9**(4), 359 (2012)
16. V. Kadyan et al., Refinement of HMM model parameters for Punjabi automatic speech recognition (PASR) system. IETE J. Res. **64**(5), 673–688 (2018)
17. S. Saraswathi, T.V. Geetha, Building language models for tamil speech recognition system. Springer **3285**, 161–168 (2004)
18. J.B. Graber, Language models. March 2011, Creative Commons Attribution-non Commercial-share Alike 3.0 United States. http://creativecommons.org/licenses/by-nc-sa/3.0/us/
19. E.W.D. Whittaker, Statistical language modelling for automatic speech recognition of Russian & English, Thesis, Trinity College, University of Cambridge, 1998

20. T.R. Niesler, P.C. Woodland, A variable-length category-based n-gram language model, in *Proceedings of the IEEE International Conference on Acoustics, Speech and Signal Processing* (Atlanta, USA, 1996)
21. HTK-3.4.1, retrieved July 7, 2012 from http://htk.eng.cam.ac.uk
22. P.P. Singh, *Sidhantak Bhasha Vigiyaan* (Madaan Publication, Patiala, 2010)
23. R. Weerasinghe, T. Nadungodage, Continuous Sinhala speech recognition, in Conference on Human Language Technology for Development (Alexandria, Egypt, 2011), 2–5

# Domain of Competency of Classifiers on Overlapping Complexity of Datasets Using Multi-label Classification with Meta-Learning

Shivani Gupta and Atul Gupta

**Abstract** A classifier's performance can be greatly influenced by the characteristics of the underlying dataset. We aim at investigating the connection between the overlapping complexity of dataset and the performance of a classifier in order to understand the domain of competence of these machine learning classifiers. In this paper, we report the results and implications of a study investigating the connection between four overlapping measures and the performance of three classifiers, namely KNN, C4.5 and SVM. In this study, we first evaluated the performance of the three classifiers over 1060 binary classification datasets. Next, we constructed a multi-label classification dataset by computing the four overlapping measures as features and multi-labeled with the competent classifiers over these 1060 binary classification datasets. The generated multi-label classification dataset is then used to estimate the domain of the competence of the three classifiers with respect to the overlapping complexity. This allowed us to express the domain of competence of these classifiers as a set of rules obtained through multi-label rule learning. We found classifiers' performance invariably degraded with the datasets having high values of complexity measures (N1 and N3). This suggested for the existence of a strong negative correlation between the classifiers' performance and class overlapping present in the data.

**Keywords** Multi-label classification · Multi-class classification · Class overlapping · Meta-learning

S. Gupta (✉)
Manipal University, Jaipur, India
e-mail: shivani.gupta@jaipur.manipal.edu

A. Gupta
Indian Institute of Information Technology, Design and Manufacturing, Jabalpur, India

© Springer Nature Singapore Pte Ltd. 2021                                              407
X.-Z. Gao et al. (eds.), *Advances in Computational Intelligence and Communication Technology*, Advances in Intelligent Systems and Computing 1086,
https://doi.org/10.1007/978-981-15-1275-9_33

# 1    Introduction

The reported results of prediction in machine learning make it impossible to know
in advance whether a given learning algorithm, the performance of classifiers will
improve or not [1]. When we focus on a particular machine learning task like classification, a significant number of learning algorithms and their improvisations are
continuously being offered, from approaches suggesting suitability of different classifiers [2], to handle imbalanced data [3], to handle the curse of dimensionality [4,
5], to refine the combination of classifiers [6] and to apply different techniques to
minimize the effects of the presence of noise [7].

   We aim to find out the domain of competence of a given classifier by connecting
its performance with the overlapping characteristics of datasets. We conjecture that
one of the main reasons for the observed variability of a classifier performance can
be the nature of the underlying dataset. And if we can characterize this very nature of
the dataset along with the observed classifier performance, we may get some useful
clues suggesting the domain of competence of that classifier.

   Accordingly, we carried out an investigation assessing the performance of various
classifiers over multiple datasets characterized by the eleven complexity measures
of data using meta-learning . In this study, we have used a set of 1060 binary classification datasets in a multi-label classification framework labeled with the observed
performance of three classifiers, namely KNN, C4.5 and SVM on the classification
datasets. The steps of the study are as follows: We first computed the performance of
the three classifiers over the binary classification datasets as well as also computed the
eleven complexity metrics for each of the 1060 binary classification datasets. These
two kinds of information are then combined to construct a multi-labeled dataset for
the intended meta-learning. Then, we extracted rules from the multi-labeled dataset
by applying multi-label rule mining [8]. These rules enable us to analyze the behavior
of the learning algorithms with respect to the characteristics of the dataset.

   The rest of this paper is organized as follows. The next section provides some
background information which includes the data complexity measures, multi-label
classification and multi-label rule mining. Section 3 describes all the details of the
experimental framework including the approach used, data collection and results of
the study. Section 4 draws implications as well as threats to validity of the findings.
Section 5 summarizes the research findings in this area and puts our results in that
perspective. Finally, Sect. 6 draws conclusions and points to interesting extensions
of this work for the future.

# 2    Related Work

It is very interesting to identify the characteristics of classifiers the help researchers
to identify the domain of competency of classifiers. With the increasing popularity

of machine learning techniques, studies started to receive attention and have become more popular since the work of [9].

Recently, idea matures, where to characterizing the complexity of classification problems is presented by a selection of several measures, along with an empirical study on the real-world problems, indicating that it is possible to find learnable structures with the geometrical measures presented [10]. These measures indicate the separability of classes, overlap of individual attribute values and geometry, topology and density of manifolds [11, 12].

Recently, multi-label classifier required by modern application such as semantic annotation of images [13] and video [14] and directed marketing [15].

## 3   Background

In this section, we provide some background information on the various elements used in this study such as the data complexity measures, multi-label classification and multi-label rule mining. As already mentioned, the performance of the learning algorithm is strongly dependent on data characteristics.

### 3.1   Complexity Measures for Classification Problems

Recently, nowadays to extract the characteristics of datasets, data complexity measures have received increasing consideration for classification [9].

Ho and Basu [4] define some complexity measures for two classes, and we have also shown in Fig. 1.

### 3.2   Multi-label Classification

Multi-label classification deals with problems where an instance may belong to more than one class label [16]. The two popular techniques of multi-label classification are problem transformation and algorithm adaptation methods. Problem transformation methods transform the multi-label problem into one or more single-label problems. After the transformation, single-label classifiers are employed, and their single-label predictions turn into a multi-label prediction [17]. Algorithm adaptation methods extend particular single-label classification algorithms to handle multi-label data directly.

**Fig. 1** Data complexity measures

## 3.3 Multi-label Rule Mining

The discovery of rules from data is an important task in data mining [18]. In recent research, the classification problem is shifting the motivation from descriptive to be predictive by applying rule mining algorithms [19]. Due to this the whole classifier algorithm is remodeled in which rule filtering is combined with the learning phase. The main role of the rule mining algorithm is not only to find the class value of consequents (CARs) but also derived new steps like filtering, rule sorting and test data classification.

### 3.3.1 MMAC

MMAC uses the associative classification in which it extracts the rule sets by applying the association rule mining in classification dataset [20]. First, it generates the rule sets using association rule mining and filter out those instances that are associated with the rules. In a second step, it repeatedly applies the above procedure the rest of the remaining instance until no frequent set left. The rules are ranked according to the support of the rule. The rule belongs to the same precondition, but different labels are merged into a single rule [20] .

### 3.3.2 CBA

The CBA algorithm is proposed by Liu et al. [21] that integrates association rule mining with classification. CBA follow three main steps:

**First step**     Discretization of attributes.
**Second step**     Discovery of frequent itemsets and rule by using apriori algorithm.
**Third step**     Identification of a subset of rules.

The discovery of frequent itemsets in CBA is the most time consuming, and it requires a large number of resources because it requires a number of iteration to generate frequent itemset from the training set.

# 4 Problem Definition

There are a number of property of data that make prediction more difficult to learn the classifier model. Recently in machine learning, researchers found that the class overlapping highly degrades the performance of classifiers [12].

A class overlapping region in $\Omega \in R_n$, an n-dimensional attribute space $R_n$, can be described as a region where at least different classes $C_a$ and $C_b$ simultaneously exhibit probability > than zero, for the instances that belongs to $\Omega$.

## 4.1 Research Questions

We focus our attention on the following research questions:

- RQ 1: How multi-label classification is suitable for meta-learning?
- RQ 2: Are the overlapping characteristics of a dataset cause the classifiers to perform considerably worse or best?
- RQ 3: How multi-label rule learning applied to find the best subset of classifiers using meta-learning?

Our aim to address the problem of classifiers' applicability on a particular domain by using multi-label classification with meta-features of data complexity measures.

# 5 The Experimental Study

In this section, we present the steps carried out in the experiments in order to identify the domain of competency of classifier in the context of meta-learning.

## 5.1 Dataset Collection

In our experiments, we have included 54 datasets [22] taken from the well-known UCI repository as shown in Table 1. UCI Machine Learning Repository is the most famous publicly available data repository. It is usually the first place to go if you are looking for datasets related to machine learning repositories [23]. First, we convert 54 multi-class problems to 1060 binary class problem. A total of 1060 binary datasets are built by the combination of their different label of classes present in the datasets. Then, we use the fivefold cross-validation method that has been employed to record the accuracy of the classifier for each dataset.

## 5.2 Classifiers Used

The classification techniques of three conceptually different methods have been used in this work: the k-nearest-neighbor classifier (KNN), a support vector machine (SVM) and a decision tree (C4.5) using the default parameters for them in the open-source software project WEKA, using the Explorer interface [24].

## 5.3 Experiment Execution

This section outlines the steps involved to perform the experiment. In order to construct the meta-dataset with the overlapping measures. We extract values from different datasets for each binary datasets. The steps to execute the experiment is described as follows:

Step 1: Evaluating the algorithm's performance: • The test performance of classifiers, on each of the 1060 datasets, is obtained by run of 5 cross-validation.

Step 2: Formation of meta-data using meta-learning: • In these, we transform the problem into multi-label problem shown in Table 2. We assign class label to each dataset according classifier performance. • Multi-label problem: For multi-label problem, the class labels are binary and correspond to the three following classifiers: KNN, SVM and C4.5: A value equal to 1 for the $i$th class on the $j$th dataset indicates that the classifier on obtaining the accuracy greater than 90% for the dataset.

Step 3: Domain of competence of classifiers: • By applying multi-label rule mining, we obtain the rule set to identify the classifier subsets for different characteristics of dataset.

**Table 1** Statistics of the datasets: number of attributes (f), classes (c) instances (m) and binary data generated (n)

| Datasets | $f$ | $c$ | $m$ | $n$ | Datasets | $f$ | $c$ | $m$ | $n$ |
|---|---|---|---|---|---|---|---|---|---|
| Balance-scale | 4 | 3 | 625 | 3 | Led7digit | 7 | 10 | 500 | 43 |
| Breast-w | 9 | 2 | 699 | 1 | Diabetes | 8 | 2 | 768 | 1 |
| Ecoli | 7 | 8 | 336 | 30 | Glass | 9 | 7 | 214 | 15 |
| Hayes-roth | 4 | 4 | 160 | 1 | Segment | 19 | 7 | 2310 | 118 |
| Heart-statlog | 13 | 2 | 270 | 1 | Yeast | 8 | 10 | 1484 | 8 |
| Iris | 4 | 3 | 150 | 3 | Marketing | 40 | 3 | 5000 | 329 |
| Sonar | 60 | 2 | 208 | 1 | Spambase | 57 | 2 | 4601 | 1 |
| Zoo | 13 | 3 | 178 | 3 | Car | 13 | 3 | 178 | 3 |
| Bupa | 6 | 2 | 345 | 1 | Glass | 9 | 7 | 214 | 15 |
| Tae | 5 | 3 | 151 | 3 | Monks | 6 | 2 | 432 | 1 |
| Texture | 40 | 11 | 5500 | 154 | Flair | 40 | 3 | 5000 | 20 |
| Abalone | 8 | 28 | 4177 | 18 | Letter | 16 | 26 | 20,000 | 288 |

**Table 2** Multi-label dataset with meta-features

| Data | F1 | F2 | N3 | N1 | KNN | C4.5 | SVM |
|------|------|------|------|------|------|------|------|
| irisc01 | 31.8 | 0.01 | 1 | 0.02 | 1 | 0 | 1 |
| irisc02 | 50.9 | 0.02 | 1 | 0.02 | 1 | 0 | 1 |
| irisc12 | 4.36 | 0.02 | 0.87 | 0.12 | 0 | 0 | 1 |
| winec01 | 4.89 | 0 | 1.97 | 0.08 | 1 | 0 | 1 |
| winec02 | 20.2 | 0 | 1 | 0.01 | 0 | 0 | 1 |
| winec12 | 3.84 | 0 | 1.74 | 0.08 | 0 | 0 | 1 |
| thyroidc12 | 10.11 | 0 | 1 | 0.03 | 1 | 0 | 1 |
| . | . | . | . | . | . | . | . |

## 5.4 Experiment Results

The experimental results are presented and discussed in three subsections which address our research questions. First, we explore the multi-label classification which is suitable for meta-learning. The second subsection is devoted to the specific characteristics of a dataset that cause the classifiers to perform considerably worse or best. The last one is focused on multi-label rule learning which helps us to find the domain of competence of classifiers using meta-learning.

### 5.4.1 RQ 1: How Multi-label Classification Is Suitable for Meta-Learning?

Each instance in multi-label classification might be associated with multiple class labels which implies a real-world scenario, as opposed to a binary classification. On the other hand, taking into account the fact that the classification ability of classifiers depends on the property of the data, the complexity of each particular dataset was analyzed. The extraction of complexity measures to be used as meta-features in meta-learning frameworks. As a set of algorithms extracted by using multi-label rule mining, the meta-learning problem can be addressed from a multi-label learning (MLL) perspective. In contrast with binary classification, in MLL multiple target labels assigned to each instance.

### 5.4.2 RQ 2: Are the Overlapping Properties of a Data Affect the Degradation in Classifier Performance?

Each classifier learning may depend on the values of the overlap measures to perform learning properly. To do this, we learned an interesting pattern with each classifier which will be applied to predict the competency of classifiers. Figures 2–4 depict the accuracy results for C4.5, SVM and KNN, plotted in ascending values. It is shown in

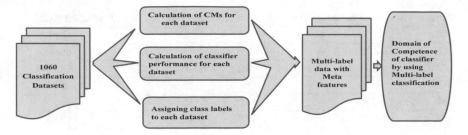

**Fig. 2** Experimental process

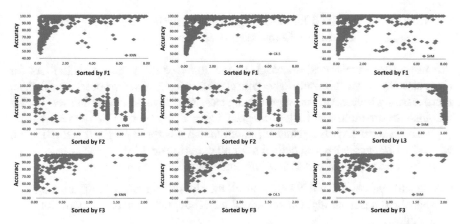

**Fig. 3** Accuracy for KNN, C4.5 and SVM sorted by F1, F2 and F3

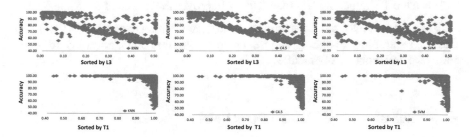

**Fig. 4** Accuracy for KNN, C4.5 and SVM L3 and T1

Fig. 3, in which the accuracy sorted by the N1 measure. The dataset with low values will yield good performance and poor performance values in terms of accuracy for high values for N1.

**Table 3** Analysis of the nature of the classifiers on chosen rule set using RIPPER and PART

| Rule | F1    | F2 | N3    | N1    | Supp | Conf  |
|------|-------|----|-------|-------|------|-------|
| R1   | ≤0.70 | –  | >0.01 | ≤0.41 | 2.00 | 65.00 |
| R2   | ≤0.70 | –  | >0.01 | >0.41 | 6.28 | 98.7  |
| R3   | >0.70 | –  | –     | ≤0.41 | 2.65 | 75.00 |
| R4   | >0.70 | –  | ≤0.01 | ≤0.41 | 4.50 | 97.50 |
| R5   | >0.70 | –  | ≤0.01 | >0.41 | 2.34 | 63.33 |

### 5.4.3   RQ 3: How Multi-label Rule Learning Applied to Find the Best Subset of Classifiers Using Meta-Learning?

We use a more real-world approach to the study of the nature of classifiers for three classifiers by using multi-label rule classification, and we propose a multi-label meta-learning to select the most promising classifiers for a particular dataset. Table 3 summarizes the multi-label rules discovered after applying the CBA and MMAC.

We have set the MinSupp and MinConf thresholds for the algorithms: CBA and MMAC to 2% and 60%, respectively, for all experiments [25].

From the analysis of these rules, we can observe the following:

- When the values N1 and N3 are low, all the classifiers perform well it means that when the dataset are highly overlapped then it degrades the performance of three classifiers.
- Increasing value of L2 improves the performance of SVM and KNN.
- When the T1 values are less than 0.99, it improves the results of C4.5 and SVM.

From these results, the dispersion of the instances within each class (N1 and N3) and the separability of classes and the complex decision boundaries(T1) are most influence performance of classifiers.

## 6   Conclusions and Future Directions

In this paper, we use the data complexity of the datasets for which the most popular classifiers provide accurate results by using multi-label classification. Multi-label classification helps us to find the set of classifiers which perform better in certain properties of datasets than to find a single best classifier.

In our study, we choose a large set of binary datasets in which we apply three classical classifiers very different in their nature. We have built descriptive rules by using multi-label rule learning, and we have obtained a final result rules which codify the domains of competence of the classifiers. From the results, we found that the increasing values of N1 and N3 measures highly affect the performance of the classifier. The results show that overlapping among classes highly affects the performance of classifiers.

The study carried out in this paper can be easily extended to different points. As future works, we plan to generate artificial datasets with different level of overlapping among classes and also add more classification algorithm to the experiments.

# References

1. D.H. Wolpert, The lack of a priori distinctions between learning algorithms. Neural Comput. **8**(7), 1341–1390 (1996)
2. J. Luengo, F. Herrera, Domains of competence of fuzzy rule based classification systems with data complexity measures: a case of study using a fuzzy hybrid genetic based machine learning method. Fuzzy Sets Syst. **161**(1), 3–19 (2010)
3. E. Ramentol et al., SMOTE-RSB*: a hybrid preprocessing approach based on oversampling and undersampling for high imbalanced data-sets using SMOTE and rough sets theory. Knowl. Inf. Syst. **33**(2), 245–265 (2012)
4. J. Derrac et al., Integrating instance selection, instance weighting, and feature weighting for nearest neighbor classifiers by coevolutionary algorithms. IEEE Trans. Syst. Man Cybern. Part B (Cybernetics) **42**(5), 1383–1397 (2012)
5. I. Vainer et al., Obtaining scalable and accurate classification in large-scale spatio-temporal domains. Knowl. Inf. Syst. **29**(3), 527–564 (2011)
6. L.I. Kuncheva, J.J. Rodruez, A weighted voting framework for classifiers ensembles. Knowl. Inf. Syst. **38**(2), 259–275 (2014)
7. J.A. Sez et al., Analyzing the presence of noise in multi-class problems: alleviating its influence with the one-vs-one decomposition. Knowl. Inf. Syst. **38**(1), 179–206 (2014)
8. F. Thabtah, P. Cowling, Y. Peng, MCAR: multi-class classification based on association rule, in *The 3rd ACS/IEEE International Conference on Computer Systems and Applications* (IEEE, 2005)
9. M. Basu, T.K. Ho (eds.) *Data Complexity in Pattern Recognition* (Springer Science and Business Media, 2006)
10. T.K. Ho, M. Basu, Complexity measures of supervised classification problems. IEEE Trans. Pattern Anal. Mach. Intell. **3**, 289–300 (2002)
11. A. Orriols-Puig, N. Macia, T.K. Ho, *Documentation for the Data Complexity Library in C++* (Universitat Ramon Llull, La Salle 196, 2010)
12. J.S. Snchez, R.A. Mollineda, J.M. Sotoca, An analysis of how training data complexity affects the nearest neighbor classifiers. Pattern Anal. Appl. **10**(3), 189–201 (2007)
13. M.-L. Zhang, Zhi-Hua Zhou, ML-KNN: a lazy learning approach to multi-label learning. Pattern Recogn. **40**(7), 2038–2048 (2007)
14. G.-J. Qi, et al. Correlative multi-label video annotation, in *Proceedings of the 15th ACM International Conference on Multimedia* (ACM, 2007)
15. Y. Zhang, S. Burer, W.N. Street, Ensemble pruning via semi-definite programming. J. Mach. Learn. Res. **7**(Jul), 1315–1338 (2006)
16. G. Tsoumakas, I. Katakis, I. Vlahavas, Mining multi-label data, in *Data Mining and Knowledge Discovery Handbook* (Springer, Boston, MA, 2009), pp. 667–685
17. S. Godbole, S. Sarawagi, Discriminative methods for multi-labeled classification, in *Pacific-Asia Conference on Knowledge Discovery and Data Mining* (Springer, Berlin, Heidelberg, 2004)
18. K. Rameshkumar, M. Sambath, S. Ravi, Relevant association rule mining from medical dataset using new irrelevant rule elimination technique, in *2013 International Conference on Information Communication and Embedded Systems (ICICES)* (IEEE, 2013)
19. B.M. Al-Maqaleh, Discovering interesting association rules: a multi-objective genetic algorithm approach. Int. J. Appl. Inf. Syst. **5**(3), 47–52 (2013)

20. F.A. Thabtah, P. Cowling, Y. Peng, MMAC: a new multi-class, multi-label associative classification approach, in *Fourth IEEE International Conference on Data Mining, 2004, ICDM'04* (IEEE, 2004)
21. Y. Ma, B. Liu, W. Hsu, Integrating classification and association rule mining, in *Proceedings of the Fourth International Conference on Knowledge Discovery and Data Mining* (1998)
22. R.S. Lynch, P.K. Willett, Classifier fusion results using various open literature data sets, in *SMC'03 Conference Proceedings. 2003 IEEE International Conference on Systems, Man and Cybernetics. Conference Theme-System Security and Assurance (Cat. No. 03CH37483)*, vol. 1 (IEEE, 2003)
23. C. Blake, *UCI Repository of Machine Learning Databases*. http://www.ics.uci.edu/~mlearn/MLRepository.html (1998)
24. I. Russell, Z. Markov, An Introduction to the Weka data mining system, in *Proceedings of the 2017 ACM SIGCSE Technical Symposium on Computer Science Education* (ACM, 2017)
25. F. Thabtah, Rules pruning in associative classification mining, in *Proceedings of the IBIMA Conference* (2005)

# Intelligent Hardware

# Comparative Analysis of FOPID and Classical Controller Performance for an Industrial MIMO Process

Sheetal Kapoor, Mayank Chaturvedi and Pradeep Kumar Juneja

**Abstract** In this paper, the fractional-order controller (FOPID) and proportional–integral–derivative (PID) controllers are designed for a selected MIMO system having a delay. FOPID controller provides two additional control parameters which may facilitate the controller to offer more stability and flexibility. The interaction of the variables is analyzed using relative gain array (RGA) and suitable loop pairing for the process under consideration recommended. The simplified decoupler is used to reduce the undesirable process interactions between the controlled and manipulated variables. These controllers are designed for the reduced SISO process achieved after decoupling using Skogestad method.

**Keywords** MIMO process · Decoupling · FOPID · PID · Nelder–Mead

## 1 Introduction

Basically, the design of multiple-input, multiple-output (MIMO) controllers is additional complex thus makes it challenging to control and compared to that of the scalar system because of the loop interactions. In multi-loop control, the MIMO processes are considered to be an assembly of multi-single loops, for which the controller is designed and executed on the loops [1]. And the major reasons for the acceptance of multi-loop control by the process control industry are their realistic performances and robustness.

The major benefits behind utilizing decoupling control methodology are that: (1) It permits the implementation of SISO controller design methods; and (2) if the actuator or sensor fails, it is reasonably simple to balance the loop automatically,

S. Kapoor (✉)
Tula's Institute, Dehradun, India
e-mail: sheetalkapoor059@gmail.com

M. Chaturvedi · P. K. Juneja
Graphic Era Deemed to be University, Dehradun, India
e-mail: mayankchaturvedi.geit@gmail.com

P. K. Juneja
e-mail: mailjuneja@gmail.com

© Springer Nature Singapore Pte Ltd. 2021
X.-Z. Gao et al. (eds.), *Advances in Computational Intelligence and Communication Technology*, Advances in Intelligent Systems and Computing 1086,
https://doi.org/10.1007/978-981-15-1275-9_34

since the effect of failure will be on only one loop. Ideally, simplified and inverted are the three types of decoupling techniques. Simplified decoupling is a wide margin by the most well-known method [2, 3].

For the purpose of measurement of interactions relative gain array (RGA), ($\Lambda$) is evaluated. In this work, a multivariable process model is controlled with FOPID along with PID controller results have been compared and have been presented [4, 5]. Foremost with the assistance of simplified decoupling, communications have been lessened and the MIMO process model has been changed over into two individual SISO processes.

## 2  Methodology

Here the industrial scale polymerization (ISP) reactor has been considered proposed by Chien et al. The transfer function model of the process is given by:

$$G_{pi}(s) = \begin{bmatrix} \frac{22.89}{4.572s+1}e^{-0.2s} & \frac{-11.64}{1.807s+1}e^{-0.4s} \\ \frac{4.689}{2.174s+1}e^{-0.2s} & \frac{5.80}{1.801s+1}e^{-0.4s} \end{bmatrix} \qquad (1)$$

For the above process, the controlled temperature is along with the level and the manipulated variables are the two input feed reactors.

### 2.1  Relative Gain Array (RGA)

RGA is used for the calculation of quantitative interaction by the parameter $\lambda_{ij}$ (relative gain) based on the steady-state information which can be calculated as:

$$\lambda = \frac{1}{\frac{K_{12}K_{21}}{K_{11}K_{22}}} \qquad (2)$$

For $2 \times 2$ case, RGA can be expressed as:

$$\Lambda = \begin{bmatrix} \lambda & 1-\lambda \\ 1-\lambda & \lambda \end{bmatrix} \qquad (3)$$

Likewise, RGA, RNGA, and Niederlinski index have been calculated for ISP reactors and are as follows [6]:

The results of RGA-NI-RNGA suggest the pair of u1−y1, u2−y2. The phase and gain margin were chosen for the controller design which are $72°$ and $5.0$, respectively.

## 2.2 Decoupling Method

Decoupling control configuration, called "simplified decoupling" by Luyben, is generally utilized in the literature [7, 8]. $P_{ij}(s)$ is the process transfer matrix of MIMO system and represented as:

$$P(s) = \begin{bmatrix} P_{11}(s) & P_{12}(s) \\ P_{21}(s) & P_{22}(s) \end{bmatrix} \tag{4}$$

The decoupler $D_{ii}(s)$ as follows:

$$D(s) = \begin{bmatrix} 1 & -\frac{D_{12}(s)}{D_{11}(s)} \\ -\frac{D_{21}(s)}{D_{22}(s)} & 1 \end{bmatrix} \tag{5}$$

The resultant transfer matrix $T(s)$ comes out to be:

$$T(s) = \begin{bmatrix} P_{11}(s) - \frac{P_{12}(s)P_{21}(s)}{P_{22}(s)} & 0 \\ 0 & P_{22}(s) - \frac{P_{12}(s)P_{21}(s)}{P_{11}(s)} \end{bmatrix} \tag{6}$$

Figure 1 represents the simplified decoupling control, where $w_1$ and $w_2$ are reference inputs; $P_{ij}(s)$ are the process transfer function; $D_{21}$ and $D_{12}$ are the two controllers designed for the decoupled SISO process model; $Y_1$ and $Y_2$ are system outputs.

After that with help of Skogestad scheme for reducing the order of the model, the decoupled model has been decreased into FOPDT on account of the extensive

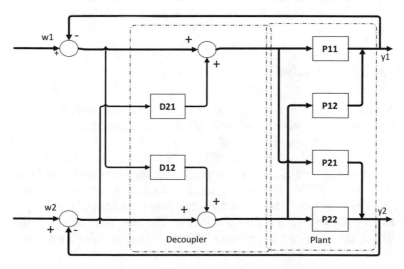

**Fig. 1** Block diagram for simplified decoupling control for 2 × 2 model

accessibility of the tuning procedure for FOPID and PID controller [9, 10]. Further, FOPID and PID controllers are obtained based on the tuning techniques: direct synthesis and Wang–Juang–Chan tuning technique [11]. The decoupler matrix is given below:

$$D_i(s) = \begin{bmatrix} 1 & \frac{0.508(4.572s+1)}{(1.807s+1)}e^{-0.2s} \\ \frac{-0.808(1.801s+1)}{(2.174s+1)}e^{0.2s} & 1 \end{bmatrix} \quad (7)$$

The resultant diagonal decoupler transfer function:

$$h_{11}(s) = \frac{167.394s^2 + 15.094s + 32.3}{17.962s^3 + 22.13s^2 + 8.553s + 1}e^{-0.2s} \quad (8)$$

$$h_{22}(s) = \frac{42.411s^2 + 38.282s + 8.184}{7.07s^3 + 11.1s^2 + 5.78s + 1}e^{-0.4s} \quad (9)$$

Using Skogestad model order reduction method, $h_{11}(s)$ and $h_{22}(s)$ are reduced to the FOPDT models [9, 12] And the equation is given below:

$$f_{i1}(s) = \frac{32.3}{(1.907s + 1)}e^{-2.2s} \quad (10)$$

$$f_{i2}(s) = \frac{10.85}{(1.80s + 1)}e^{-0.4s} \quad (11)$$

For the purpose of the controller design of the above SISO model, FOPID and PID controllers are designed. Nelder–Mead algorithm has been utilized for the calculation of extra two parameters ($\lambda, \mu$) along with IMC-Maclaurin PID tuning, direct synthesis, and Wang–Juang–Chan tuning technique [5, 13, 14].

## 3  Result and Analysis

Presently, different sorts of tuning techniques have been considered for the purpose of designing controller for both FOPID and PID, and the examination has been done on the basis of closed-loop performance characteristics. With the help of various tuning techniques, all the parameters of FOPID and PID controllers have been obtained. Figure 1 shows the simulated closed-loop response for First Order Plus Dead Time (FOPDT) (11) model, and IMC–MAC tuning technique has been utilized here.

From Table 3, it can be seen that the percentage overshoot provided by FOPID controller is 6.77 which is less than provided by PID controller, that is, 15.108.

In Fig. 2, the simulated response of FOPDT (11) model has been shown with

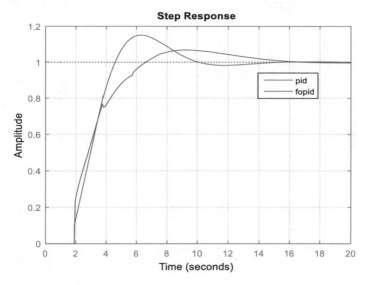

**Fig. 2** Comparing closed-loop response with FOPID and PID controllers with IMC-MAC technique for FOPDT (11)

FOPID and PID controllers for Wang–Juang–Chan tuning technique. The performance characteristics have been shown in Table 3. Simulated closed-loop performance has been analyzed with direct synthesis tuning technique, where the manipulated variable is $\tau_c = 1$; in Fig. 3, it can be depicted very clearly that the overshoot has been reduced too much extent for FOPDT (11) process model.

Figure 4 represents the simulating step response with FOPID and PID controllers tuned with direct synthesis with $\tau_c = 0.5$ for FOPDT (11), and it can be seen that with PID controller response is showing oscillations which is highly undesirable where the FOPID controller percentage overshoot is 6.79 as shown in Table 3.

The simulated response comparison has been presented in Fig. 5 and is implemented with IMC-Maclaurin tuning technique for which the performance characteristics have shown in Table 1 for FOPDT (22) process model. Figure 6 presents the performance analysis with direct synthesis for FOPDT (22) and comparative performance characteristics have been compared in Table 3. Closed-loop simulated step response for FOPDT (22) model has been shown in Fig. 7, and changing $\tau_c = 1$. Controller designed with Direct Synthesis and Wang–Juang–Chan tuning technique, the simulated response has been shown in Figs. 8 and 9.

All the performance characteristics have been shown in Tables 2 and 3, where the comparison has been carried out among closed-loop performance of FOPID and PID controllers for the two reduced SISO models, i.e., FOPDT (11) and FOPDT (22).

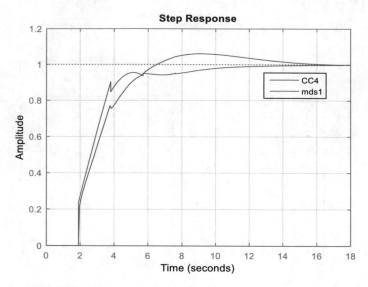

**Fig. 3** Comparing closed-loop response with FOPID and PID controllers with Wang–Juang–Chan tuning technique for FOPDT (11)

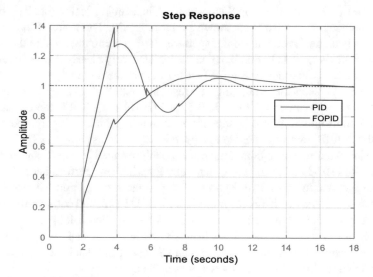

**Fig. 4** Comparing closed-loop response with FOPID and PID controllers with direct synthesis for FOPDT (11)

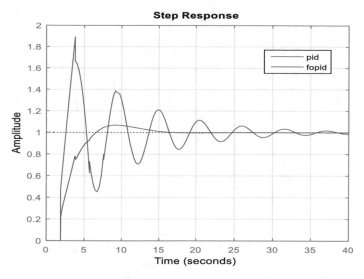

**Fig. 5** Comparing closed-loop response with FOPID and PID controllers with direct synthesis ($t_c$ = 0.5) technique for FOPDT (11)

| **Table 1** RGA, RNGA, NI values of diagonal paring | Control configuration | RGA | RNGA | NI |
|---|---|---|---|---|
| | (1, 1)–(2, −2) | 0.7087 | 0.5482 | 1.4111 |
| | (1, −2), (2, −1) | 0.2913 | 0.4518 | −3.4323 |

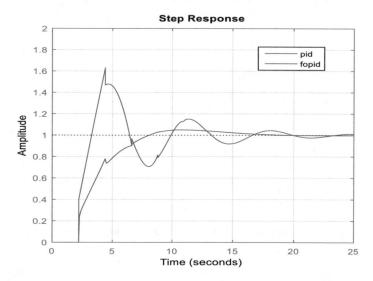

**Fig. 6** Comparing closed-loop response with FOPID and PID controllers with IMC–MAC for FOPDT (22)

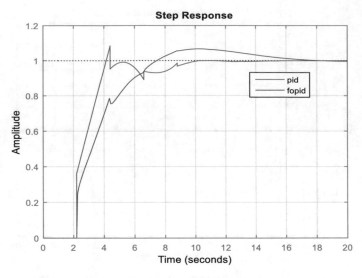

**Fig. 7** Comparing closed-loop response with FOPID and PID controllers with direct synthesis for FOPDT (22)

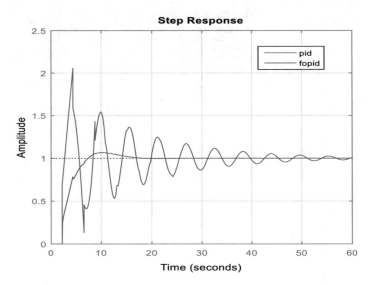

**Fig. 8** Comparing closed-loop response with FOPID and PID controllers with DS ($t_c = 0.5$) for FOPDT (22)

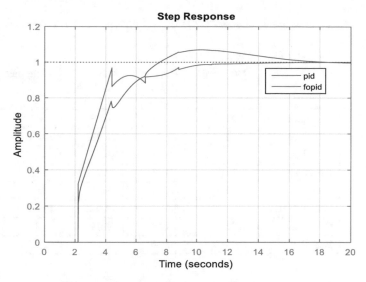

**Fig. 9** Comparing closed-loop response with FOPID and PID controllers with Wang–Juang–Chan tuning technique for FOPDT (22)

**Table 2** Performance characteristics for various types of PID and FOPID controller designed for FOPDT (11) process model

| Controller | Characteristics | Tuning techniques | | | |
|---|---|---|---|---|---|
| | | IMC–MAC | Direct synthesis $t_c = 1$ | Direct synthesis $t_c = 0.5$ | Wang–Juang–Chan |
| PID | Rise time | 0.87 | 1.63 | 0.34 | 1.96 |
| | Settling time | 22.03 | 9.23 | 55.82 | 9.84 |
| | Overshoot | 63 | 8.42 | 105.74 | 0 |
| | Peak time | 4.40 | 4.40 | 10.26 | 19.28 |
| FOPID | Rise time | 4.30 | 3.73 | 3.67 | 3.92 |
| | Settling time | 15.43 | 15.14 | 15.06 | 15.93 |
| | Overshoot | 4.95 | 6.66 | 6.60 | 6.92 |
| | Peak time | 10.5 | 10.28 | 10.26 | 10.36 |

## 4 Conclusion

In this work, the results of the performance of the FOPID for a multivariable process model are compared with PID controller. The parameters of FOPID controller have been obtained utilizing the Nelder–Mead algorithm. From the simulated response, it can be concluded that overshoot has largely been reduced when the controller is designed with FOPID controller. Thus, the flexibility and stability of the controller

**Table 3** Performance characteristics for FOPDT (22) for PID and FOPID designed with various tuning techniques

| Controller | Characteristics | Tuning techniques | | | |
|---|---|---|---|---|---|
| | | IMC–MAC | Direct synthesis $t_c = 1$ | Direct synthesis $t_c = 0.5$ | Wang–Juang–Chan |
| PID | Rise time | 2.25 | 0.97 | 0.55 | 1.889 |
| | Settling time | 9.42 | 13.56 | 34.86 | 10.11 |
| | Overshoot | 15.11 | 38.59 | 89.07 | 0 |
| | Peak time | 6.27 | 3.80 | 3.79 | 22.75 |
| FOPID | Rise time | 3.29 | 3.35 | 3.28 | 3.09 |
| | Settling time | 13.90 | 14.04 | 13.92 | 13.48 |
| | Overshoot | 6.77 | 6.86 | 6.79 | 6.04 |
| | Peak time | 9.13 | 9.12 | 9.14 | 9.07 |

have been improved with a fractional-order-based PID controller. The overshoot of the considered MIMO process has been reduced by a considerable amount.

# References

1. D. Chen, D.E. Seborg, Multiloop PI/PID controller design based on Gershgorin bands, in *IEE Proceedings—Control Theory and Applications* (2002), pp. 68–73
2. M. Waller, J.B. Waller, K.V. Waller, Decoupling revisited. Ind. Eng. Chem. Res. **42**, 4575–4577 (2003)
3. C. Rajapandiyan, M. Chidambaram, Controller design for MIMO processes based on simple decoupled equivalent transfer functions and simplified decoupler. Indus. Eng. Chem. Res. **51**, 12398–12410 (2012)
4. T.N.L. Vu, M. Lee, Independent design of multi-loop PI/PID controllers for interacting multivariable processes. J. Process. Control. **20**, 922–933 (2010)
5. D.K. Maghade, B.M. Patre, Decentralized PI/PID controllers based on gain and phase margin specifications. ISA Trans. **51**, 550–558 (2012)
6. C.A. Smith, A.B. Corripio, *Principles and Practice of Automatic Process Control* (John Wiley, New York, 1985)
7. Q.G. Wang, B. Huang, X. Guo, Auto-tuning of TITO decoupling controllers from step tests. ISA Trans. **39**(4), 407–418 (2000)
8. R. Jangwan, P.K. Juneja, M. Chaturvedi, S. Sunori, P. Singh, Comparative analysis of controllers designed for pure integral complex delayed process model, in *International Conference on ICACCI,* (IEEE, 2014), pp. 712–717
9. D. Naithani, M. Chaturvedi, P.K. Juneja, Integral error based controller performance comparison for a FOPDT model, in *Image Information Processing (ICIIP)* (IEEE, 2017)
10. P. Verma, P.K. Juneja, M. Chaturvedi, Various mixed approaches of model order reduction (CICN), in *2016 8th International Conference on IEEE* (2016)
11. S. Kapoor, M. Chaturvedi, P.K. Juneja, Design of fractional order PID controller for a SOPDT process model, in ICIIP (IEEE, 2017), pp. 215–218
12. Sigurd Skogestad, Simple analytic rules for model reduction and PID controller tuning. J. Process. Control. **13**(4), 291–309 (2003)

13. P. Kholia, M. Chaturvedi, P.K. Juneja, S. Kapoor, Effect of padè first order delay approximation on controller capability designed for IPDT process model, in *International Conference on Power Energy, Environment & Intelligent Control (PIEEC)* (IEEE, 2018)
14. S. Kapoor, M. Chaturvedi, P.K. Juneja, Design of FOPID controller with various optimization algorithms for a SOPDT model, in *International Conference on Emerging Trends in Computing and Communication Technologies (ICETCCT),* (IEEE, 2017)

# Reduced-Order Modeling of Transient Power Grid Network with Improved Basis POD

Satyavir Singh, Mohammad Abid Bazaz and Shahkar Ahmad Nahvi

**Abstract** Power systems dynamics is represented by coupled equations and influenced by several elements. Most of these elements exhibit nonlinear behavior, and hence, computational efforts required for solving these coupled nonlinear equations are expensive. To simplify computational efforts, reduced-order model can be used. This reduced-order model is formulated by projecting a large-dimensional state vector onto a small-dimensional subspace spanned by an orthonormal basis which is conventionally obtained through a technique, called proper orthogonal decomposition (POD). However, this technique involves simulation of the high-dimensional nonlinear system to obtain the orthonormal basis for the projection. In this paper, we have presented an improvement of POD technique with the approximate snapshot ensemble bases extraction to avoid the need to simulate full-order dynamics of the large power grid network. The POD basis is generated from approximate trajectory which reduces simulation time with insignificant or no error in the employed model.

**Keywords** Model order reduction · Snapshots · Proper orthogonal decomposition · Power grid network

## 1 Introduction

Simulation of nonlinear dynamics of large power grid networks has heavy computational cost and large simulation time. An example of such modeling and simulation methods is the energy-based methods [1] which investigate the dynamical behavior of large, nonlinear power system models. The large size of power grid network makes

S. Singh (✉) · M. A. Bazaz
Department of Electrical Engineering, National Institute of Technology, Srinagar, India

S. A. Nahvi
Department of Electrical Engineering, IUST, Awantipora, India
e-mail: shahkar.nahvi@islamicuniversity.edu.in

© Springer Nature Singapore Pte Ltd. 2021
X.-Z. Gao et al. (eds.), *Advances in Computational Intelligence and Communication Technology*, Advances in Intelligent Systems and Computing 1086,
https://doi.org/10.1007/978-981-15-1275-9_35

simulation of system dynamics numerically expensive, and hence, the requirement for model order reduction (MOR) is arising. Reduced-order models are numerically cheaper and preserve system physics with an acceptable accuracy. In the literature, many approximation approaches are addressed for MOR and the selection of an appropriate approach depends upon the type of problem to be solved [2]. Popular MOR schemes are balanced truncation [3], moment matching [2], and proper orthogonal decomposition (POD) [4].

Many researchers have applied MOR strategies to reduce the size of large power engineering models, but most of these methods involve linearization of the power system model about an equilibrium point [5]. Hence, the model order reduction is applicable only for small excursions about the point of linearization. There are also examples [6–8] of application of nonlinear MOR strategies to power system problems. For example, [6] addresses the application of POD to reduce order of the nonlinear model of power grid network failures where POD is validated on the coupled swing equation model representing the cascading dynamics of the alternators.

The aim of POD is to obtain a compact system by projecting a large-dimensional model into a small-dimensional subspace while retaining the dominant features of state evolution dynamics. The low-dimensional subspace is obtained from state snapshots in response to certain inputs to which the full-order model (FOM) is subjected [9]. The conventional POD was introduced to simplify the nonlinearity of the system in a power grid model. However, this scheme required snapshots of full-order nonlinear model. This work proposed a scheme to improve basis extraction time by generating the approximate snapshot ensemble to further reduce the size of original model. This is an attempt to validated this scheme on the large power grid network.

This work is organized in the following sections: in Section I modeling of power grid network has been discussed. This is accompanied by the section on approach of MOR, POD, and the generation of approximate snapshot ensembles and its subsequent application in POD for creating the reduced-order model. In the numerical validation section, the proposed approach is implemented on a nonlinear power grid network, showing improved simulation time with little or no loss in accuracy. The final section concludes the work. Pseudo codes are reproduced wherever necessary.

## 2 Modeling of Power Grid Network

A mathematical approach is adapted to describe the transient dynamics of an alternator, known swing equation. It have a second-order differential equation showing the alternator node or bus. This equation originates from the rotor dynamics of the alternator, involved with algebraic equation of electrical power [10, 11]. The dynamical behavior of the $i$th alternator is given by following equation

$$\frac{2H_i}{\omega}\frac{d^2\delta_i}{dt^2} + D_i\frac{d\delta_i}{dt} = P_{m_i} - P_{e_i}, for, i = 1, \dots, N \qquad (1)$$

This is a swing equation which describes the power flow of the $i$th bus [12]. The symbol $H_i$ is the inertia constant of the alternator, $\omega$ is the reference frequency of the model, $D$ is damping factor, $\delta_i$ is rotor angle, $P_{m_i}$ is the mechanical power provided to the alternator, and $P_{e_i}$ is the power demanded on the alternator by the grid network (including the power lost to damping). Under equilibrium, $P_{m_i} = P_{e_i}$ and if system experience fault or fluctuation, the power demand $P_{e_i}$ varies. In (1), any variation in the difference between the power demand and power supply is compensated by the angular momentum of the rotor [13].

The power grid network voltage is given by $V_i = |V_i|e^{j\delta_i}$, where $j = \sqrt{-1}$. Complex admittance of $\ell$th line connected to $i$th alternator is expressed as $y_i = g_{i\ell} + ib_{i\ell}$, with $b_{i\ell}$ is the line susceptance and $g_{i\ell}$ is the line conductance. It is assumed that voltage magnitude is kept constant and transmission lines are purely reactive, i.e., $g_{i\ell} = 0$. The transient state of the power system is described by rotor dynamics of $i$th alternator in the following form [11],

$$m_i\ddot{\delta} + d_i\dot{\delta} = P_{m_i} - \Sigma_{\ell=1}^{N}|V_i||V_\ell|b_{i\ell}\sin(\delta_i - \delta_\ell) \qquad (2)$$

$i = 1, 2, \dots, N$. Equation (2) has solution, $\delta_i(t)$ and it describes the transient behavior of the power grid network [6]. All the nodes are considered as generator or alternator (or PV) nodes. Equation (2) will take the following form [11]:

$$M\ddot{\delta} + D\dot{\delta} = P_m - P(\delta(t)) \qquad (3)$$

where $M$ and $D$ are $N \times N$ diagonal matrices, and $\ddot{\delta}$, $\dot{\delta}$, $\delta$ are vectors in $R^N$. The number of alternators is $N$ in the grid network. All these products are matrices vector product. The function $P(\delta)$ is a nonlinear behavior of $\delta$. Equation (3) is showing the high fidelity model of the power grid network. Electrical power depends on $\delta$ as well as $\dot{\delta}$ that means it depends on relative angular velocity as compared with a synchronously rotating system.

The nonlinear power grid network model (3) is formulated in state space as below:

$$\begin{bmatrix} \dot{\delta} \\ \ddot{\delta} \end{bmatrix} = \begin{bmatrix} \dot{\delta} \\ -M^{-1}D\dot{\delta} + M^{-1}(P_m - P(\delta(t))) \end{bmatrix} \qquad (4)$$

The computational requirement to solve (4) is expensive due to large size ($2N$) and nonlinearity of the model. Therefore, requirement of MOR is raised. Numerical techniques of MOR are elaborated in the next sections to reduce the model size and simulation time.

# 3 Approach of MOR

## 3.1 POD

To the given input or an initial condition applied to a dynamical model have a state vector evolves in $R^n$. This state vector is approximated by a trajectory in $R^r$ with $r < n$. The problem stated as a set of snapshots $(x^1, x^2, \ldots, x^{n_s}) \in R^n$ and let $\chi = span(x^1, x^2, \ldots, x^{n_s}) \subset R^n$, be smaller dimension with $r = rank(\chi)$. To get orthonormal basis vectors $(\phi_1, \phi_2, \ldots, \phi_k)$ is approximated $\chi$ for given $k < r$. It can be expressed as $X = [x^1 \ x^2, \ldots, x^{n_s}] \in R^{n \times n_s}$, and best-fitted $\hat{X}$ with $rank(\hat{X}) = k$ such that the error $E = ||X - \hat{X}||_F^2$ is minimized. The result is given by, $\hat{X}^*$, [9]:

$$\hat{X}^* = V_k \Sigma_k W_k^T \tag{5}$$

where $\Sigma$ is the singular values (SVs) of $X$ with the column vectors $V$ and $W$ are left and right orthogonal vectors, respectively. The $\Sigma \in R^{n \times n_s}$ have SVs $\sigma_1 \geq \sigma_2, \ldots, \geq \sigma_r$ on main diagonal and other elements zero, $\Sigma_k = diag(\sigma_1, \sigma_2, \ldots, \sigma_k) \in R^{k \times k}$, $V_k$ consists of the first $k$ columns of $V$, and $W_k^T$ consists of the top $k$ rows of $W^T$. Let smallest singular value is $\sigma_k = \epsilon$ (numerical value). The approximation error (5) is given as follows

$$||X - \hat{X}^*||_F^2 = \sum_{i=k+1}^r \sigma_i^2 \tag{6}$$

Model order reduction has approximation error results from the Galerkin projection in POD. The projection matrix $V_k$ is obtained to generate reduced-order model based on SVD of the snapshot matrix $X$. The projection matrix can be expressed as follows:

$$V_k = \{v_i\}_{i=1}^k \tag{7}$$

In particular, consider the following nonlinear system with input $u(t)$, state vector $x(t)$, and nonlinear function $f(x(t))$ where state trajectory $x(t)$ evolves in $R^n$:

$$\dot{x}(t) = f(x(t)) + Bu(t) \tag{8}$$

The size of reduced-order model is $k \ll n$ which is obtained from a subspace spanned by a smaller dimension $k$ in $R^n$. It is obtained by orthogonal projection matrix obtained in (7) and projected the system in (8) onto $V_k$. The reduced-order system of order-$k$ takes the following form:

$$\dot{\tilde{x}}(t) = V_k^T f(V_k \tilde{x}(t)) + V_k^T Bu(t) \tag{9}$$

Here $\tilde{x}(t) \in R^k$ and $x(t) \approx V_k \tilde{x}(t)$.

## 3.2   Generation of Approximate Snapshot Ensembles

The work [14] had addressed the reduced basis shown that an acceptable basis $V_k$, obtained from the approximate trajectory of states. To perform this task an algorithm was presented Algorithm 1. The snapshot matrix of nonlinear model by successive linearization is formulated with this algorithm. This is performed by the simulation of the successive linearized model instead of nonlinear model. The objective is to reduce the requirement of priori-heavy numerical computation of large order model. It has been carried out by simulating approximate ensemble obtained from full-order model (FOM). This was done by eliminating the need for computationally heavy, prior simulations of the high-dimensional problem to generate state snapshot ensembles by proposing to generate an approximate snapshot ensemble obtained from simulations of successive linearization of the nonlinear system [15].

To generate the approximate snapshot ensemble for the states, to an applied input $u(t)$ or an initial condition $x_0$, the nonlinear model is linearized around $x_0$. The $x(t)$ can be estimated from $\dot{x} \approx \tilde{\dot{x}}(t) = A_i x + (f(x_i) - A_i x_i))$ along the approximate trajectory. The approximate state ensemble, $x(t)$, is obtained by stacking the estimates $\tilde{x}(t)$ at each time step. The linearized model is simulated at further points of state vector $x$, till it reaches a threshold $\alpha$ from initial vector $x_0$, i.e., till $||x - x_0|| < \alpha$. If inequality violates, another linearization is done. The process is continued till distance of all previous linearized models greater than $\alpha$. This gives the linearized model of nonlinear system, and it will be stacked at every time step, and as a result, approximate snapshot ensemble matrix obtained. The SVD is applied on the obtained snapshot ensemble matrix, and subsequently, POD performed. This approach is experimented in the next section to get cheaper reduced-order model. The pseudo-code for the addressed strategy is presented in Algorithm 1.

---

**Algorithm 1** Approximate Snapshot trajectory generation

---

1: $i \leftarrow 0, j \leftarrow 0, X = [x_j]$, $x_0$: starting state, $T$: simulation time-steps, $\alpha$ is an appropriate constant.
2: Linearized model at $x_i$

$$\left. \begin{cases} \dot{x} = A_i x + (f(x_i) - A_i x_i)) + Bu \\ y = Cx \end{cases} \right\} \tag{10}$$

3: Simulate (10) with state vector $x = x_j$ for single time step
4: $X \leftarrow [X \quad x_{j+1}]$;                          ▷ *Store the snapshot of states*
5: **if** $min_{(0 \le k \le i)} \frac{||x_{j+1} - x_k||}{||x_k||} > \alpha$ **then**
6:    $i \leftarrow i + 1, \quad j \leftarrow j + 1$. go to (2)
7: **else**
8:    $j \leftarrow j + 1$ go to (3)
9: **end if**
10: Matrix of state snapshots $X$, SVD of $X$, $X = V \Sigma W^T$
11: Retain few columns of $V$ corresponding to dominant SVD.
12: Projection matrix, $V_k = \{v_i\}_{i=1}^k$.

---

## 7  Numerical Validation on Grid Network

The power grid network studied is having only the alternators with one reference node connected to all the alternators, as indicated in Fig.1 [7]. The dynamics of machines in power grid network can be addressed as the collective motion of it. The emerging areas are interconnected network of wind farm, group of parallel generators, and micro-grids toward the development of the smart grid model. This work has the same configuration of the proposed model. To make calculations easy, certain assumptions have been taken for the study of transient power grid network.

- The power system model is lossless.
- The length of transmission lines between two consecutive alternators is very small than the line connecting alternators with the infinite bus.
- The length of transmission lines between infinite bus and all the alternators is identical.
- All alternators are mutually connected with same-sized transmission lines.

These assumptions give us a mathematical model of the motion of alternators where $\delta_i$ is the rotor angular position with respect to the reference frame of the $i$th alternator.

$$\frac{1}{P_r} P_{e_i}(\delta_1, \ldots, \delta_n) = \underbrace{\frac{V V_{\text{ref}}}{X_{\text{ref}}}}_{b} \sin \delta_i + \underbrace{\frac{V^2}{X_{\text{int}}}}_{b_{\text{int}}} \{\sin(\delta_i - \delta_{i-1}) + \sin(\delta_i - \delta_{i+1})\} \quad (11)$$

where $P_r$ is rated active power in watts. Parameters $V$, $V_{\text{ref}}$, $X_{\text{ref}}$, and $X_{\text{int}}$ are constant per unit values and respectively terminal voltage of the alternator, voltage of reference bus, the reactance of transmission lines joining alternator with reference bus, and

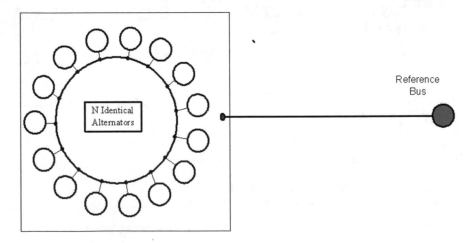

**Fig. 1**  Large power grid network

**Table 1** Data table

| Symbol | Parameter name | value |
|---|---|---|
| $m_i$ | Mass of alternator | 1.0 (pu) |
| $d_i$ | Damping of alternator | 0.25 (pu) |
| $P_m$ | Power demanded by alternator | 0.95 (pu) |
| $b$ | Susceptance between alternator and reference bus | 1 (pu) |
| $b_{int}$ | Susceptance between consecutive alternators | 100 (pu) |
| $N$ | Number of alternators | 1000 |

reactance of transmission lines joining alternator $i$ and $i+1$ in the grid network [7]. The nonlinear term $P(\delta)$ with four assumptions can be represented to (11) in per unit values for $i$th bus as follows:

$$P_{e_i} = b\sin\delta_i + b_{int}\left[\sin(\delta_i - \delta_{i-1}) + \sin(\delta_i - \delta_{i+1})\right] \tag{12}$$

where $b$ is the critical value of susceptance between the reference bus and the $i$th alternator, and $b_{int}$ is the critical value of susceptance between generator $i$ and $i+1$. The constants $b$ and $b_{int}$ are inversely proportional to the lengths of associated transmission lines.

The swing equation formulated in the standard form as given in (8), where $x$ is rotor angle, $f(x)$ is given by (12), and $u$ is mechanical power input $u(t) = P_m$. The output is represented by an average value of rotor angle of alternators $y(t) = \delta_{avg}$. The average value of $\delta$ is defined as follows:

$$\delta = \frac{1}{N}\sum \delta_i \tag{13}$$

The used parameter per unit values for the study is given in Table 1 [7, 11]. To address steady-state stability of the power grid, a damping factor is included. The reduced-order model size is truncated, $k = 35$ POD modes. The model simulation time and approximation error with the proposed algorithm is investigated with perturbation of nodes. For this purpose, two distinct cases are tested as given below:
1. All nodes starting from over-perturbation
2. All nodes starting from synchronously equilibrium condition.

**First Test Case: All nodes starting from over-perturbation**
In this case, assumed initial value of all the nodes $\delta_i = 1.12$, $\forall$ $i$ with $D = 0.25$. The approximate snapshot trajectory is generated using Algorithm 1, and snapshots stacked at regular interval to capture the system dynamics. The SVD is performed to obtain the dominant orthonormal basis vectors. Henceforth, simulation time reduces in basis extraction with proposed approach as compared to conventional POD. The

orthonormal basis in the projection is applied to generate a reduced-order model. It observed that the simulation time for the successive linearization is significantly reduced as compared to conventional POD, as summarized in Table 2, and error is also very less as shown in Table 3.

The system dynamical profile, error profiles in MOR, and approximation error profile are shown in Figs. 2, 3, and 4, respectively. The approximation error between exact–approximate ensemble is also very small 0.099%.

**Second Test Case: All nodes starting from synchronously equilibrium condition**
This case assumed initial value of all the nodes $\delta_i = 1, \quad \forall \; i$ with $D = 0.25$. The approximate snapshot trajectory is generated using Algorithm 1, and snapshots stacked at regular interval to capture the system dynamics. The SVD is performed to obtain the dominant orthonormal basis vectors. Henceforth, simulation time reduces in basis extraction with proposed approach as compared to conventional POD. The orthonormal basis in the projection is applied to generate a reduced-order model. It observed that the simulation time for the successive linearization is significantly reduced as

**Table 2** Comparison of simulation time for POD and APOD

| – | Full model sim | POD | APOD |
|---|---|---|---|
| Time (s) | 187.09 | 139.79 | 98.34 |

**Table 3** Quantitative comparison of errors

| – | POD | APOD |
|---|---|---|
| % error ($\delta_{avg}$) | $4.34 \times 10^{-10}$ | $6.35 \times 10^{-10}$ |

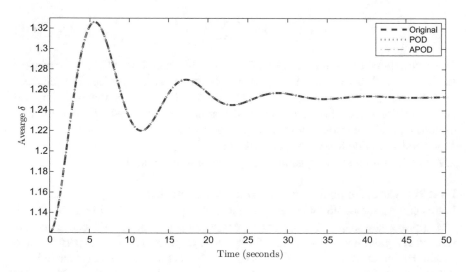

**Fig. 2** Dynamical profiles of over-perturbation

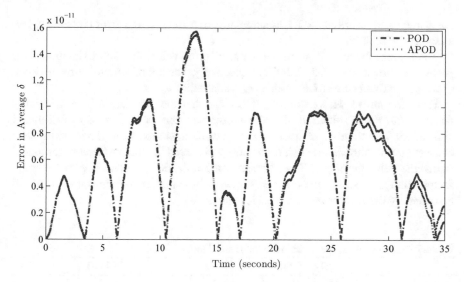

**Fig. 3** Error profiles

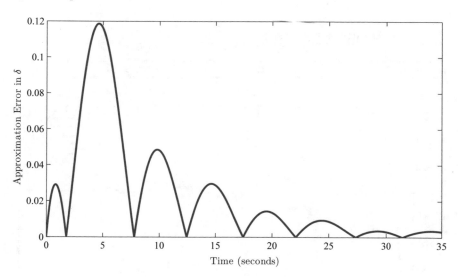

**Fig. 4** Error in approximation

compared to conventional POD, as summarized in Table 4, and error is also very less as shown in Table 5.

The system dynamical profile, error profiles in MOR, and approximation error profile are shown in Figs. 5, 6, and 7, respectively. The approximation error between exact–approximate ensemble is also very small 2.30%.

From the above demonstration, it is clear that the simulation is considerably faster due to the elimination of residual computations as compared to full model simulation. The number of mathematical evaluations also went down and results cheaper computational demand. It intuitively observed the that reduced-order model mimic the behavior of the FOM. The error minimization associated with the many factors such as modes, numbers, and step size. For smaller size steps, the error will be bound by the truncation of the basis vectors.

**Table 4** Comparison of time taken to find POD and APOD

| –        | Full model sim | POD    | APOD  |
|----------|----------------|--------|-------|
| Time (s) | 182.63         | 141.51 | 97.22 |

**Table 5** Quantitative comparison of errors

| –                     | POD                    | APOD                   |
|-----------------------|------------------------|------------------------|
| % error ($\delta_{avg}$) | $3.84 \times 10^{-10}$ | $4.23 \times 10^{-10}$ |

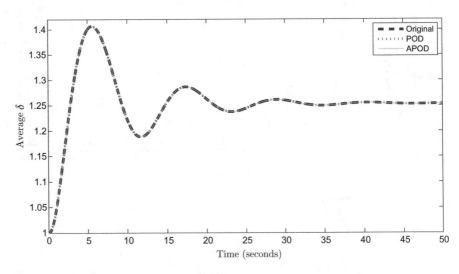

**Fig. 5** Dynamical profile of critical perturbation

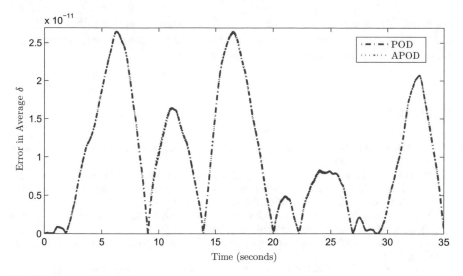

**Fig. 6** Error profiles

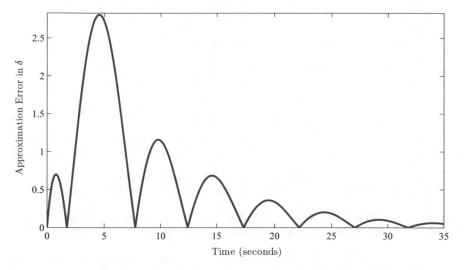

**Fig. 7** Error in approximation

## 8 Conclusion

This work presents a successful attempt to implement approximate snapshot ensemble bases extraction in POD to reduce the computational efforts of large power grid network. The approximate snapshot ensemble algorithm is addressed to improve the simulation time of conventional POD technique. The dynamical behavior of the systems is verified under different test cases to present the effectiveness of the proposed approach. The addressed model preserved system stability and accuracy in all test cases. The proposed approach can be extended to improve the simulation time of the available MOR techniques with an insignificant error.

# References

1. P.C. Magnusson, The transient-energy method of calculating stability. Trans. Am. Inst. Electr. Eng. **66**(1), 747–755 (1947)
2. D. Harutyunyan, R. Ionutiu, E.J.W. ter Maten, J. Rommes, W.H. Schilders, M. Striebel, Advanced topics in model order reduction, in *Coupled Multiscale Simulation and Optimization in Nanoelectronics* (Springer, Berlin, 2015), pp. 361–432
3. W.H. Schilders, H.A. Van der Vorst, J. Rommes, *Model Order Reduction: Theory, Research Aspects and Applications*, vol. 13 (Springer, 2008)
4. A. Chatterjee, An introduction to the proper orthogonal decomposition. Curr. Sci. **78**(7), 808–817 (2000)
5. D. Chaniotis, M. Pai, Model reduction in power systems using krylov subspace methods. IEEE Trans. Power Syst. **20**(2), 888–894 (2005)
6. P.A. Parrilo, S. Lall, F. Paganini, G.C. Verghese, B.C. Lesieutre, J.E. Marsden, Model reduction for analysis of cascading failures in power systems, in *Proceedings of the 1999 American Control Conference, 1999* (IEEE, 1999), vol. 6, pp. 4208–4212
7. Y. Susuki, I. Mezić, T. Hikihara, Coherent swing instability of power grids. J. Nonlinear Sci. **21**(3), 403–439 (2011)
8. N. Kashyap, S. Werner, T. Riihonen, Y.-F. Huang, Reduced-order synchrophasor-assisted state estimation for smart grids, in *2012 IEEE Third International Conference on Smart Grid Communications (SmartGridComm)*, (IEEE, 2012), pp. 605–610
9. A.C. Antoulas, *Approximation of Large-Scale Dynamical Systems* (SIAM, 2005)
10. T. Gonen, *Modern Power System Analysis* (CRC Press, 2013)
11. M.H. Malik, D. Borzacchiello, F. Chinesta, P. Diez, Reduced order modeling for transient simulation of power systems using trajectory piece-wise linear approximation. Adv. Model. Simul. Eng. Sci. **3**(1), 31 (2016)
12. A.R. Bergen, *Power Systems Analysis* (Pearson Education India, 2009)
13. D.P. Kothari, I. Nagrath, et al., *Modern Power System Analysis* (Tata McGraw-Hill Education, 2011)
14. S.A. Nahvi, M.A. Bazaz, M.-U. Nabi, S. Janardhanan, Approximate snapshot-ensemble generation for basis extraction in proper orthogonal decomposition. IFAC Proc. Volumes **47**(1), 917–921 (2014)
15. S. Singh, M.A. Bazaz, S.A. Nahvi, A scheme for comprehensive computational cost reduction in proper orthogonal decomposition. J. Electr. Eng. **69**(4), 279–285 (2018)

# Role of IoT and Big Data Support in Healthcare

Vikash Yadav, Parul Kundra and Dhananjaya Verma

**Abstract** Healthcare is one of the major sectors in any country, both in terms of the revenue that it generates and the services that it is providing to mankind, and also, it provides employment opportunities to millions of people. Healthcare sector comprises of technology-enabled health services, multi-facility hospitals, on board healthcare services and telemedicine, multifunction medical devices and specialized machines. In this paper, the role of Internet of things with the help of new technologies involved and big data has been shown towards the field of healthcare.

**Keywords** Internet of things · Big data · Bioinformatics · Healthcare · Medical imaging

## 1 Introduction

In recent years, exponential growth and a large amount of investments have been noticed in this sector, and taking into consideration, the huge amount of revenue that it generates and various types of medical services that it provides. So it becomes the need of the hour to power this sector with advanced blooming technologies, and for our reference, we are focussing on big data analytics and IoT in healthcare.

V. Yadav · D. Verma (✉)
Department of Computer Science & Engineering, ABES Engineering College,
Ghaziabad, India
e-mail: dhananjaya.15bcs1170@abes.ac.in

V. Yadav
e-mail: vikash.yadav@abes.ac.in

P. Kundra
Department of Computer Applications, ABES Engineering College, Ghaziabad, India
e-mail: parul.kundra@abes.ac.in

© Springer Nature Singapore Pte Ltd. 2021                                                          445
X.-Z. Gao et al. (eds.), *Advances in Computational Intelligence and Communication
Technology*, Advances in Intelligent Systems and Computing 1086,
https://doi.org/10.1007/978-981-15-1275-9_36

## 1.1  Big Data in Healthcare

Big data analytics (BDA) has brought together by two different branches of computer science that are big data and analytics and is collectively designed to deliver a data management approach.

Big data is mostly described by the industry professionals as extremely a large amount of unstructured and structured data an organization generate. Big data (BD) has feature a vast size that exceeds the handling capability of traditional information management technologies.

Big data has driven the need for the development of state-of-the-art technological infrastructure and sophisticated tools that are capable of holding, storing and analysing huge amounts of structured and unstructured data, such as data from biometric applications [1], medical and text reports, and medical and research images. These data are being generated exponentially from data profound technologies like surfing through the Internet for various activities such as accessing information, visiting and posting on social networking sites, mobile computing, cloud uploads, electronic commerce.

Analytics in reference to big data is the process of investigating and analysing huge amounts of data, coming from various sources of data in all different data formats, and in order to provide useful insights that can power decisions at present or in near future. Various analytical technologies and tools such as data mining, deep learning algorithms, natural language processing tools (NLP), artificial intelligence and predictive analytics are used to analyse and contextualize the data; also, various analytical approaches are used to identify inherent patterns in the data, correlations and anomalies may also get discovered in the integration process of huge amounts of data from different data sets [2].

Big data analytics helps to draw important conclusions from the data to provide solutions to many important and newly discovered problems. Within the health sector, it provides stakeholders with a blooming new technology and advanced tools that have tremendous potential to move healthcare to a much better level and to make healthcare more economically feasible.

Big data in healthcare apart from being voluminous can also be described through three main characteristics that are as follows Fig. 1:

**Volume** can be defined as the amount of data and information generated and consumed by various organizations and individual users. For example, medical imaging (e.g., magnetic resonance imaging, X-rays, ultrasounds, computed tomography scans) generates a large variety of data with high level of complexity and much broader scope for further analysis.

**Velocity** can be defined as the incidence and velocity at which data is being created, hold, stored and shared among various devices like new technologies available in healthcare are producing billions of data every single minute for a day long.

**Variety** can be defined as the prevalence of new data types including those coming from machine and mobile sources. Like some healthcare technologies produce

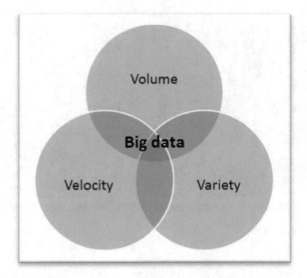

**Fig. 1**   3V's of big data [7]

"Omics" data which is scientifically produced by various sequencing technologies at almost every level of cellular components, from genomics, proteomics and metabolomics to protein interaction and phenomics.

## 1.2   Internet of Things in Healthcare

Internet of things (IoT) is an arrangement of multiple physical, electronic and various sensor-based devices connected to each other which enables them to store and exchange data among each other (Fig. 2).

For medical devices and applications to connect with each other to collect exchange data among each other through various IT solutions developed for healthcare, Internet of Healthcare Things (IoHT) has been developed. It comprises consumerization of various wearable devices which enables personalized monitoring of one's health and many more professional medical devices.

## 2   Need for New Technologies

Research and developments in the healthcare sector lead to the following factors which in turn accelerated the need for the implementation of new technologies like big data analytics and Internet of things as follows:

**Fig. 2** IoT in healthcare [8]

(i)   **Higher cost of medical and healthcare services**: Big data may reduce the cost of fraud, unnecessary tests and abuse in the healthcare sector.
(ii)  **Higher demand for overall health coverage**: Big data can power various predictive models for better diagnosis and treatment of various diseases quite efficiently by using the data that is being aggregated many times for better results ranging from DNA, cells to protein-related data.
(iii) **Growing popularity of wearable devices in healthcare**: The modern development in the wearable devices has gained much popularity among youngsters and demand much accurate results.
(iv)  **Growing popularity of handy health monitors for various healthcare issues**: Various developments in the monitoring devices have enabled aged patients and handicapped patients to regularly track and monitor their health in real time.
(v)   **Maintaining electronic health records (EHRs)**: For the diagnosis of many diseases, the trend of the patients must needed to be maintained and kept for further research purposes, so EHRs are created which comprises of patient-related demographics, medical history and test-related history which is greatly powered by big data.

## 3   IoT and Big Data Support

### 3.1   IoT Support

The IoT provides support to the healthcare sector in the following ways:

**Fig. 3** Research aid [9]

**Research aid**: IoT powers the medical research with real-world information through real-time data coming from IoT devices and can be analysed and tested to derive much accurate results (Fig. 3).

**Moodables**: These devices are intended to enhance or change the mood of a person; these devices are basically designed and developed through research in neurosciences powered with IoT (Fig. 4).

**Emergency care**: Through IoT-based applications and devices, the medical-aid providers are alerted before any mishappening to occur and also to take appropriate precautions to boost healthcare services (Fig. 5).

**Tracking and alerts**: Regular tracking can be possible manually or automated as the IoT-based devices to collect and share data in real time so that alerts can be provided according to the varying health of a person (Fig. 6).

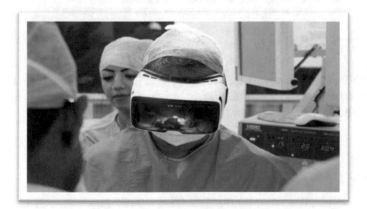

**Fig. 4** Moodables [10]

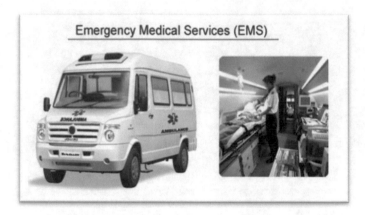

**Fig. 5** Emergency car [11]

**Fig. 6** Tracking and alerts [12]

**Ingestible sensors**: These are the most unique innovation powered by IoT can help to monitor and detect any irregularities in body as they are pill-sized and can remain inside the body without any harm and benefit healthcare (Fig. 7).

## 3.2 Big Data Support

The conceptual framework can use any of the business intelligence tools on any regular system and can be used for a project on healthcare analytics; the reason is that big data processes data across various nodes in small pieces of data that are broken down from large data sets. These data sets are analysed to explore important

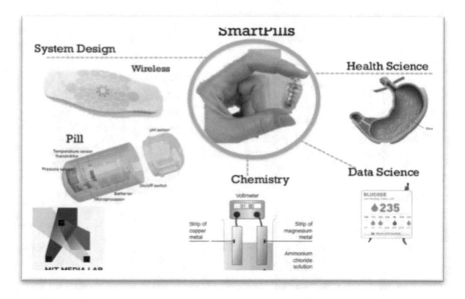

**Fig. 7** Ingestible sensors [13]

insights from them that have the potential to influence the healthcare sector with better decision-making capabilities and many more cost-effective diagnosis techniques.

**Architectural framework**: Big data analytics employs various open-source platforms like Hadoop, MapReduce. These platforms are programming oriented, complex and require the deployment of complex infrastructure, and technocrats with specialized skills are required to use these tools. The shortcoming with these technologies are that they are not user-friendly and are different from various service-based proprietary tools as they are provided with additional support which is not present in big data tools and platforms (Fig. 8).

Big data provides support to the healthcare sector through the following ways:

1. **Bioinformatics applications**: Research in bioinformatics analyses various variations in the biological system up-to molecular level. It is necessary to store and analyse data in a time efficient manner to align with the current trends of healthcare. With the help of modern sequencing technologies, the genomic data can be acquired in very less time. Big data analytics helps bioinformatics applications to create data repositories, and also it provides infrastructure for computing, and other data manipulation tools to efficiently analyse biological data and draw valuable insights from it [3].

2. **Improved security and reduce fraud**: Some research results also indicate that the data in the healthcare sector are 200% more likely to be breached than any other sector and any harm to the personal and medical data would result in catastrophic consequences because it possesses great market value for the cyber criminals in the black markets. By the use of big data analytics, many threats can be prevented by identifying and analysing any change in the network traffic and

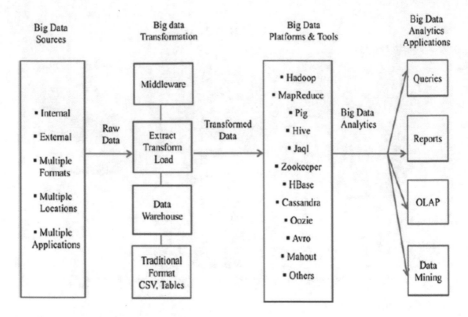

**Fig. 8** Conceptual infrastructure of BDA [14]

any other type of attack; also, it prevents many inaccurate insurance claims in a systematic way. It empowers various encryption techniques, network firewalls, processing of insurance claims, enabling patients to get better returns on their claims and caregivers are paid faster. For example, according to the Centres for Medicare and Medicaid Services big data analytics helped them to save more than 210.7 million USD in just a year by preventing frauds and boosting their security standards.

3. **Predictive analysis**: Big data helps to analyse used cases and tested data for BDA in healthcare that have the potential to assist researchers to discover causes and treatments for certain diseases. Such that doctors may actively monitor patients for traits of certain diseases, and adequate precautions should be taken before any malefic event to occur and the patient may be provided with personalized care and resources required for the treatment may be gathered before its too late.

4. **Human errors prevention**: Sometimes healthcare professionals may unintentionally prescribe a wrong medicine or wrong doses for a drug. These type of errors can be minimized by the use of big data as it can be used to analyse data and the corresponding prescribed treatment; by this, the loss due to human errors can be minimized and many lives can be saved. Hence, by the use of big data many busy healthcare professionals may work more efficiently and reliability towards healthcare is boosted.

5. **Building diagnostic machines**: These are the innovations in which the medical machines diagnose the disease and intelligently interpret the results by itself, so that the doctors are alerted for any abnormality if found in the results. For

example, in case of X-rays, MRIs and ultrasound images, the machine with the aid of big data can analyse results as well as report for any abnormalities.

6. **Big data in image processing**: One of the important sources of medical data is medical images generated through various imaging techniques like computed tomography (CT), magnetic resonance imaging (MRI), X-rays, molecular imaging, ultrasound, photoacoustic imaging, fluoroscopy, positron emission tomography-computed tomography (PET-CT) and requires large data storage infrastructures. These images hold a wide spectrum of different image acquisition methodologies. For example, structures of blood vessels structure can be visualized using magnetic resonance imaging (MRI), computed tomography (CT), ultrasound and photoacoustic imaging, etc.

## 4 Challenges in Introducing New Technologies in Health Sector

1. **Data security and privacy**: There are much chances in IoT that the data security and privacy may get compromised as the data being sent and received among various IoT-enabled devices lack various standard data transmission protocols, and also ambiguity may arise regarding data ownership and its regulation. These factors make data prone to hackers and crackers, and a result of which the personal health information (PHI) may get compromised and both patient and doctor may have to suffer from the loss due to the same [4].

2. **Hindrance in integration of multiple devices**: The developers of various IoT-enabled devices do not use standard protocols for communication, and thus, the devices may face hindrance while transmitting and storing data which, as a result, slow down the process of data aggregation among various devices and thus reduces the scalability factor of IoT-enabled devices in healthcare.

3. **Data accuracy and overload**: Due to the non-uniformity of protocols for data transmission among various IoT-enabled devices, data accuracy is compromised and some devices also face issues like data overload, fact being that IoT devices may result in the accumulation of terabytes of data which can be further analysed to gain insights.

4. **Costs**: The higher cost of healthcare services in developed countries makes it difficult for common man to access them so it gives rise to the concept of medical tourism in which the person travels to developing countries for treatment and can save up-to 90% of its expenditure on healthcare services. But there is a huge scope through IoT and big data-powered devices to cut off medical expenditures and healthcare costs.

5. **Closing the loop for drug delivery**: The loop for drug delivery can be closed by combining data from IoT-enabled devices and data analytics and developing devices that can automatically respond to the corresponding changes, for example applications like artificial pancreas. Also, it has the potential to develop

automated drug delivery systems and other sensor data-based and data-driven applications. Medical devices or smartphone-based health applications powered with big data analytics can automate the systems for drug delivery and give recommendations for various diseases.

6. **Lack of digitalized health data**: A very small amount of medical data and clinical information are available in digital format and are ready to be used for analytics [5].
7. **Lack of unified health data**: The health data should be unified among various data sets for patients across different hospitals.
8. **Data storage**: Data stored in different and distributed data silos make it difficult for further analytics to happen and make analytics highly unstable [6].

## 5 Conclusion

Healthcare sector is of uttermost importance in any country in terms of both the revenue that it generates and the services that it is providing to mankind. It comprises of technology-enabled health services, multi-facility hospitals, on-board healthcare services, telemedicine, and specialized machines, etc., and also the large-scale employment that this sector provides also makes this sector of prime consideration. So this sector needs to be powered with technology-enabled health services and employs the usage of new technologies like big data analytics and Internet of things (IoT) in healthcare.

In this paper, we have majorly focused on two technologies that are big data analytics and Internet of Things in Healthcare. Big data analytics (BDA) has brought together by two different branches of computer science that are big data and analytics and is collectively designed to deliver a data management approach. Big data is mostly described by the industry professionals as extremely large amount of unstructured and structured data an organization may hold, generate and share.

Internet of things (IoT) is an arrangement of multiple physical, electronic and various sensor-based devices connected to each other which enables them to store and exchange data among each other.

## References

1. V. Yadav et al., A biometric approach to secure big data, in *International Conference on Innovation and Challenges in Cyber Security* (ICICCS-2016), 03–05 Feb 2016, pp. 75–79. ISBN: 978-93-84935-69-6
2. V. Yadav et al., Big data analytics for health systems, in *IEEE International Conference on Green Computing and Internet of Things* (*ICGCIOT-15*), 08–10 Oct 2015, pp. 253–258. ISBN: 978-1-4673-7909-0
3. https://www.i-scoop.eu/internet-of-things-guide/internet-things-healthcare/

4. https://www.peerbits.com/blog/internet-of-things-healthcare-applications-benefits-and-challenges.html
5. https://econsultancy.com/internet-of-things-healthcare/
6. 12 Examples of big data analytics in healthcare that can save people. https://www.datapine.com/blog/big-data-examples-in-healthcare/
7. https://bigdataldn.com/big-data-the-3-vs-explained
8. https://readwrite.com/2018/01/13/internet-things-healthcare-possibilities-challenges
9. https://yourstory.com/2018/08/market-research-importance-healthcare
10. https://www.solutionanalysts.com/blog/5-iot-applications-that-will-change-the-face-of-healthcare
11. https://www.researchgate.net/figure/Distribution-on-the-forms-of-healthcare-records-in-this-review_tbl2_318897881
12. https://www.indiamart.com/proddetail/emergency-medical-services-4795931133.html
13. https://gajitz.com/touchscreen-monitor-lets-doctors-monitor-patients-remotely
14. W. Raghupathi, V. Raghupathi, Big data analytics in healthcare: promise and potential. Health Inf. Sci. Syst. **2**(3) (2014). http://www.hissjournal.com/content/2/1/3

# Compendious and Succinct Data Structures for Big Data

Vinesh Kumar, Akhilesh Kumar Singh and Sharad Pratap Singh

**Abstract** Recent growth of cloud data and cloud computing has been expediter and predecessor to the appearance of big data. Cloud computing has improved data storage by calculating and saving time with the means of identical and relevant technologies. While cloud computing provides important benefits over conventional physical deployments, its platform has also originated in numerous forms from time to time (Gog and Petri in Softw Pract Exp 44:1287–1314 [1]). In this proposed paper, the main data structure used in big data is tree. Quad tree is used for graphics and spatial data in the main memory. Traditional sub-linear algorithms that are used to handle Quad tree were inefficient. Also, SDS can be optimized for query handling and space. Optimized SDS can improve functionality of different SDS like rank and select, FM index (Blandford et al. in Proceedings of the fourteenth annual ACM-SIAM symposium on discrete algorithms, pp 679–688, 2003 [2]). As geometric data, proteins database, Gnome data, DNA data are large databases for main memory, an efficient and simple representation is required in main memory of computer system. However, overall quantity of storing area is not a vital problem in recent times, considering the fact that external memory can store large quantity of data and may be inexpensive, time needed to get access to information is a vital blockage in numerous programs. Number of access for hitting outside of memory is conventionally lower than number of access for hitting into main memory which has caused examine of recent compressed demonstrations of information that might be capable to save identical data in a reduced area.

**Keywords** SDS · Big data · CT · RMQ · XML

V. Kumar (✉) · A. K. Singh · S. P. Singh
GLAU, Mathura, UP, India
e-mail: vinesh.kumar@gla.ac.in

A. K. Singh
e-mail: akhileshkr.singh@gla.ac.in

S. P. Singh
e-mail: sharad.singh@gla.ac.in

© Springer Nature Singapore Pte Ltd. 2021
X.-Z. Gao et al. (eds.), *Advances in Computational Intelligence and Communication Technology*, Advances in Intelligent Systems and Computing 1086,
https://doi.org/10.1007/978-981-15-1275-9_37

457

# 1 Introduction

Among all the various applications of big data [3, 4], the main application of this paper is to represent raster information in geographic information systems, where information is measured. In spatial information it is a very common property and is exploited through typical demonstrations in this area. Nonetheless, configurations of general data are similar to K2-tree and do no longer take profits of this form of symmetries in spatial data [5]. The message is demonstrated equally a chain of source symbols $x_1x_2, \ldots, x_n$. Coding procedure of message contains making use of cipher to every sign in message and concatenating all of codeword resultant. Output of encoding is series of target symbols $C(x_1)C(x_2), \ldots, C(x_n)$ [6]. Decrypting technique is opposite method that acquires source symbol consistent to every code word to reconstruct real note. Compressed data is a universal trouble in computer science. Solidity method is utilized approximately universally to permit efficient storage and management of big datasets [1, 7]. Large quantity of data (in kind of text, image, video, and so forth.) that needs to be managed and conveyed daily makes flawless need of solidity methods that minimize scale of data for a large effective loading and communication. Compression is powerfully connected with entropy [8]. Given a message, objective of solidity is to decrease its mass while preserving all of data it consists of. Entropy signifies common area need to keep a sign for the available data source [9]. Therefore, to decrease space vital in count to the entropy of source that indicates notional smallest is the main goal of solidity. Distinction among distance of a given code and source of entropy is known as redundancy.

## 1.1 Succinct Data Structure

On the basis of rank and select, succinct data structure is faster in runtime performance and compression than traditional data structure [2, 10]. The basic aim behind the usage of different data structures is to improve memory consumption of dataset. Space required for succinct data structure is less as compared to other data structure. It has been used for information retrieval and bioinformatics [10, 11]. Then succinct dataset is compared with uncompressed suffix array, and it requires $2n + O(n)$ bits for tree representation, whereas later requires *klogn* bits per node which consumes huge memory (Fig. 1).

## 1.2 Big Data

The growth of cloud data stores and cloud computing has been expediter and predecessor to the appearance of big data. A cloud computing is co-modification of

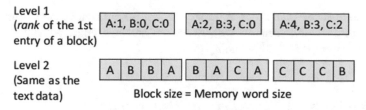

Level 1
(*rank* of the 1st
entry of a block)

| A:1, B:0, C:0 | A:2, B:3, C:0 | A:4, B:3, C:2 |

Level 2
(Same as the
text data)

| A | B | B | A | B | A | C | A | C | C | C | B |

Block size = Memory word size

**Fig. 1** Succinct data structure implementation on hardware

data storage and calculating time by means of identical technologies. It has important benefits over conventional physical deployments. Nevertheless, cloud platforms originated in numerous forms, and from time to time, it is combined with traditional architectures [2, 12].

## 1.3 SDS in Big Data

Today, the big data has become a buzz word and still in developing stage. Weather forecasting, basically the problem of initial value, is considered by researcher as a case of big data, which will help to improve the accuracy of forecasting. For handling this huge data need for weather forecasting, there is a requirement of a well-organized data structure [5, 13].

Through this section researcher discuss process of weather forecasting, different approaches used for forecasting, review of big data and role of big data in weather forecasting, review of data structures used for big data as well as weather forecasting. Numerical weather prediction (NWP) is the desirable technique for weather forecasting. The data structures available till now have some limitations to apply for weather data, and hence, researcher plans to design a new data structure which will store the weather data efficiently [14].

The present implemented data [15] arrangement for massive facts is an arrangement of data in a tree form for big facts units, which saves summary of statistics, having low value of degree and able to filling most of the demands of person as well as unique facts till now. The tree is an extension of quad tree statistics shape [16]. The data and information produced by the satellites and supercomputers are very difficult to handle by the simple databases, and they need tree like structure to handle these types of data. Global climate model is applied for studies for reason first of all its far MDD, and climate version information is big in length and did not have clean get admission to it. Data structure can be discussed with four different names such as transformation function, subdivision structure, subdepartment criteria, and location codes. SDS basics [12] are mentioned in this section. A compression based data in which the offspring of each node regarded as BFS or DFS [13]. Variable-length encoding is used for compression in this data shape. A tree is used to solve issues of looking minimal and most in a variety. It can resolve records shape trouble in integers in lexicographic order. Another tree that is nonlinear records forms for set of code strings. In this scenario, we will add on characters edges with the route from parent node to separately descending leaf.

## 1.4 Big Data—A Challenge and Opportunity

Data is composed at the extreme level. If we focus on a large variety of application areas, data is being collected at brilliant level. On the authenticity of model, the model of judgment is based on the estimation. It is also based on self-authenticated data. Investigation of big data provides every feature of users [9, 17]. These features are composed of mobile applications, life sciences, marketing, etc.

This large data structure has apparent to convert not abandoned research into learning phase. A newest comprehensive measurable evaluation of audible approaches affianced by 35 allotment schools in NYC has authorize that one of apical 5 rules accompanying with assessable enlightening account abounding angry into application figures to adviser tutoring. Imagine a cosmos in which we accept get admission to a gigantic database where we accrue anniversary assertive a measurement of every scholar's bookish all-embracing performance. Moreover, there is an able tendency for all-inclusive Web deployment of advisory actions, and this can actualize an added huge abundance of abundant abstracts about academy scholars' all-embracing routine. It is abundantly intended that application information processing can reduce payment process of healthcare sector while enslaving its superiority, finished authoritative affliction added arresting and founding it on added accepted connected observing. McKinsey estimates accumulation of three billion dollar's anniversary year in the USA alone.

In a matching of band, abundant affairs fabricated for amount of big abstracts for city-limit's planning, active bus line, careful clay, ability extenuative, beautiful substances. Abundant bodies acutely cognizance artlessly on analysis/modeling phase: at the aforementioned time as that appearance is vital, it is distant of slight advance after adverse levels of abstracts assay pipeline.

## 2  Review of Literature

Grossi et al. [7] recommended a new execution of suffix arrays of compacted techniques which represents new transactions between time and space complexity for a given text of n codes along with every text of the alphabet, where every sign was programmed by $\log|\Sigma|$ bits. This form represents complex arrays and their usage while conserving wide-ranging text indexing functionalities, and its length adjust according to the size $O(m \log|\Sigma| + \text{poly} \log(n))$ time. Term $M_h \log|\Sigma|$ signifies $m$th—order observed of the text. This means that their key changed and uses optimal space other than lower-order terms. Gottlob et al. [11] defined the ability to remember the site of ontological database admittance, in the form of relational database R, A-box is defined and Boolean conjunctive query is evaluated toward R. This condition can be rewritten on recursive data and can be accessed over database R. Conversely, D Lite version is used to authorize for role presence, altering methods are the result of non-recursive approaches. This bounces rising stab to stimulate inquiry of whether such reworking basically needs to be of larger size. In this article, they show that its just likely to interpret $(\Sigma, q)$ into equivalent non-recursive polynomial size of Datalog program. Ladra [18] represents the data retrieval efficiency issues they deal with issue of efficiency displayed by the data structure of compressed for and various algorithms that can be applied in various fields and applications and hold various alike properties.

In this paper, they discussed the following concepts:

(i)   For the integer sequence with encoding system of variable length which allows quick access to the system and gives results in a good way containing techniques for the prediction having low intensity and space.

(ii)  Word-based and text-based methods of compression that allow quick searches for words and phrases; words and phrases on the text of the compressed form and utilized equal space and gave better than the traditional methods for the small occupying space.

(iii) Web graphs are the well-organized techniques that require forward and back-ward for the smaller area. Muthukrishnan [17], they discussed the latest algo-rithms for the data streams and connected requests which are useful for the research purposes. Basically, they work upon the three puzzles Puzzle 1: Dis-cover missing numbers Puzzle 2: Spinning, Puzzle 3: Pointer and Chaser. Aaronson [19] defined various techniques such as soap bubbles, quantum com-puting, for computing, and entropic calculating. Steps of soap bubbles also have some output. He has not idea that how these algorithms helps to solve NP-Problems. He also suggested that by studying them deeply, we are able to solve the computational problems and also added that it could be helpful in physics. Richard et al. [9], they measured concise or presentation of trees using space criteria that and support various functions related to the naviga-tions. Mainly they focused on static ordinal trees where every node of children is well ordered. These set of operations are combined with the previous results. Their protest takes $2n + O(n)$ bits to construct n-node tree, that's inside $o(n)$ bits of information-theoretic minimum and supports all operations in $O(1)$ time on RAM model.

Munro et al. [2] present suffix of the tree that uses $(n \lg n) + O(n)O(m)$ time, where n is size of text and m is pattern period. The output of structure is easy to understand, and using Muthukrishnan answers are evaluated. Past compact illustrations of suffix trees had also complex lower-order assumptions and want more time for searching. With fixed size, the alphabet it do not considered of this structure and takes similar time $O(m \lg k)$ time for string searching.

Munro et al. [10], they focused on the static objects such as trees like binary, root tree with order, and a balanced series of parentheses. Their symbols exploited an amount of space inside the algorithm and needs lesser space and time. Further, it is compared with the previous researches an comparison work. It goes from root then left to right child to determine its time in case of binary tree.

Jacobson [16] data structures that constitute stationary unlabeled trees and simple graphs are generated. This arrangement of data was more space proficient as compared to conventional pointer-based illustrations; however, they are just as time proficient for traversal processes. For trees, the whole arrangement states asymptotically most wanted. It is possible with the help of n-node trees with fewer bits per node for encoding, as $N$ grows without bound.

Static unlabeled trees and planar graphs of data structure were proposed by Jacobson [16] After comparison with conventional pointer-based structures that were time proficient for traversal processes, these structures were more space proficient. Tries as data structures are most desirable because they have the efficiency of encoding n-node trees with less bits per node. As per as value of n grows without bound for planar graphs, this data structure utilizes linear space.

For representation of graphs, Blandford [20] considered an issue. So they defined a new data structure for representing n-vertex unlabeled graphs. It was able to overcome problem of previous output for graphs. They gave some experimental output after using "real-world" graphs which includes three-dimensional finite elements, Internet router graphs, link graphs from Web, VLSI, and street map graphs. This method uses less space as adjacency lists with some order of magnitude in support to depth—first traversal in same time duration as in running time.

Blandford et al. [21] proposed a method for efficiently representing sets $S$ having size $n$ of an ordered universe $U = \{0, ..., m - 1\}$. Let an ordered dictionary, structure $D$ has $O(n)$ pointers. A simple blocking method was proposed in which an ordered set of data structure was created which performed equal operations in equal time bound with $O(n \log (m + n)/n)$ bits while information-theoretic lower bound remains constant. The unit cost of RAM model was chosen with word size $\log \vee U \vee @ \Omega$) and a table of size $O(m^a \llbracket \log \rrbracket^2 m)$ bits for this constant $\alpha > 0$. Time bound for their operations carried $1/\alpha$ component. They gave experimental output for Standard Template Library (STL) execution of red–black trees, and for an execution of Traps whose execution was associated with blocking and without blocking. Blocking versions utilize a factor among 1.5 and 10 less space depending on the density of set.

Kumar et al. [13–15, 22] data structures for big data representation have been shown. Three different tree data structures had been proposed.

# 3 Proposed Methodology

SDS has been especially considered in a theoretical setting. There are various articles published that define the strength of SDS in a broader way and some authors explained binary sequences, trees, and layout of this design are also connected to bit probe involvedness of arrangement of data and complexity of time–space, and there are SDS sets of functions, threads, trees, charts, associations, sequences, permutations, integers, geometric data, unique formats (XML, JSON), and many more. Although the cost of memory is decreasing and the processor speeds are increasing day by day, the amount of textual data to be processed (such as dictionaries, encyclopedias, newspaper archives, Web and genetic databases) is also increasing at a much higher rate. These representations have useful applications in portable devices (like mobile phones and smart cards) where the amount of memory is limited. The latest technique for the data structures indexing is to compress and file information in a single shot. Main motive of these filtered techniques indexes is to save best query times and consumes best space also. Improvement in these techniques compartmentalization has conjointly delayed to additional combined structures, for instance, trees and subsets. Compressed text compartmentalization makes serious utilization of summary information structures for set info, or dictionaries. Succinct data is mostly deal with the static data. Saving of space is a big principle for exploitation of succinct data structures for dynamic processes.

*Datasets*: In our prior work that was based on rank and select, bit vectors of different densities and size were created to evaluate the data structures used there. The range of instance size varies from 1 MB to 64 GB, multiplying four times of previous size each time. The FM indexes are evaluated on the basis of following datasets.

- The biggest available instance file of size 200 MB for all categories including DBLP, DNA, PROTEINS, SOURCES, ENGLISH of the Web site (pizza and chili) are considered as standard benchmark in text indexing provides texts from varying applications.
- A 3600 MB text file containing genome sequence (DNA) created by appending the soft masked assembly sequence of the human and cat DNA. All the comment separators were removed and are exchanged with separator token to adjust the alphabet size.

*Optimization*: The performance of SDS is optimized by showing the effect of feature set on varying indexes. As far as result discussion is considered, the implementation proposed runs approximately 50% times faster than highly optimized feature sets. Although in few cases of the test set the runtime of the implementation setup reduces slightly compared to original experimental setup, the results were found to be good.

## 4 Results and Discussions

*Experimental Setup*: To calculate the value of the future Top-p completion methods, Completion Tree (CT), Score-Decomposed Tree and baseline RMQ Tree, its main features are with the datasets and application of varying situations on an Intel i5-2640M 2.9 GHz processor with a 32 GB of DDR3 RAM attached to 2 sockets indicating non uniform memory access and 12 MB L3 Cache, assembled with associated software. All experiments are performed single threaded including two-level translation look aside buffer. The machine setup is also used for sanity checkup on smaller experimental data. Ubuntu linux is used as operating environment, g++ as application environment.

Queries P: 10,454,552 searching queries those are from the Google query log (20). This dataset is representative of the style, and frequency of queries users may enter into the search box of a search engine or large Web site.

Queries Q: We have filtered more than 300 M search queries from Google search engine for scalability evaluation. Figures 4 and 5 show the graphical representation of space complexity of SDS and optimized SDS in comparison with size of stored big data. From below graph, it is clear that optimized SDS takes very less space as compared to SDS. Figure 5 shows the performance of SDS and optimized SDS with dataset from Google.

*Results*: As discussed Fig. 2, CT **utilizing indicator math is fundamentally speedier than information arrangements utilizing adjusted brackets for traversal, particularly in discovery the underlying locus hub**.

Figure 3 shows the graphical representation of **time complexity of SDS and**

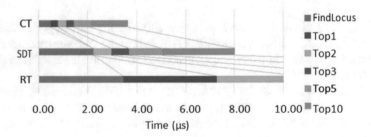

**Fig. 2** Completion time breakdowns

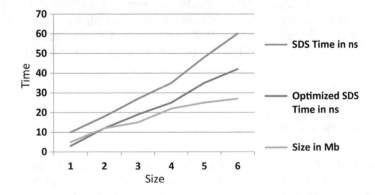

**Fig. 3** Time complexity for big data

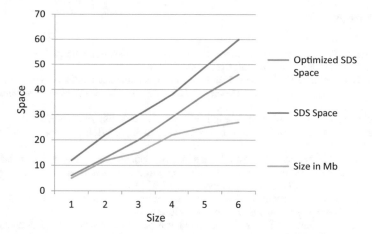

**Fig. 4**  Space complexities for big data

**optimized SDS in comparison with size of stored big data**. From below graph, it is clear that optimized SDS takes very less time as compared to SDS.

Figure 4 shows the graphical representation of **space complexity of SDS and optimized SDS in comparison with size of stored big data**. From below graph, it is clear that optimized SDS takes very less space as compared to SDS.

*Time and Space Complexity analysis*: The space complexity of the experiment discussed is bounded by $O$ $(n + n/k)$ bits. As the main aim was improvement of implementation by changing feature set, the main highlight was improvement of space rather than time.

## 5  Conclusion

In this paper, we introduced **three information constructions to speech the issue of Top-p completion**; each through various interplanetary/time/many-sided quality exchanges offs. **Trials on expansive scale datasets demonstrated** that Completion Tree, in light of established information structures, requires generally two fold the span of Score-Decomposed Tree, in light of compact primitives. Nonetheless, it is about twice as quick. Things being what they are sorting out the information in a territory delicate requesting are important to the execution additions of these dual constructions over the easier RMQ Tree. If we wish to increase scalability in big data, then we have to make the proposed data structures practically. We are not implemented these SDS due to space constraints. We can design such pseudocode that has minimum time complexity. We have to choose programming language to implement these data structures for big data.

# References

1. S. Gog, M. Petri, Optimized succinct data structures for massive data. Softw. Pract. Exp. **44**(11), 1287–1314 (2003). https://doi.org/10.1002/spe.2198
2. D.K. Blandford, G.E. Blelloch, I.A. Kash, Compact representations of separable graphs, in *Proceedings of the Fourteenth Annual ACM-SIAM Symposium on Discrete Algorithms* (2003), pp. 679–688
3. S. Rajbhandari, A. Singh, Big data in healthcare, in *International Conference on Innovative Computing and Communications* (Springer, Berlin, 2019), p. 261. https://doi.org/10.1007/978-981-13-2354-6_28
4. A. Saxena, A. Chaurasia, N. Kaushik, Handling big data using map reduce over hybrid cloud, in *International Conference on Innovative Computing and Communications* (Springer, Berlin, 2019), p. 135. https://doi.org/10.1007/978-981-13-2354-6_16
5. D.K. Blandford, G.E. Blelloch, Compact representations of ordered sets, in *Proceedings of the ACM-SIAM Symposium on Discrete Algorithms*, Jan 2004
6. G. Gottlob, T. Schwentick, Rewriting ontological queries into small non recursive data log programs, in *KR* (2012)
7. R. Raman, V. Raman, S.S. Rao, Succinct indexable dictionaries with applications to encoding kary trees and multisets, in *SODA* (2002), pp. 233–242
8. J.I. Munro, V. Raman, A.J. Storm. Representing dynamic binary trees succinctly, in *Proceedings of the Twelfth Annual ACM-SIAM Symposium on Discrete Algorithms (SODA-01)*, 7–9 Jan 2001 (ACM Press, New York), pp. 529–536
9. J.I. Munro, V. Raman, Succinct representation of balanced parentheses and static trees. SIAM J. Comput. **31**(3), 762–776 (2002)
10. R. Raman, S.S. Rao, Succinct dynamic dictionaries and trees. in *Annual International Colloquium on Automata, Languages and Programming (CALP), Lecture Notes in Computer Science*, vol. 2719 (Springer, Berlin, 2003), pp. 357–368
11. S. Ladra, Algorithms and compressed data structures for information retrieval. Ph.D. thesis, University of A Coruña, 2011
12. P. Ferragina, G. Manzini, On compressing and indexing data. J. ACM **52**(4), 552–581 (2005). (Also in IEEE FOCS 2000)
13. V. Kumar et al., Data representation in big data via succinct data structures. Int. J. Eng. Sci. Technol. (IJEST) **10**(1), 21–28 (2018). ISSN 0975-5462. (UGC Approved Journal till 01/05/2018, ICI indexed)
14. V. Kumar, J. Shekhar, S. Kumar, Comparative study of sensible compendious data structures for massive data store, in *Proceedings of International Conference on Research and Innovation in Engineering (ICRIE-2016)*, 12–13 Feb 2016, Gr. Noida, India, pp. 128–131. ISBN:978-93-5254-3
15. V. Kumar et al., Compendious and optimized data structures for big data store, in *Proceedings of International Conference on IOT (ICIoTCT-2018)*, 26–27 Mar 2018, MNIT Jaipur, India, SSRN (Elsevier). (Scopus indexed)
16. R. Raman, V. Raman, S.S. Rao, Succinct indexable dictionaries with applications to encoding k-ary trees and multisets, in *ACM—SIAM Symposium on Discrete Algorithms*, (2002), pp. 233–242
17. Scott Aaronson, NP-complete problems and physical reality. SIGACT News **36**(1), 30 (2005)
18. B. Chazelle, Who says you have to look at the input? The brave new world of sublinear computing, in *Plenary talk at the 15th Annual ACM-SIAM Symposium on Discrete Algorithms (SODA 2004)* (2004)
19. R.F. Geary, R. Raman, V. Raman, Succinct ordinal trees with level-ancestor queries, in *SODA'04: Proceedings of the fifteenth annual ACM-SIAM symposium on Discrete algorithms* (Society for Industrial and Applied Mathematics, 2004), p. 110
20. J.I. Munro, R. Raman, V. Raman, S.S Rao. Succinct representations of permutations. in *Annual International Colloquium on Automata, Languages and Programming (CALP), Lecture Notes in Computer Science*, vol. 2719 (Springer, Berlin, 2003), pp. 345–356

21. J.I. Munro, S.S. Rao, Succinct representations of functions, in *Annual International Colloquium on Automata, Languages and Programming (CALP)*, *Lecture Notes in Computer Science*, vol. 3142 (Springer, Berlin, 2004), pp. 1006–1015
22. V. Kumar, J. Shekhar, Design of sensible compendious data structures for Top p completion in Massive data store, in *Proceedings of National Conference on Research and Innovation in Engineering (RSCTA-2016)*, 12–13 Mar 2016, Ramanujan College University of Delhi, India, pp. 256–131. ISBN 978-93-5254-320
23. R. Grossi, A. Gupta, J.S. Vitter, High-order entropy-compressed text indexes, in *SODA* (2003), pp. 841–850
24. S. Muthukrishnan, Data streams: algorithms and applications, in *Plenary Talk at the 14th Annual ACM-SIAM Symposium on Discrete Algorithms (SODA 2003)* (2003)
25. D. Benoit, E.D. Demaine, J.I. Munro, R. Raman, V. Raman, S. Rao, Representing trees of higher degree. Algorithmica **43**(4), 275–292 (2005)
26. J.I. Munro, V. Raman, S.S. Rao, Space efficient suffix trees. J. Algorithms **39**, 205–222 (2001)

# Neural Network Techniques in Medical Image Processing

Sonika Nagar, Mradul Jain and Nirvikar

**Abstract** The paper is a critical review of the uses of neural network in medical imaging process. First, a thorough review of machine-learning-based artificial neural network is introduced, with its historical background. Inhibiting reasons why ANN was almost dropped altogether in research centers is given and an explanation why it has become so important in modern medical image processing is given. Types of neural networks including a brief description of CNN are given before an introduction of the application in the medical image processing before the paper concludes with an emphasis that ANN has not only improved medical image processing, it is likely to take a leading role in medication in the coming years.

**Keywords** Artificial neural network (ANN) · Image processing · Image segmentation · Image detection and classification · Convolutional neural network (CNN)

## 1 Introduction

Medical imaging plays a prominent role in diagnosing the diseases and in over the past few decades, the interests in medical image processing has significantly increased [1–3]. Especially machine-learning-based artificial neural networks (ANNs) methods have grabbed more attention. An ANN is developed for a particular application, such as data classification or pattern recognition after undergoing a learning process (Wang, Heng & Wahl). The purpose of this article is to cover the techniques of ANN used in medical imaging. Rather than covering all the research aspects and related issues of ANNs in medical image processing, we target three major topics of medical

S. Nagar (✉) · M. Jain
ABES Engineering College, Ghaziabad, India
e-mail: nagar.sonika1994@gmail.com

M. Jain
e-mail: mradul.jain@abes.ac.in

Nirvikar
Department of CSE, College of Engineering Roorkee, Roorkee, India
e-mail: nirvikarlohan@yahoo.com

© Springer Nature Singapore Pte Ltd. 2021
X.-Z. Gao et al. (eds.), *Advances in Computational Intelligence and Communication Technology*, Advances in Intelligent Systems and Computing 1086,
https://doi.org/10.1007/978-981-15-1275-9_38

images: first, medical image pre-processing (i.e. enhancement and de-noising) [3, 4], second, medical image segmentation [5–8] and third, object detection, recognition and classification [11–14]. We do not intend to go any deeper into any particular algorithm or technique, rather summarizing the major approaches and pointing out some interesting factors of the neural networks especially CNN in medical imaging is done, furthermore, we want to give some ideas to answer what are the major applications of neural networks techniques in medical image processing now and in the nearby future.

## 1.1   Historical Background of Neural Network

The development of neural networks and their simulations is not a very much recent development. However, evidence exists to show that the same field was in existence before the advent of computers, and is known to have overcome at most a major setback and a number of eras. The Warren McCulloch and Walter Pits as a neuro-physiologist and logician, respectively, designed the first artificial neuron in 1943. Unfortunately, the technology that was available then could not afford them to do much. A number of advances have been made possible using some of the inexpensive computer simulations. In this period, professional support and support were extremely low, resulting in few important discoveries in the field by researchers. In spite, of the short backs, the few pioneers managed to build up convincing technology that surpassed the limitations that had been noted by Minsky and Papert. In 1969, Minsky and Papert published a book that detailed the main hindrances of frustration on neural networks by researchers, which made it to be accepted without a critical analysis, leading to the near collapse of the ANN field.

## 1.2   Significance of Using Neural Networks

The neural networks, have the capabilities of deriving meaning from a complex set of data, extract patterns and be in a position to show trends that are not easy to detect by humans or other computer methods. A neural network that is well trained is seen as an "expert" in the specific area it is designed to analyze. The same expert is then used in the interpolation of new projection whose aim is to answer "what if" questions.

Other advantages of ANNs include its capability of adaptive learning, which is the ability to adopt by learning how to carry out tasks that based on the data provided from the initial experience or training. An ANNs is reliable because of its self-organization capabilities through the creation of its own representation or organization of the data it receives during the learning process. An inherent advantage of neural networks over other is that it has capabilities to operate in real time through computations that are either in parallel or by use of special hardware devices to take advantage of the prevailing capability. ANNs are also preferred because of what is termed as fault tolerance through redundant information coding. In normal circumstances, any damage on the network often leads to its poor performance, however, some ANN capabilities are still maintained even after the network has undergone major destruction [1].

## 2 Architecture of Neural Network

An artificial neuron is a designed model system having several inputs and one output. In engineering, such systems are referred to as multiple input single output (MISO). Both the training and the testing modes of operations are within the neuron. The training mode empowers the neuron to either fire or not to fire whenever a specific input pattern is realized. In the testing mode, when a trained input pattern from the training mode is detected, its related output becomes the present output. If the input pattern detected is not associated with the any of the input patterns of trained list, then the rules of firing are used to determine whether to fire or not.

### 2.1 Basic Structure

For a neuron, if the inputs are denoted as $x_1$, $x_2$ and $x_3$, and the synaptic weights applied to the input are symbolized as $w_1$, $w_2$ and $w_3$.

Simply, a matrix multiplication between input $x_i$ and weights $w_i$ arrive at the weighted sum. Bias acts as an intercept that is added to the linear equation. It is added to the weighted sum, which is used to adjust the output along with the weighted sum of the inputs to the neuron. The processing done by a neuron is thus denoted as:

$$y = \sum_{i=1}^{n} W_i X_i + b$$

where $n$ is the number of inputs.

An activation function is applied on this output which determines the firing of neurons in a neural network Fig. 1.

### 2.2 Feed-Forward Network

Feed-forward ANNs shown in Fig. 2 makes it possible for signals to move in one direction only, which is from the inputs to the output of the network. In other words, there is no loop for feedback. What this means is that the output of a given layer does not influence the same layer. A characteristic of feed-forward ANN is that it

**Fig. 1** The basic model of a neuron

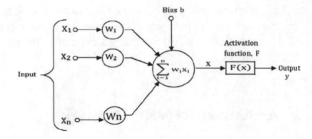

**Fig. 2** A simple feed-forward network

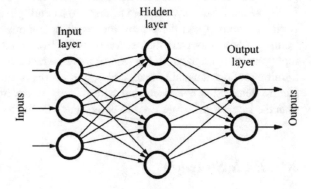

is a network that is straightforward because it only associating all its inputs to the outputs [2]. Feed-forward ANNs are also referred to as top-down or bottom-down networks.

## 2.3 Feed-Back Network

Unlike, feed-forward networks, the feed-back networks shown in Fig. 3 allow signals to move in both directions. This type of network introduces loops within the network, making them extremely powerful and often can be overly complicated. One key characteristic of feed-back networks is that they are dynamic, meaning that they easily change from one state to another continuously until they attain their optimum point of operation. This optimum position is the equilibrium point, which they retain

**Fig. 3** A simple feed-back network

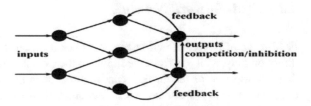

until the input sources are changed again and another optimum point is reached. These types of neural networks are also known as recurrent, which meant for the feed-back connections in the single-layer network.

## 2.4 Convolutional Neural Network

CNNs are a type of biologically inspired feed-forward networks characterized by a sparse local connectivity and weight sharing among its neurons. A CNN can also be seen as a sequence of convolutional and subsampling layers in which the input is a set of $H \times W \times D$ images, where $H$ is the height, $W$ is the width and $D$ is the number of channels which, in the case of RGB images corresponds to $D = 3$.

A typical convolutional layer (volume) is formed by K filters (kernels) of size $F \times F \times D$, where $F \leq H$ and $F \leq W$ These filters are usually randomly initialized and are the parameters to be tuned in the training process. Since the size of the filter is generally strictly smaller than the dimensions of the image, this leads to a local connectivity structure among the neurons. Each of this convolutional volumes has an additional hyper-parameter, S, which corresponds with the stride that the filter is going to slide spatially in the image Fig. 4.

## 3 ANN in Medical Image Processing

In recent years, medical image processing has been increasing used and analyzed using artificial neural networks (ANNs). The main components of medical image processing that heavily depend on ANNs are medical image recognition and object detection, medical image pre-processing and medical image segmentation [3].

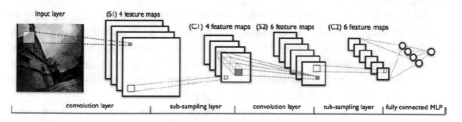

**Fig. 4** Example of CNN architecture (LeNet, 1998–5 layers) [5]

## 3.1 Pre-processing

Image reconstruction and image restoration are the two categories under which neutral networks are applied in image processing. A specific type of ML-based ANN used in image reconstruction is the Hopfield neural network [1]. Hopfield neural network has a major advantage that the problem of reconstruction of medical image can always be conceptualized as an optimization problem making the convergence of the network to a stable position and also minimize the energy functions.

A high number of neural network applications in medical image pre-processing are concentrated in medical reconstruction. In the most simple approach of medical image restoration, filtering is used to remove noise from the image neural network filters (NF) for this problem was developed by Suzuki et al. who also proposed the first neural edge enhancer (NEE) which is based on modified multilayer neural network used to enhance edges that are desired [3]. By training using teaching edges and input noisy images, the NEE is able to attain the function of desired edge enhancer. NEE is robust against noise, when compared with conventional edge enhancers and it is also able to enhance continuous edges found in images that are noisy. NEE has an also superior capability to the conventional edge enhancers with the same desired edges.

## 3.2 Image Segmentation

Neural networks that are formulated with feed-forward capabilities are the most used in medical image segmentation. Feed-forward neural networks perform much better than their traditional maximum likelihood classifier (MLC) counterparts do [3]. The segmental images using neural networks appear to have less noisy than MLC counterparts do and they are not severely affected to the selection of the set of training used than their MLC counterparts.

The main aim of introducing Hopfield neural networks was for the purpose of finding satisfactory solutions associated with problems of optimization that are complex. This is the reason why they have come of edge as alternatives to the traditional optimization algorithms for reconstruction of medical image that can formulate as a problem of optimization. For example, Hopfield neural network has been used for segmentation of some organs from the medical images [3].

Author in [6] has proposed a method by which whole brain can be segmented within 25 s to 3 min. In [7], CNN model used for segmentation for pancreas achieves an average dice score of 46.6–68.8%. The approach used for segmentation of the ultrasound images of placenta's first trimester gives almost equivalent performance compared to the results acquired through MRI [8]. In [9], an automated segmentation approach used for liver, heart and great vessels MRI images have great potential for clinical applications. CNN used for retina vessel segmentation by author in [10] reduces the occurrences of false-positives thus increases the specificity of the model.

## 3.3 Object Detection, Recognition and Classification

Back propagation neural network has been applied in several cases for image detection, recognition and classification. Back propagation in image recognition has been used in interpreting mammograms cold lesion detection in SPECT image, diagnosis of liver disease classes based on ultra sonographic reduction in the occurrences of false-positives (higher specificity) in the computerized detection of modules of the lungs in LDCT, differentiating between interstitial lung diseases, as well as in radiography of the chest [3]. Other neural networks, such as ART neural network, Hopfield neural network, fuzzy neural networks [3]. Radial Basis Function Neural Network (RBFNN) and Probabilistic Neural Network (PNN) are extensively used in medical image detection and recognition [4].

In [11, 12], breast cancer diagnosis CNN models are proposed, in [11], MRI images are used and accuracy 96.39% accuracy was obtained with 97.73% sensitivity, and 94.87% specificity, while in [12], pectoral muscles were detected from the mammographic images used and 83% accuracy is obtained. In [13], a CNN model proposed for automatic detection for ECG images of myocardial infarction attains 93.53% average accuracy with noise while 95.22% without noise. An approach was proposed for the malignancy classification for CT scan images for lung cancer nodule and a high level of accuracy was achieved 99% [14]. A CNN model used for Gastric cancer classification attained high accuracy of 97.93% [15].

## 3.4 Fuzzy Neural Networks in Medical Imaging

Body part medical images are analyzed to find any anomalies and it is the key objective of every medical imaging application. The task of identifying abnormalities from medical images can be done using many methods including fuzzy logic, genetic algorithms or neural network applications. However, a hybrid approach that combines all the three is also possible. By combining the strengths of the three fields and independent algorithms to improve the use of medical imaging, treating of many diseases becomes easy. The hybrid fuzzy neural network has been used to show that aspects of a priori realized from the expertise if humans can be instrumental in decision-making [12]. It is also important to note that the algorithm used the advantages of parallel neuron architecture to realize reliable results. The network delivers correct image interpretations with respect to the capabilities of hybridizing from the different algorithms used.

The hybrid algorithm, in short, utilizes the benefits of the three unique factors, by combining them into one architecture. The responsibility of the fuzzy system in the combined architecture is very important as unpredictable and non-numerical fuzzy to decisions made by physicians and doctors in health facilities is the most vital part. Another use of fuzzy is the fusion of images based on the multimodal neural networks. The technique is used to fuse several images from a single source to achieve enhanced performance and accuracy. Pattern recognition and image fusion deliver high-end services in medical imaging field. The approach can attain better results when compared to a number of algorithms within the same domain.

The ratio of noise and data available is used in the measure of performance of the input images applied to the system. Research has shown that blurred images give quality results from the fusion method. As a disease diagnosis system, the results are marginally higher. The relatively new method is anticipated to offer a milestone in medical imaging by going beyond the limitations of the existing systems found in the same domain. The operator or doctor carries out image retrieval through parameters and handles image retrieval control. The quality is affected by adjusting the parameters. For example, tongue image algorithm retrieval often has more than 300 images as test runs coming from different sources [12]. The resulting comparisons are computed and evidence shows that the fuzzy CMAC realizes better outcomes for each color type of tongue image at an average of 91% retrievals.

The field of tissue classification has increasingly been using neural networks based on fuzzy logic. In this concept, different medical images are introduced to the algorithm as inputs. The algorithm then is used to generate and work as data of fuzzed classification. The outcome is the prominent features of the corresponding data images displayed as outputs.

## 4   Discussion and Comparative Analysis

The performance of neural networks when compared to the conventional ones for medical image processing in terms of quality was much higher. However, it should be noted that neural network training time is much more and the computations are quite complex and in order to realize reliable and high performance for eases that do not require training, often very large amount of training cases are needed [3]. In cases where ML-based ANN is trained with much less cases, the reliability of non-training cases is often low. For example, the ML-based ANN is only likely to fit the training cases. The disadvantages of most feed-forward neural network methods are that they are extremely slow when it comes to convergence rate and will always need a prior learning parameter. These disadvantages have halted the effective applications of feed-forward neural networks in segmentation of medical images.

With each passing day, we have overcome these limitations with the availability of more computation power, increasing number of digitally stored medical images and by improving the data storage facilities.

Tables 1 and 2 highlight the potential applications of CNN in medical image processing. By these tables, we conclude that CNN has been successfully applied to many tasks in medical image processing such as image segmentation, image detection and classification domain and provides great results in almost every case.

**Table 1** A table highlighting application of CNN in medical image segmentation

| Author | Application | Method | Dataset |
|--------|-------------|--------|---------|
| Havaei [6] | Brain tumor segmentation | DNN | MRI |
| Roth [7] | Pancreas segmentation | CNN | CT |
| Looney [8] | Ultrasound segmentation of first-trimester placenta; | CNN | The 3D volumetric ultrasound data |
| Dou [9] | Automated segmentation of heart, liver, heart and great vessel | CNN | MRI |
| Fu [10] | Retina vessel segmentation | CNN | Fundus images |

**Table 2** A table highlighting application of CNN in medical image classification

| Author | Application | Method | Dataset used | Accuracy obtained |
|--------|-------------|--------|--------------|-------------------|
| Rasti [11] | Breast cancer diagnosis | CNN | MRI | 96.39% |
| Dubrovina [12] | Automatic breast tissue classification | CNN | Mammography | 83% |
| Acharya [13] | myocardial infarction detection | CNN | ECG | 93.53% with noise 95.22% without noise |
| Zhang [14] | Lung cancer nodule malignancy classification | CNN | CT | 99% |
| Shen [15] | Gastric cancer identification | CNN | Gastric images | 97.93% |

## 5 Conclusion

From the historical background, working and application of ANN, its widespread use in medical image processing has energized the need for further research in solving centuries-old problems in medicine. A detailed review of the neural network techniques and its application in the field of medical image analysis is presented. The ANN will continue to play a central role in medical image processing where CNN-based method outperforms other conventional methods and networks. Already, its use in this field has made it possible for quick diagnosis of ailments and its expansion into other medical is already happening and is greatly beneficial.

## References

1. A. Cichocki, R. Unbehauen, M. Lendl, K. Weinzierl, Neural networks for linear inverse problems with incomplete data especially in applications to signal and image reconstruction. Neurocomputing **8**(1), 7–41 (1995)

2. Y. Wang, P. Heng, F.M. Wahl, Image reconstructions from two orthogonal projections. Int. J. Imaging Syst. Technol. **13**(2), 141–145 (2003)
3. S. Wei, W. Yaonan, Segmentation method of MRI using fuzzy gaussian basis neural network. Neural Inf. Process. **8**(2), 19–24 (2005)
4. S. Zhenghao, H. Lifeng, Application of neural networks in medical image processing, in *Proceedings of the Second International Symposium on Networking and Network Security* (Academy Publishers, Jinggangshan, China, 2010), pp. 23–26
5. LISA Lab, My LeNet, Retrieved 2016-5-26. [Online]. Available: http://deeplearning.net/ tutorial/ images/mylenet.png
6. M. Havaei, A. Davy, D. Warde-Farley, A. Biard, A. Courville, Y. Bengio, H. Larochelle, Brain tumor segmentation with deep neural networks. Med. Image Anal. **35**, 18–31 (2017)
7. H.R. Roth, A. Farag, L. Lu, E.B. Turkbey, R.M. Summers, Deep convolutional networks for pancreas segmentation in CT imaging. Med. Imaging Image Process. **9413**, 94131G (2015)
8. P. Looney, G.N. Stevenson, K.H. Nicolaides, W. Plasencia, M. Molloholli, S. Natsis, S.L. Collins, Automatic 3D ultrasound segmentation of the first trimester placenta using deep learning. in *Proceedings of the 2017 IEEE 14th International Symposium on Biomedical Imaging, Biomedical Imaging (ISBI 2017)* (Melbourne, Australia, 2017), pp. 279–282
9. Q. Dou, L. Yu, H. Chen, Y. Jin, X. Yang, J. Qin, P.A. Heng, 3D deeply supervised network for automated segmentation of volumetric medical images. Med. Image Anal. **41**, 40–54 (2017)
10. H. Fu, Y. Xu, D.W.K. Wong, J. Liu, Retinal vessel segmentation via deep learning network and fully-connected conditional random fields. in *Proceedings of the Biomedical Imaging (ISBI), 2016 IEEE 13th International Symposium on Biomedical Imaging* (Prague, Czech, 2016), pp. 698–701
11. R. Rasti, M. Teshnehlab, S.L. Phung, Breast cancer diagnosis in DCE-MRI using mixture ensemble of convolutional neural networks. Pattern Recognit. **72**, 381–390 (2017)
12. A. Dubrovina, P. Kisilev, B. Ginsburg, S. Hashoul, R. Kimmel, Computational mammography using deep neural networks. Comput. Methods Biomech. Biomed. Eng. Imaging Vis. **6**, 243–247 (2016)
13. U.R. Acharya, H. Fujita, S.L. Oh, Y. Hagiwara, J.H. Tan, M. Adam, Application of deep convolutional neural network for automated detection of myocardial infarction using ECG signals. Inf. Sci. **415**, 190–198 (2017)
14. J.L. Causey, J. Zhang, S. Ma, B. Jiang, J.A. Qualls, D.G. Politte, F. Prior, S. Zhang, X. Huang, Highly accurate model for prediction of lung nodule malignancy with CT scans. Sci. Rep. **8**, 9286 (2018)
15. Y. Li, X. Li, X. Xie, L. Shen, Deep learning based gastric cancer identification, in *Proceedings of the 2018 IEEE 15th International Symposium on Biomedical Imaging, Biomedical Imaging (ISBI 2018)* (Washington, DC, USA, 2018), pp. 182–185

# A Detection Tool for Code Bad Smells in Java Source Code

Aakanshi Gupta, Bharti Suri and Bimlesh Wadhwa

**Abstract** Code bad smells are the indication of an unstructured code snippet that can be a problem in software design. Code bad smell deteriorates the source code quality that resulting in unnecessary maintenance cost. This paper presents a tool to automatically detect bad smells in Java software. This tool can detect eight bad smells, namely God class/large class, data class and empty catch block, comments, nested try statement, exception thrown in finally block, unprotected main program by understanding their definitions, empty catch block, dummy handler, nested try statement, exception thrown in finally block, dummy handler, and unprotected main program considered as exception handling bad smells, and these bad smells received less attention from researchers. No single tool is available to detect all considered eight code bad smells in one go.

**Keywords** Bad smell · Software maintenance · Exception handling · Java software

## 1 Introduction

Code bad smells are the absurdity of the developed code in the software projects. These code smells are obstruction in understanding the code and increase the effort and cost of maintenance process. They can result in software failures also. Researchers have developed many detection tools like Checkstyle, Infusion, iPlasma

A. Gupta (✉)
ASET, Guru Gobind Singh Indraprastha University, New Delhi, India
e-mail: aakankshi@gmail.com

B. Suri
USICT, Guru Gobind Singh Indraprastha University, New Delhi, India
e-mail: bhartisuri@gmail.com

B. Wadhwa
Department of Computer Science School of Computing, NUS, Singapore, Singapore
e-mail: dcsbw@nus.edu.sg

© Springer Nature Singapore Pte Ltd. 2021
X.-Z. Gao et al. (eds.), *Advances in Computational Intelligence and Communication Technology*, Advances in Intelligent Systems and Computing 1086,
https://doi.org/10.1007/978-981-15-1275-9_39

479

[1], PMD [2], Jdeodarant [3], Stench Blossom [4], and so on. These available detection tools can detect either three or four bad smells out of these eight bad smells considered in this research: God class/large class, comments, data class, empty catch block, nested try statement, exception thrown in finally block, dummy handler, and unprotected main program (Table 1).

The motivation of this research is to reduce the maintenance cost and enhance the quality of the software. If the code smell detection is performed manually, it will not reduce the cost of the software; instead will increase the effort and time also. In this research, eight bad smells are considered and only for Java software the reason behind that; the Java is an free and open-source implementation language so any developer can edit the code at the software repository which is the major cause of the bad smell generation. The eight bad smells which are considered are very popular in the software and can be harmful in the future. Thus, this is a limitation of the tool and working on all the bad smells in a single go takes time so we will consider the remaining bad smells in the future work.

To attain this aim, a preliminary version of a tool has been proposed. Fowler et al. [5] have listed 22 bad code smells in the code snippet of object-oriented languages. This paper considers ten bad smells detection in Java source code, namely God class/large class, comments, data class, empty catch block, nested try statement, exception thrown in finally block, dummy handler, and unprotected main program. The exception handling bad smells named empty catch block, nested try statement, exception thrown in finally block, dummy handler, and unprotected main program distinguish the tool from other available tools. Bad code smell detection has been stated as a pre-work to the refactoring. Refactoring is the process to restructure the developed code without creating new functionality [5]. To enhance the internal quality and non-functional attribute of the software, refactoring plays a vital role. As the bad smells are detected, the developer evaluates the relevant decay in the code

**Table 1** Bad smells with their detection tools

| Tools name | Bad smells | | | | | | | Availability | Type | Language |
|---|---|---|---|---|---|---|---|---|---|---|
| | CS1 | CS2 | CS3 | CS4 | CS5 | CS6 | CS7 | | | |
| Jdeodarant | * | * | | | * | | * | Yes | Eclipse Plugin | Java |
| Décor | * | * | * | | | | | No | Standalone | Java |
| Stench | * | * | | | * | | | No | Standalone Blossom | Java |
| PMD | * | * | * | | | | | Yes | Eclipse Plugin/Sandalone | Java |
| iPlasma | * | * | * | | * | | | No | Standalone | Java, C, C++ |
| InFusion | * | | | | * | * | | No | Standalone | Java, C++ |
| Robusta | | | | * | | | | Yes | Eclipse Plugin | Java |
| DT | * | * | * | | * | | | No | Standalone | Java |
| Checkstyle | * | * | * | | | | | No | Eclipse Plugin | Java |

structure and identifies the required refactoring technique. Refactoring techniques may heal the associated problem and eliminate their indication.

The paper is organized as follows. Section 2 describes the related work of this research. Section 3 describes experimental setup for the detection method of the implemented tool. In Sect. 4, the industrial application of the research work is elaborated. Finally, Sect. 5 summarizes the research work and effort for the future extension.

## 2 Related Work

The concept of code bad smells was coined by Fowler et al. [5]. This can result in symptoms of bad code design or disobey the principles of object-oriented programming concepts. Zhang et al. [6] stated the improved definition of bad smells more precisely. Zhang et al. [7] also revealed in their systematic review that the code clone smell was the most studied. Gupta et al. [8] presented a systematic literature review on code bad smells and another survey on the most famous code smell: Code clone [9] which helps in deciding detection techniques for bad smells. Göde and Koschke [10], Thummalapenta et al. [11], and Rahman et al. [12] presented a exploratory study about the code clone bad smell. Brown et al. [13] described a category of code bad smells documented as antipatterns in the literature. Though Ouni et al. [14] stating the Blob and spaghetti code antipatterns as a type of code smells. Tufano et al. [15, 16] analyzed circumstances behind the bad smells introduction in the system and also explained when the smells got induced in the software.

Chatzigeorgiou and Manakos [17] noticed that the bad smells increases during the time. Peters and Zaidman [18] reported that developers do not spend time in removing the bad smells although they are aware with the presence of bad smells in the code base. Arcoverde et al. [19] studied a survival age of the bad smells in the source code. Olbrich et al. [20] analyzed two bad smells God class and Shotgun surgery that, in a specific period, the number of bad smells increases as well as decreases and it does not depend on the size of the software. Vaucher et al. [21] examined about the God class bad smell evolution in the code designing phase or unintentionally introduction in the code base and discriminated between them.

Li and Shatnawi [22] observed the detection process of code bad smells could be a method to identify and restructure the problematic classes. Khomh et al. [23] analyzed bad smells which were correlated than others to change-proneness which in turn affects maintenance effort. Tufano et al. [15, 16] described the two major reasons behind the code bad smells: lack of time and too much pressure on software developers. Ouni and Kessentini et al. [24] revealed that bad smells increase maintenance effort. Yamashita and Moonen [25] realized that interaction between bad smells also affected e-maintenance. Mäntylä et al. [26] grouped bad smells into six categories based on their similarities. Hall et al. [27] proposed a relationship between defects and bad smells, based on three open-source softwares. Rapu et al. [28] stated the use of historical information of the divergent code base and increase the automatic detection accuracy. Palomba et al. [29] presented that using histori-cal data not only intrinsically characterized bad smells Divergent change, Shotgun surgery, and Parallel inheritance can be identified but also Blob and Feature envy type of bad smells. Sharma and Spinellis [30] tabulated the forty-six methods of bad smell detection with the respective smells in their survey and analyzed the methods according to platforms and languages. Maiga et al. [31] provided SVMDetect, an approach based on SVM for antipatterns detection, namely Blob, Functional decom-position and Swiss army knife on three open-source systems: ArgoUML, Azureus, and Xerces. Kreimer [32] described about the design flaws Lazy class and Long method on the basis of metrics with learning decision trees. Fontana et al. [33] com-pared the sixteen different machine learning techniques on seventy-four systems for four bad smells: divergent change, lazy class, feature envy, and long method and obtained the highest performance by J48 and random forest technique.

Moha et al. [34, 35] developed a tool which placed bad smells in a domain-specific language definition of code smells. Sjoberg et al. [36] stated that the quality of a code decays due to bad smells. The size of a code in industrial projects is enormous and hence detecting bad smells manually is not practical. That is why code bad smell detection tools are needed. Van Emden and Moonen [37] released their first code bad smell detection tool. Marinescu et al. [1] introduced a tool iPlasma which detects four bad smells. It is a subversion of the Infusion detection tool. Liu and Zhang [38] presented a SLR on detection tools. They presented a comparison between various available detection tools.

## 3  Experimental Setup

Automatic detection of bad smells is necessary today and so researchers have shown a keen interest in developing a tool for bad smell detection. Bad smells considered in this research are detected using their threshold values. Table 1 describes the difference between other available tools and this work. Bad smells have been identified by codes CS1 to CS7 which are described further in the same section. There are few tools available for implementing the detection process and with only one tool being available for Java to handle bad smells; it is named Robusta (https://marketplace. eclipse.org/content/robustaeclipse-plugin). These reasons motivate us to implement a tool for bad smell detection. Fowler et al. [5] identified twenty-two code bad smells and suggested that every code bad smell needs a refactoring technique to improve source code structure. The following is a brief description about the bad smells under consideration.

- God Class/Large Class (CS1): A class with too many responsibilities leading to a huge no. of variables and functions in the system is referred to as a God class [33]. The traditional way to detect God class is with a specific number of line of code (LOC) of one class, which is used as a metric as to whether a project has a God class or not. The threshold value for God class detection is 1000 [33]. Figure 1

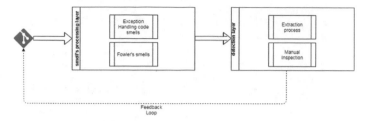

**Fig. 1** Architecture of the tool

presents a snapshot of the large class bad smell detection with our implemented tool.

- Data class and Long Method (CS2): Data class bad smell is generated when the classes contain the data for another classes and not able to control their data independently. Long method bad smell has many lines and implements more than one task. The threshold value for long method detection is 20 lines of code [38, 39].

- Comments and Long parameter list (CS3): When the source code contains the comments more than the threshold value that is 10, then comments bad smell got introduced in the software. Long parameter list code bad smell can be detected by counts of parameters for each method. This bad smell makes invoking methods hard. Threshold value for long parameter list is 10 variables [39]. Figure 2 presents a snapshot of the long parameter list bad smell detection with our implemented tool.

- Java Exception handling bad smells (CS4). Empty catch block, nested try statement, exception thrown in finally block, dummy handler, and unprotected main program are considered as Java exception handling bad smells. These are an exceptional type bad smells in Java. The code block in Java with empty catch blocks is called empty catch block bad smell. The traditional way to detect empty catch block is to scan each catch block and look for a number of lines of code in each catch block. Nested try statement bad smell is generated when one try statement populated with one or more try statements. Handling the exceptions that are thrown inside the finally block of the another try catch statement is termed as exception thrown in finally block bad smell. Dummy handler bad smell is introduced when the exception handling code does not actually handling the exception rather viewing the exceptions only. The bad smell unprotected main generated in the source code when the outer exception can be only handled in a subprogram; not in the main program.

- Other bad Smells: Few bad smells like feature envy(CS5), code clone(CS6), and type checking(CS7) can be detected by their respective tools and are presented in Table 1.

**Fig. 2** Detection of large class code smell

## 3.1 Tool's Architecture and Detection Approach

In Fig. 3, the architecture of the tool is illustrated. Tool is composed of five modules. The input file is a source code of Java with code smells that are taken as input for the extraction procedure. Afterward, code smells related to exception handling and Fowler's smells that are being considered in this research are shown in the architecture. These smells are found using extraction procedure which is written with the help of their definitions. Then, the smells are calculated by the tool and we can observe them from the output window.

Large class bad smell is detected by measuring the count of coding lines called LOC metrics. Figure 3 shows a code snippet for computation of LOC using LOC value to detect large class bad smell. Bad smell empty catch block is detected by counting the number of lines in every exception handling block in Java code.

## 4 Result and Discussion

In this paper, a bad smell detection tool has been proposed. Although many tools like jdeodarant, stench blossom, decor, robusta, and pmd etc., are available in the literature, yet they are not sufficient to detect all the bad smells. The proposed tool detects the bad smells in the Java code snippets. This tool is different from the other available tools as it can detect the more number of smells as shown in Table 2; a comparison between proposed tool and other available tools. The output window and other screenshots also have been shown in Fig. 3. The Fig. 4 also shows a code snippet for the detection of bad smells; God class and Long parameter list. The bad smell detection is a vital step in the software development process as the bad smells increase the maintenance cost. In the industries, the bad smell free code also helps

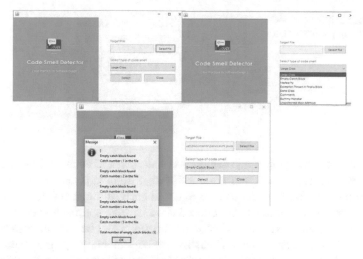

**Fig. 3** Screen shots of the tool with output window

**Table 2** Comparison between proposed tool and other tools

| Tools | Bad smells | | | |
|---|---|---|---|---|
| | CS1 | CS2 | CS3 | CS4 |
| Proposed tool | Yes | Yes | Yes | Yes |
| Jdeodarant | Yes | Yes | No | No |
| Stench Blossom | Yes | Yes | No | No |
| PMD | Yes | Yes | Yes | No |
| Robusta | No | No | No | Yes |

for the employees performance evaluation. Further section includes the industrial application of the bad smells.

## 4.1 Industrial Application

Softwares are the prerequisite for today's digital world. Good quality software with low maintenance cost is industry requirement. The maintenance cost of the software is about 60–80% of the total cost [40–42]. Bad smells are the major factor which increases maintenance cost and reduces software quality. Detection and removal of bad smells are very important aspect for industries. Manual detection of bad smell is time-consuming and tedious job. Developing tools for automatic detection of bad smells are comparatively better solution. This tool will help industries to detect code bad smells in the code of the software. These smells can be eliminated before inducing complications, like reduced quality and readability of the software.

**Fig. 4** Code snippet for God class and Long parameter list detection

```
public class Reflection{
    public static void main(String args[]) throws Exception {
        int count= 0;
//      Parameter parameter[] = new Parameter;
        Class c = Class.forName("TestClass");
        System.out.println("\nMethods:");
        Method methods[] = c.getDeclaredMethods();
//      Method methods2[] = c.getDeclaredMethod()
//      String[] array = new String[methods.length];
        for (Method meth : methods)
        {
            System.out.println(" " + meth);
            System.out.println(meth.getParameterCount());
            if(meth.getParameterCount()>5){
                System.out.println("Code Smell: \"Long Parameter List\" found");
            } else{
                System.out.println("Code Smell: \"Long Parameter List\" not found");
            }
```

In the usual software cycle, smell detection is undertaken after completing the coding step; if detection is performed after the software designing phase, it will benefit the software industry. Bad smells can be removed early through refactoring process. Development process time can be reduced if we have an approach to detect the bad smells earlier. The approach is useful for industry managers who manage the software code phase. The detected code bad smells will help in analyzing the quality of the code. A developer can participate in the development process of the open source. It is the responsibility of industry manager to judge the quality of the code developed by a specific developer. The code developer, who develops the code with a few smells can be used as a vital resource by the project manager. Thus, this research will also help to decide employee performance in the industry.

## 5   Conclusion

It can be asserted that the code bad smells detection processes are certainly a necessity to know which code snippet should be enhanced. The code bad smells automatic detection is the practical solution as compared to manual detection process. This paper presented a tool for bad smells (God class/large class, data class, comments and empty catch block, nested try statement, exception thrown in finally block, unprotected main program, dummy handler) detection. Detection of the Java exception handling code smell also named as empty catch block, nested try statement, exception thrown in finally block, unprotected main program, and dummy handler distinguish this tool with the other tools.

It is not easy to use and choose an available tool. Often there is a lack of proper documentation or no documentation, or no manual available regarding some tools. According to Table 1, many tools are not available for detection process. There is also being observed various difficulties in deciding the deduction rules used on the threshold values of the software metrics in the detection.

In future, aims to detect those code bad smells that have not drawn much attention so far. To accommodate other languages that are object oriented like C++, Ruby, Python, more work needs to be contributed. Additionally, long method code bad smell can be further improved by considering other logical aspects of a method.

# References

1. P.F.M.C. Marinescu, R. Marinescu, R. Wettel, *iPlasma: An Integrated Platform for Quality Assessment of Object-Oriented Design*
2. S. Slinger, *Code Smell Detection in Eclipse* (Delft University of Technology)
3. N. Tsantalis, T. Chaikalis, A. Chatzigeorgiou, Jdeodorant: identification and removal of type-checking bad smells, in *12th European Conference on Software Maintenance and Reengineering, 2008. CSMR 2008* (IEEE, 2008), pp. 329–331
4. E. Murphy-Hill, A.P. Black, An interactive ambient visualization for code smells, in *Proceedings of the 5th International Symposium on Software Visualization* (ACM, 2010), pp. 5–14
5. M. Fowler, K. Beck, J. Brant, W. Opdyke, D. Roberts, *Refactoring: Improving the Design of Existing Code* (Addison-Wesley Professional, 1999)
6. M. Zhang, N. Baddoo, P. Wernick, T. Hall, Improving the precision of Fowler's definitions of bad smells, in *Software Engineering Workshop, 2008. SEW'08. 32nd Annual IEEE* (IEEE, 2008), pp. 161–166
7. M. Zhang, T. Hall, N. Baddoo, Code bad smells: a review of current knowledge. J. Softw. Maint. Evol.: Res. Pract. **23**(3), 179–202 (2011)
8. A. Gupta, B. Suri, S. Misra, A systematic literature review: code bad smells in java source code, in *International Conference on Computational Science and Its Applications* (Springer, 2017), pp. 665–682
9. A. Gupta, B. Suri, A survey on code clone, its behavior and applications, in *Networking Communication and Data Knowledge Engineering* (Springer, 2018), pp. 27–39
10. N. Göde, R. Koschke, Studying clone evolution using incremental clone detection. J. Softw: Evol. Process **25**(2), 165–192 (2013)
11. S. Thummalapenta, L. Cerulo, L. Aversano, M. Di Penta, An empirical study on the maintenance of source code clones. Empir. Softw. Eng. **15**(1), 1–34 (2010)
12. F. Rahman, C. Bird, P. Devanbu, Clones: what is that smell? Empir. Softw. Eng. **17**(4–5), 503–530 (2012)
13. W.H. Brown, R.C. Malveau, H.W. McCormick, T.J. Mowbray, *AntiPatterns: refactoring software, architectures, and projects in crisis*. Wiley (1998)
14. A. Ouni, M. Kessentini, S. Bechikh, H. Sahraoui, Prioritizing code-smells correction tasks using chemical reaction optimization. Softw. Qual. J. **23**(2), 323–361 (2015)
15. M. Tufano, F. Palomba, G. Bavota, R. Oliveto, M. Di Penta, A. De Lucia, D. Poshyvanyk, When and why your code starts to smell bad, in *Proceedings of the 37th International Conference on Software Engineering*, vol. 1 (IEEE Press, 2015), pp. 403–414
16. M. Tufano, F. Palomba, G. Bavota, R. Oliveto, M. Di Penta, A. De Lucia, D. Poshyvanyk, When and why your code starts to smell bad (and whether the smells go away). IEEE Trans. Softw. Eng. **43**(11), 1063–1088 (2017)
17. A. Chatzigeorgiou, A. Manakos, Investigating the evolution of bad smells in object-oriented code, in *International Conference on the Quality of Information and Communications Technology* (IEEE, 2010), pp. 106–115
18. R. Peters, A. Zaidman, Evaluating the lifespan of code smells using software repository mining, in *16th European Conference on Software Maintenance and Reengineering (CSMR), 2012* (IEEE, 2012), pp. 411–416
19. R. Arcoverde, A. Garcia, E. Figueiredo, Understanding the longevity of code smells: preliminary results of an explanatory survey, in *Proceedings of the 4th Workshop on Refactoring Tools* (ACM, 2011), pp. 33–36
20. S. Olbrich, D.S. Cruzes, V. Basili, N. Zazworka, The evolution and impact of code smells: a case study of two open source systems, in *3rd International Symposium on Empirical Software Engineering and Measurement, 2009. ESEM 2009* (IEEE, 2009), pp. 390–400
21. S. Vaucher, F. Khomh, N. Moha, Y.-G. Guéhéneuc, Tracking design smells: lessons from a study of god classes, in *16th Working Conference on Reverse Engineering, 2009. WCRE'09* (IEEE, 2009), pp. 145–154

22. W. Li, R. Shatnawi, An empirical study of the bad smells and class error probability in the post-release object-oriented system evolution. J. Syst. Softw. **80**(7), 1120–1128 (2007)
23. F. Khomh, M. Di Penta, Y.-G. Guéhéneuc, G. Antoniol, An exploratory study of the impact of antipatterns on class change- and fault-proneness. Empie. Softw. Eng. **17**(3), 243–275 (2012)
24. M. Kessentini, R. Mahaouachi, K. Ghedira, What you like in design use to correct bad-smells. Softw. Q. J. **21**(4), 551–571 (2013)
25. A. Yamashita, L. Moonen, To what extent can maintenance problems be predicted by code smell detection? An empirical study. Inf. Softw. Technol. **55**(12), 2223–2242 (2013)
26. M.V. Mäntylä, C. Lassenius, Subjective evaluation of software evolvability using code smells: an empirical study. Empirical Softw. Eng. **11**(3) 395–431 (2006)
27. T. Hall, M. Zhang, D. Bowes, Y. Sun, Some code smells have a significant but small effect on faults. ACM Trans. Softw. Eng. Methodol. (TOSEM) **23**(4) (2014). Article No. 33
28. D. Rapu, S. Ducasse, T. Gîrba, R. Marinescu, Using history information to improve design flaws detection, in *Eighth European Conference on Software Maintenance and Reengineering, 2004. CSMR 2004. Proceedings* (IEEE, 2004), pp. 223–232
29. F. Palomba, G. Bavota, M. Di Penta, R. Oliveto, A. De Lucia, D. Poshyvanyk, Detecting bad smells in source code using change history information, in *Proceedings of the 28th IEEE/ACM International Conference on Automated Software Engineering* (IEEE Press, 2013), pp. 268–278
30. T. Sharma, D. Spinellis, A survey on software smells. J. Syst. Softw. **138**, 158–173 (2018)
31. A. Maiga, N. Ali, N. Bhattacharya, A. Sabané, Y.-G. Guéhéneuc, G. Antoniol, E. Aïmeur, Support vector machines for anti-pattern detection, in *2012 Proceedings of the 27th IEEE/ACM International Conference on Automated Software Engineering (ASE)* (IEEE, 2012), pp. 278–281
32. J. Kreimer, Adaptive detection of design flaws. Electron. Notes Theor. Comput. Sci. **141**(4), 117–136 (2005)
33. F.A. Fontana, P. Braione, M. Zanoni, Automatic detection of bad smells in code: an experimental assessment. J. Object Technol. **11**(2), 5:1 (2012)
34. N. Moha, Y.-G. Guéhéneuc, A.-F. Le Meur, L. Duchien, A. Tiberghien, From a domain analysis to the specification and detection of code and design smells. Formal Asp. Comput. **22**(3–4), 345–361 (2010)
35. N. Moha, Y.-G. Guéhéneuc, A.-F. Le Meur, L. Duchien, A domain analysis to specify design defects and generate detection algorithms, in *International Conference on Fundamental Approaches to Software Engineering* (Springer, 2008), pp. 276–291
36. D.I. Sjoberg, A. Yamashita, B.C. Anda, A. Mockus, T. Dyba, Quantifying the effect of code smells on maintenance effort. IEEE Trans. Softw. Eng. **39**(8), 1144–1156 (2013)
37. E. Van Emden, L. Moonen, Java quality assurance by detecting code smells, in *Ninth Working Conference on Reverse Engineering, 2002. Proceedings* (IEEE, 2002), pp. 97–106
38. X. Liu, C. Zhang, DT: a detection tool to automatically detect code smell in software project. Adv. Comput. Sci. Res. **71**
39. N. Roperia, *JSmell: A Bad Smell Detection Tool for Java Systems* (California State University, Long Beach, 2009)
40. S.S. Yau, J.J.-P. Tsai, A survey of software design techniques. IEEE Trans. Softw. Eng. **SE-12**(6), 713–721 (1986)
41. R.S. Arnold, *Software Reengineering* (IEEE Computer Society Press, 1993)
42. B.P. Lientz, E.B. Swanson, G.E. Tompkins, Characteristics of application software maintenance. Commun. ACM **21**(6), 466–471 (1978)

# Performance Comparison of Graphene Terahertz Antenna and Copper Terahertz Antenna

Subodh Kumar Tripathi and Ajay Kumar

**Abstract**  Terahertz frequency, also called terahertz gap, ranges from 0.1 to 10 THz is the area of the most up-to-date research spectrum useful for wireless communication. In this paper, performance of terahertz antenna using graphene and copper patch are analyzed. MATLAB codes are used to characterize the properties of graphene and are applied to high-frequency electromagnetic software which is used for modeling and simulation of the proposed antenna. Paper highlights the use of counterpart of copper, i.e. graphene as a patch material in terahertz antenna and also shows the performance comparison of graphene-based terahertz antenna and copper-based terahertz antenna. Return loss, directivity, and gain of the proposed antenna are compared. Results analysis shows that graphene is better option for the design of terahertz antenna.

**Keywords**  Graphene · Copper · Terahertz frequency · Terahertz antenna

## 1  Introduction

Graphene is a single-layer carbon crystal made from graphite, which has subjugated in recent times the research society due to its incomparable properties [1–3]. Among many researchers, only little research work found the antenna applications of graphene [4–8]. Due to outstanding electromagnetic, mechanical properties, and tunable characteristics of the material in reconfigurable designs, graphene is extensively used in many THz applications. Foremost challenge in using graphene material as a nano component is to mathematically model the material that would give better THz frequency properties. THz frequency range has great potential in applications such as imaging, sensing, spectroscopy, and detection [9].

S. K. Tripathi (✉)
IKGPTU, Jalandhar, Punjab, India
e-mail: subodh.lama@gmail.com

A. Kumar
ECE Department, BECT, Gurudaspur, Punjab, India
e-mail: ajaykm_20@yahoo.co.in

© Springer Nature Singapore Pte Ltd. 2021
X.-Z. Gao et al. (eds.), *Advances in Computational Intelligence and Communication Technology*, Advances in Intelligent Systems and Computing 1086,
https://doi.org/10.1007/978-981-15-1275-9_40

In this letter, terahertz antenna with graphene which can be made tunable is simulated and analyzed. Here, tunable behavior of the graphene has been utilized to form tunable terahertz antenna. Many tuning schemes have been used in past such as micro electro mechanical switch (MEMS), PIN diode, and tunable materials (ferroelectric materials) for achieving tunable behavior of the reconfigurable antenna, and their performance comparison has been shown in [10], their performance is limited by operation of frequency and drawbacks such as MEMS requires high control voltage, poor reliability, and mechanical movement of the switch. Ferroelectric material is another choice for the design of reconfigurable terahertz antenna [11–16] as the permittivity of these materials varies with change in applied dc bias voltage but has the drawback of requirement of large bias voltage for their operation, while graphene provides better platform for the reconfigurable antenna. This is important to note that graphene-based terahertz antenna is not limited by frequency of operation and does not require high bias voltage.

THz communication can be very beneficial for increase of fast data transfer in short distance communication, and for these purposes, we require an efficient antenna operating in THz range. For the future applications such as wireless networks on chip and wireless nano sensor networks, antenna is a very essential element and size of the antenna is also a very crucial. Here, in this letter, we have proposed a small size efficient terahertz antenna which can be useful for above-mentioned applications.

From literature understanding, it is obvious that the electromagnetic fields for metallic antenna are governed by classical Maxwell's equations, the graphene, however, is represented by a surface conductivity arising from a semi-classical intraband mode and quantum-dynamical interband mode [17]. Surface conductivity of graphene has been calculated by Kubo's formula.

The surface conductivity expression of an infinite graphene sheet specified in Eq. (1), includes two parts, intraband as a first term and interband as a second term [18]. The graphene conductivity model is given by [19].

$$
\sigma(i\omega, \mu_c, T, \Gamma) = \frac{q_e^2 (\omega + 2i\Gamma)}{\pi\hbar^2} \left[ \frac{q_e^2}{(\omega + 2i\Gamma)^2} \int_0^\infty \xi \left( \frac{\partial f_d(\xi)}{\partial \xi} - \frac{\partial f_d(-\xi)}{\partial \xi} \right) d\xi - \int_0^\infty \frac{f_d(-\xi) - f_d(\xi)}{(\omega + 2i\Gamma)^2 - \left(\frac{4\varepsilon}{\hbar}\right)^2} d\xi \right]
$$

(1)

Surface conductivity of the graphene is plotted using MATLAB and is shown in Figs. 1 and 2. Inductive nature of conductivity can be observed from Figs. 1 and 2 and that allows graphene to support surface plasmon polariton (SPP) [20].

Graphene conductivity $\sigma$ or surface impedance ($Zs = 1/\sigma$) can be changed by DC bias. This behavior of graphene is utilized to design reconfigurable small size antennas. In this paper, we have designed and analyzed inset-fed rectangular microstrip antenna using copper as a patch material and graphene as a patch material. We have compared the results of both the antenna for return loss, gain, and directivity of the antenna. In second section, we have illustrated traditional design procedure and calculated the design parameters for terahertz antenna resonating at 0.3 THz. In beginning of this section, we have modeled surface conductivity of the graphene

**Fig. 1** Real part of surface conductivity of graphene for different chemical potential

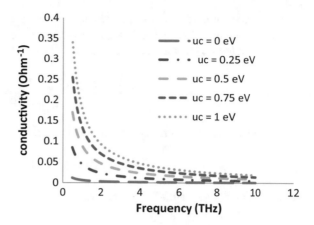

**Fig. 2** Imaginary part of surface conductivity of graphene for different chemical potential

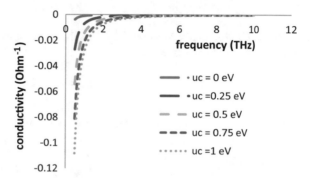

to incorporate graphene in electromagnetic simulator such as HFSS which does not include this material in the library. Using HFSS, we have performed simulation of the copper-based microstrip terahertz antenna and graphene-based microstrip terahertz antenna.

## 2  Theory

The studied inset-fed microstrip terahertz antenna consists of a rectangular patch arranged on a dielectric substrate of permittivity ($\varepsilon_r = 2.2$). A perfect ground plane is placed below the substrate that participates in the radiation of the fields. Patch is made of graphene. To determine the width $W_p$ of the patch antenna, we use the following equation [21]

$$W_p = \frac{C}{2f_r\sqrt{\frac{\varepsilon_{r+1}}{2}}} \tag{2}$$

where $f_r$ is desired resonant frequency and $c$ is the speed of light.

The effect of fringing fields acting at the outside of the radiating patch. The effective dielectric constant can be given by

$$\varepsilon\text{reff} = \frac{\varepsilon_r + 1}{2} + \frac{\varepsilon_r - 1}{2}\left(1 + \frac{12 * h}{W_p}\right)^{-1/2} \tag{3}$$

The length of the patch, $L_p$ can be calculated as:

$$L_p = L_{\text{eff}} - (2 * \Delta L) \tag{4}$$

where the effective length $L_{\text{eff}}$ is calculated as

$$L_{\text{eff}} = \frac{C}{2 f_r \sqrt{\varepsilon\text{reff}}} \tag{5}$$

where $\Delta L$ is the fields overflow. The length, width of the ground plane and width of the feed line are calculated as

$$L_g = L_p + (6 * h) \tag{6}$$

$$W_g = W_p + (6 * h) \tag{7}$$

$$W_f = h\left(\frac{Z_c}{50\sqrt{\varepsilon_r}} - 2\right) \tag{8}$$

where $Z_c$ is intrinsic impedance of free space and value is 377 $\Omega$.

The position of the inset feed point where the input impedance is 50 $\Omega$ is calculated as:

$$\text{Zin}(x) = \frac{1}{2(G_1 \pm G_{12})}\cos^2\left(\frac{\pi}{L_p}x\right) \tag{9}$$

where $G_1$ and $G_{12}$ are given by

$$G_1 = \begin{matrix} \frac{1}{90}\left(\frac{W_p}{\lambda_0}\right)^2 & \text{for } W_p \ll \lambda_0 \\ \frac{1}{120}\left(\frac{W_p}{\lambda_0}\right) & \text{for } W_p \ll \lambda_0 \end{matrix} \tag{10}$$

$$G_{12} = \frac{1}{120\pi^2}\int_0^\pi \left[\frac{\sin\left(\frac{K_0 W_p}{2}\cos\theta\right)}{\cos\theta}\right]^2 J_0(K_0 L \sin\theta)\sin^3\theta d\theta \tag{11}$$

**Table 1** Parameters of the proposed terahertz antenna

| Parameters | Symbol | Value |
|---|---|---|
| Resonant frequency | $f_r$ | 0.3 THz |
| Length of the substrate | $L_s$ | 445.4 μm |
| Width of the substrate | $W_s$ | 513.76 μm |
| Thickness of the substrate | $T_s$ | 19.76 μm |
| Length of the patch | $L_p$ | 326.86 μm |
| Width of the patch | $W_p$ | 395.2 μm |
| Length of the feed line | $L_f$ | 56 μm |
| Width of the feed line | $W_f$ | 60.86 μm |
| Notch gap | A | 12 μm |
| Inset feed position | B | 112.32 μm |

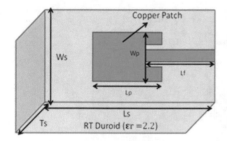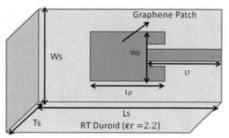

**Fig. 3** Copper patch and graphene patch terahertz antenna

The calculated design parameters are tabulated in Table 1 and proposed design is shown in Fig. 3.

# 3 Proposed Antenna Design

Proposed antenna using copper patch and using graphene patch are shown in Fig. 3. Both the antennas are same in all design parameters except patch material. Antenna consists of rectangular patch on RT Duroid substrate and is fed by inset microstrip line. The parameters detail are given in Table 1.

# 4 Proposed Antenna Result Analysis

Proposed antenna is designed and simulated on high-frequency electromagnetic software HFSS. Firstly, antenna using copper patch is simulated and the return loss plot is shown in Fig. 4. Antenna resonates at 0.3 THz and return loss is −12 dB. Gain of the copper patch THz antenna is 5.68 dB and directivity of the antenna is 6.49 dB. Gain and directivity plots are shown in Fig. 5.

Proposed antenna using graphene as a patch material is simulated and return loss plot is shown in Fig. 6. Antenna resonates at 0.49 and 0.55 THz. And return loss values are −14 dB and −16 dB, respectively. These results are obtained at chemical voltage (0.5 eV) of graphene. Resonant frequency can be set by setting proper chemical voltage as chemical voltage is also used for reconfiguration of the THz antenna. Gain and directivity of the graphene terahertz antenna are also improved and are shown in Fig. 7 and values are 6.27 dB and 7.47 dB, respectively. Comparison of return loss plot of copper patch antenna and graphene patch antenna is shown in Fig. 8. It shows the resonate condition of both the antennas.

Comparison of performance parameters of the proposed antenna and work reported earlier is given in Table 2.

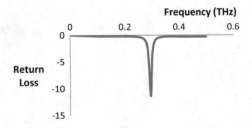

**Fig. 4** Return loss plot of metal terahertz antenna

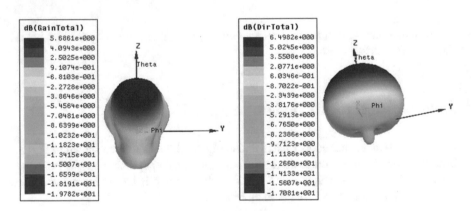

**Fig. 5** Gain and directivity of the metal terahertz antenna

**Fig. 6** Return loss plot of graphene terahertz antenna

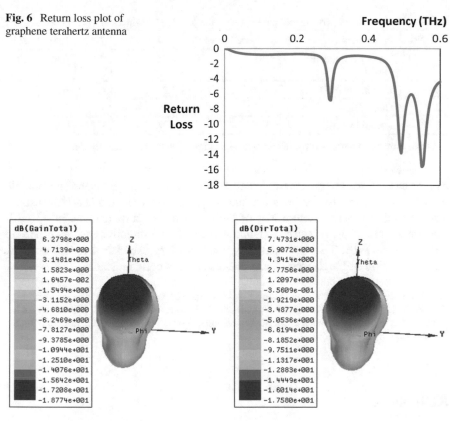

**Fig. 7** Gain and directivity of the graphene terahertz antenna

**Fig. 8** Comparison of return loss plot of copper and graphene terahertz antenna

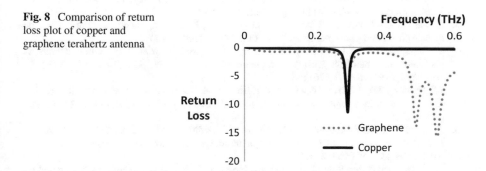

## 5 Conclusion

Presented work utilizes exceptional surface conductivity of graphene in design of terahertz antenna. Terahertz antenna using copper as a patch material and graphene

**Table 2** Comparison of performance parameters of the proposed antenna and work reported earlier

| SI. No. | Parameters | Proposed antenna | [22] | [23] | [24] | [25] | [26] |
|---------|-----------|------------------|------|------|------|------|------|
| 1 | Resonant frequency (THz) | 0.3 | 0.75 | 1.3 | 2.8 | 0.69 | 1.03 |
| 2 | Directivity (dB) | 7.47 | 5.71 | – | – | 6.57 | 1.8 |
| 3 | Gain (dB) | 6.27 | – | – | – | 5.74 | – |
| 4 | Return loss (dB) | −16 | −36 | – | −31 | −27 | −35 |

– (dash) shows that particular parameter is not investigated in the reference article

as a patch material are compared. Proposed work shows that graphene is a suitable material than copper for terahertz antenna design as copper is fragile and unsustainable in terahertz range. Return loss of the graphene terahertz antenna is −16 dB, maximum gain of the proposed antenna is 6.27 dB and directivity of the proposed antenna is 7.47 dB. The proposed design can be very useful in wireless network communication and high-speed short distance indoor communication.

**Acknowledgements** I convey my sincere thanks for the support provided by Thapar University, Patiala, Punjab, IKGPTU, Jalandhar, Punjab, India, IIT Indore, India and MIET, Meerut, UP, India. I also convey my sincere thanks to Dr. Mukesh Kumar, Associate professor, IIT, Indore, India for his kind support and guidance.

# References

1. K. Geim, K.S. Novoselov, The rise of graphene. Nat. Mater. **6**, 183–191 (2007)
2. P. Blake, E.W. Hill1, A.H.C. Neto, K.S. Novoselov, D. Jiang, R. Yang, T.J. Booth, A.K. Geim, Making grapheme visible. Appl. Phys. Lett. **91**, 063124 (2007)
3. C. Stampfer, J. Güttinger, F. Molitor, D. Graf, T. Ihn, K. Ensslin, Tunable coulomb blockade in nano structured graphene. Appl. Phys. Lett. **92**, 012102 (2008)
4. M. Tamagnone, J.S. Gomez-Diaz, J.R. Mosig, J. Perruisseau Carrier, Reconfigurable terahertz plasmonic antenna concept using a grapheme stack. Appl. Phys. Lett. **101**, 214102 (2012). https://doi.org/10.1063/1.4767338
5. J.S. Gomez-Diaz, J. Perruisseau Carrier, P. Sharma, A. Ionescu, Non-contact characterization of graphene surface impedance at micro and millimeter waves. J. Appl. Phys. **111**, 114908 (2012)
6. H.S. Skulason, H.V. Nguyen, A. Guermoune, V. Sridharan, M. Siaj, C. Caloz, T. Szkopek, 110 GHz measurement of large-area graphene integrated in low-loss microwave structures. Appl. Phys. Lett. **99**, 153504 (2011)
7. M. Dragoman, A.A. Muller, D. Dragoman, F. Coccetti, R. Plana, Terahertz antenna based on graphene. J. Appl. Phys. **107**, 104313 (2010)
8. I. Frigyes, J. Bito, B. Hedler, L.C. Horvath, in Applicability of the 50–90 GHz frequency bands in feeder networks, in *Proceedings European Antennas Propagation Conf* (Berlin, Germany, 2009 March), pp. 36–40
9. I.F. Akyildiz, M. Jornet, C. Han, Terahertz band: next frontier for wireless communication. Phys. Commun. **12**, 16–32 (2014)
10. N. Haider, A.G. YArovoy, Recent developments in reconfigurable and multiband antenna technology. Int. J. Antenna Propag. **2013**, 869170 (2013)

11. J. Modelski, Y. Yashchyshyn, Semiconductor and ferroelectric antennas. in *Proceedings of the Asia-Pacific Microwave Conference (APMC '06)* (2006 Dec), pp. 1052–1059
12. Y. Yashchyshyn, J. Marczewski, K. Derzakowski, J.W. Modelski, P.B. Grabiec, Development and investigation of an antenna system with reconfigurable aperture. IEEE Trans. Antenna Propag. **57**(1), 2–8 (2009)
13. A.E. Fathy, A. Rosen, H.S. Owen et al., Silicon-based reconfigurable antenna concepts, analysis, implementation, and feasibility. IEEE Trans. Microw. Theory Tech. **51**(6), 1650–1661 (2003)
14. L. Liu, R.J. Langley, Liquid crystal tunable microstrip patch antenna. Electron. Lett. **44**(20), 1179–1180 (2008)
15. Y. Yashchyshyn, J. Modelski, Reconfigurable semiconductor antenna, in *Proceedings of the 9th International Conference: The Experience of Designing and Application of CAD Systems in Microelectronics (CADSM '07)* (2007 Feb), pp. 146–150
16. A. Gaebler, A. Moessinger, F. Goelden et al., Liquid crystal reconfigurable antenna concepts for space applications at microwave and millimeter waves. IEEE Trans. Antenna Propag. **2009** (2009)
17. J.R. Lima, Controlling the energy gap of graphene by Fermi velocity engineering. Phys. Lett. A **379**(3), 179–182 (2015)
18. G.W. Hanson, Dyadic Green's functions and guided surface waves for a surface conductivity model of graphene. J. Appl. Phys. **103**(6), 064302 (2008). https://doi.org/10.1063/1.2891452
19. J.M. Jornet, I.F. Akyildiz, Graphene based nano antennas for electromagnetic nano communications in the terahertz band. in *Proceedings of the Fourth European Conference on Antennas and Propagation* (Barcelona, Spain, 2010), pp. 1–5. https://doi.org/10.1016/j.nancom.2010.04.001
20. G.W. Hanson, Dyadic Green's functions and guided surface waves for a surface conductivity model of graphene. Appl. Phys. **103**, 064302 (2008)
21. C.A. Balanis, *Antenna Theory, Analysis and Design*, 3rd edn. (Wiley Publications, 2011)
22. S. Anand, D.S. Kumar, R.J. Wu, M. Chavali, Graphene nano ribbon based terahertz antenna on polyimide substrates. Optik **125**(2014), 5546–5549 (2014). Elsevier
23. I. Llaster, C. Kremers, A. Cabellos-Aparicio, J.M. Jornet, E. Alarcon, D.N. Chigrin, Graphene-based nano-patch antenna for terahertz radiation. Photonics Nanostruct. Fundam. Appl. **10**, 353–358 (2012)
24. H.A. Abdulnabi, R.T. Hussiein, R.S. Fyath, Design and performance investigation of tunable UWB THz antenna based on graphene fractal artificial magnetic conductor. Int. J. Electron. Commun. Eng. & Technol. (IJECET) **6**(9), 39–47 (2015)
25. S.K. Tripathi, A. Kumar, High gain highly directive graphene based terahertz antenna for wireless communication. I-Manag. J. Commun. Eng. Syst. **6**(4) (2017 Aug–Oct). https://doi.org/10.26634/jcs.6.4.13804
26. T. Zhou, Z. Cheng, H. Zhang, M. Le Berre, L. Militaru, F. Calmon, Miniaturized tunable terahertz antenna based on graphene. Microw. Opt. Technol. Lett. **56**(8) (2014 Aug). https://doi.org/10.1002/mop

# Computer Graphics and Vision

# A Review of Various State of Art Eye Gaze Estimation Techniques

Nandini Modi and Jaiteg Singh

**Abstract** Eye gaze tracking is a technique to track an individual's focus of attention. This paper provides a review on eye gaze trackers (EGTs) classified on the basis of intrusive and non-intrusive techniques. According to the numerous applications of EGTs in human–computer interface, neuroscience, psychology and in advertising and marketing, this paper brings forward a deep insight into recent and future advancements in the field of eye gaze tracking. Finally, comparative analyses on various EGT techniques along with its applications in various fields are discussed.

**Keywords** Eye gaze trackers (EGTs) · Computer vision · Intrusive techniques · Eye gaze · Video oculography

## 1 Introduction

Eye gaze tracking is a technique to estimate the focus of attention of a user. Eye gaze tracking has its applications in human–computer interaction, virtual reality and in diagnosis of eye diseases [1, 2]. Eye gaze tracking can help the disabled person and is helpful in controlling mouse pointer with one's eyes [3]. The distinct photometric, geometric and motion attributes of eyes also yield valuable visual cues for face recognition, face detection and for interpreting facial expressions [4]. This paper presents a review on various eye gaze tracking techniques used to detect the gaze of an eye and brings forward a comparative analysis on eye gaze tracking techniques.

To understand, record and project eye gaze, a number of approaches have been proposed by research community. Majority of these eye gaze tracking techniques are dependent on intrusive devices like contact lenses, head-mounted equipment and

N. Modi
Department of Computer Science and Engineering, Chitkara University Institute
of Engineering and Technology, Chitkara University, Patiala, Punjab, India
e-mail: nandini.modi@chitkara.edu.in

J. Singh (✉)
Department of Computer Applications, Chitkara University Institute of Engineering
and Technology, Chitkara University, Patiala, Punjab, India
e-mail: jaiteg.singh@chitkara.edu.in

© Springer Nature Singapore Pte Ltd. 2021  501
X.-Z. Gao et al. (eds.), *Advances in Computational Intelligence and Communication
Technology*, Advances in Intelligent Systems and Computing 1086,
https://doi.org/10.1007/978-981-15-1275-9_41

electrodes. Primary disadvantage of such techniques is that they require physical contact with the users and cannot be used without their consent. Further, such devices are also a bit inconvenient to use [5]. Subsequently, video-based gaze tracking techniques were proposed as a non-intrusive alternative to existing intrusive methods.

Video-based gaze tracking approach commonly deployed two types of imaging techniques, namely infrared imaging and visible imaging as shown in Table 1. As infrared imaging techniques rely upon invisible infrared light sources to obtain the controlled light and a better contract image, it is expected to reduce the effect of light conditions to produce sharp contrast iris and pupil image. On the other hand, visible imaging methods do not require sensitive equipment for tracking eye gaze and are suitable to work in a natural environment where ambient light is uncontrolled [6].

Various intrusive and non-intrusive techniques for eye gaze estimation are detailed as under.

**Table 1** Comparative analysis of eye gaze estimation techniques

| Evaluation parameters | EOG | Scleral search coil | Infrared oculography | Video oculography |
|---|---|---|---|---|
| Tracking Approach | Intrusive | Intrusive | Intrusive | Non-intrusive |
| Sensing technology | Electrodes | Coil of wire in contact lens | IR light sources, corneal reflection, purkinje image | Images captured using two or more video cameras |
| Application area | Clinical applications | Medical and psychological research | Saccadometer research systems | Human–computer interaction and medical field |
| Accuracy | High | High | Medium | High |
| Cost | Low | Intermediate | High | High |
| System setup complexity | High | High | Intermediate | Low |
| Comfort of use | Uncomfortable | Uncomfortable | Uncomfortable | Uncomfortable |
| Eye movement direction detection | ±25° vertical | Vertical | Vertical and horizontal | Limited due to head movement restriction |
| Real-time implementation | Yes | No | No | Yes |
| Example | BIOPAC MP | Chronos vision | IntelliGaze IG-30 | ERICA and tobii |
| Implicit illumination conditions | Robust | – | Robust | Vulnerable |

## 1.1   Intrusive EGTs

**Electrooculography (EOG).** EOG technique consists of sensors which are placed on the skin surrounding the eyes for measuring the electric field which tends to exist when the eyes rotate. In this method, eye position can be determined by noting small deviations in the skin surrounding the eye. Horizontal as well as vertical movements can be distinctly gauged by placing the electrodes appropriately. This method is not recommended for day to day use, due to its requirement of proximity to the user. It is often used by clinicians due to its ease of use and cost effectiveness. Andreas et al. proposed an algorithm using EOG technique to assess and recognize human activity based upon recording of eye movements [7].

**Scleral Search Coil.** Eye gaze tracking through scleral search coil is done by producing an eye position signal. This is done by inserting contact lens in the eyes. This contact lens has small coils of wire embedded in it. Whenever eye moves, the coil of wire in the lens moves in a magnetic field, thereby inducing voltage in the coil. It then produces eye position signal. This technique is one of the most accurate techniques available so far, but being an invasive method it has not been used extensively for gaze detection except for medical and psychological research. The method discussed by Robinson et al. used scleral search coil along with contact lens having high resolution and low drift. The author measured user's gaze points from the voltage induced from the coil. The approximate accuracy provided by this method is $0.08°$ [8, 9].

**Infrared Oculography.** This technique makes use of infrared light reflected from an eye and using that information, eye positions are inferred [10]. Based on the difference in eye positions, change in eye gaze is estimated. This is intrusive because it makes use of sensors and light source which are placed on spherical glasses. This technique measures eye gaze more accurately only in the horizontal direction.

## 1.2   Non-intrusive Eye Gaze Trackers

**Video Oculography.** Video oculography is most widely used in commercial eye trackers. Video-based eye gaze tracking can be intrusive or non-intrusive depending upon the usage of light, i.e., infrared or visible light. This technique makes use of single or multiple cameras to obtain gaze position which is estimated using images captured from camera [11, 12]. Eye gaze tracking using camera depends upon certain properties of eye, i.e., pupil size, eye blinks, etc. Most of these techniques have the potential to be implemented in a non-intrusive way. Video-based gaze tracking systems are divided into two categories.

*Single camera video oculography.* In this technique, single camera system and a single illumination source are used to estimate the direction of gaze. It works by eye illumination using an infrared light source which produces glint on the eye cornea, which is also known as corneal reflection [13].

*Multiple camera video oculography.* Using multiple camera eye trackers, high resolution eye images are captured in order to give accurate gaze estimation results. This eye tracker makes use of separate camera, one for eye image and other for head pose estimation [14]. In the method proposed by Zhu et al., two video cameras have been used wherein infrared illuminator is placed on the front part of camera to obtain the image of an eye with glint. Pupil-glint vector is then obtained from the captured eye images for estimating gaze [15].

# 2 Literature Review

Eye gaze detection techniques could broadly be classified as model-based, appearance-based and hybrid methods. The literature review undertaken covers gaze estimation technologies proposed from year 2010 to 2018 and published at reputed platforms.

The literature review is organized into three subsections. The first section comprises of model-based gaze estimation techniques. The second section provides work done by researchers using appearance-based gaze estimation technique. Third section gives hybrid technique combining model-based and appearance-based techniques for gaze estimation.

## 2.1 Model-Based Gaze Estimation Techniques

Model-based eye gaze estimation deploys 3D eye models for tracking gaze directions, based upon geometric eye features, i.e., iris center and eye contours. The model-based approach generally calculates average offset between optical and visual axis of eyes for estimating gaze points. Multiple cameras have been used to capture the optical axis from both the eyes and calculate horizontal and vertical gaze angle from the screen to the camera [16]. Valenti et al. combine head pose and eye location information in effectively estimating eye gaze. The authors infer eye gaze from eye shapes such as pupil centers and iris edges and thereby suffer from low image quality and variation in illumination conditions [14]. Wood et al. used limbus model for detecting iris edges from camera-based eye images that works effectively on portable devices. Despite their usage of 3D eye model for inferring gaze, their method still relies on eye appearance and is affected by slight appearance and head pose variations [17]. Some other researchers combine 3D eye model with visual saliency in order to predict eye parameters and eye gaze in a natural manner [18]. Ince et al. presented a low-cost 2D gaze estimation system using shape and intensity-based pupil center detection method which detects eye gaze accurately without any explicit calibration setup [19].

## 2.2  Appearance-Based Gaze Estimation Techniques

In appearance-based methods, eye gaze is detected on the basis of photometric appearance of eyes. Appearance-based methods use eye image to estimate gaze points mapped onto screen coordinates. Appearance-based methods use neural networks, linear regression or other forms of learning to establish a mapping between the image of the face and the user's gaze point [20, 21]. Lu et al. proposed an adaptive linear regression method calibrated with small number of training samples for effectively measuring gaze. The framework developed by the authors handles head pose variations and eye blinks when change in illumination occurs [21]. Sugano et al. proposed a novel gaze estimation method using visual saliency maps using MPIIGaze data set for both training and testing purpose thereby eliminating the need of personal calibration for each user [22]. Lu et al. used learning-based model and geometric properties of eyes for estimating eye gaze. These appearance-based methods achieve an average accuracy of around 3 degrees by using only a single camera [23]. Vicente et al. presented a low-cost system capable of detecting eye gaze under different conditions, for example, a user wearing eye glasses [24]. The two publically available gaze data sets (MPIIGaze and EYEDIAP) have been used by the researchers to train the model to show increased variability in head pose and illumination conditions [25].

## 2.3  Hybrid Techniques

Hybrid techniques combine benefits of both model-based and appearance-based techniques for gaze estimation. Cheung et al. used model-based approach that handles head movement using ordinary camera but requires large number of training samples having different head poses to learn a correction model. The other hybrid techniques detect eye corners using a three-phase feature-based approach for estimating the eye locations using ordinary camera [26]. Rizwan et al. proposed a near infrared camera sensor-based gaze classification system suitable for vehicular environment. The authors had used convolutional neural network and deep learning techniques to solve gaze tracking in an unconstrained environment [27].

Although model-based methods can estimate gaze directions accurately, its requirement of specialized equipment or head-mounted eye tracker limits its application. Therefore, they are more suitable for use in a controlled environment such as in the laboratory. Appearance-based methods estimate gaze directions using only a single common camera, and thus, they attract increasing focus of attention [23]. A generic comparison of existing eye gaze tracking solutions regarding various evaluation criteria is summarized in Table 2.

**Table 2**  Generic comparison of eye gaze estimation techniques

| Evaluation criteria | Model-based | Appearance-based | Hybrid |
|---|---|---|---|
| Setup complexity | High | Low | Low |
| System calibration | Fully calibrated | ✗ | Intermediate |
| Hardware (camera) requirements | Two or more | ordinary camera | Ordinary web camera |
| Implicit robustness to head movements | medium–high | Low | Low |
| Implicit robustness to varying illumination | medium–high | Low | Low |
| Gaze estimation accuracy error | low (<1°) | (>2°) high | (1–3°) high |

# 3  Application Areas of Eye Gaze Tracking

## 3.1  Marketing

Marketing is all about convincing customer to purchase the product. Various factors contribute in tilting customer's decision in their favor, for example, product packaging, placement of product at stores and promotional offers, etc. All these indications can be gauged by the direction in which the customer is looking, and that is where eye gaze tracking proves its worth. It not only helps in tracking the direction in which the user is looking, but also provides useful inputs to the product analysts which help them in fabricating and designing their products in a better way [28].

## 3.2  Disabled People/Medical

Eye gaze tracking is proving to be nothing short of blessing for disabled person. With the help of eye gaze tracking technique, these people can interact with others using eye movements. Through their eye movements, now they can send input to specially designed computers using EGT which converts it into words for communication [29].

## 3.3  Simulator

Another EGT application is in analyzing the attention of pilots in realistic conditions. On the basis of scan patterns generated by experienced professionals where they generally look out of glass cockpit at critical devices, an EGT can help them in understanding the focus of attention of novice pilots in emergency situations [30].

## 3.4  Video Games

Gaming is an ever evolving field. Every now and then new innovative ideas are being introduced to make gaming experience all the more interesting. Now gaming consoles can not only be controlled by hands but also through eye movements. Eye movements are tracked using EGTs which perform corresponding actions on the gaming screen. For example, in a racing game, car can be controlled through eye gaze, or in a puzzle game the puzzle pieces can be controlled on-screen through eye movements and can be brought together to complete the puzzle [31].

## 3.5  E-Learning/Education

In the past few years, various techniques like virtual classroom, cloud computing and devices like mobile phones and tablets have been used for the development and accessibility of e-learning environment. The use of eye tracking technique in e-learning has made possible to estimate the focus of learner in real time. From various measures used in eye tracking like the duration of fixation, gaze point and blink duration, it has been observed that the stress level, focus of attention and problem-solving ability of the learners can be estimated effectively [32].

Review summary sheet containing eye gaze application areas categorized into intrusive and non-intrusive techniques usage is presented in Table 3. From this review, it can be said that EGTs have proven themselves to be valuable in e-learning study, psychology and many other commercial industrial applications.

## 4  Conclusion

This paper concludes to the fact that eye gaze tracking could be used with numerous (intrusive and non-intrusive) technologies. From this comparative analysis, it was observed that owing to high cost and required expertise, they are not yet available for generic research and usage. Earlier techniques used for eye gaze tracking were limited to laboratory research and are not suitable for real-time applications. Now as the new technology evolves with the usage of desktops and mobile devices, new video-based and camera-based techniques for estimating eye gaze have been used these days. So, in order to effectively measure eye gaze, cost-effective solution with limited system setup complexity is required. This paper discussed application areas where eye gaze tracking has proven its worth. Use of eye gaze tracking in various fields without using costly equipment can be done, which helps in future enhancements in the field of computer vision.

**Table 3** Review summary on application areas of eye gaze tracking

| Paper Ref. | Method category | | | Application area | Intrusive | Non-intrusive | Using webcam | Using eye tracker |
|---|---|---|---|---|---|---|---|---|
| | *Model-based* | *Hybrid-based* | *Appearance-based* | | | | | |
| [33] | ✓ | | | Ad interface design | ✓ | | | ✓ |
| [31] | ✓ | | | Gaming | | ✓ | | ✓ |
| [34] | ✓ | | | Human–robot interaction | | ✓ | | ✓ |
| [27] | | ✓ | ✓ | Vehicular environment | | ✓ | ✓ | |
| [32] | ✓ | | | E-learning | ✓ | | | ✓ |
| [35] | ✓ | | | Sports | | ✓ | | ✓ |
| [36] | | | ✓ | Security and authentication | | ✓ | ✓ | |
| [24] | | | ✓ | Driver drowsiness detection | | ✓ | | ✓ |
| [37] | | | ✓ | Medical field | | ✓ | | ✓ |
| [38] | ✓ | | | Marketing | | ✓ | | ✓ |

# References

1. S. Chandra, G. Sharma, S. Malhotra, D. Jha, A.P. Mittal, Eye tracking based human computer interaction: applications and their uses. in *Man and Machine Interfacing (MAMI), 2015 International Conference on* (IEEE, 2015 Dec), (pp. 1–5)
2. S. Wibirama, H.A. Nugroho, K. Hamamoto, Evaluating 3D gaze tracking in virtual space: a computer graphics approach. Entertain. Comput. **21**, 11–17 (2017)
3. T.O. Zander, M. Gaertner, C. Kothe, R. Vilimek, Combining eye gaze input with a brain–computer interface for touchless human–computer interaction. Intl. J. Hum.-Comput. Interact. **27**(1), 38–51 (2010)
4. X. Zhu, D. Ramanan, Face detection, pose estimation, and landmark localization in the wild. in *Computer Vision and Pattern Recognition (CVPR), 2012 IEEE Conference on* (IEEE, 2012 June), pp. 2879–2886
5. B. Noris, J.B. Keller, A. Billard, A wearable gaze tracking system for children in unconstrained environments. Comput. Vis. Image Underst. **115**(4), 476–486 (2011)
6. K.W. Choe, R. Blake, S.H. Lee, Pupil size dynamics during fixation impact the accuracy and precision of video-based gaze estimation. Vision. Res. **118**, 48–59 (2016)
7. A. Bulling, D. Roggen, G. Tröster, Wearable EOG goggles: seamless sensing and context-awareness in everyday environments. J. Ambient. Intell. Smart Environ. **1**(2), 157–171 (2009)
8. D.A. Robinson, A method of measuring eye movement using a scleral search coil in a magnetic field. IEEE Trans. Bio-med. Electron. **10**(4), 137–145 (1963)
9. R.S. Remmel, An inexpensive eye movement monitor using the scleral search coil technique. IEEE Trans. Biomed. Eng. **4**, 388–390 (1984)
10. C. Anderson, A.M. Chang, J.P. Sullivan, J.M. Ronda, C.A. Czeisler, Assessment of drowsiness based on ocular parameters detected by infrared reflectance oculography. J. Clin. Sleep Med. **9**(09), 907–920 (2013)
11. E. Skodras, V.G. Kanas, N. Fakotakis, On visual gaze tracking based on a single low cost camera. Sig. Process. Image Commun. **36**, 29–42 (2015)
12. J. Turner, A. Bulling, H. Gellersen, Extending the visual field of a head-mounted eye tracker for pervasive eye-based interaction. in *Proceedings of the Symposium on Eye Tracking Research and Applications* (ACM, 2012 March), pp. 269–272
13. Y. Sugano, Y. Matsushita, Y. Sato, H. Koike, An incremental learning method for unconstrained gaze estimation. in *European Conference on Computer Vision* (Springer, Berlin, Heidelberg, 2008 Oct), pp. 656–667
14. R. Valenti, N. Sebe, T. Gevers, Combining head pose and eye location information for gaze estimation. IEEE Trans. Image Process. **21**(2), 802–815 (2012)
15. Z. Zhu, Q. Ji, K.P. Bennett, Nonlinear eye gaze mapping function estimation via support vector regression. in *Pattern Recognition, 2006. ICPR 2006. 18th International Conference on*, vol. 1 (IEEE, 2006 Aug), pp. 1132–1135
16. T. Nagamatsu, R. Sugano, Y. Iwamoto, J. Kamahara, N. Tanaka, User-calibration-free gaze estimation method using a binocular 3D eye model. IEICE Trans. Inf. Syst. **94**(9), 1817–1829 (2011)
17. E. Wood, T. Baltrušaitis, L.P. Morency, P. Robinson, A. Bulling, A 3D morphable eye region model for gaze estimation. in *European Conference on Computer Vision* (Springer, Cham, 2016 Oct), pp. 297–313
18. J. Chen, Q. Ji, A probabilistic approach to online eye gaze tracking without explicit personal calibration. IEEE Trans. Image Process. **24**(3), 1076–1086 (2015)
19. I.F. Ince, J.W. Kim, A 2D eye gaze estimation system with low-resolution webcam images. EURASIP J. Adv. Signal Process. **2011**(1), 40 (2011)
20. X. Fan, K. Zheng, Y. Lin, S. Wang, Combining local appearance and holistic view: dual-source deep neural networks for human pose estimation. in *Proceedings of the IEEE Conference on Computer Vision and Pattern Recognition* (2015), pp. 1347–1355
21. F. Lu, Y. Sugano, T. Okabe, Y. Sato, Adaptive linear regression for appearance-based gaze estimation. IEEE Trans. Pattern Anal. Mach. Intell. **36**(10), 2033–2046 (2014)

22. Y. Sugano, Y. Matsushita, Y. Sato, Appearance-based gaze estimation using visual saliency. IEEE Trans. Pattern Anal. Mach. Intell. **35**(2), 329–341 (2013)
23. F. Lu, T. Okabe, Y. Sugano, Y. Sato, Learning gaze biases with head motion for head pose-free gaze estimation. Image Vis. Comput. **32**(3), 169–179 (2014)
24. F. Vicente, Z. Huang, X. Xiong, F. De la Torre, W. Zhang, D. Levi, Driver gaze tracking and eyes off the road detection system. IEEE Trans. Intell. Transp. Syst. **16**(4), 2014–2027 (2015)
25. X. Zhang, Y. Sugano, M. Fritz, A. Bulling, It's written all over your face: full-face appearance-based gaze estimation. in *2017 IEEE Conference on Computer Vision and Pattern Recognition Workshops (CVPRW)* (IEEE, 2017 July), pp. 2299–2308
26. Y.M. Cheung, Q. Peng, Eye gaze tracking with a web camera in a desktop environment. IEEE Trans. Hum.-Mach. Syst. **45**(4), 419–430 (2015)
27. R.A. Naqvi, M. Arsalan, G. Batchuluun, H.S. Yoon, K.R. Park, Deep learning-based gaze detection system for automobile drivers using a NIR camera sensor. Sensors **18**(2), 456 (2018)
28. R.D.O.J. Dos Santos, J.H.C. de Oliveira, J.B. Rocha, J.D.M.E. Giraldi, Eye tracking in neuromarketing: a research agenda for marketing studies. Int. J. Psychol. Stud. **7**(1), 32 (2015)
29. M. Miyamoto, Y. Shimada, M.A.K.I. Yasuhiro, K. Shibasato, Development of eye gaze software for children with physical disabilities. in *Advanced Informatics: Concepts, Theory And Application (ICAICTA), 2016 International Conference On* (IEEE, 2016 Aug), pp. 1–6
30. P. Biswas, J. DV, Eye gaze controlled MFD for military aviation. in *23rd International Conference on Intelligent User Interfaces* (ACM, 2018 March), pp. 79–89
31. P.M. Corcoran, F. Nanu, S. Petrescu, P. Bigioi. Real-time eye gaze tracking for gaming design and consumer electronics systems. IEEE Trans. Consum. Electr. **58**(2) (2012)
32. C.C. Wang, J.C. Hung, S.N. Chen, H.P. Chang, Tracking students' visual attention on manga-based interactive e-book while reading: an eye-movement approach. Multimed. Tools Appl. 1–22 (2018)
33. R.N. Khushaba, C. Wise, S. Kodagoda, J. Louviere, B.E. Kahn, C. Townsend, Consumer neuroscience: Assessing the brain response to marketing stimuli using electroencephalogram (EEG) and eye tracking. Expert Syst. Appl. **40**(9), 3803–3812 (2013)
34. A. Zaraki, D. Mazzei, M. Giuliani, D. De Rossi, Designing and evaluating a social gaze-control system for a humanoid robot. IEEE Trans. Hum.-Mach. Syst. **44**(2), 157–168 (2014)
35. B. Pires, M. Hwangbo, M. Devyver, T. Kanade, Visible-spectrum gaze tracking for sports. in *Proceedings of the IEEE Conference on Computer Vision and Pattern Recognition Workshops* (2013), pp. 1005–1010
36. V. Cantoni, M. Musci, N. Nugrahaningsih, M. Porta, Gaze-based biometrics: an introduction to forensic applications. Pattern Recogn. Lett. **113**, 54–57 (2018)
37. S. Wyder, F. Hennings, S. Pezold, J. Hrbacek, P.C. Cattin, With gaze tracking toward noninvasive eye cancer treatment. IEEE Trans. Biomed. Eng. **63**(9), 1914–1924 (2016)
38. J. Mundel, P. Huddleston, B. Behe, L. Sage, C. Latona, An eye tracking study of minimally branded products: hedonism and branding as predictors of purchase intentions. J. Prod. & Brand. Manag. **27**(2), 146–157 (2018)

# Gabor Filter and ICA-Based Facial Expression Recognition Using Two-Layered Hidden Markov Model

Mayur Rahul, Rati Shukla, Puneet Kumar Goyal, Zaheer Abbas Siddiqui and Vikash Yadav

**Abstract** This paper introduces the framework based on the extraction of features using Gabor filters and modified hidden Markov model for classification. The three regions of interest (nose, mouth and eyes) are extracted using Gabor filters and dimensions are reduced by independent component analysis. Then these reduced dimensions are input to our two-layered HMM for training and testing. Seven facial expressions are recognized using publicly available JAFFE dataset. Experimental data shows the efficient and robust nature of our framework and shows its uniqueness on comparing it with other existing available methods.

**Keywords** Hidden Markov model · JAFFE · Feature extraction · Classifier · Facial expressions · Pattern recognition

## 1 Introduction

As we all know communication is the main medium to exchange our thoughts, idea, views, etc. It is of different types out of which facial expression is one of the main and essential means for communication. Ekman et al. in their research introduces it and found six fundamental facial expressions, i.e., anger, sadness, surprise, disgust, fear

M. Rahul
Department of Computer Applications, UIET, CSJMU, Kanpur, India
e-mail: mayurrahul209@gmail.com

R. Shukla
GIS Cell, MNNIT, Allahabad, India
e-mail: mca.rati@gmail.com

P. K. Goyal (✉) · Z. A. Siddiqui · V. Yadav
Department of Computer Science and Engineering, ABES Engineering College, Ghaziabad, India
e-mail: puneet.goyal@abes.ac.in

Z. A. Siddiqui
e-mail: zaheer.abbas@abes.ac.in

V. Yadav
e-mail: vikash.yadav@abes.ac.in

© Springer Nature Singapore Pte Ltd. 2021                                                  511
X.-Z. Gao et al. (eds.), *Advances in Computational Intelligence and Communication Technology*, Advances in Intelligent Systems and Computing 1086,
https://doi.org/10.1007/978-981-15-1275-9_42

and joy [1]. These six expressions are used to convey the inner situation of human emotions and intentions. Meharabian et al. state that 7% of total communication is by verbal, 38% uses sound and 55% uses facial expressions and gestures [2].

We can apply automatic facial expression recognition (FER) in various fields like human–computer interaction, surveillance system, e-learning and security systems in offices [3]. The FER systems generally have pre-processing, feature extraction and classification as three basic steps. Pre-processing, the first step, is basically the removal of noise present in images. Feature extraction, the second step, is used to fetch the essential features, and classification step is the recognition of facial expression in which training and testing of classifier take place.

This paper uses the filter and ICA for feature extraction, whereas modified HMM is used for the classification. Facial expression recognition used in the security purposes in the offices and shopping complexes to read the facial expression of the suspect peoples. Our system is capable to identify facial expressions efficiently and robustly. The main objectives of our paper are as follows: (1) To investigate our work with other existing methods in terms of recognition rate and ROC curve. (2) To investigate the processing speed of our modified HMM classifier. (3) To investigate the error rates of our FER system.

The organization of the paper has made in such a way that the next section will cover related work. Section 3 will cover the proposed system. Section 4 covers experimental results and their discussion. Section 5 covers the conclusion and future scope of the research.

## 2   Related Work

Geometric-based and appearance-based is two types of FER systems. Ahmad et al. extract the texture from image to represent facial changes is the example of appearance-based method [4]. Lei et al. use angle and displacement to extract geometric features are the example of geometric-based method [5]. A Lot of researches has been done in appearance-based methods by LBP and its several variants [6]. Some uses MTP [7], LTP [8], CLBP [4], PCA [9] and HOG [10]. Shan et al. introduce the concept of LBP for low-resolution images and shows its superiority with the others existing available methods. Gritti et al. also show the performance of various local features, such as HOG, LTP, LBP and Gabor, and show that LBP performs better than all the available features [11].

Caragni et al. introduce the setting parameters of HOG descriptor and prove that good parameter setting makes the HOG more efficient for the recognition of facial expressions [3]. Ahmed et al. propose the Compound Local Binary Pattern (CLBP) which is an extended version of LBP, which combines the information between gray values and normal LBP [1]. Bashar et al. introduce the Medium Ternary Pattern, which is the variation between gray values and median filter [2].

**Table 1** Related works

| Authors | Database | Methods | Accuracy |
|---------|----------|---------|----------|
| Rivera [16] | 29 subjects | Local directional number pattern | 92.9 |
| Li [17] | PCA, LDA and SVM | 29 subjects | 3D–90%<br>2D–80% |
| Shan [18] | Local binary patterns, adaboost LDA, SVM | CK | 89.14 |
| Jain [19] | Latent-dynamic conditional random fields(LDCRFs) | CK | 85.4 |
| Wu [20] | Gabor motion energy filters | CK | 78.6 |
| Zhang [21] | NN-based facial emotion recognition | CK | 75.83 |

Gunawan et al. propose the active appearance model (AAM) to extract essential features and fuzzy logic for classification of facial expressions [11]. Zhou et al. created an active shape model (ASM) for extraction of important features points and then displacement and projections of these feature points used to recognized facial expressions [12]. Andrade et al. combine two techniques to recognize facial expressions. They used Empirical Normalized Distances (END) and Correlation Features Selection (CFS) to extract important features. Both are in the category of geometric-based feature extraction [13]. Both techniques are based on point distribution model (PDM) for the purpose of feature extraction under the category of geometric-based feature extraction.

Appearance-based feature extraction relies on facial points, entire face and patches around some specific regions to extract appearance-based information. Wang et al. proposed a framework for extraction of facial regions using active appearance model (AAM) and then Gabor wavelet transformation is applied to get extracted features [14]. Chen et al. created a framework for facial regions. These facial regions are extracted using HOG descriptors [15]. Most of the above frameworks use the SVM for classification Table 1.

# 3 Proposed System

The proposed system has been depicted with the help of block diagram in Fig. 1. The Gabor filter is used to extract feature. Independent component analysis is used to reduce the dimensions of the features from the Gabor filter. Further, we are using these features to train and test the HMM classifier. Our new modified HMM classifier is the extension of normal HMM which consists of individual expressions (mouth, nose) and combination of individual expressions (anger, surprise). This HMM handles both types of expressions. N-cross validation rule is applied for avoiding overfitting and underrating of the classifier.

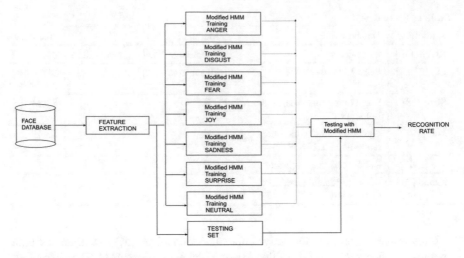

**Fig. 1** Block diagram of the proposed system

## 4 Experiments and Results

JAFFE dataset has been used in MATLAB2016 environment. The extracted feature from Gabor filter has very high dimensions. To reduce the dimension, ICA has been used. N-cross validation rule is used to avoid overfitting. In this rule, all the images are categorized into ten-folds. In which each fold consists of 90% images used for training and 10% for testing purpose. Final recognition rate is the mean of all folds Table 2.

Feature extraction using Gabor filter is shown in Fig. 2. The recent research shows that Gabor filter is the best feature extraction technique. After feature extraction, dimensions are reduced using ICA. The reduced dimensions give the final feature vector, which is used to train and test the classifier. The results in the form of confusion matrix are shown in Table 3.

**Table 2** Instances of seven expressions from JAFFE

| Expression | No. of images |
|------------|---------------|
| Anger | 16 |
| Disgust | 26 |
| Fear | 19 |
| Joy | 33 |
| Sadness | 29 |
| Surprise | 27 |
| Neutral | 29 |

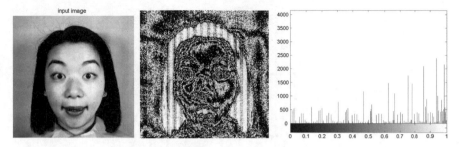

**Fig. 2** Feature extraction using Gabor filter

Table 3 shows the improvement in results. The disgust expression is recognized well by our system as compared to work of Jun et al. Except recognition rate of anger and fear, all the recognition rate is better as compared to work by Jun et al.

The performance of our HMM classifier is shown in Fig. 3. The figure depicts the goodness of our classifier. The plot between true positive and false positive rates above the line of no-discrimination shows the best performance of our classifier.

The complete algorithm is divided into two parts: average feature extraction processing time and average rest-of-the-algorithm processing time. Our proposed system processing speed of 210 and 935 ms are better than the Jun et al. processing speed of 234 and 1050 ms (Table 4).

The error rates of Jun et al. is 18.6 and 20.23 as compared to our 11.2 and 16.64 for the 10 and 170 training samples, respectively, required for the classification (Table 5).

## 5 Conclusions and Future Works

In this research work, the introduction of Gabor filter and ICA with the new modified HMM has been made. Our proposed system is able to work better in terms of recognition rate, processing speed and error rates. Our HMM classifier works fastly and efficiently as compared to other existing available methods. Further, in the future, we also incorporate contempt facial expression and spontaneous expression in our system. We also apply our proposed system to big-size environment.

**Table 3** Confusion matrix for the proposed method

| | Anger | Disgust | Fear | Joy | Sadness | Surprise | Neutral | Recognition rate (our method) | Jun et al. |
|---|---|---|---|---|---|---|---|---|---|
| Anger | 13 | 0 | 1 | 0 | 0 | 2 | 0 | 81.25 | 83 |
| Disgust | 0 | 21 | 0 | 0 | 2 | 3 | 0 | 80.7 | 75.6 |
| Fear | 1 | 0 | 12 | 4 | 2 | 0 | 0 | 63.1 | 79 |
| Joy | 0 | 0 | 0 | 32 | 0 | 1 | 0 | 96.96 | 89 |
| Sadness | 2 | 0 | 0 | 0 | 25 | 0 | 2 | 86.20 | 75 |
| Surprise | 0 | 0 | 0 | 5 | 0 | 22 | 0 | 81.48 | 77 |
| Neutral | 0 | 0 | 0 | 0 | 1 | 0 | 18 | 94.73 | – |

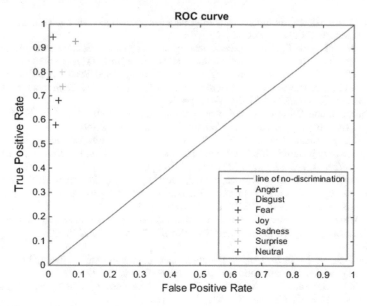

**Fig. 3** ROC curve for the confusion matrix

**Table 4** Comparison of processing time

| Average feature extraction processing time (in ms) | | Average rest-of-the-algorithm processing time (in ms) | |
|---|---|---|---|
| Proposed system | Jun et al. | Proposed system | Jun et al. |
| 210 | 234 | 935 | 1050 |

**Table 5** Comparison of error rates

| No. of training samples required | | 10 | 170 |
|---|---|---|---|
| Error rates (%) | Jun et al. | 18.6 | 20.23 |
| | Proposed framework | 11.2 | 16.64 |

**Acknowledgements** Japanese Female Facial Expression (JAFFE) is publicly available datasets. It is available free of charge from Web site http://www.kasrl.org/jaffe.html. The database was planned and assembled by Michael Lyons, Miyuki Kamachi and Jiro Gyoba.

# References

1. P. Ekman, An argument for basic emotions. Cogn. Emot. **6**(3–4), 169–200 (1992)
2. A. Mehrabian et al., *Silent Messages*, vol. 8 (Wadsworth Belmont, CA, 1971)
3. G. Molinari, C. Bozelle, D. Cereghetti, G. Chanel, M. Bétrancourt, T. Pun, Feedback emotionnel et collaboration médiatisée par ordinateur: Quand la perception des interactions est liée aux

traitsémotionnels. In Environnements Informatiques pour l'apprentissage humain, Actes de la conférence EIAH (2013), pp. 305–326

4. F. Ahmed, H. Bari, E. Hossain, Person-independent facial expression recognition based on compound local binary pattern (clbp). Int. Arab J. Inf. Technol. **11**(2), 195–203 (2014)

5. G. Lei, X.-h. Li, J.-l. Zhou, X.-g. Gong, Geometric feature based facial expression recognition using multiclass support vector machines, in *IEEE International Conference on Granular Computing, 2009, GRC'09* (IEEE, 2009), pp. 318–321

6. C. Shan, S. Gong, P.W. McOwan, Facial expression recognition based on local binary patterns: a comprehensive study. Image Vis. Comput. **27**(6), 803–816 (2009)

7. F. Bashar, A. Khan, F. Ahmed, M.H. Kabir, Robust facial expression recognition based on median ternary pattern (mtp), in *2013 International Conference on Electrical Information and Communication Technology (EICT)* (IEEE, 2014), pp. 1–5

8. T. Gritti, C. Shan, V. Jeanne, R. Braspenning, Local features based facial expression recognition with face registration errors. in *8th IEEE International Conference on Automatic Face & Gesture Recognition, 2008. FG'08*, (IEEE, 2008) pp. 1–8

9. S.S. Meher and P. Maben, Face recognition and facial expression identification using pca, in *Advance Computing Conference (IACC), 2014 IEEE International* (IEEE, 2014), pp. 1093–1098

10. P. Carcagnì, M. Coco, M. Leo, C. Distante, Facial expression recognition and histograms of oriented gradients: a comprehensive study. SpringerPlus **4**(1), 1 (2015)

11. A.A. Gunawan et al., Face expression detection on kinect using active appearance model and fuzzy logic. Procedia Comput. Sci. **59**, 268–274 (2015)

12. R. Shbib, S. Zhou, Facial expression analysis using active shape model. Int. J. Sig. Process. Image Process. Pattern Recognit. **8**(1), 9–22 (2015)

13. J. de Andrade Fernandes, L.N. Matos, M.G. dos Santos Arag˜ao, Geometrical approaches for facial expression recognition using support vector machines, in *Conference on Graphics, Patterns and Images (SIBGRAPI), 2016 29th SIBGRAPI* (IEEE, 2016), pp. 347–354

14. L. Wang, R. Li, K. Wang, A novel automatic facial expression recognition method based on AAM. J. Comput. **9**(3), 608–617 (2014)

15. J. Chen, Z. Chen, Z. Chi, H. Fu, Facial expression recognition based on facial components detection and hog features, in *International Workshops on Electrical and Computer Engineering Subfields* (2014), pp. 884–888

16. A.R. Rivera, J.R. Castillo, O.O. Chae, Local directional number pattern for face analysis: face and expression recognition. IEEE Trans. Image Process. **22**, 1740–1752 (2013). https://doi.org/10.1109/TIP.2012.2235848

17. C. Li, A. Soares, Automatic facial expression recognition using 3D faces. Int. J. Eng. Res. Innov. **3**, 30–34 (2011)

18. C. Shan, S. Gong, P.W. McOwan, Facial expression recognition based on local binary patterns: a comprehensive study. Image Vis. Comput. **27**, 803–816 (2009). https://doi.org/10.1016/j.imavis.2008.08.005

19. S. Jain, C. Hu, J.K. Aggarwal, Facial expression recognition with temporal modeling of shapes, in *Proceedings of the 2011 IEEE International Conference on Computer Vision Workshops (ICCV Workshops)*, Barcelona, Spain. 6–13 Nov 2011 (IEEE Piscataway, NJ, USA, 2011), pp. 1642–1649

20. T. Wu, M.S. Bartlett, J.R. Movellan, Facial expression recognition using Gabor motion energy filters, in *Proceedings of the 2010 IEEE Computer Society Conference on Computer Vision and Pattern Recognition Workshops (CVPRW)*, San Francisco, CA, USA, 13–18 June 2010, (IEEE, Piscataway, NJ, USA, 2010), pp. 42–47

21. L. Zhang, M. Jiang, D. Farid, M.A. Hossain, Intelligent facial emotion recognition and semantic-based topic detection for a humanoid robot. Expert Syst. Appl. **40**, 5160–5168 (2013). https://doi.org/10.1016/j.eswa.2013.03.016

# FPGA Implementation of Modified Clipping Method for Impulsive Noise Cancellation

Priyank H. Prajapati, Digvijay R. Khambe and Anand D. Darji

**Abstract** Noise is an unwanted signal that distorts or creates interference in the measurement of the desired signal. Impulsive noise is an important noise because of its uncertainty, which adds severe distortion in the signal and makes the algorithms to fail or may cause a performance reduction. In this work, a modified clipping method has been presented to remove the impulsive noise. The analysis and comparison of different methods have been carried out for impulsive noise cancellation and shown that the proposed modified clipping method performs better than the conventional methods. Moreover, the hardware architecture of modified clipping methods has been proposed to remove impulsive noise and implemented using VHDL language. The proposed architecture can able to operate at the maximum clock frequency of 87.71 MHz when implemented on the Virtex 4 FPGA platform. The proposed impulsive noise removal architecture improves the maximum clock frequency by 75% and has 2% less LUTs utilization as compared to existing MRMN architecture. Also, the real-time implementation of the impulsive noise cancellation system using proposed architecture has been tested on the Spartan 3E FPGA board.

**Keywords** FPGA · Impulse noise · Modified clipping · Adaptive filter · Relative average

## 1 Introduction

Signal processing systems suffer from mainly two noises, they are Gaussian noise and impulsive noise. The noise gets generated by many natural sources such as thermal noise, Johnson–Nyquist noise, black body radiation from the earth and other warm objects, shot noise and from celestial sources such as the Sun [1, 2]. In various fields of the signal processing application like noise cancellation in system identification, active noise cancellation system and wireless communication system demands, the

P. H. Prajapati (✉) · D. R. Khambe · A. D. Darji
SVNIT, Surat 395007, India
e-mail: add@eced.svnit.ac.in

© Springer Nature Singapore Pte Ltd. 2021
X.-Z. Gao et al. (eds.), *Advances in Computational Intelligence and Communication Technology*, Advances in Intelligent Systems and Computing 1086,
https://doi.org/10.1007/978-981-15-1275-9_43

signal-to-noise ratio should be improved by minimizing the effect of noise on the desired signal. The main objective of the work is to remove the impulsive noise since it creates the burst error in pre-processing of the signal processing system.

Adaptive filters can remove the Gaussian noise using least mean square (LMS) and normalized LMS (NLMS) algorithms [3]. But, removing the impulsive noise using LMS and NLMS may cause the algorithms failure since, the direct dependency of their weight update function on the error signal [3]. Researchers have developed numerous adaptive algorithms, like robust mixed-norm (RMN) [4], normalized RMN (NRMN) [5], modified RMN (MRMN) [6] algorithm, etc. to remove the impulsive noise using the adaptive filters. In RMN-based adaptive algorithms, two different adaptive algorithms are merged to get better performance in the noisy environment. Adaptive filters are generally designed using active noise cancellation (ANC) filters, but these filters require an additional reference signal for active noise cancellation. Further, the adaptive algorithms may suffer for stability since its structure is similar to the closed-loop system. There are other methods too, to remove the impulsive noise such as clipping and modified clipping method as discussed in [7–10] are mostly targeting the PAPR noise reduction in a communication system which is not an adaptive method. The clipping method has been presented by Mirza [7, 9], removes impulsive noise to some extent but not completely. Moreover, it adds distortions in the processed signal. The hardware implementation of the impulsive noise cancellation method using adaptive filters is shown in [6, 11]. However, these hardware implementations consume more hardware due to the adaptive algorithm complexity.

In this work, a modified clipping method has been proposed and shown that this method can remove impulsive noise. The distortion due to this method while processing the signal is also less as compared to the conventional clipping method. The proposed method is not an adaptive ANC noise cancellation method; hence, it does not need a reference signal, and the hardware complexity of the proposed method is also less.

The work has been organized as follows: Sect. 2 describes the different clipping methods to remove the impulsive noise and also, their MATLAB simulation results have been presented. In Sect. 3, analysis and comparisons of modified clipping have been shown. Section 4 describes the proposed hardware architecture for modified clipping method for impulsive noise cancellation. Moreover, in Sect. 5, the real-time hardware implementation and its results have been shown, and also comparisons have been carried out with the existing implementation for impulsive noise cancellation. The final section concluding the work.

## 2  Clipping Method for Impulsive Noise Cancellation

The clipping method [9] is the simplest method to implement wherein the amplitude range of the input signal has priorly known, and the comparison of the threshold and the input signal has been carried out to clipped out unwanted signal. This method is work for those impulses which have a higher amplitude than the signal's amplitude. It has seen that this method cannot remove the impulse, which is within the range of the signal amplitude and still the processed signal has abrupt changes. Hence, the modified clipping method has been proposed to remove the impulsive noise from the signal.

## 2.1 Modified Clipping Method for Impulsive Noise Cancellation

In the proposed method, the impulse has been detected using the concept of the relative average of signal and replacing that impulse by a constant value or gradually changing the amplitude of the signal. There are two steps involved in this method;

1. Calculating relative average of signal and deciding thresholds.
2. Detecting impulse and clipping it.

### 2.1.1 Calculating Relative Average of Signal and Deciding Thresholds

In this method, threshold calculation is estimated first based on the information about impulsive noise, which is extracted using the relative average value. To calculate the relative average of the signal, window (the number of samples used for processing the signal) of length $L$ has been selected. This window has been slid along the signal to calculate relative averages using (1).

$$Relativeavg(i) = \frac{r(i) + r(i-1) + r(i-2) + \cdots + r(i-L)}{L * \frac{(r(i)+r(i-L))}{2}} \tag{1}$$

where $r(i)$ is an input signal, and $L$ is a length of the window. After applying the relative average equation (1), for the window of length $L$ and sliding this window along the signal, the array of relative average values have been obtained.

### 2.1.2 Detecting and Clipping Impulse

The relative average is very high where the impulse is present and have lower values to other parts of the signal. Hence, it can be possible to decide thresholds values $V_{tl}$ and $V_{th}$ for impulse detection. For that, the computed relative average of the signal has been compared with a threshold. If the relative average is less than $V_{tl}$ or greater than $V_{th}$, then impulse has been detected and can be clipped out. The following equation has been used for clipping method.

$$y(i-k) = \begin{cases} r(i-k); & \text{if } ravg < V_{th} \text{ or } ravg > V_{tl} \\ r(i-L-1); & \text{if } ravg > V_{th} \text{ or } ravg < V_{tl} \end{cases} \quad (2)$$

where $k = L, L-1, L-2 \cdots 1$ and $ravg$ is relative average value for a particular window. Figure 1 (a) shows the output of modified clipping method using Eqs. (1) and (2) for window $L = 50$. The impulses have been clipped out from the signal, but it creates a staircase wave at impulse position. Moreover, while using this method, if thresholds get slightly changed, then method gives distorted output.

In order to make the algorithm robust to thresholds variations, Eq. (2) has been modified to give Eq. (3).

$$y(i-k) = \begin{cases} r(i-k); & \text{if } ravg < V_{th} \text{ or } ravg > V_{tl} \\ r(i-k-1) + ks; & \text{if } ravg > V_{th} \text{ or } ravg < V_{tl} \end{cases} \quad (3)$$

where $k = L, L-1, L-2 \cdots 1$ and $ravg$ is relative average value for a particular window. $s$ is the step size given as

$$s = \frac{r(i) - r(i-L)}{L} \quad (4)$$

Figure 1b shows the output of modified clipping method using Eqs. (1) and (3) for window $L = 50$.

It is evident that output of function given by Eq. (3) is smoother at clipping point as compared to the output given by Eq. (2). Even if there has been the change in the thresholds value, the output would not get distorted.

## 3   Analysis and Comparison of Modified Clipping

To check the performance of modified clipping method, analysis and comparison with conventional clipping method and adaptive filter have been carried out. All the simulations have been performed using MATLAB (R2015b) software on Intel i5-CPU having 2.50 GHz clock speed and 6 Gb memory.

### 3.1   Modified Clipping Method for Non-sinusoidal and Non-periodic Signals

The non-periodic signal with Gaussian and impulsive noise has been processed using the modified clipping method as shown in Fig. 2. The output of modified clipping proves that it can be able to process the different input signals having different shapes.

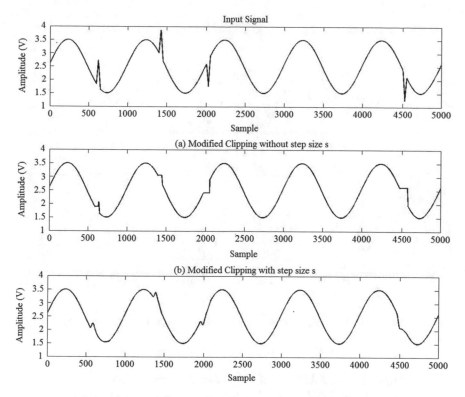

**Fig. 1** Output of modified clipping method **a** using Eqs. (1) and (2): $V_{th} = 1.02$, $V_{tl} = 0.87$; **b** using Eqs. (1) and (3): $V_{th} = 1.01$, $V_{tl} = 0.87$

## 3.2 Effect of Window on Modified Clipping Method

In the modified clipping method, selection of the window length is a crucial part. In this section, the effect of the change in the window length of the modified clipping method has been discussed. The sine wave has been generated using MATLAB with a triangular impulse of width 50 has been added to the sine wave. After that, window of length L has been varied from 10 to 100. The effect of the varying length of the window on impulsive noise is shown in Fig. 3. It has been proved that as the length of the window increases towards the width of an impulse, it would reduce the impulse noise. At window length equal to the impulse width, there has been reasonable noise reduction as shown in Fig. 3b. Hence, to decide the length of the window, there has been a need for information about the impulsive noise prior to filtering operation performed by this method.

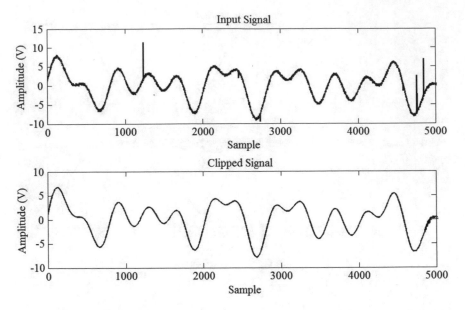

**Fig. 2** Output of modified clipping method for non-periodic signal with Gaussian and impulsive noise

## 3.3 Comparison of Modified Clipping Method with Conventional Clipping Method in Frequency Domain

In this section, the conventional clipping method and modified clipping method [9] have been compared in a frequency domain. As shown in Fig. 4, the modified clipping has less abrupt changes in its output as compared to the conventional clipping method, because the modified clipping reduces the impulsive noise present in the amplitude range of the signal, which is not in the case of conventional clipping method as discussed in Sect. 2.

## 3.4 Comparison of Modified Clipping Method with Adaptive Filter Method

In this section, the sine wave with impulsive noise has been given to the adaptive filter and modified clipping method. The adaptive filter of 9 taps with MRMN algorithm has been used as given in [6, 12], for impulse noise cancellation. The window length $L = 10$ has been used for modified clipping method. The results have been plotted in Fig. 4b, and observed that the outputs of adaptive filter and modified clipping method

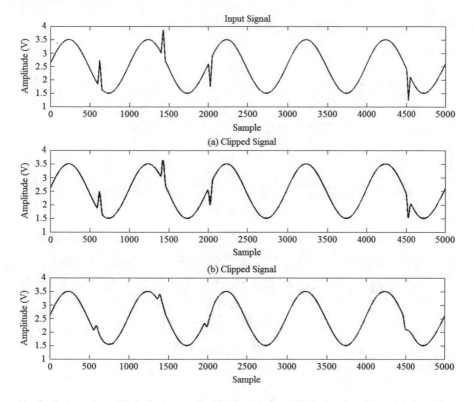

**Fig. 3** Output of modified clipping method for the window of length **a** $L = 10$, and **b** $L = 50$

are nearly same. However, the modified clipping method has very less computation complexity because of its simple relative average calculation as compared to complex adaptive weight update functions of adaptive filter.

# 4 Proposed Hardware Architecture of Modified Clipping Method

As proven in Sect. 3, the impulse noise reduction performance of the modified clipping method is better, and also its complexity is less as compared to other existing methods. So, it has been used for hardware implementation. There are various advantages of FPGAs over DSP processors and ASIC [11]. Hence, FPGA has been used for the hardware implementation of modified clipping method. The modified clipping method has been implemented using the Spartan 3E FPGA starter kit for the real-time noise cancellation application.

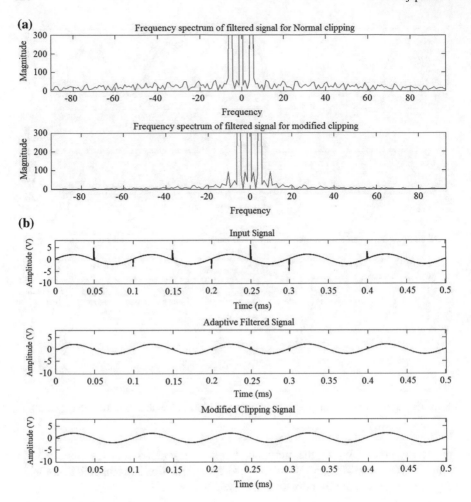

**Fig. 4** Comparison of modified clipping with **a** conventional clipping method [9] in frequency domain, and **b** adaptive filtering method [3]

Figure 5 shows the architecture of modified clipping method. In this architecture, input samples have been passed through register array of length L in order to compute the relative average of the input signal. The relative average value has been calculated with the use of register array, adders, accumulator, shifter, and one division module and the impulse is detected using the comparator module as shown in Fig. 5. In relative average calculation, the numerator term of Eq. (1) which is an accumulation of input data has been calculated using the iterative method given by Eq. (5).

$$accsq_{new} = accsq_{old} + r_L - r_1 \tag{5}$$

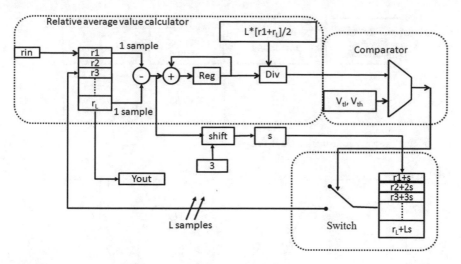

**Fig. 5** Architecture of modified clipping method

The old value of accumulation $accsq_{old}$ has been stored in the register. This old value is used to calculate a new value of accumulation as shown in Eq. (5). This technique requires only one accumulator, one 2's complement adder and L registers.

The output of relative average calculator block is compared with the threshold. When relative average value crosses threshold values, it is considered as the impulse, since the impulse has high relative average value as compared to the desired signal. If the impulse is detected, then register array of switch module is computed simultaneously with the relative average calculation to replace the register array of the relative average value calculator by new values as shown in the architecture. The last value of the register array gives a processed output.

## 5 Hardware Implementation Results of Modified Clipping Method

Figure 6 shows the hardware implementation block diagram of proposed impulse noise cancellation system. For the real-time implementation, Spartan 3E FPGA board has been used. To generate the noisy signal, arbitrary function generator is used and sampled using ADC of Spartan 3E FPGA board. The sampled data is given to modified clipping block, which clips the impulses from the input signal. The processed data is transferred to the remote desktop machine using Ethernet interfacing. The remote desktop receives data transmitted by Ethernet, for that the wire-shark tool has been used. The received data is displayed using MATLAB. The noisy input signal given to the system is shown in Fig. 7a and output of test bench is given in Fig. 7b. The real-time received data has been plotted as shown in Fig. 8. Figure 9 shows, the

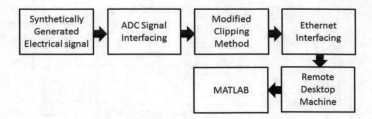

**Fig. 6** Hardware experimentation block diagram

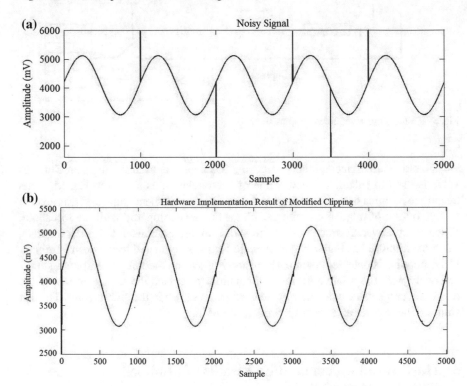

**Fig. 7** **a** Noisy input signal for RTL design verification. **b** Hardware implementation simulation result of modified clipping

hardware experimentation set-up of the noise cancellation system. Modified clipping method works in the frequency range of 3–40 KHz when window and threshold are set for 10 KHz input signal frequency.

Table 1 gives the comparison of modified clipping method with MRMN adaptive filter implementation and modified clipping method using Spartan 3E FPGA platform. MRMN [6] adaptive filter has operating frequency of 41.81 MHz and it utilizes 893 LUTs, where as proposed architecture of modified clipping method has a maximum operating frequency of 71.88 MHz and it utilize 365 LUTs only. It has

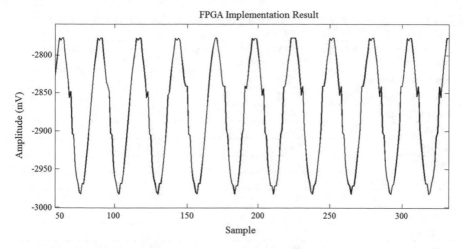

**Fig. 8** FPGA hardware implementation result of modified clipping method

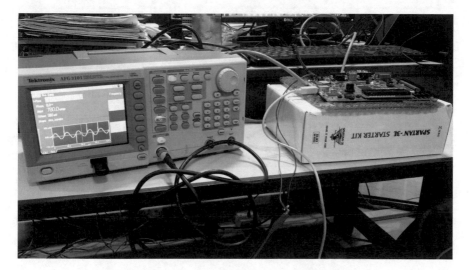

**Fig. 9** Experimental set-up

been shown that since the complexity of modified clipping is less as a compared to adaptive filter the proposed architecture occupies less resource utilization.

Table 2 gives the comparison of different methods for impulse removal using the Virtex 4 FPGA platform. Since modified clipping method has less complexity and registers array utilized in proposed architecture break the critical path and improves the maximum operating clock frequency of architecture, the proposed architecture has 87.71 MHz maximum clock frequency. Moreover, the slice utilization of proposed architecture gives the difference of 44% and 2% of the robust and MRMN architecture, respectively. Also, noted that the proposed architecture has not utilized

**Table 1** Comparison of proposed modified clipping architecture with MRMN architecture on Spartan-3E (XC3S500E FG320-4) FPGA

| Algorithms/timing and logic utilization | MRMN [6] (Gaussian and impulsive noise) | Modified clipping (impulsive noise) |
|---|---|---|
| Synthesis delay (ns) | 20.754 | 14.28 |
| Max. clock freq. (MHz) | 41.81 | 71.88 |
| Slice Flipflops | 288 | 216 |
| LUTs | 893 | 365 |
| Slices | 545 | 204 |
| Multipliers | 18 | 2 |
| Adders/subtractor | 27 | 15 |
| Counters | 0 | 2 |
| Accumulators | 0 | 2 |
| Registers | 379 | 202 |
| Comparators | 18 | 4 |
| Logic shifter | 1 | 0 |

**Table 2** Comparison of existing impulsive noise cancellation method on Virtex 4 (XC4VFX12 ff668-12) for PAR timing and resource utilization

| FPGA logic | Robust [11] | MRMN [6] | Modified clipping |
|---|---|---|---|
| Slices | 2586 (47%) | 277 (5%) | 206 (3%) |
| Multipliers | 9 | 18 | 0 |
| 4 input LUTs | 4777 (43%) | 438 (4%) | 372 (3%) |
| Max. clock freq. (MHz) | 18.97 | 49.863 | 87.71 |

a single multiplier, since the architecture used shifter instead of a DSP multiplier for multiplication, which is shown in Table 2.

# 6 Conclusion

The conventional clipping method is not an adaptive method and it has less complexity. But, this method fails to remove complete impulse from the signal. Adaptive filters can remove complete impulse, but have higher hardware complexity and need a reference signal for active noise cancellation. Hence, modified clipping method has been proposed which can remove impulses. The analysis and comparison shows that the modified clipping gives better output as compared to the existing noise removal methods. Moreover, the hardware implementation of the proposed modified clipping method is carried out. The proposed architecture utilizes only 365 LUTs and has a maximum operating frequency of 71.88 MHz when implemented on the Spartan3E FPGA. The proposed architecture leads to a low area and high-speed realization for fast signal preprocessing applications. Hardware implementation comparison of proposed architecture shows the resource utilization reduce by 44% and 2% of the robust and MRMN architecture, respectively, when implemented on the Virtex 4 FPGA platform.

**Acknowledgements** The research work carried out in this paper has been supported by Special Manpower Development Programme Chip to system Design, Miety Government of India. Authors are thankful to them for providing research facilities.

# References

1. S.V. Vaseghi, *Advanced Digital Signal Processing and Noise Reduction* (Wiley, 2000)
2. A. Zaknich, *Principles of Adaptive Filters and Self-Learning Systems* (Springer, 2006)
3. S. Haykin, *Adaptive Filter Theory* (Pearson Education, 2008)
4. J. Chambers, A. Avlonitis, A robust mixed-norm adaptive filter algorithm. IEEE Signal Process. Lett. **4**(2), 46–48 (1997)
5. E.V. Papoulis, T. Stathaki, A normalized robust mixed-norm adaptive algorithm for system identification. Signal Process. Lett. IEEE **11**(1), 56–59 (2004)
6. C.A. Parmar, B. Ramanadham, A.D. Darji, FPGA implementation of hardware efficient adaptive filter robust to impulsive noise. IET Comput. Digit. Tech. **11**(3), 107–116 (2016)
7. S.M. Kabir, A. Mirza, S.A. Sheikh, Impulsive noise reduction method based on clipping and adaptive filters in AWGN channel. Int. J. Future Comput. Commun. **4**(5), 341 (2015)
8. K. Sultan, H. Ali, Z. Zhang, Joint SLM and modified clipping scheme for PAPR reduction, in *13th International Bhurban Conference on Applied Sciences and Technology (IBCAST), 2016* (IEEE, 2016), pp. 710–713
9. A. Mirza, S.M. Kabir, S.A. Sheikh, Reduction of impulsive noise in OFDM systems using a hybrid method. Int. J. Signal Process. Syst. **4**(3) (2016)
10. S. Singh, A. Kumar, A modified clipping algorithm for reduction of PAPR in OFDM systems, in *2015 IEEE International Conference on Computational Intelligence and Computing Research (ICCIC)* (IEEE, 2015), pp. 1–4
11. A. Rosado-Muñoz, M. Bataller-Mompeán, E. Soria-Olivas, C. Scarante, J.F. Guerrero-Martínez, FPGA implementation of an adaptive filter robust to impulsive noise: two approaches. IEEE Trans. Ind. Electron. **58**(3), 860–870 (2011)
12. E. Soria, J. Martin, J. Calpe, A. Serrano, J. Chambers, Robust adaptive algorithm with low computational cost. Electron. Lett. **42**(1), 60–80 (2006)

# Autonomous Cars: Technical Challenges and a Solution to Blind Spot

Hrishikesh M. Thakurdesai and Jagannath V. Aghav

**Abstract** Automotive industry is progressing forward toward the future, where the role of driver is becoming smaller and leading to become ideally driverless. Designing a fully driverless car (DC) or self-driving car is a most challenging automation project, since we are trying to automate complex processing and decision-making of driving a heavy and fast-moving vehicle in public. There are many scenarios where self-driving cars are not able to perform like human drivers. There are a lot of technical, non-technical, ethical and moral challenges to be addressed. Furthermore, two recent accidents caused by self-driving cars of Uber [1] and Tesla [2] have raised a concern toward the readiness and safety of using these cars. Therefore, it is necessary to address these challenges and issues of DC's. In this paper, we have surveyed various technical challenges and scenarios where DCs are still facing issues. We have also addressed an issue of blind spots and proposed a systematic solution to tackle the issue. Before self-driving cars go live on road, we have to overcome these challenges and work on technology barriers so that we can make the DCs safe and trustworthy.

**Keywords** Self-driving cars · Machine learning · LIDAR · Automotive

## 1 Introduction to Self-driving Cars

A Self-driving car is also known as an "autonomous car (AC)" or "robot car" which has the capability to sense its environment and move ahead on its own. DC contains various sensors and cameras such as radio detection and ranging (RADAR), light detection and ranging (LIDAR) and sound navigation and ranging (SONAR) to understand the surrounding environment. Radar is a detection system that makes use of radio waves to determine various parameters like distance, speed or angle of the object. SONAR makes use of sound waves to determine presence of object. LIDAR

H. M. Thakurdesai (✉) · J. V. Aghav
College of Engineering, Wellesley Rd, Shivajinagar, Pune, Maharashtra 411005, India
e-mail: thakurdesaihm17.is@coep.ac.in

J. V. Aghav
e-mail: jva.comp@coep.ac.in

© Springer Nature Singapore Pte Ltd. 2021
X.-Z. Gao et al. (eds.), *Advances in Computational Intelligence and Communication Technology*, Advances in Intelligent Systems and Computing 1086,
https://doi.org/10.1007/978-981-15-1275-9_44

is considered as an eye of self-driving car. It uses pulsed laser light for detection. These sensors and other actuators generate a huge amount of data in real time. This data is processed by central computer for driving decisions. DCs have huge benefits in various domains, mainly in military and surveillance. These vehicles will be extremely useful for elderly and disabled people, also to children as they will be safer and secured without human interventions [3]. DCs will reduce the number of accidents which are caused by human errors in driving (94% of total accidents are because of human errors). Other benefits include fuel efficiency, car, car availability for everyone (e.g., person without a license, small child, old persons, etc.) and efficient parking. Experimental work is going on in the field of driverless cars since 1920. In 1977, the first truly automated car was invented by Tsukuba Laboratories in Japan. That car travelled with the speed of 30 km per hour with the help of two cameras. In 2015, the US government gave clearance to test the DCs on public roads. In 2017, Audi A8 was proved to be the first Level 3 automated car which travelled with the speed of 60 km per hour using Audi AI. In 2018, Google Waymo started to test its autonomous car and completed around 16,000,000 km of road test by October 2018. Waymo is the first company which has launched fully autonomous taxi service in USA from December 2018 [4].

## 1.1   Key Components in Self-driving Car

One of the major components in self-driving car is a light detection and ranging (LIDAR) which is considered as an eye of the vehicle. Main objective of LIDAR is to create 3D map of the surrounding world in real time. It emits laser beams which are invisible to human eyes and calculates the time taken to come back after hitting the nearby objects. This calculation gives distances of the surrounding objects along with their identification and hence helps the car to guide "How to Drive." LIDARs are capable of giving high-resolution images with minute details and exact distances. In addition to LIDAR, DC also uses video camera and other cameras to identify traffic lights and road signs and to maintain safety distance between other vehicles and pedestrians. Radar sensors are used to monitor position of nearby vehicles, especially in bad weather conditions. These sensors are also used in adaptive cruise control. Cameras can only give the object image but cannot give the depth of the object. Hence, we need LIDAR and RADAR. LIDAR sometimes fails in foggy conditions. RADAR is used in those cases to get the perception of surrounding. There is a global positioning system (GPS) antenna on the top of DC which gives car position on the road [5]. Another major component of DC is central computer unit which collects data from all sensors and manipulates steering, acceleration and monitoring control (Refer Fig. 1 for working of components: summary). Movement of self-driving car from one point to another point includes perseverance of environment, planning of path and controlled movements on the path. Environment perseverance includes tracking of objects, lanes and identification of self-position. This is achieved by using various medium and short-term sensors and cameras. Radars are proven to be more effective

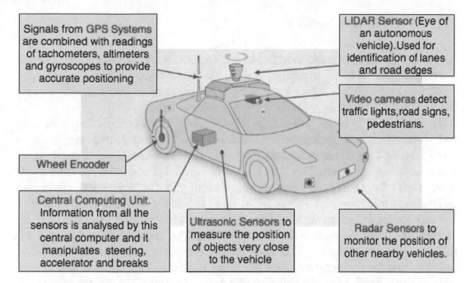

**Fig. 1** Components of driverless car. *Source* The economist, "how the self-driving car works"

than cameras for objects tracking in vehicles. Autonomous vehicles use LIDARs. One LIDAR mounted on the top of the car gives 360° view of surrounding with long range. The data obtained from this device is used to maintain speed and to apply breaks when necessary. Navigation and path planning are majorly done using GPS. These paths are initially calculated based on the destination, later dynamically changed based on the road scenarios like blocks or traffic. Accelerometers and gyroscopes are used along with GPS as the satellite signals sometimes may not reach in tunnels or underground roads. Data processing is done for taking actions like lane keeping, braking, maintaining distance, stopping and overtaking. Various control systems, electronic units and CAN networks are used for this purpose (Fig. 2).

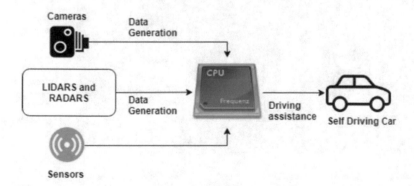

**Fig. 2** Flow of working for DC

## 1.2 Levels of Automation

As per the standards organization Society of Automotive Engineers (SAE) International, there are six levels in driving system. These are defined based on automation involved in the driving, which ranges from "no automation" to "full automation" (Refer Fig. 3 for summary of levels).

In **Level 0**, there is no automation. All operations like steering control, braking and speed control are performed by human driver. **In Level 1**, this level involves some functions for driver assistance. Most of the main tasks like steering, braking and monitoring are handled by driver only. **In Level 2**, there is partial automation level. Most of the companies are working currently in this level where driver gets assistance for steering and acceleration. Here, the driver must always be attentive and monitoring to take the control in case of safety-critical issues. **Level 3** includes monitoring assistance along with steering and braking. The vehicles in this level use LIDAR as an eye of the car. This level does not require human attention when the speed is moderate (around 37 miles per hour) and conditions are safe. But in case of higher speed and unusual scenarios, still drivers control plays critical role. Major players in the industry like Audi and Ford have announced the launch of Level 3 cars in 2018–2019. **Level 4** is known as high automation. Here, the vehicle is capable of taking steering control, braking, acceleration and monitoring. Human attention is required only for certain critical scenarios like complex traffic jams or merges into highways. **Level 5** This level involves complete automation. Here, absolutely no human attention is required for driving. All complex situations like traffic jams are also handled by vehicle, allowing driver to sit back without paying attention. Still the research is going on in this level and there are many technical and moral challenges which are to be addressed before these cars are made available to public.

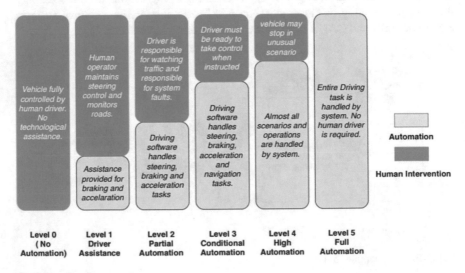

**Fig. 3** Levels of automation

With the given background of levels of automation, in the next section, we will survey the technical challenges.

## 2    Technical Challenges in Self-driving Cars

Extensive research is going on in the field of self-driving cars and all big companies in the world like Google, BMW, Audi, Uber and many others are constantly working to make these cars available to the common public [6]. In 2018, Google Waymo has already launched a commercial taxi in Arizona, USA in four suburbs of Phoenix-Chandler, Tempe, Mesa and Gilbert. The day is not far when entire world will travel driverless. Still, there are a lot of technical, legal and moral challenges to be addressed before self-driving cars can be made available [7–9]. It is necessary to develop the trust among the people for user acceptance and to make the cars ready to hit the roads [10–12]. Hence, it is essential to work in this field by analyzing current technical challenges faced by DCs so that we can contribute to enhance the safety and mobility of these cars.

Below is the list of some key challenges which are to be addressed before making the DC live on the road [13–15].

### 2.1    Unpredictable Road Conditions

Road conditions vary from place to place. Some roads are very smooth and contain proper lane markings, whereas some roads are very deteriorated which can contain potholes and are mountainous [16]. It becomes challenging for driverless car to drive itself on such road where there are no signs and lane markings [17]. Lane markings can also be disappeared because of quick snowfall. Humans can still identify lanes by natural road curves but DCs may not. Even the roads with slight water or flood can also confuse DC. Bosch has announced a system designed to give feel of road conditions to DCs. Advance information about wet roads or snow will help the AV to decide exactly where it can drive autonomously. This system aims to increase driving safety and availability of autonomous driving function.

### 2.2    Unusual Traffic Conditions

DCs can be made available on the roads only when they can handle all sorts of traffic conditions. Roads will contain other autonomous vehicles as well as human-driven vehicles. The situations may arise when humans are breaking the traffic rules and unexpected situation may occur for which DC is not trained to handle [18]. In case of dense traffic, even a few centimeters of movement matter a lot. If the DC is waiting for the traffic to automatically clear, it may have to wait indefinitely. Deadlock conditions may also arise where all cars are waiting for others and no one is moving ahead.

## 2.3 Radar Interference

As stated above, DC uses radars and sensors for navigation which are mounted on the top and on the body of the car. DC emits the radio waves which strike the object and reflects back. The time taken for reaction is used to calculate the distance between car and object. When hundreds on cars on the road will use this technology, the specific car will not be able to distinguish between the waves of its own and the waves emitted by other cars. Even if we have range of bands for the waves, still they can be insufficient in case of large number of vehicles.

## 2.4 Crosswalk

There are some situations which are very easy to handle for the humans but not for the machine. For example, driving the car between the two walking persons on the crosswalks and take immediate left turn. This situation is common for humans but very difficult for machine as they do not have perception power and cannot read from human faces. This is known as Moravec's Paradox where it becomes difficult to teach or engineer the machine. Current algorithms which are used in DCs will become policy of the future. Humans are not actually good drivers as they drive with emotions. One of the surveys has found that in specific crosswalk, black pedestrians waits more than white ones. Human driver decides to yield at crosswalk based on pedestrian's age or speed. But this becomes challenging for DC to analyze and take driving step ahead.

## 2.5 Bugs and Attacks

As DC uses machine learning algorithms, we cannot be 100% confident about any result when human safety comes in, as it learns from experience. Even a small error in the code can cause huge consequences [19]. The example of Elaine Herzberg death can be given to prove the point. This was the first death case of killed pedestrian by autonomous car on March 18, 2018. This lady in Tempe, Arizona was struck by Uber's car which was operating in self-driving mode. As a result, Uber suspended testing of self-driving car in Arizona. It is obvious that this death was caused by bugs in the system of DC. These bugs can be missing lines of codes or external attack in the system which ignored the sensor data. DCs can also suffer from intrusion attacks like denial of service which can affect the working of the system [20]. Research is going on for developing intrusion detection systems (IDS) for self-driving cars [21].

## 2.6 Need of Extensive Classification

Machine learning and AI systems learn from datasets and past experience which results in detection of various objects like cyclists, pedestrians, road markings, signs, etc. Hence, if something is missing in the dataset using which the car is trained, there is a risk of misinterpretation resulting is malfunctioning. For example, the STOP sign dropped on the side of the road which is not supposed to be followed, may confuse DC. Another example is a person walking with bicycle in hand or speedy boat came out of river and landed on the road. Zombie Kangaroo Costume Challenge can also be an example of the same where the small child is wearing kangaroo costume or halloween costume. This situation is too unusual to be included in the risk database. Therefore, the vehicle will either wrongly categorize it or will freeze and wait for human instructions. Hence, it is important to work on extensive classifier which will be able to identify unusual scenarios for taking respective actions like human capabilities.

## 2.7 Issues Due to Co-existence of Manual and Self-driving Cars

At some point, we are going to come across a scenario where there will be considerable number of driverless cars and human-driven cars [22, 23]. This may lead to a lot of problems as drivers from different country will have different etiquettes. For example, Indian drivers will show hand to turn on the road which may not be identified by the DC. Hence, it is essential to develop some sort of universal signaling language which can be followed by all to avoid failures.

## 2.8 Platooning Problem

Platooning means driving very close to another vehicle and together. This reduces road space consumption and wind resistance resulting in reduction on fuel consumption [24, 25]. Along with these advantages, there is also a risk of car crashes, especially if human driver tries to come and merge into platoon space. One of the solutions is to maintain dedicated lanes for the DCs but it is not practical in all areas.

## 2.9 Sharing Cost

At present, the cost of DC is high, hence, it is predicted that most of the autonomous vehicles will be shared [26, 27]. Hence, efficient methods must be developed to pick up the passengers on road so that other passengers in the car will not have to wait for huge time. Shared cars will also help to increase per-vehicle occupancy and decrease number of cars on the road, thereby helping in reduction of traffic.

## 2.10 Building Cost Effectiveness

As stated above, DC uses LIDAR, various cameras and sensors which are very costly. In particular, the cost of LIDAR is huge which increases overall cost of DC. At present, a DC setup can cost up to $80,000 in which LIDAR cost itself can range from $30,000 to $70,000. One strategy to reduce cost is to use LIDAR with fewer lasers. Audi has claimed that LIDAR with only four lasers will be sufficient for safe driving on highways. Further, research is going on in companies like "Valeo" and "Ibeo" to make LIDARs in less than $1000.

## 2.11 Providing Prediction Power and Judgmental Calls Taking Capability to DC

To drive safely on the roads, it is essential to see, interpret and predict the behavior of humans [28]. The car must understand if it is in the blind spot of some other car. Human drivers are able to pass through high-congestion areas because of eye contact with the humans. Eye contact also helps us to predict the state of mind. For example, we can identify if the person is distracted or not realizing the vehicle movement or the other driver is not attentive for lane change, etc. This prediction capability has to be established in the DCs to make them safe for humans. Sometimes, driver has to take instant decision when a particular object suddenly comes in front. Most easy solution is instant braking but it may lead to a hit from behind which may result in huge accidents on the highway. Another situation is to decide whether to hit a child on the road or to hit an empty box on the side which is blocking the lane [29]. These scenarios are familiar to humans hence they can take right calls. The call taken by DC may be technically correct but may fail in real world. Obviously, huge trial and error experimentation is required to make sure that DCs can handle these tasks.

## 2.12 Challenge of Identifying Animals Which Do Not Stay on Ground

DC needs to identify and respond to dozens on animals like horses, moose and kangaroos. Wild animals can suddenly jump onto the DCs path. Here, the animals especially like kangaroos represent a unique problem. Most identification systems use ground as a reference to identify the location. But kangaroos hop off the ground and it becomes challenging for the system to identify the location and where they will land. As per National Roads and Motorists' Association of Australia, 80% of animal collisions in the country are with kangaroos. More than 16 thousand kangaroo strikes each year with the vehicles. These accidents also create millions of dollars of insurance claims. Major companies like Volvo are taking this issue seriously. The safety engineers from Volvo have started filming kangaroos and they have used this data to developed kangaroo detection system in their vehicle (Table 1).

**Table 1** Summary of technical challenges and possible solutions

| Summary of technical challenges | Possible solutions |
| --- | --- |
| Unpredictable road conditions | To develop a system which will give advanced feel of road conditions like wet roads based on weather reports and maps |
| Unusual traffic conditions | Advanced warning system based on GPS and maps along with dynamic path changes to avoid the traffic areas |
| Radar interference | Maintaining sufficient distance between the vehicles to avoid radar interference. Use of multiple radars and validations for results |
| Bugs and attacks | Secured designs, development of IDS for possible attacks, ML-based solutions for attacks |
| Unusual scenarios and identification of rare objects | Designing a classifier by taking heterogeneous datasets of unusual scenarios and objects |
| Platooning and cost effectiveness | Efficient algorithms for car sharing service to reduce the wait time, designing cheap LIDARs, platooning systems to drive closely and together, dedicated lanes to avoid crashes |
| Prediction power to vehicle | Development of human behavioural identification system, speed prediction system for driverless cars |
| Animals identification | Designing exclusive animal detection system for animals like kangaroos by collecting the video data and training to predict the landing (Volvo is working on these systems already) |

# 3 Blind Spots in Self-driving Car

In manual driving, "blind spots" are the areas around the car which are not directly visible from the rear and side mirrors. These areas can be seen by driver by little extra efforts like turning the head on sides. Blind spots also consist of areas which are not visible because of body of the cars which includes pillars which are used to join top and body of the car [30]. Like human-driven cars, DCs are also suffering from blind spot issues. DC uses LIDAR as an eye to get overall 360° view of surrounding in 3D. Detection of object for a DC is done using LIDAR sensors. Hence, the blind spot area depends on the number of sensors used. Uber have recently encountered big trouble because of this issue [15]. Uber's self-driving car has stroked the lady in Arizona which resulted in her death. The reason for this is said to be the blind spot due to the reduction of sensors from five to one [31]. In 2016, Uber shifted to Volvo instead of using Fords autonomous cars. This leads to large changes in sensors design. The sensors were reduced from five to just one which is mounted on the top of the roof. To compensate this change, they increased RADAR sensors from seven to ten. Removal of number of LIDARs reduced the cost but increased the blind spots. One LIDAR on the top results in blind spot area low to the ground all around the car. Hence, there is a chance of accident if the object is not detected in the blind spot area. Therefore, it is necessary to design a system which will guide the autonomous vehicle while changing lanes or taking turns.

## 3.1 Existing Solutions to Blind Spots Problem

For human-driven cars, numbers of solutions are there for elimination of blind spots. Most simple solution to reduce blind spots is to adjust the mirrors of the car. Blind spots are eliminated in heavy vehicles by using special seat designs. Special blind spot mirrors are developed which gives wide view to the driver. Some high-end cars also use electronic-based systems which detect the cars in the blind spots and warning is given on the mirrors using LED light. Vision-based blind spot detection systems (BSD) are there for daytime and nighttime driver assistance which makes use of lasers and cameras [32–34]. At present, LIDAR is used in autonomous vehicle to see around the world. LIDARs used at the sides of the car to reduce the blind spots. But the cost of LIDAR is huge which increases the overall cost of driverless car. Only LIDAR can cost up to $80,000. If we reduce the number of LIDARs to one (mounted on the top), there can be some areas around the car which may not be in the visibility. Any small object like cyclist can remain undetected because of this. In 2018, Uber already had an accident in Arizona due to this reason. One of the Uber cars in Arizona stroked the pedestrian woman who died due to this accident. As a result, Uber had to terminate testing in Arizona. Further, even if we detect the object by using sensors instead of LIDAR, the driverless car will not get enough assistance for further steps (to reduce or increase the speed) which is done easily by human driver using eye contact. Hence, it is necessary to develop a system for blind spot detection which will be cost effective as well as will assist the vehicle for driving actions.

## 3.2   Proposed System for Blind Spot Detection

We are proposing a novel approach using machine learning to tackle the problem of blind spots. The aim of this approach is to reduce the accidents caused by visibility and to improve human's safety. This BSD system will consist of two modules. First module will be using sensors to identify any object in the blind spot area of vehicle [35–37]. Second module will use machine learning techniques to identify relative speed of that vehicle. First module is object detector module and second module is assistance module. Based on the speed calculations, assistance can be given to the DC. If the object in the blind spot is moving with relatively same speed as our DCs speed, it has to vary its speed so that the object will be out of blind spot area [38]. Post that it can change the lane or take the turn. If the speed of detected object is relatively slow or fast, our DC will wait for that vehicle to move out. Note that we are not using any extra LIDAR for detection; hence, the solution will not add huge cost in the design. Here, the use of machine learning to calculate relative speed is essential to avoid long wait. The scenario may occur where the car behind our car is also moving with relatively same speed and hence maintaining same distance with our car. Therefore, our car cannot indefinitely wait for that car to vary its speed. In case of human drivers, the driver will judge the speed of that car and vary its speed to take it out of blind spot. But this is not that simple for the machine. Also, we are taking "speed" into account instead of distance because vehicles with same speed will maintain same distance for longer time but the calculation of relative speed will give instant assistance. We have surveyed the existing methods and their limitations [23, 32, 35, 36]. The table below gives summary of known solutions for DCs (Table 2).

## 3.3   Objective of the Solution

Blind spots are one of reasons for accidents on the roads. These accidents are caused due to unrecognized vehicles in the blind spot areas of the car. Hence, it is necessary to find the solution by which we can recognize the vehicles and take them according to their action. The objective of given solution is to reduce the accidents caused by visibility and effectively to improve human safety.

**Table 2** Survey of existing solutions and limitations

| Existing solutions | Limitations |
|---|---|
| Use of light detection and ranging (LIDAR) | LIDARs are very costly. If we increase the LIDARs, overall cost of the vehicle will increase. If we reduce the LIDARs, the vision will be affected and blind spots will be there |
| Blind spot monitoring system which gives warning and alerts | Useful for human-driven cars where driver can quickly act upon the indication. Not effective for driverless car as the system will not assist the car about the action |
| Use of RADARs and video cameras for detection of objects in combination with single LIDAR mounted on the top | By combining RADARs, LIDAR and video cameras, even if we detect the object, we cannot identify its speed. Hence, it is challenging for DC to take the action |

## 3.4 Proposed Algorithm

*Step 1*: *Detect the object using Object Detection Module in the blind spot area. Object Detection Module will contain sensors to detect the presence of object.*
*Step 2*: *Calculate the relative speed of the object using Machine Learning Techniques (Refer Output in Fig. 3).*
*Step 3*: *If the relative speed is Zero (Moving with same speed), Decrease the speed slowly so that it will move out of blind spot.*
*Step 4*: *If the relative speed is Negative (Moving slowly with respect to our DC), Increase the speed and give the appropriate indicator.*
*Step 5*: *If the relative speed is Positive (Moving fast with respect to our DC), Wait for the car to move out without any efforts.*
*Step 6*: *Once the blind spot area detector gives green signal, indicating Safe to Move, give the proper indicator (left or right) and change the lane or take the turn.*
*Step 7*: *Perform above steps continuously while driving (Figs. 4 and 5).*

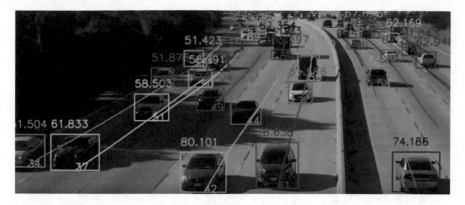

**Fig. 4** Vehicle speed estimation. *Source* Zheng Tang, Nvidia AI City challenge workshop

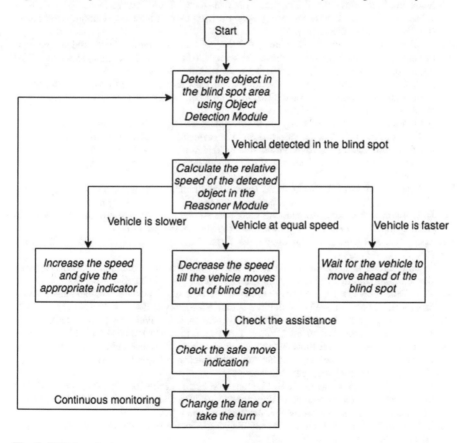

**Fig. 5** BSD flowchart

# References

1. K. Naughton, Ubers fatal crash revealed a self-driving blind spot: night vision. Bloomberg, 29 May 2018
2. J. Stewart, TESLA'S autopilot was involved in another deadly car crash. Wired, 30 Mar 2018
3. J. Meyer, H. Becker, P.M. Bsch, K.W. Axhausen, Autonomous vehicles: the next jump in accessibilities. Res. Transp. Econ. **62**, 80–91 (2017)
4. Wikipedia contributors, History of self-driving cars. Wikipedia, The Free Encyclopedia. Wikipedia, The Free Encyclopedia, 26 Jan 2019. Web 29 Jan 2019
5. Y. Zein, M. Darwiche, O. Mokhiamar, GPS tracking system for autonomous vehicles. Alex. Eng. J. Available Online. 13 Nov 2018
6. T. Kanade, C. Thorpe, W. Whittaker, Autonomous land vehicle project at CMU, in *Proceedings of the 1986 ACM Fourteenth Annual Conference on Computer Science*, *CSC'86* (ACM, New York, 1986), pp. 71–80
7. K. Kaur, G. Rampersad, Trust in driverless cars: investigating key factors influencing the adoption of driverless cars. J. Eng. Technol. Manag. **48**, 87–96 (2018)
8. J. De Bruyne, J. De Werbrouck, Merging self-driving cars with the law. Comput. Law Secur. Rev. **34**(5), 1150–1153 (2018)
9. R. Okuda, Y. Kajiwara, K. Terashima, A survey of technical trend of adas and autonomous driving. in *Proceedings of Technical Program—2014 International Symposium on VLSI Technology, Systems and Application (VLSI-TSA)*, Apr 2014, pp. 14
10. N. Adnan, S.M. Nordin, M.A. Bahruddin, M. Ali, How trust can drive forward the user acceptance to the technology? In-vehicle technology for autonomous vehicle. Transp. Res. Part A: Policy Pract. **118**, 819–836 (2018)
11. P. Marks, Autonomous cars ready to hit our roads. New Sci. **213**(2858), 19–20 (2012)
12. Y.-C. Lee, J.H. Mirman, Parents perspectives on using autonomous vehicles to en-hance childrens mobility. Transp. Res. Part C Emerg. Technol. **96**, 415–431 (2018)
13. Biggest challenges in driverless cars, https://9clouds.com/blog/what-are-thebiggest-driverless-car-problems
14. I. Barabs, A. Todoru, N. Cordo, A. Molea, Current challenges in autonomous driving, in *IOP Conference Series: Materials Science and Engineering*, vol. 252, (2017) p. 012096
15. R. Hussain, S. Zeadally, Autonomous cars: research results, issues and future challenges. in *IEEE Communications Surveys & Tutorials,* (10 Sept 2018), pp. 1–1
16. S.A. Cohen, D. Hopkins, Autonomous vehicles and the future of urban tourism. Ann. Tour. Res. **74**, 33–42 (2019)
17. L. Ye, T. Yamamoto, Impact of dedicated lanes for connected and autonomous vehicle on traffic flow throughput. Phys. A Stat. Mech. Appl. **512**(15), 588–597 (2018)
18. W.X. Zhu, H.M. Zhang, Analysis of mixed traffic flow with human-driving and autonomous cars based on car-following model. Phys. A Stat. Mech. Appl. **496**, 274–285 (2018)
19. N.E. Velling, From the testing to the deployment of self-driving cars: legal challenges to policymakers on the road ahead. Comput. Law Secur. Rev. **33**(6), 847–863 (2017)
20. J. Cui, L.S. Liew, G. Sabaliauskaite, F. Zhou, A review on safety failures, security attacks, and available countermeasures for autonomous vehicles. Ad Hoc Networks Available online (7 Dec 2018 In Press). Accepted Manuscript
21. K.M.A. Alheeti, A. Gruebler, K.D. McDonald Maier, An intrusion detection system against malicious attacks on the communication network of driverless cars, in *2015 12th Annual IEEE Consumer Communications and Networking Conference (CCNC)* 2015
22. E.R. Teoh, D.G. Kidd, Rage against the machine? Google's self-driving cars versus human drivers. J. Saf. Res. **63**, 57–60 (2017)
23. Uber self driving car fatality. NewScientist **237**(3170), 5–57 (2018)
24. M. Lad, I. Herman, Z. Hurk, Vehicular platooning experiments using autonomous slot cars. IFAC-PapersOnLine **50**(1), 12596–12603 (2017)
25. J. Yu, L. Petng, Space-based collision avoidance framework for autonomous vehicles. Procedia Comput. Sci. **140**, 37–45 (2018)

26. R. Iacobucci, B. McLellan, T. Tezuka, Modeling shared autonomous electric vehicles: potential for transport and power grid integration. Energy **158**, 148–163 (2018)
27. J.P. Hanna, M. Albert, D. Chen, P. Stone, Minimum cost matching for autonomous carsharing. IFAC-PapersOnLine **49**(15), 254–259 (2016)
28. M. Zhu, X. Wang, Y. Wang, Human-like autonomous car-following model with deep reinforcement learning. Transp. Res. Part C Emerg. Technol. **97**, 348–368 (2018)
29. P. Bohm, M. Kocur, M. Firat, D. Isemann, Which factors influence attitudes towards using autonomous vehicles, in *Automotive UI, Proceedings of the 9th International Conference on Automotive User Interfaces and Interactive Vehicular Applications Adjunct* (2017), pp. 141–145
30. Y.H. Cho, B.K. Han, Application of slim A-pillar to improve driver's field of vision. Int. J. Autom. Technol. **11**, 517–524 (2010)
31. K. McCarthy, Uber self-driving car death riddle: was LIDAR blind spot to blame? Emergent Tech, 28 Mar 2018
32. G. Liu, M. Zhou, L. Wang, H. Wang, X. Guo, A blind spot detection and warning system based on millimeter wave radar for driver assistance. Sci. Direct **135**, 353–365 (2017)
33. B.F. Wu, H.Y. Huang, C.J. Chen, Y.H. Chen, C.W. Chang, Y.L. Chen, A vision based blind spot warning system for daytime and nighttime driver assistance. Comput. Electr. Eng. **39**, 846–862 (2013)
34. M.W. Park, K.H. Jang, S.K. Jung, Panoramic vision system to eliminate driver's blind spots using a laser sensor and cameras. Int. J. ITS Res. **10**, 101–114 (2012)
35. Y.L. Chen, B.F. Wu, H.Y. Huang, C.J. Fan, A real-time vision system for nighttime vehicle detection and traffic surveillance. IEEE Trans. Ind. Electron. **58**(5), 2030–2044 (2011)
36. Y.C. Kuo, N.S. Pai, Y.F. Li, Vision-based vehicle detection for a driver assistance system. Comput. Math. Appl. **61**, 2096–2100 (2011)
37. C.T. Chen, Y.S. Chen, Real-time approaching vehicle detection in blind-spot area, in *Proceedings IEEE International Conference Intelligence Transport System* (2009), pp. 1–6
38. V. Milanes, D.F. Llorca, J. Villagra, J. Perez, C. Fernandez, I. Parra, C. Gonzalez, M.A. Sotelo, Intelligent automatic overtaking system using vision for vehicle detection. Expert. Syst. Appl. **39**, 3362–3373 (2012)

# Upper Half Face Recognition Using Hidden Markov Model and Singular Value Decomposition Coefficients

Pushpa Choudhary, Ashish Tripathi, Arun Kumar Singh
and Prem Chand Vashist

**Abstract** Face recognition is the process of recognizing a special face from a set of different faces. A human face is a very complex object/structure with varying features from childhood to old age. The first main objective of this paper is to recognize a person even if the person is not in a neutral state or with any facial expression. The second main objective of this paper is to recognize those people whose upper half face is available. In this paper, a sequence of blocks has been created for an image of each face and each block is featured by singular value decomposition parameters. Seven-state Hidden Markov Model has been used to cover whole face details. The system has been evaluated on 400 face images of Olivetti Research Laboratory face database, in MATLAB with different illuminations, lighting conditions and Cohn-Kanade database with different expressions. For both databases, recognition rate of approximately 100% has been achieved with the increase of training images. The major achievement of this algorithm is that it recognizes the person correctly even if an upper half face of the person is provided as test image and it recognizes the person with correct expression in CK+ database. This result is achieved by using a small number of feature describing blocks and resizing the images into a smaller size.

**Keywords** Face recognition · Hidden Markov Model (HMM) · Singular value decomposition (SVD)

P. Choudhary · A. Tripathi (✉) · A. K. Singh · P. C. Vashist
Department of Information Technology, G. L. Bajaj Institute of Technology & Management,
Greater Noida, India
e-mail: ashish.mnnit44@gmail.com

P. Choudhary
e-mail: pushpak2728@gmail.com

A. K. Singh
e-mail: arun.k.singh.iiit@gmail.com

P. C. Vashist
e-mail: pcvashist@gmail.com

© Springer Nature Singapore Pte Ltd. 2021                                   549
X.-Z. Gao et al. (eds.), *Advances in Computational Intelligence and Communication Technology*, Advances in Intelligent Systems and Computing 1086,
https://doi.org/10.1007/978-981-15-1275-9_45

# 1 Introduction and Background

Face detection and face recognition from still and video images is an emerging field with numerous commercial applications, such as an ATM machine or controlling the entry of people in security system into restricted areas, etc. In real-time application, face recognition is a challenging area. It is a very useful technique in biometric system to find out identity of a particular person by applying face recognition technique.

The face recognition is performed by matching the test face image with the trained face images of known persons which are stored in the database. If the test face image is matching with any one of the trained face images of database, then corresponding person is identified; otherwise, person is an unknown to the system.

The first main objective of this paper is to recognize a person even if the person is not in a neutral state or with any facial expression. The second main objective of this paper is to recognize those people whose upper half face is available. Expression occurs on a person's face depending upon his/her mental state. Following are the main challenges, which have been addressed.

- There are many difficulties in face recognition due to skin colour, face shape, eye colour, eye closing, etc. Further, due to various occlusions like beard, jewellery, and eyeglasses, the detection process may be hindered.
- The head pose of a person can lead to difficulty in extracting the features of the face.
- Intensity among expressions is also different from context to context and time to time. This may also lead to failure in face recognition.

In 2001, Viola–Jones [1] constitutes the fundamental set of ideas for face detection algorithm. This algorithm is suitable for frontal upright faces. Later in 2003, Viola and Jones [2], improved version of algorithm has been introduced in that detection of profile and rotated views images was verified. In 2004, by Viola and Jones [3] presented a concept of real-time face detection method which is very successful for solving face detection tasks. This is a vast improvement in speed compared to previous face detectors methods, even taking into account improvements in computer hardware.

In 1994, on the ORL database, one-dimensional Hidden Markov Model has been used by Samaria and Harter [4], by this method recognition rate was 87% obtained. After some time, upgraded version of algorithm was produced by them. In that, one-dimensional HMM was converted to a pseudo-two-dimensional HMM and achieved a recognition rate from 87 to 95% on the same database by using half the images for testing and the other half for training.

In 2004, Zhang et al. [5] applied a similarity function to describe the person confidence level for two more images belong to the same person or not. The recognition rate of this system is 97.9%.

In 2005, Li and Yin [6] proposed an algorithm in which a face image with wavelet transform decomposed into three levels. Then, Fisher face method is applied to these levels.

Kharat and Dudul [7, 8] used support vector machine (SVM) method. To extract the facial feature, they used various feature extraction techniques.

## 1.1 Preliminaries Concepts

### 1.1.1 Face Detection

For the detection of face region, widely used Viola–Jones real-time face detector method has been used in the proposed method. Though the training through this method is slow, detection is very fast. AdaBoost algorithm is used to train the cascading classifiers and integral image filters used for each classifier, and this technique is based on Haar functions. This process is essential to speed up the detection rate of face.

### 1.1.2 Cascade Classifier

The cascade classifier [9] is a combination of stages, each stage containing a strong classifier. A given sub-window is a face or not a face is determined in every stage of classifier. The sub-window is immediately discarded if it is declared as non-face. Conversely, sub-window is passed on to the next stage when it is classified as a face by a given stage. The chances of a sub-window containing a face increase, as it moves on to next stage in the cascade. The cascading classifier process is illustrated below in Fig. 1.

### 1.1.3 Hidden Markov Models

One-dimensional Hidden Markov Models [10–17] applied to find out face recognition and object recognition system. In this model, some hidden states are associated and an observation sequence is generated by these hidden states individually.

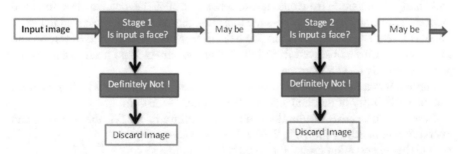

**Fig. 1** Cascade classifier

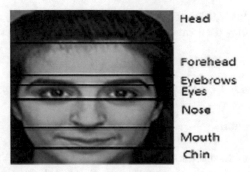

**Fig. 2** Seven regions of face image

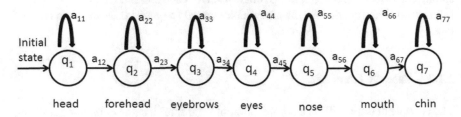

**Fig. 3** Block diagram of one-dimensional HMM model with seven-state face regions

Seven different regions, i.e. head, forehead, eyebrows, eyes, nose, mouth and chin constitute a face image, as shown in Fig. 2. For an each person face, significant facial regions such as head, forehead, eyebrows, eyes, nose, mouth and chin occur in natural order from top to bottom. Therefore, each of these regions assigned to a state, to constitute one-dimensional HMM in left-to-right order as shown in Fig. 3.

## Assumption

In the experiment, it has been assumed that the probability of transition from one state to next is 50% (i.e. from head to forehead, forehead to eyebrows and so on) and the probability of transition from one state to self is also 50% (i.e. head to head, forehead to forehead and so on). The seven-state transition diagrams are represented in Fig. 3. There is a sequence of movements like head to forehead, forehead to eyebrows and so on, and it can be observed from Fig. 3, that movement from head to eyes is not possible directly or vice versa.

Further, it is assumed that the probability of the initial state is 1 which is the head and the probability of the final state is also 1 which is the chin.

Based on the above assumption, initial probability of HMM, the matrix $A$ and matrix $B$ is defined, as shown in Tables 1, 2 and 3, respectively.

Initially, observation probability matrix OP is defined as

**Table 1** Initial probability ($\pi$)

| Head | Forehead | Eyebrow | Eye | Nose | Mouth | Chin |
|------|----------|---------|-----|------|-------|------|
| 1 | 0 | 0 | 0 | 0 | 0 | 0 |

**Table 2** Matrix A (transition probability)

|  | Head | Forehead | Eyebrows | Eyes | Nose | Mouth | Chin |
|------|------|----------|----------|------|------|-------|------|
| Head | 0.5 | 0.5 | 0 | 0 | 0 | 0 | 0 |
| Forehead | 0 | 0.5 | 0.5 | 0 | 0 | 0 | 0 |
| Eyebrow | 0 | 0 | 0.5 | 0.5 | 0 | 0 | 0 |
| Eyes | 0 | 0 | 0 | 0.5 | 0.5 | 0 | 0 |
| Nose | 0 | 0 | 0 | 0 | 0.5 | 0.5 | 0 |
| Mouth | 0 | 0 | 0 | 0 | 0 | 0.5 | 0.5 |
| Chin | 0 | 0 | 0 | 0 | 0 | 0 | 1 |

**Table 3** Matrix OP (observation probability)

|  | 1 | 2 | 3 | ... | 1260 |
|------|------|------|------|-----|------|
| Head | 0.0008 | 0.0008 | 0.0008 | ... | 0.0008 |
| Forehead | 0.0008 | 0.0008 | 0.0008 | ... | 0.0008 |
| Eyebrow | 0.0008 | 0.0008 | 0.0008 | ... | 0.0008 |
| Eye | 0.0008 | 0.0008 | 0.0008 | ... | 0.0008 |
| nose | 0.0008 | 0.0008 | 0.0008 | ... | 0.0008 |
| mouth | 0.0008 | 0.0008 | 0.0008 | ... | 0.0008 |
| chin | 0.0008 | 0.0008 | 0.0008 | ... | 0.0008 |

$$OP = \frac{1}{L} \times ones(K, L)$$

Here, $L$ is defined as number of all possible observation symbols obtained from quantization procedure (i.e. $L = 1260$) and $K$ is the number of states (i.e. $K = 7$).

According to literature, 'HMMs normally work on sequence of symbols, is called observation vectors. As we know that an image is represented by a 2D matrix. Therefore, the images were interpreted as a one dimensional sequence'. To solve this problem, singular value decomposition (SVD) has been used.

### 1.1.4 Singular Value Decomposition

In signal processing and statistical data analysis, singular value decomposition [18–21] method is an important tool. SVD method is applicable to any real matrix as shown below:

Let us consider a matrix $A$ with r rows and c columns. Then, the $A$ can be factorized into three matrices $R$, $S$ and $P^T$ such that $A = RSP^T$

- Matrix $R$ is an $r \times r$ orthogonal matrix.
  $R = [r_1, r_2...r_i, r_{i+1}... r_r]$
  Column vectors $r_i$, for $i = 1, 2, ..., r$, form an orthonormal set.
  $R_i^T R_j = \delta_{ij} = 1$ when $i = j$ and $\delta_{ij} = 0$ when $i \neq j$
- matrix $P$ is an $c \times c$ orthogonal matrix.
  $P = [p_1, p_2,... p_i, p_{i+1},..., p_c]$
  Column vectors $p_i$ for i $= 1, 2... c$, form an orthonormal set.
  $V_i^T V = \delta_{ij} = 1$ when $i = j$ and $\delta_{ij} = 0$ when $i \neq j$
- $S$ is $r \times c$ values with components $\sigma_{ij} = 0$, $i \neq j$ and $\sigma_{ij} > 0$

## 2 Proposed System

Face detection gives us the location of the face. But face recognition is used for identification of the person. In the proposed methodology, ORL and CK+ face database has been used. Facial feature is extracted for each training image and stored in the database. Same facial feature extraction method has also been applied to test images, and if test image matches with one of the images stored in database, then recognition process has been performed. The steps involved in the proposed methodology are depicted in Fig. 4 and detail descriptions of methodology are as follows:

For the improvement of face recognition system, some pre-processing is required like filtering, block extraction and quantization as described below and shown in Fig. 5. Detailed description of these processes which involved in training and testing

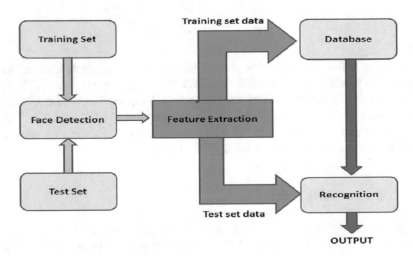

**Fig. 4** Flow chart of proposed face recognition methodology

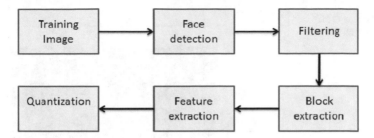

**Fig. 5** Training and testing processes of image

images is shown in Fig. 6.

**Face Detection**: For the detection of face region, widely used Viola–Jones real-time face detector has been used in the proposed method.

**Filtering**: In the proposed method, order statistic filter has been used for filtering, which is nonlinear spatial filters. This filter is basically used for speed up the algorithm. In this algorithm, order statistic filter has been applied with $3 \times 3$ windows. In this algorithm, centred pixel is replaced by minimum element in the sliding window. This is represented by following equation

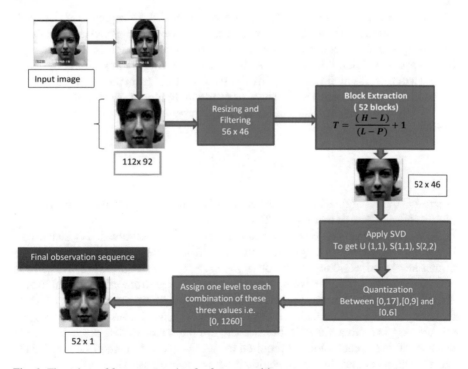

**Fig. 6** Flow chart of feature extraction for face recognition

**Fig. 7** Sequence of
overlapping blocks

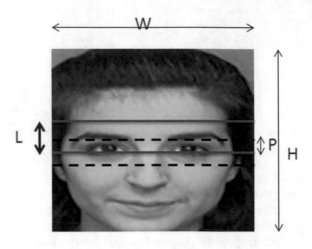

$$f(x, y) = \min_{(s,t) \in S_{xy}} \{g(s, t)\}$$

In the above equation, $g(s, t)$ is the grey level of pixel $(s, t)$ and $S_{xy}$ is the mentioned window.

**Block Extraction**: Observation sequence in HMM is represented by one-dimensional model and face image is represented by two-dimensional models; therefore, images must be converted into one-dimensional model. For that reason, each face image is dividing into width **w** and height **h** with overlapping blocks of height $\mathbf{h_1}$ and width **w**. A patch window with size **h1** × **w** slides from the top to the bottom of the image and generates a sequence of overlapping blocks. Each time, the patch moves only one pixel, as shown in Fig. 7. The number of blocks (**t**) can be evaluated as follows

$$\mathbf{t} = \frac{(\mathbf{h} - \mathbf{h}1)}{(\mathbf{h}1 - \mathbf{p})} + 1$$

where $h$ = height of face image, $w$ = width of face image, $h1$ = height of overlapped blocks and $p$ = size of overlapped blocks.

With the help of the above formula, 52 blocks are extracted. As stated earlier, HMM model needs single discrete value. But face images are stored in two-dimensional, and one-dimensional observation sequence is required in HMM; in this case, conversion is required from two-dimensional image to one-dimensional image. To cater this, SVD has been applied on this block to get three matrices $(R, S, P)$. Only three single values, i.e. $R(1,1)$, $S(1,1)$ and $S(2,2)$ are retained from these matrices and rest are skipped. Now, for each block, there are three real values, but still a single discrete value for each block is required to use as observation sequence for HMM model. Again to cater this, quantization process has been used.

**Quantization**: Since SVD coefficients have continuous values, these cannot be modelled by discrete HMM and hence require quantization. Now, select the features from diagonal elements of $S$. An image with $56 \times 46$ resolution has an $S$ matrix with same dimensions, which means 56 distinct singular values. In this model, the first feature $S(1,1)$ into 10, the second feature $S(2,2)$ into 7 and the third one $R(1,1)$ into 18 levels have been quantized, leaving 1260 ($10 \times 7 \times 18$) possible distinct vectors for each block. The detail process of feature extraction for face recognition is shown in Fig. 6. Now, apply the HMM model with this single value.

## 3 Experimental Results

The proposed recognition system is tested on the Olivetti Research Laboratory (ORL) face database and Extended Cohn-Kanade database (CK+). For the reduction of computational complexity and memory consumption, PGM format images of this database have been resized from $112 \times 92$ to $56 \times 46$ bmp format images as shown in Fig. 8. This model is trained for each person in the database. For each person, there are ten images; in that, five face images of the same person are used in training and the remaining five face images are used for testing. Recognition rate of both the databases, i.e. ORL and CK+ are 96.5% and 95.33%, respectively. As numbers of training images grow, the recognition rate also increases, as shown in Table 4.

From Cohn-Kanade database (as shown in Fig. 9), seven folders were created based on the basic facial expression such as sad, disgust, happy, angry, neutral, surprise and fear. This database is different from ORL database. In this database, face images with expressions are stored. This proposed algorithm recognizes the

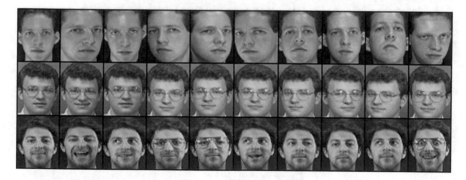

**Fig. 8** Three examples of ORL face database

**Table 4** Recognition rate as compared to number of training images

| No. of training images | 3 | 4 | 5 | 6 | 7 | 8 |
|---|---|---|---|---|---|---|
| ORL | 83.21 | 91.66 | **96.5** | 98.75 | 98.95 | 99.25 |
| Cohn-Kanade | 90.47 | 94.44 | **95.33** | 95.66 | 97.22 | 100 |

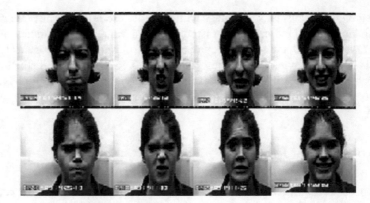

**Fig. 9** Examples of CK+ face database

person with correct expression. As shown in Table 4, recognition rate of algorithm is 95.33% in case of five training images of this database.

This proposed system also recognizes a person with correct expression. It is not necessary that a person should be in neutral pose. Because in CK+ database, images are stored with different expressions, folders are created based on the expression of a person rather than personwise, even though this experiment gives 95.33% correct result with low computational cost as compared to ORL database as shown in Table 5.

As shown in Table 6, the number of methodology of face recognition is depicted with computational cost on the ORL face database. It can be observed from Table 6 that proposed system has recognition rate of 96.5% (ORL) with low computational cost. This methodology also takes less memory space consumption due to reduction in the size of the face image. If the sizes of image are not reduced, then it will take more space in the memory as well as more processing time. As described in Table 6, this system takes less recognition time as compared to other systems.

Most important point in the proposed system is it also recognizes person correctly even if an upper half face of a person is provided as a test image. By providing the

**Table 5** Result of experiments of our proposed method on ORL and CK+ database

| Database | ORL | CK+ |
|---|---|---|
| Total images | 400 | 60 |
| No. of trained images | 200 | 30 |
| No. of tested images | 200 | 30 |
| Image size | $56 \times 46$ | $56 \times 46$ |
| No. of trained images (per person) | 5 | 5 |
| Training time per image (s) | 0.32 | 0.5 |
| Recognition time per image (s) | **0.15** | **0.06** |
| Recognition rate (%) | **96.5** | **95.33** |

Bold specifies the outcome of the experiment

**Table 6** Comparative recognition rate with other methods on ORL face database

| Method | Recognition rate in % | Training time per image | Recognition time per image (s) |
|---|---|---|---|
| PDBNN [13] | 96 | 20 min | $\leq$0.1 |
| 1DHMM + Wavelet [21] | 100 | 1.13 s | 0.3 |
| DCT-HMM [12] | 99.5 | 23.5 s | 3.5 |
| 1D HMM + SVD [1] | 96.5 | 0.63 s | 0.28 |
| Proposed method | 96.5 | 0.32 s | 0.15 |

upper half face only, the recognition time as well as computational cost will also be reduced. As shown in Table 7, recognition time of the upper half face is very less as compared to the whole face of ORL database. In this proposed system, if partial data of face are available then also recognition is possible. It is not compulsory to have the complete face of person to recognize. This is the major advantage of the proposed system. Some of the upper half face of ORL database is shown in Fig. 10, which is used as a test image for this proposed system. In this proposed system, lower half face was also tested, but the result was not good as compared to upper half face. The reason behind this is number of face region is more in upper half face as compared to lower half face. Upper half face contains following face region like head, forehead, eyebrows and eyes, and lower half face contains only mouth and chin.

**Table 7** Comparative recognition results between whole face and upper half face on ORL face database

| Database | ORL (whole face) | ORL (upper half face) |
|---|---|---|
| Total images | 60 | 60 |
| No. of trained images | 30 | 30 |
| No. of tested images | 30 | 30 |
| Image size | 56 × 46 | 56 × 46 |
| No. of trained images (per person) | 5 | 5 |
| Training time per image (s) | 0.30 | 0.30 |
| Recognition time per image (s) | 0.05 | 0.03 |
| Recognition rate (%) | 100 | 100 |

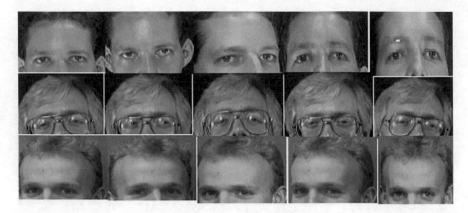

**Fig. 10** Upper half face of ORL database (Given as test image)

## 4 Conclusion

In the proposed method, following points have been observed: (i) For both databases, recognition rate of approximately 100% has been achieved with the increase in images used for training. (ii) This proposed algorithm recognizes the person with correct expression with recognition rate of 95.33% in case of five training images of CK+ database. (iii) Further, it also recognizes person correctly even if an upper half face of a person is provided as a test image, which in turn reduces the recognition time as well as computational cost. (iv) Recognition rates of 96.5% on ORL database and 95.33% on CK+ database with low computational cost as compared to other researcher's work have been achieved.

## References

1. P. Viola, M. Jones, Rapid object detection using a boosted cascade of simple features, in *Proceedings of the 2001 IEEE Computer Society Conference on Computer Vision and Pattern Recognition, CVPR* 2001, vol. 1 ( IEEE, 2001)
2. P. Viola, M.J. Jones, D. Snow, Detecting pedestrians using patterns of motion and appearance (IEEE, 2003)
3. P. Viola, M.J. Jones, Robust real-time object detection. Int. J. Comput. Vis. **57**, 137–154 (2004)
4. F.S. Samaria, A.C. Harter, Parameterisation of a stochastic model for human face identification, in *Proceedings of the Second IEEE Workshop on Applications of Computer Vision* 1994 (IEEE, 1994)
5. G. Zhang et al., Boosting local binary pattern (LBP)-based face recognition, in *Advances in biometric person authentication* (Springer, Berlin, Heidelberg, 2004), pp. 179–186
6. Y.L. Tian, R.M. Bolle, System and method for automatically detecting neutral expressionless faces in digital images. U.S. Patent No. 6,879,709, 12 Apr 2005
7. V. Bettadapura, Face expression recognition and analysis: the state of the art (2012). arXiv preprint arXiv:1203.6722

8. G.U. Kharat, S.V. Dudul, Neural network classifier for human emotion recognition from facial expressions using discrete cosine transform, in *First International Conference on Emerging Trends in Engineering and Technology ICETET'08* 2008 (IEEE, 2008)
9. S. Soo, Object detection using Haar-cascade classifier. Institute of Computer Science, University of Tartu (2014)
10. F. Samaria, F. Fallside, Face identification and feature extraction using Hidden Markov models, in *Image Processing: Theory and Application*, ed. by G. Vernazza (Elsevier, 1993)
11. F. Samaria, Face recognition using Hidden Markov Models. Ph.D. thesis, Engineering Department, Cambridge University, Oct 1994
12. M. Bicego, U. Castellani, V. Murino, Using Hidden Markov Models and wavelets for face recognition, in *Proceedings IEEE International Conference on Image Analysis and Processing (ICIAP)* (2003), pp. 7698–1948
13. A.V. Nefian, M.H. Hayes, Hidden Markov Models for face recognition, in *Proceedings IEEE International Conference on Acoustics, Speech and Signal Processing (ICASSP)* (1998), pp. 2721–2724
14. H.R. Farhan, M.H. Al-Muifraje, T.R. Saeed, A novel face recognition method based on one state of discrete Hidden Markov Model, in *Annual Conference on New Trends in Information & Communications Technology Applications (NTICT)*, 2017 (IEEE, 2017)
15. S.D. Ruikar, A.A. Shinde, Face recognition using singular value decomposition along with seven state HMM. Int. J. Comput. Sci. Telecommun. **4**(6) (2013)
16. H. Miar-Naimi, P. Davari, A new fast and efficient HMM-based face recognition system using a 7-state HMM along with SVD coefficients. Iran. J. Electr. Electron. Eng. **4**(1), 46–57 (2008)
17. C. Anand, R. Lawrance, Algorithm for face recognition using HMM and SVD coefficients. Artif. Intell. Syst. Mach. Learn. **5**(3), 125–130 (2013)
18. Q. Zhen et al., Muscular movement model-based automatic 3D/4D facial expression recognition. IEEE Trans. Multimed. **18**(7), 1438–1450 (2016)
19. H. Chen-Chiung et al., Effective semantic features for facial expressions recognition using SVM. Multimed. Tools Appl. **75**(11), 6663–6682 (2016)
20. H. Farhan, M. Al-Muifraje, T. Saeed, Face recognition using maximum variance and SVD of order statistics with only three states of hidden Markov model. Int. J. Comput. Appl. **134**(6), 32–39 (2016)
21. Hung-Hsu Tsai, Yi-Cheng Chang, Facial expression recognition using a combination of multiple facial features and support vector machine. Soft. Comput. **22**(13), 4389–4405 (2018)

# Comparison of Different Metaheuristic Algorithms for Multilevel Non-local Means 2D Histogram Thresholding Segmentation

**Garima Vig and Sumit Kumar**

**Abstract** Multilevel image segmentation technique segregates an image into disjoint regions and has application in many real-world problems like object recognition, boundary estimation of motion systems, image compression, etc. Conventional image segmentation does not consider the spatial correlation of image's pixels and lack in providing better post-filtering efficiency. This paper performs an analysis of results obtained from different metaheuristic algorithms using an efficient technique of 2D histogram multilevel thresholding based on non-local means filter and Renyi entropy. Further, this study aims to compare newly proposed whale optimization algorithm with some prominent algorithms in recent past and some conventional metaheuristic algorithms to achieve an efficient image segmentation.

**Keywords** Image segmentation · Metaheuristic algorithms · Multilevel thresholding · 2D histogram

## 1 Introduction

Image segmentation is the widely adopted technique for object extraction and recognition, where the major approaches can be categorized as edge-based segmentation, thresholding, clustering, and region-based segmentation. Multilevel thresholding technique is one of the prominent approaches of image segmentation as it extracts more than one region of interest in a simple and efficient manner, majorly applicable in the field of pattern recognition [1] and medical image analysis [2]. Moreover, entropy-based multilevel thresholding helps to achieve the appropriate partition of the object image as it provides information about the distribution of pixel levels of an image [3].

G. Vig (✉) · S. Kumar
CS&E Department, Amity University, Noida, UP, India
e-mail: garimavig311@gmail.com

S. Kumar
e-mail: sumitkumarbsr19@gmail.com

© Springer Nature Singapore Pte Ltd. 2021
X.-Z. Gao et al. (eds.), *Advances in Computational Intelligence and Communication Technology*, Advances in Intelligent Systems and Computing 1086,
https://doi.org/10.1007/978-981-15-1275-9_46

Conventional multilevel thresholding algorithms require exhaustive computation and hence are time-consuming [4]. This drawback is addressed by using threshold level as a spatial dimension of metaheuristic algorithms. Metaheuristic algorithms have the capability to produce near-optimal solution for all the problems that do not have any problem-specific algorithm known as their solution. Over the last few years, many researchers have shown interest in solving multilevel thresholding segmentation problem using different metaheuristic algorithms like ant colony optimization [5], genetic algorithm [6], firefly algorithm [7], particle swarm optimization [8], etc.

In this study, we have performed a comparison of different metaheuristic algorithms using an efficient technique of 2D histogram multilevel thresholding method based on non-local means filter. Results obtained from newly proposed whale optimization algorithm has been analyzed with some prominent algorithm in recent past like cuckoo search, gravitational search algorithm, artificial bee colony, and some conventional algorithms like genetic algorithm, particle swarm optimization, and differential evolution. Images used in this study are from image data of MATLAB R2015a software.

The remainder of the paper is organized in the following sections: Sect. 2 describes the methodology used in the study, Sect. 3 gives a brief description of the methodology used in the paper, Sect. 4 elaborates comparative result of implemented metaheuristic techniques and Sect. 5 concludes the study.

## 2 Related Work

Multilevel thresholding being the vital part of image analysis process is a prominent research area and some of the salient research work in the field is described below:

Wong and Sahoo extended posteriori entropy under inequality constraint for characterizing shape and uniformity of regions [9]. Abutaleb, developed the concept of a 2D histogram that contains spatial information in addition to gray-level distribution information [10]. Sahoo and Arora presented a 2D histogram and Reny's entropy-based technique but that ignore information corresponding to the edges [11]. To overcome this, Zheng et al. presented a local variance histogram-based technique depending on pixels' gray level [12]. Abutaleb developed a technique to perform thresholding using the gray level and local mean value of pixels of the image [10]. Xue-guang and Shu-hong developed a local gradient-thresholding technique that exploited edge-related information of image but it does not give better results for real applications [13]. Mittal and Saraswat presented a non-local means filter-based thresholding technique that results in better post filter clarity of image with better results [14]. Metaheuristic algorithms are widely used to get an optimal solution of multilevel thresholding based on the 2D histogram with less computation. Fengjie and Jieqing used an improved version of the genetic algorithm using simulated annealing based on Otsu method [15]. Sarkar and Das performed segmentation using differential evolution (DE) and Tsallis entropy method [16]. Shen et al. presented an adaptive ant colony optimization (ACO) algorithm where both ant and pheromone are used for the construction of objective function [17]. Panda et al. developed an adaptive swallow swarm optimization (SSO) technique along with evolutionary gray gradient algorithm (EGGA) for brain MR image containing many regions of interest [18].

## 3 Methodology

The methodology adopted for image segmentation in this paper is depicted through the flow diagram in Fig. 1. In this study, a 2D histogram is created using non-local means filtered image to estimate multilevel thresholds. Optimal threshold levels are estimated using metaheuristic algorithms and Renyi entropy as a fitness function. Threshold level so obtained are then used for image segmentation process and as final output segmented and color-mapped images are obtained. For further insight into the proposed methodology, Renyi entropy, non-local means filtering, 2D histogram, and metaheuristic algorithms are illustrated below.

– **Renyi Entropy**:

Entropy is the measure of the global amount of information contained in an image histogram, as stated by Shanon in his initiated research work known as information theory. Renyi entropy is a generalization of Shanon entropy proposed by Renyi. Renyi entropy is a more adaptable and flexible function due to its dependence on parameters. It is an extended version of Shanon entropy which is applied to the continuous family of entropy measures. Renyi entropy can be stated as:

$$R_q(I) = \frac{1}{1-q} \ln \sum_{i=1}^{E} p_i^q \tag{1}$$

– **Non-local means filter**:

This method calculates the arithmetic mean of pixels by performing a weighted comparison of each center pixel of a filter with the target pixel of an image to get better post filter clarity by including edge information in the right amount. Consider

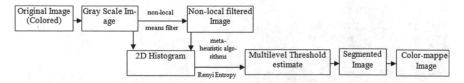

**Fig. 1** Flow diagram of image thresholding segmentation method

$Y(n)$ and $X(m)$ as two pixels of input image then non-local means output image pixels can be computed as:

$$Y(n) = \frac{\sum_{m \in X} X(m) \cdot w(n, m)}{\sum_{m \in X} w(n, m)} \tag{2}$$

where $w(n, m) = e^{-\frac{|\mu(n)-\mu(m)|^2}{\sigma(n)\cdot\sigma(m)}}$, $\mu$ is mean and $\sigma$ is the standard deviation for the pixels corresponding to the filter mask of size $k \times k$. In this paper square of $k = 3$.

– **2D Histogram**:

2D Histogram for an image is obtained by mapping non-local means image pixel with grayscale image pixels to provide better exploitation of spatial information of the image. Consider a spatial coordinate $(u, v)$ of the pixel of the grayscale image $f(u, v)$ and of the non-local mean filtered image $g(u, v)$. 2D histogram matrix value can be calculated by mapping the occurrence of $f(u, v)$ with $g(u, v)$. Mathematically,

$$H(x, y) = O_{xy} \tag{3}$$

where $O_{xy}$ is the total number of occurrence of pair $x, y$; $x = f(u, v)$ and $y = g(u, v)$

The normalized histogram is calculated by dividing occurrence value with the size of the image $M \times N$. Mathematically,

$$N(x, y) = O_{xy}/M \times N \tag{4}$$

Histogram depicted in the table show that object and background information can be obtained through diagonal elements; hence, only diagonal elements are considered for further obtaining optimal threshold values.

Table 1 shows the gray image, non-local means image, and 2D histogram obtained from the original image for sample images considered in this study.

**Table 1** Non-local means image and the 2D histogram of sample images

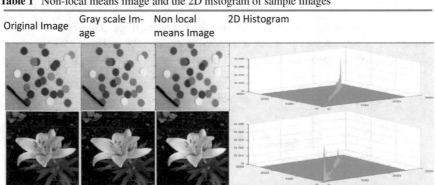

| Original Image | Gray scale Image | Non local means Image | 2D Histogram |
| --- | --- | --- | --- |

– **Metaheuristic Algorithms**:

In this work, the comparison of results obtained from seven algorithms is done to find the best optimal threshold values for performing non-local means multilevel thresholding segmentation. Table 2 describes all the algorithms and Table 3 describes the value of parameters of the algorithms used in the study.

**Table 2** Description of metaheuristic algorithms

| Algorithm | Author/year | Description |
|---|---|---|
| GA | John Holland, 1992 | GA technique is inspired by biological evolution process and is used for solving complex optimization problems. Selection, crossover, and mutation are three main steps of GA. Trivially in GA the coding is done in binary format but in some cases, the continuous coded version of GA is used [19] <br> Uniform Crossover : $y_i = \alpha * x_i + (1 - \alpha) * x_j$ <br> (5) |
| PSO | Eberhart and Kennedy, 1995 | PSO algorithm is a type of optimization algorithm that finds a global minimum or maximum within the search space of the given problem. Particles in PSO have velocities associated with them to travel through the search space. This technique maintains personal best position for each particle and the best position globally [20] <br> Velocity update: <br> $v(l + 1) = w * v(l) + r_1 * c_1 * (p\text{best}(l) - x(l))$ <br> $\quad + r_2 * c_2 * (g\text{best}(l) - x(l))$ <br> (6) |
| DE | Storn and Price, 1997 | DE helps to achieve the robust optimum solution for a given problem faster. DE performs crossover by mixing mutated vector generated with some predefined vector called target vector. In the selection process, the target vector is replaced with the trial vector if the latter is strictly superior [21] <br> Crossover : $U = \begin{cases} V \text{ if rand}_j \leq \text{Cr} \| j = j_{\text{rnd}} \\ X \text{ otherwise} \end{cases}$ <br> (7) |
| ABC | Karaboga, 2007 | ABC simulates the behavior of real bees for solving multidimensional and multimodal optimization problems. In ABC, bee colony is divided into three groups: the first group is called employed bees (those which go to find the food source), the second group is called onlookers (make the decision to select food source) while the last group is scouts (search randomly for the food source [22] <br> Position Update : $V = X_{ij} + \varphi * (X_{ij} - X_{kj})$ (8) |

(continued)

**Table 2** (continued)

| Algorithm | Author/year | Description |
|---|---|---|
| CS | Yang and Deb, 2009 | In CS algorithm, each set of nests with one egg is placed at random locations in the algorithm's search space where the egg represents a candidate solution. Search pattern for discovering the best nest is inspired by Levy flights which is an efficient technique as compared to other random walks [23]<br>Position Update : $x(t+1) = x(t) + \alpha \oplus \text{levy}(\lambda)$ (9) |
| GS | Rashedi, 2009 | The agents interact through gravitational force and their performance is measured by the mass of each agent. The optimal solution is with the heaviest mass. $K$best agent value proceeds toward optima with every iteration in a time-bound linear manner [24]<br>Mass Update : $M_i(t) = \dfrac{m_i(t)}{\sum_{j=1}^{N} m_i(t)}$; (10)<br>where: $m_i = \dfrac{f_i(t) - \text{Wrst}(t)}{\text{Bst}(t) - \text{Wrst}(t)}$ (11)<br>Acceleration Update : $a_i(t) = \dfrac{f_i(t)}{M_i(t)}$; (12)<br>where: $f_i(t) = \sum_{j=1}^{N} \text{rand}_j f_{ij}(t)$ (13) |
| WO | Mirjalili and Lewis, 2016 | WO works around the current best solution and tries to find the best candidate. Exploitation of the search space is performed using a bubble net method which constitutes of two mechanisms: Shrinking encircling mechanism and spiral updating position [25]<br>Position Update:<br>$X(t+1) =$<br>$\begin{cases} X^*(t) - A.D & \text{if } p < 0.5 \\ D'.e^b.\cos(2\pi l) + X^*(t) & \text{if } p \geq 0.5 \end{cases}$ (14) |

**Table 3** Parameters' setting of metaheuristic algorithms

| Algorithm | Parameters |
|---|---|
| GA | CrPercent = 0.7, MutPercent = 0.3, MutRate = 0.1 |
| PSO | $c_1 = c_2 = 1.494$, $w = 0.72$ |
| DE | $C_r = 0.5$, $F = 0.8$ |
| ABC | $\varphi = 1$ |
| CS | $\alpha = 1$ |
| GS | Go = 100, $\alpha = 1$, Rpower = 1, Rnorm = 2 |
| WO | randRate = 0.5, $b = 1$ |

# 4 Results

The optimal thresholds value for sample images was obtained using seven meta-heuristic algorithms. MATLAB was used to implement all algorithms and to perform segmentation. The optimal threshold value, fitness function value, and time elapsed by each algorithm is illustrated in Table 4. Figures 2 and 3 show the comparison of the result of all metaheuristic algorithms for yellow lilly and colored chips images, respectively. Table 5 shows segmented and colored map images obtained after applying threshold valued obtained through different metaheuristic algorithms.

**Table 4** Metaheuristic algorithms' result comparison for sample images

| Metaheuristic algorithm | Yellow lily | | | | | Colored chip | | | | |
|---|---|---|---|---|---|---|---|---|---|---|
| | Optimal thresholds | | | Optimum value | Time elapsed | Optimal thresholds | | | Optimum value | Time elapsed |
| GA | 96 | 162 | 207 | 35.5579 | 11.6088 | 163 | 215 | 237 | 33.3041 | 20.595 |
| PSO | 96 | 164 | 236 | 35.516 | 12.7846 | 155 | 212 | 256 | 33.2162 | 18.6455 |
| DE | 96 | 162 | 207 | 35.5579 | 12.7343 | 155 | 209 | 236 | 33.3211 | 17.3271 |
| ABC | 96 | 162 | 207 | 35.5576 | 22.7731 | 155 | 210 | 236 | 33.3207 | 27.7496 |
| CS | 98 | 164 | 206 | 35.5516 | 21.9905 | 155 | 207 | 235 | 33.3136 | 36.6717 |
| GS | 99 | 170 | 220 | 35.4959 | 13.8851 | 71 | 145 | 199 | 31.6597 | 15.8032 |
| WO | 96 | 162 | 204 | 35.5491 | 14.1781 | 154 | 209 | 235 | 33.3158 | 24.4027 |

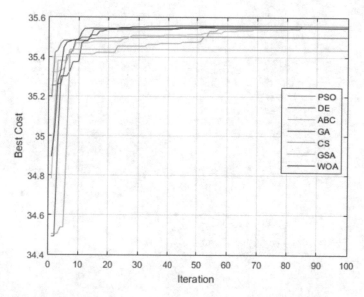

**Fig. 2** Yellow lilly comparison

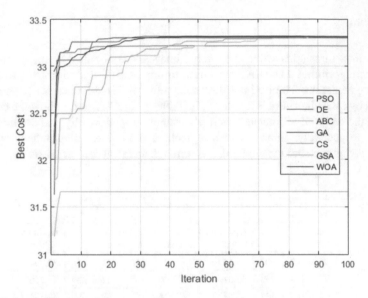

**Fig. 3** Colored chip comparison

**Table 5** Segmented and color-mapped image of sample images

| Algo-rithms | Segmented Image | Color-mapped Image | Segmented Image | Color-mapped Im-age |
|---|---|---|---|---|
| GA | | | | |
| PSO | | | | |
| DE | | | | |
| ABC | | | | |
| CS | | | | |
| GS | | | | |
| WO | | | | |

# 5 Conclusion

This study performed a comparative analysis of different metaheuristic algorithms for multilevel image segmentation using non-local mean 2D histogram thresholding method. In this paper, we have compared results for three-level threshold obtained from seven metaheuristic algorithms. Results show that all metaheuristic algorithms perform well but differential evolution was able to consistently outperform with less time elapsed. Recently developed metaheuristic algorithm whale optimization algorithm was also able to achieve good results in less time. Particle swarm optimization algorithm converged very quickly for both the sample images processed. All metaheuristic algorithms were able to differentiate objects from the background with less computational complexity for multilevel image segmentation process.

# References

1. H.F. Ng, Automatic thresholding for defect detection. Pattern Recognit. Lett. **27**(14), 1644–1649 (2006)
2. J.A. Noble, D. Boukerroui, Ultrasound image segmentation: a survey. IEEE Trans. Med. Imaging **25**(8), 987–1010 (2006)
3. A.L. Barbieri, G.F. De Arruda, F.A. Rodrigues, O.M. Bruno, L. da Fontoura Costa, An entropy-based approach to automatic image segmentation of satellite images. Phys. A Stat. Mech. Its Appl. **390**(3), 512–518 (2011)
4. J.N. Kapur, P.K. Sahoo, A.K. Wong, A new method for gray-level picture thresholding using the entropy of the histogram. Comput. Vis. Graph. Image Process. **29**(3), 273–285 (1985)
5. X.N. Wang, Y.J. Feng, Z.R. Feng, Ant colony optimization for image segmentation, in *Proceedings of 2005 International Conference on Machine Learning and Cybernetics*, Aug 2005, vol. 9 (IEEE, 2005), pp. 5355–5360
6. U. Maulik, Medical image segmentation using genetic algorithms. IEEE Trans. Inf. Technol. Biomed. **13**(2), 166–173 (2009)
7. M.H. Horng, R.J. Liou, Multilevel minimum cross entropy threshold selection based on the firefly algorithm. Expert Syst. Appl. **38**(12), 14805–14811 (2011)
8. P.Y. Yin, Multilevel minimum cross entropy threshold selection based on particle swarm optimization. Appl. Math. Comput. **184**(2), 503–513 (2007)
9. A.K. Wong, P.K. Sahoo, A gray-level threshold selection method based on maximum entropy principle. IEEE Trans. Syst. Man Cybern. **19**(4), 866–871 (1989)
10. A.S. Abutaleb, Automatic thresholding of gray-level pictures using two-dimensional entropy. Comput. Vis. Graph. Image Process. **47**(1), 22–32 (1989)
11. P.K. Sahoo, G. Arora, Image thresholding using two-dimensional Tsallis–Havrda–Charvát entropy. Pattern Recognit. Lett. **27**(6), 520–528 (2006)
12. X. Zheng, H. Ye, Y. Tang, Image bi-level thresholding based on gray level-local variance histogram. Entropy **19**(5), 191 (2017)
13. W. Xue-guang, C. Shu-hong, An improved image segmentation algorithm based on two-dimensional Otsu method. Inf. Sci. Lett. **1**(2), 77–83 (2012)
14. H. Mittal, M. Saraswat, An optimum multi-level image thresholding segmentation using non-local means 2D histogram and exponential Kbest gravitational search algorithm. Eng. Appl. Artif. Intell. **71**, 226–235 (2018)
15. S. Fengjie, W. He, F. Jieqing, 2D Otsu segmentation algorithm based on simulated annealing genetic algorithm for iced-cable images, in *IFITA'09 International Forum on Information Technology and Applications*, May 2009, vol. 2 (IEEE, 2009), pp. 600–602

16. S. Sarkar, S. Das, Multilevel image thresholding based on 2D histogram and maximum Tsallis entropy—a differential evolution approach. IEEE Trans. Image Process. **22**(12), 4788–4797 (2013)
17. X. Shen, Y. Zhang, F. Li, An improved two-dimensional entropic thresholding method based on ant colony genetic algorithm, in *GCIS'09 WRI Global Congress on Intelligent Systems,* May 2009, vol. 1 (IEEE, 2009), pp. 163–167
18. R. Panda, S. Agrawal, L. Samantaray, A. Abraham, An evolutionary gray gradient algorithm for multilevel thresholding of brain MR images using soft computing techniques. Appl. Soft Comput. **50**, 94–108 (2017)
19. J.H. Holland, *Adaptation in natural and artificial systems*: *an introductory analysis with applications to biology, control, and artificial intelligence,* MIT press, 1992
20. R. Eberhart, J. Kennedy, A new optimizer using particle swarm theory, in *MHS'95 Proceedings of the Sixth International Symposium on Micro Machine and Human Science,* Oct 1995, (IEEE, 1995), pp. 39–43
21. R. Storn, K. Price, Differential evolution–a simple and efficient heuristic for global optimization over continuous spaces. J. Glob. Optim. **11**(4), 341–359 (1997)
22. D. Karaboga, B. Basturk, A powerful and efficient algorithm for numerical function optimization: artificial bee colony (ABC) algorithm. J. Glob. Optim. **39**(3), 459–471 (2007)
23. X.S. Yang, S. Deb, Cuckoo search via Lévy flights. in *World Congress on Nature & Biologically Inspired Computing NaBIC,* Dec 2009, (IEEE, 2009), pp. 210–214
24. E. Rashedi, H. Nezamabadi-Pour, S. Saryazdi, GSA: a gravitational search algorithm. Inf. Sci. **179**(13), 2232–2248 (2009)
25. S. Mirjalili, A. Lewis, The whale optimization algorithm. Adv. Eng. Softw. **95**, 51–67 (2016)

# Author Index

© Springer Nature Singapore Pte Ltd. 2021
X.-Z. Gao et al. (eds.), *Advances in Computational Intelligence and Communication Technology*, Advances in Intelligent Systems and Computing 1086,
https://doi.org/10.1007/978-981-15-1275-9

Printed in the United States
By Bookmasters